Electrical Engineering: Concepts and Applications

Electrical Engineering: Concepts and Applications

Edited by **Marko Silver**

Willford Press
New York

Published by Willford Press,
118-35 Queens Blvd., Suite 400,
Forest Hills, NY 11375, USA
www.willfordpress.com

Electrical Engineering: Concepts and Applications
Edited by Marko Silver

International Standard Book Number: 978-1-68285-011-4 (Hardback)

Printed in the United States of America.

Contents

Preface

The purpose of the book is to provide a glimpse into the dynamics and to present opinions and studies of some of the scientists engaged in the development of new ideas in the field from very different standpoints. This book will prove useful to students and researchers owing to its high content quality.

Most of the gadgets and devices we use in our day to day life are made of various electrical components. The scope of electrical engineering is vast, as it branches out into significant sub-fields like electronics, digital computers, power engineering, telecommunications, etc. Latest researches and developments pertaining to electrical engineering have been covered in this book, such as power generation, microelectronics, signal processing, instrumentation, etc. The extensive content of this book provides the readers with a thorough understanding of the subject. Students, researchers, professionals and anyone else engaged in electrical and electronics engineering, communication engineering, and associated fields will benefit alike from this book.

At the end, I would like to appreciate all the efforts made by the authors in completing their chapters professionally. I express my deepest gratitude to all of them for contributing to this book by sharing their valuable works. A special thanks to my family and friends for their constant support in this journey.

Editor

1

Designing quantum-dot cellular automata circuits using a robust one layer crossover scheme

Sara Hashemi, Keivan Navi

Department of Electrical and Computer Engineering, Shahid Beheshti University, G.C., Tehran, Iran
E-mail: navi@sbu.ac.ir

Abstract: Quantum-dot cellular automata (QCA) is a novel nanotechnology which is considered as a solution to the scaling problems in complementary metal oxide semiconductor technology. In this Letter, a robust one layer crossover scheme is introduced. It uses only 90° QCA cells and works based on a proper clock assignment. The application of this new scheme is shown in designing a sample QCA circuit. Simulation results demonstrate that using this new scheme, significant improvements in terms of area and complexity can be achieved.

1 Introduction

To date different coplanar quantum-dot cellular automata (QCA) crossover schemes have been introduced [1–4]. The first structure which is shown in Fig. 1a is implemented using 90 and 45° QCA cells [1]. This crossover scheme has a high sensitivity to manufacturing faults and needs a high precision for QCA cell placement [2, 5]. The presented solutions in [2–4] are implemented using only 90° QCA cells. The coplanar crossover scheme in [2] (as shown in Fig. 2*b*) uses an eight-phase clocking mechanism with three types of clock signals (normal, crossing and converting signals). In this scheme, four-phase shifted versions of the normal clock signal are used for general logic flow in QCA cells *a*, *c*, *p* and *r*. A crossing signal is used in QCA cell *x* and two-phase shifted versions of the converting signal are required in QCA cells which are shown by *b* and *q*. Regarding this structure, it is clear that this crossover scheme leads to a QCA circuit with a complex clocking mechanism. The crossover scheme in [3] and its clocking waveforms are shown in Figs. 1*c* and *d*. This solution in contrast to the previous scheme [2] uses only four regular clock signals. As shown in Fig. 1*c*, the horizontal wire is segmented into two parts. All of the four clock signals must be elapsed at the left segment, whereas the QCA cells of the right segment are controlled by the clocking signal *C*2. At the first clock cycle, the input signal *B* propagates to the 'output' *B* (at $T=1$ all of the QCA cells of the vertical wire *B* are in the switch phase). At clock cycle 5, the input signal *A* has been propagated to QCA cell *x*. This cell is in the hold phase and drives the QCA cells of the vertical wire *B* which are in the switch phase (the input *B* has been considered unpolarised at this time). Finally, at $T=6$ the input signal *A* is propagated to the right using the QCA cells of the vertical wire (at this time, the QCA cells of the right side of the vertical wire are in the switch phase, whereas the QCA cells of the vertical wire *B* are in the hold phase). However, as mentioned in [3], in this crossover scheme, the input signals *A* and *B* are only applied during the first clock cycle and then removed (the input *B* is considered unpolarised at $T=5$). If this condition is not satisfied at clock cycle 5, the input signal *A* cannot correctly be transmitted to the vertical wire *B*. Regarding this condition, this scheme is not suitable for pipeline structure of QCA circuits.

Based on the Landauer clocking mechanism, four clocking zones can be used to design QCA circuits [5]. The schematic of a QCA wire composed of four clocking zones is shown in Fig. 2*a*. Each of these clocking zones is comprised of a group of QCA cells which are controlled by the same QCA clock. The four QCA clock waveforms [5] are shown by Fig. 2*b*. As shown in this figure, in each of the clocking zones, four clock phases can occur; a switch, a hold, a release and a relax phase. During the switch phase, the input cells affect the unpolarised QCA cells and some values are reached. In the hold phase, the polarisation of QCA cells is locked for transmission to their neighbouring cells. During the release phase, QCA cells lose their values. Finally, the relax phase means that QCA cells are completely unpolarised [5]. As shown in Fig. 2*b*, there is a 90° phase delay between clocking zones. For example, there is a 180° phase delay from the first clocking zone (zone0) to the third one (zone2). Regarding these phase delays, two routing elements with simple structures were introduced in [6]. These elements are constructed using five QCA cells and are used to implement two edges and straight crossings in a QCA programmable switch matrix (PSM) interconnection element [6]. These elements also can be considered as a suitable solution for crossing wires in a QCA circuit.

Recently, another coplanar crossover scheme has been introduced [4]. This scheme works same as the presented elements in [6]. The only difference is the clocking zone of the middle QCA cell. In [6], the middle QCA cell is not connected to the QCA clock, whereas in [4] a QCA clock is assigned to this cell. As shown in Fig. 2*c*, the crossover scheme in [4] uses two different clocking zones with a 180° phase delay to crossover wires. In Fig. 2*c*, when the QCA cells of zone0 are in the switch and hold phases, the QCA cells of zone2 are in the release and relax phases; therefore the QCA cells of zone0 cannot be affected by QCA cells of zone2 and vice versa. In the next section, the robustness of this scheme is investigated and it is changed to obtain a robust structure against sneak noise paths.

2 Robust QCA crossover scheme

In [7, 8], the robustness of the basic coplanar crossover scheme (Fig. 3*a*) is investigated. The results demonstrate that the input signal of the vertical wire can affect the output of the horizontal wire and this leads to incorrect operation of the crossover scheme. Based on calculations [7], the kink energy between QCA cells {*A*1, *C*2}, {*A*2, *C*2} and {*A*3, *C*2} is zero and therefore the polarisation of QCA cells *A*1, *A*2 and *A*3 do not affect the polarisation of QCA cell *C*2. Unfortunately, the kink energy between the cell pairs {*C*2, *A*} and {*C*2, *X*} is non-zero and these cells (*A* and *X*) affect the polarisation of QCA cell *C*2 [7]. This results in sneak noise paths and the noise is conducted from input signal *Sv* to QCA cell *C*2 [7]. It is worth mentioning that these sneak noise paths can lead to incorrect operation when small QCA blocks are used to design larger QCA circuits. To solve this problem, as shown in Fig. 3*b*, in a robust crossover scheme the horizontal QCA cells on the right side of the vertical wire are controlled

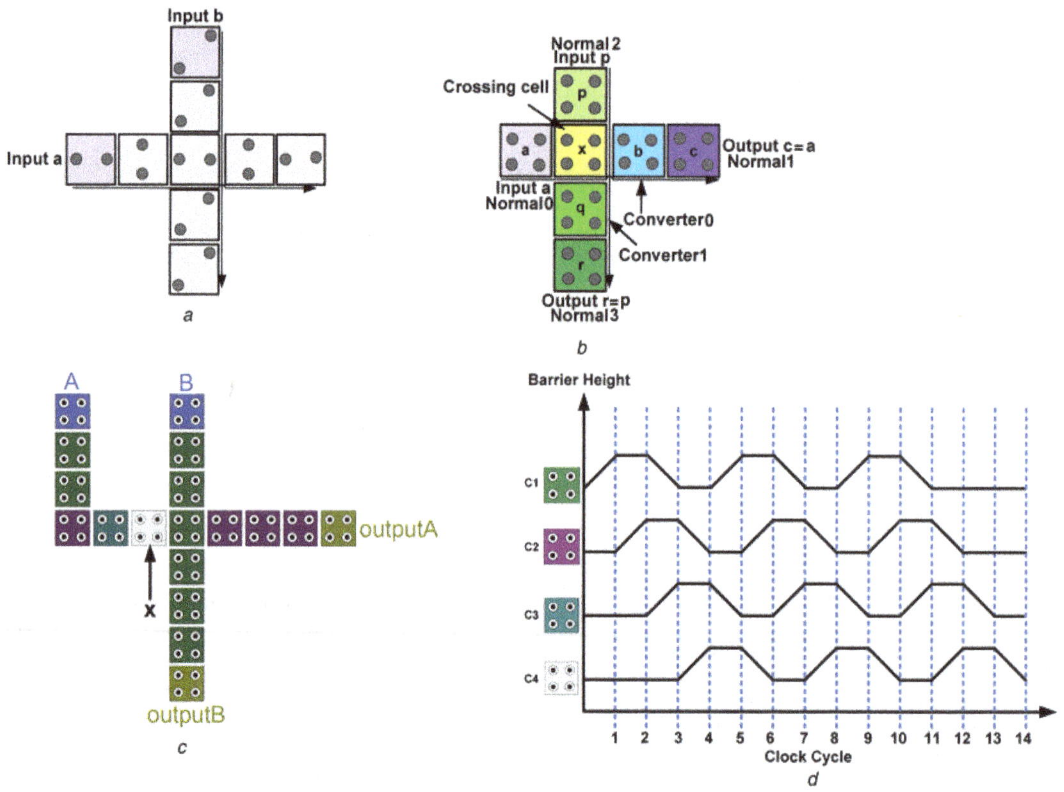

Fig. 1 *First structure*
a Basic coplanar crossing scheme constructed using 90 and 45° QCA cells [1]
b Coplanar crossing scheme constructed using only 90° QCA cells presented in [2]
c and *d* Coplanar crossover scheme in [3] with its clocking waveforms

using a QCA clock which has a 90° phase delay with the cells on the left side of the vertical wire [7, 8]. Using this proper clock assignment, during the switch phase of QCA cells *A* and *X* (controlled using clock0 in Fig. 3*b*), the QCA cells *C*2, *C*3 and *Z* are in the relax phase and they are not affected by the vertical wire. During the switch phase of clock1, the QCA cells *C*2, *C*3 and *Z* are affected by QCA cell *C*1 and the noise from cells *A* and *X* (the QCA cells *C*1, *A* and *X* are in the hold phase). In this condition, the polarisation of QCA cells *C*2, *C*3 and *Z* eventually settle down to the input signal *Sh*. In fact, using this proper clock assignment the input signal *Sh* does not arrive to the output no later than any noise signal and this leads to a robust QCA crossover scheme against sneak noise paths.

In this section, robustness of the coplanar crossover scheme in Fig. 2*c* against sneak noise paths is investigated and a robust structure is introduced. Based on Fig. 4*a*, it is clear that this scheme same as the basic design (Fig. 3*a*) is vulnerable to the sneak noise paths. As shown in this figure, the output QCA cells of the horizontal wire

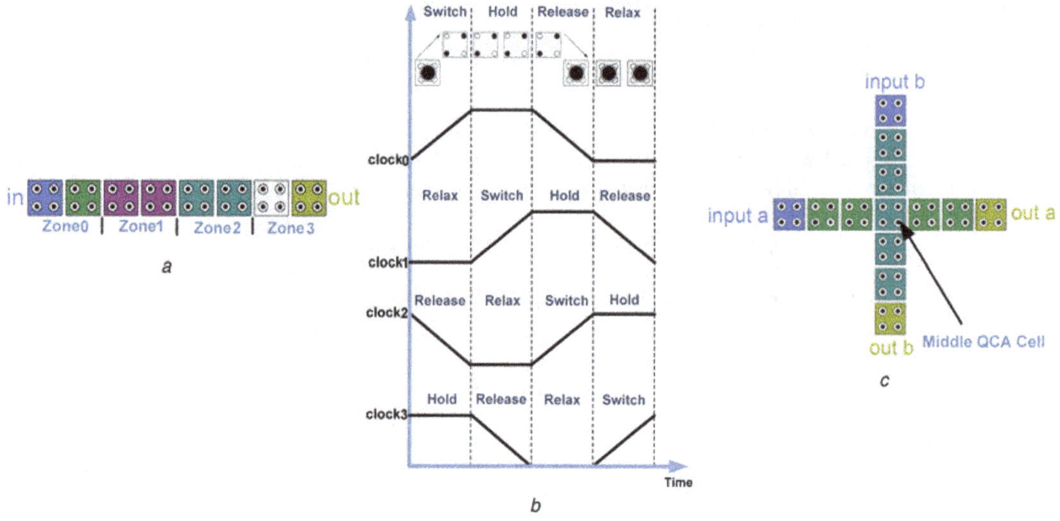

Fig. 2 *Coplanar crossover scheme*
a QCA wire constructed using four clocking zones
b Landauer QCA clocking waveforms [5]
c QCA crossover scheme in [4] which works same as the routing elements in [6]

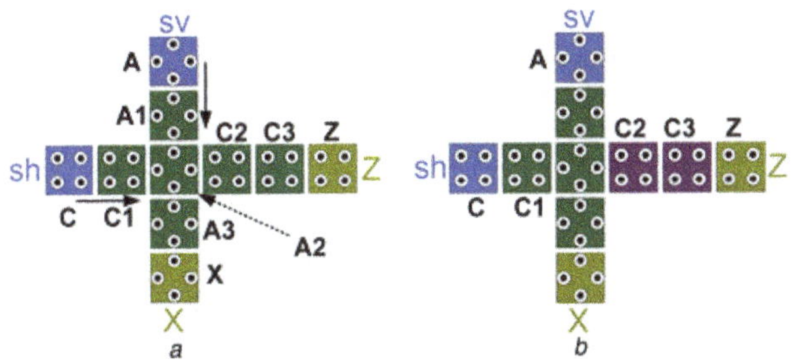

Fig. 3 *Robustness of the basic coplanar crossover scheme*
a Basic coplanar QCA crossover scheme
b Robust crossover against sneak noise paths [7, 8]

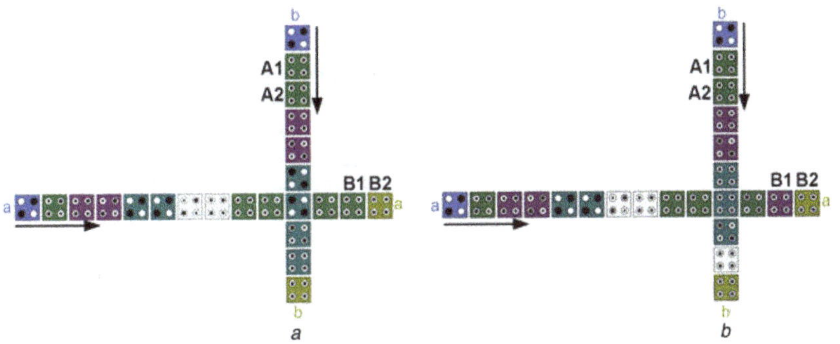

Fig. 4 *Output QCA cells of the horizontal wire*
a Coplanar crossover scheme in [4]
b Robust crossover scheme against sneak noise paths

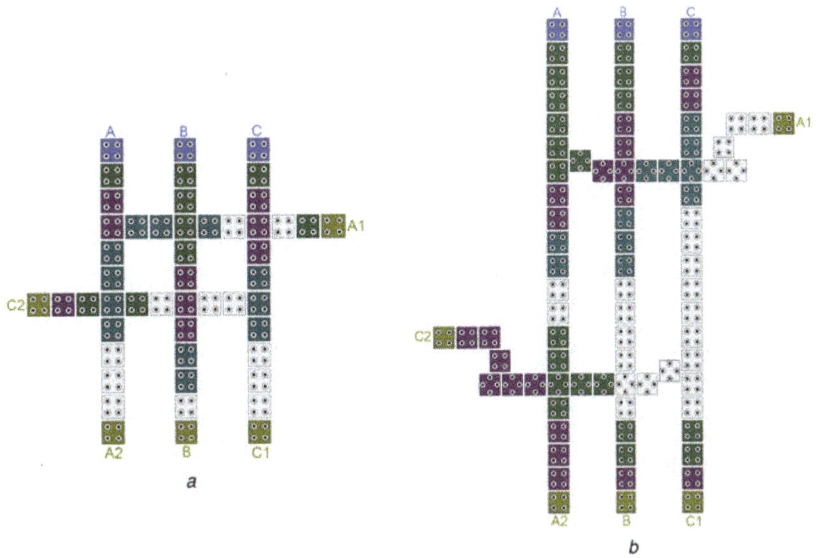

Fig. 5 *Application of the new crossover scheme*
a QCA circuit constructed using the new robust coplanar crossing scheme
b QCA circuit constructed using the coplanar crossing scheme in Fig. 3*b*

(*B*1 and *B*2) in the switch phase are affected by the QCA cells of the vertical wire which are in the same phase (QCA cells *A*1 and *A*2). The values of the kink energy between QCA cells can demonstrate this problem. The kink energy [7] between QCA cells *A*1, *A*2 and *B*1, *B*2 is presented in Table 1 (the kink energy can be considered as the difference in energy between two cells which have an opposite polarisation and those same two cells with a same polarisation [7]). In this table, the kink energy is calculated using the four possible interaction patterns between a pair of QCA cells (the length of each cell has been considered equal to 18 nm and there is a space of 2 nm between each of the two neighbour QCA cells). Based on these results, it is clear that the kink energy between QCA cells

Table 1 Kink energy between QCA cells in Fig. 4*a*

QCA cells		≃Energy, J
*B*1	*A*1	-0.08×10^{-23}
	*A*2	-2.11×10^{-23}
*B*2	*A*1	-0.96×10^{-23}
	*A*2	-3.03×10^{-23}

Table 2 Comparison results of the proposed structures in Figs. 5*a* and *b*

Design types	Complexities (QCA cells)	Area, μm^2	Maximum delay (clock cycles)
new design in Fig. 5*a*	50	0.06	1.5
presented design in Fig. 5*b*	84	0.13	1.5

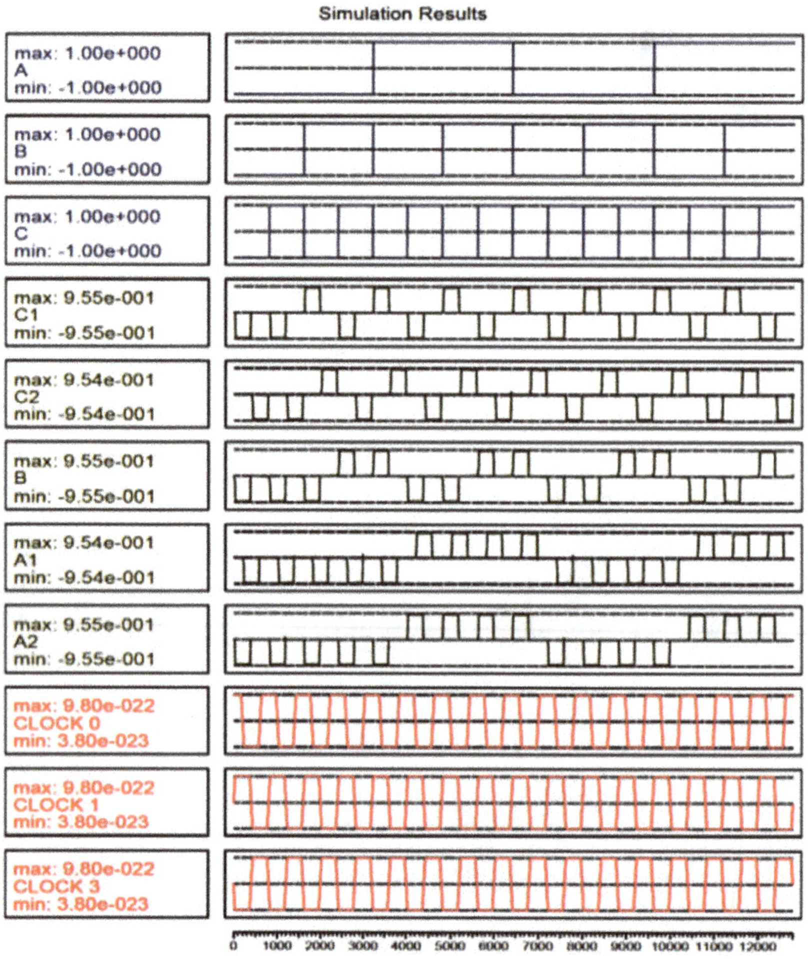

Fig. 6 *Simulation results of the proposed structure in Fig. 5a*

*A*1, *A*2 and *B*1, *B*2 is not zero. In other words, the polarisation of the horizontal outputs is affected by QCA cells of the vertical wire which are in the same clocking zone. To solve this problem, an additional clocking zone (with a 90° phase delay) must be added to the output of horizontal wire (Fig. 4*b*). In this case, same as the presented robust design in Fig. 3*b*, the horizontal input does not arrive after the noise signals. In this structure, when the QCA cells *A*1 and *A*2 are in the switch phase the horizontal outputs (*B*1 and *B*2) are in the relax phase and therefore the vertical wire does not affect the horizontal outputs. In addition, it is clear that an additional clocking zone (clocking zone 3 in Fig. 4*b*) must be added to the output of the vertical wire. This is done due to the fact that by adding a clocking zone to the horizontal wire, the output QCA cells of the horizontal and vertical wires are positioned in two clocking zones with a 90° phase delay (clocking zones 1 and 2 in this example). It is clear that in this case, the output QCA cells of the vertical wire in the switch phase (positioned in clocking zone 2) are affected by QCA cells of the horizontal wire which are in the hold phase (*B*1 and *B*2 which are positioned in clocking zone 1). To solve this problem, the output QCA cells of the vertical wire are positioned in a clocking zone which has a 180° phase delay with the outputs of horizontal wire (clocking zone 3 in Fig. 4*b*).

An application of the new crossover scheme is shown in Fig. 5*a*. As shown in this figure, using the new scheme the input signals *A* and *C* are transmitted easily (the outputs of these signals are shown by *A*1, *A*2, *C*1 and *C*2). The structure of this circuit using the basic robust crossover scheme (Fig. 3*b*) is shown in Fig. 5*b*. Based on the proposed designs, it is clear that the new scheme leads to a very dense structure and is a suitable solution for designing robust QCA circuits (Table 2). The exhaustive simulations of the proposed structure (Fig. 5*a*) using the default parameters of both bi-stable and coherence vector engines of QCADesigner version 2.0.3 [9, 10] is shown in Fig. 6. It is worth mentioning that using both engines same results have been achieved which indicate the accuracy of the new crossover scheme. Regarding the proposed results, this scheme is a suitable solution for designing efficient and robust QCA circuits.

3 Conclusion

In this Letter, a robust one layer QCA crossover scheme was introduced. This design uses only 90° QCA cells and in contrast to its counterparts does not need a complex clocking mechanism. This scheme was used to design a sample QCA circuit. Simulation results indicate that using this scheme very dense structure and robust against sneak noise paths can be achieved.

4 References

[1] Tougaw P.D., Lent C.S.: 'Logical devices implemented using quantum cellular automata', *J. Appl. Phys. Am. Inst. Phys.*, 1994, **75**, pp. 1818–1824

[2] Devadoss R., Paul K., Balakrishnan M.: 'Coplanar QCA crossovers', *IET Electron. Lett.*, 2009, **45**, pp. 1234–1235

[3] Tougaw D., Khatun M.: 'A scalable signal distribution network for quantum-dot cellular automata', *IEEE Trans. Nanotechnol.*, 2013, **12**, pp. 215–224

[4] Shin S.H., Jeon J.C., Yoo K.Y.: 'Wire-crossing technique on quantum-dot cellular automata'. Second Int. Conf. Next Generation Computer and Information Technology (NGCIT), 2013, vol. **27**, pp. 52–57

[5] Hänninen I., Takala J.: 'Binary adders on quantum-dot cellular automata', *J. Signal Process. Syst. (Springer)*, 2010, **58**, pp. 87–103
[6] Jazbec A., Zimic N., Bajec I.L., Pe'ar P., Mraz M.: 'Quantum-dot field programmable gate array: enhanced routing'. Conf. Optoelectronic and Microelectronic Materials and Devices, 2006, pp. 121–124
[7] Kim K., Wu K., Karri R.: 'The robust QCA adder designs using composable QCA building blocks', *IEEE Trans. Comput. Aided Des. Integr. Circuits Syst.*, 2007, **26**, pp. 176–183
[8] Kim K., Wu K., Karri R.: 'Towards designing robust QCA architectures in the presence of sneak noise paths'. Proc. Design, Automation and Test in Europe Conf. Exhibition, 2005, pp. 1214–1219
[9] QCADesigner Documentation [Online]. Available at http://www.qcadesigner.ca
[10] Walus K., Dysart T.J., Jullien G.A., Budiman R.A.: 'QCADesigner: a rapid design and Simulation tool for quantum-dot cellular automata', *IEEE Trans. Nanotechnol.*, 2004, **3**, pp. 26–31

2

Investigation of wound rotor induction machine vibration signal under stator electrical fault conditions

Sinisa Djurović, Damian S. Vilchis-Rodriguez, Alexander Charles Smith

School of Electrical and Electronic Engineering, The University of Manchester, Power Conversion Group, Sackville Street Building, M13 MPL Manchester, UK
E-mail: sinisa.durovic@manchester.ac.uk

Abstract: This paper investigates wound rotor induction machine torque and vibration signals spectra for operation with and without a stator short-circuit or open-circuit winding fault. Analytical expressions that enable the healthy and faulty machine pulsating electromagnetic torque frequencies to be related to shaft speed are derived and validated for operating conditions of interest. A coupled-circuit machine model is used to investigate the healthy and faulty electromagnetic torque signal. Shaft torque and stator frame vibration are measured on a laboratory test rig comprising a 30 kW wound rotor induction machine. It is shown that the existence of a stator winding inherent electrical unbalance or that arising from fault gives rise to a range of pulsating torque frequencies that are transmitted to the machine frame and can be detected in the measured vibration signal. The magnitudes of the resulting vibration components are demonstrated to be largely determined by the unbalance severity and the mechanical system response. The presented experimental results clearly validate the analytical and simulation analysis for the operating range of the investigated industrial machine design.

1 Introduction

Wind energy remains the dominant form of sustainable electricity generation. With a significant increase in wind energy penetration levels projected in the near future for a number of national markets [1], the development of effective wind turbine (WT) condition monitoring (CM) techniques is becoming crucial for minimising the downtime and maintenance cost of production systems. A considerable proportion of installed and currently manufactured megawatt size WTs use wound rotor induction generator (WRIG) drives for electromechanical energy conversion. Recent fault surveys [2–6] report high rates of winding failures in WT generators and indicate that these make a large contribution to WT downtime.

The common approach to academic research on ac machinery non-invasive electrical fault detection is based on techniques utilising spectral analysis of machine electrical signals [7]. In this context, a number of authors have recently reported that WRIG winding fault specific spectral changes can be identified in WRIG currents for different industrial designs [8–10]. When WT drive train CM in the field is concerned the commercially available monitoring systems (CMSs) today are almost exclusively focused on generator vibration signal monitoring [11]; this technique can provide mechanical fault detection and general indications of the generator electrical fault. The existing commercial CMS solutions have however been reported to suffer from a high degree of diagnostic unreliability [12]. This indicates a distinct need for a better understanding of the fault information contained in the vibration signal, and in particular whether it can be used to better recognise and assess the generator electrical condition. In addition, the success of future CM platforms in delivering the required improvements in diagnostic reliability will largely depend on how effectively fault signatures can be identified in multiple electrical and mechanical signals.

Vibration and torque signals monitoring and analysis for electrical fault detection was shown to be feasible in the literature [13–17]. However, these methods have received considerably less attention in comparison with electrical signal monitoring because of reported lower sensitivity of mechanical signals to electrical fault signature. Li and Mechefske [13] showed that a rotor broken bar fault will induce sidebands in the frame vibration spectrum; monitoring of these was however reported to provide lower sensitivity fault detection when compared with that achieved by current analysis. The estimated air-gap torque observation requires multiple generator signals to be monitored, but was shown in [14] to be an effective tool for recognition of cage rotor machine faults. The research in [15] presents a theoretical study that indicates that vibration signal spectral changes can occur with a stator winding fault in wound rotor induction machines. In [17] finite element analysis methods are applied to investigate the changes in air-gap force stress waves through cage rotor induction machine vibration measurement. This paper reports stator winding and rotor bar fault induced changes in low-frequency vibration signal components.

Given the extensive use of vibration analysis-based CMS in WT installations and their reported limitations in effective recognition of generator electrical fault, this paper investigates the electrical unbalance fault induced spectral effects in the WRIG vibration signal. The aim of this research is to clarify the nature of the WRIG vibration signal spectrum and the electrical fault and unbalance information it contains. This paper extends the work presented by Djurovic *et al.* [18] by providing an in-depth examination of the effects of a range of electrical faults on the shaft torque and frame vibration signals of an industrial WRIG. This paper also experimentally investigates the influence of the rotor converter on the WRIG vibration signature for operation with stator electrical fault in a doubly fed induction generator (DFIG) configuration. The work presents a generalised theoretical analysis of the WRIG electromagnetic torque signal spectrum that provides analytical expressions linking possible pulsating torque frequencies to machine operating conditions. A harmonic modelling technique [19] is employed to develop a WRIG model capable of predicting pulsating electromagnetic torque frequencies. The model is then used to investigate the torque signal spectrum under healthy and faulty operating conditions. The predicted pulsating torques are shown to be clearly detectable in shaft torque and stator frame vibration signals measured on the laboratory WRIG test rig. Both model and experimental results for the investigated machine design indicate the existence of a range of pulsating torques and vibration signal components that are stator winding unbalance specific. The magnitudes of these are shown to significantly increase with the presence of electrical fault. Propagation of winding fault harmonic effects in the shaft torque signal through to the frame vibration signal is clarified by undertaking impact testing to characterise the mechanical response of the examined system. The consistency of the identified fault effects and developed theoretical principles are validated for the laboratory machine's rated operating range.

2 Electromagnetic torque signal

Considering a general case of a three-phase p pole-pair wound rotor induction machine with no rotor excitation, a single frequency excited three-phase stator winding system will give rise to a range of air-gap fields with pole numbers v determined by the existing stator windings and supply arrangement. These can be expressed in general form by

$$b_{gs}(\theta, t) = \mathrm{Re}\left[\sum_{v} \sqrt{2}\bar{B}_{gs} e^{j(\omega t - v\theta)}\right] \quad (1)$$

where θ is the angular coordinate in the stator reference frame, $\omega = 2\pi f$ is the angular frequency and f is the stator supply frequency. For the rotor mechanical speed n_r, the transformation between the rotor coordinate system (θ') and the stationary stator reference frame coordinate system (θ) is given by: $\theta = n_r t + \theta'$. Substituting the above in (1) and considering the fundamental slip relationship, $n_r = \omega/p(1-s)$ allows for the stator current driven air-gap field to be expressed in the rotor reference frame

$$b_{gs}(\theta', t) = \mathrm{Re}\left[\sum_{v} \sqrt{2}\bar{B}_{gs} e^{j\left[\omega t\left(1 - \frac{v}{p}(1-s)\right) - v\theta'\right]}\right] \quad (2)$$

Each individual stator harmonic field in (2) will induce an electromagnetic force of corresponding frequency in the rotor three-phase windings. The stator driven air-gap fields in (2) can therefore set up a range of harmonic rotor currents. Each induced rotor current will in turn create a range of harmonic fields in the rotor reference frame, with pole numbers μ defined by the rotor winding arrangement. The resulting rotor current driven air-gap field will therefore take the following form in the rotor coordinate system

$$b_{gr}(\theta', t) = \mathrm{Re}\left[\sum_{\mu}\sum_{v} \sqrt{2}\bar{B}_{gr}^{\mu,v} e^{j[\omega t(1-(v/p)(1-s)) - \mu\theta']}\right] \quad (3)$$

Transferring (3) to the stator reference frame gives

$$b_{gr}(\theta, t) = \mathrm{Re}\left[\sum_{\mu}\sum_{v} \sqrt{2}\bar{B}_{gr}^{\mu,v} e^{j[\omega t(1-((v-\mu)/p)(1-s)) - \mu\theta]}\right] \quad (4)$$

The induced rotor current driven air-gap fields' angular frequencies in the stationary reference frame are defined by (4) as

$$\omega_i = \left(1 - \frac{v-\mu}{p}(1-s)\right)2\pi f \quad (5)$$

In the stator reference frame these rotating harmonic fields will interact with a range of stator current driven air-gap fields of supply angular frequency

$$\omega = 2\pi f \quad (6)$$

Each stator driven field with pole number v at frequency given by (6) can interact with a range of rotor driven air-gap fields with an identical pole number rotating at frequencies given by (5) and created by the induced rotor currents. The interaction of any two stator and rotor driven magnetic fields with identical pole numbers will establish an electromagnetic torque component. The angular frequency of the torque component created by two fields with identical pole numbers travelling in the same direction is given by the absolute value of the relative difference between their respective angular speeds, that is,

$$|\omega - \omega_i| \quad (7)$$

Similarly, for identical pole number fields travelling in opposite directions, the resulting torque angular frequency will be equal to the absolute value of the sum of the individual field angular velocities, that is,

$$|\omega + \omega_i| \quad (8)$$

$|\omega + \omega_i|$ (8). Considering (7) and (8), the resulting electromagnetic torque signal angular frequencies can be established directly from expressions in (5) and (6) by considering the balanced or unbalanced winding operating conditions of interest and introducing appropriate constraints for possible values of μ and v. This analysis neglects the effects of stator supply higher-order harmonics, which need to be investigated separately. The presented derivations also neglect the influence of the machine's mechanical system and consider only the air-gap field origins of the torque signal spectrum. An unbalanced stator/grid supply is assumed in the analysis, as is generally the case for online operating ac machinery. Electromagnetic torque frequencies can now be derived for the operating conditions investigated in this paper.

2.1 *Balanced windings and unbalanced supply*

When the machine operates with an unbalanced stator supply, for each stator driven harmonic air-gap field there will exist a forwards ($^+$) and a backwards ($^-$) rotating component. These stator field components will give rise to forwards and backwards rotating components of the rotor driven harmonic fields. Consequently, for every harmonic order, there will exist a range of field waves having the same pole number and travelling in forwards and backwards directions in the air-gap at speeds ω^+, ω^-, ω_i^+ and ω_i^-. Therefore, for each considered harmonic order in the stator reference frame, forwards with backwards rotating field components interaction will be possible, as well as the interaction between field components travelling in the same direction. The resulting torque frequencies can therefore be determined from the following interactions

$$|\omega^+ - \omega_i^+|;\ |\omega^+ + \omega_i^-|;\ |\omega^- - \omega_i^-|;\ |\omega^- + \omega_i^+| \quad (9)$$

For unbalanced supply conditions, the pole numbers μ and v are given by: $\mu = \pm p(1-6m)$ and $v = \pm p(1-6n)$, where $m = 0, \pm 1, \pm 2, \pm 3$... and $n = 0, \pm 1, \pm 2, \pm 3$.... Substituting these expressions into (5) and considering (9), the possible torque angular frequencies for operation with unbalanced supply and balanced windings are derived as

$$|6k(1-s)|2\pi f \quad (10)$$

and

$$|2 \pm 6k(1-s)|2\pi f \quad (11)$$

where $k = 0, 1, 2, 3$....

2.2 *Unbalanced stator windings and unbalanced supply*

A stator winding fault will give rise to air-gap field components and interactions identical to those described in (9). However, for the investigated faulty winding configurations, shown in Fig. 1, the considered short-circuit and open-circuit faults, as well as any electrical unbalance between individual phase winding parallel groups,

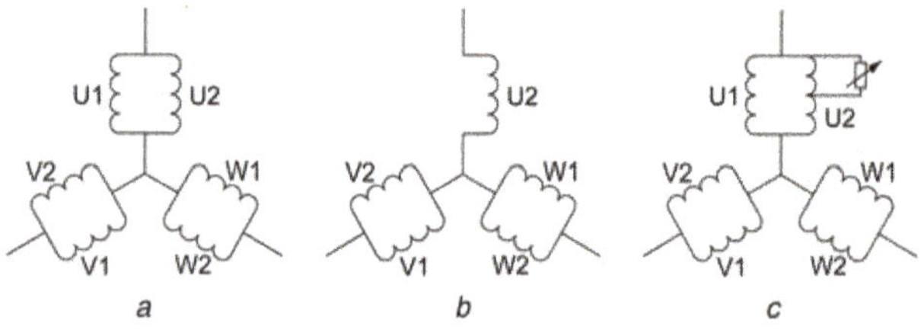

Fig. 1 *Machine winding configurations*
a Healthy stator and rotor
b Open-circuit fault
c Short-circuit fault

will create an air-gap field distortion resulting in the loss of field half-cycle symmetry and therefore periodicity over the pole-pair pitch [20]. The harmonic effects of such a field distortion will be manifested in the possible presence of all harmonic orders over the air-gap perimeter of the analysed four-pole machine. Consequently, μ and v will take a different form to that in Section 2.1, and all pole values can be assumed to exist, that is, μ, v, ± 1, ± 2, $\pm 3 \ldots$ A general expression for torque angular frequencies of a machine operating with a stator winding electrical unbalance can now be derived by substituting the above condition into (5) and evaluating the field interactions given by (9)

$$\left|\frac{k}{p}(1-s)\right|2\pi f \tag{12}$$

and

$$\left|2 \pm \frac{k}{p}(1-s)\right|2\pi f \tag{13}$$

where $k = 0, 1, 2, 3 \ldots$ The expressions in (12) and (13) are derived under the general assumption that all winding unbalance induced harmonic fields will be equally manifested in the air-gap and predict all possible torque signal frequencies that can originate from the considered conditions. However, not all frequencies given by (12) and (13) will necessarily be present in the torque signal, as how each of these is manifested will be highly dependent on a particular machine winding layout and how it responds to the unbalance induced air-gap field distortion. The derived expressions for possible torque signal frequencies are summarised in Table 1.

3 Torque signal model study

3.1 Modelling tools

The WRIG model used in this paper is based on the principles of generalised harmonic analysis and is presented in [19]. The model accounts for harmonic air-gap field effects when evaluating machine parameters and is convenient for frequency domain analysis of machine quantities, as demonstrated in [8, 19]. To simulate the behaviour of the industrial machine used for experimental research in this paper, the manufacturer's design data, winding distribution (shown in Appendix 1) and measured operational data were used as model inputs in the calculations. The model allows for numerical simulation of arbitrary healthy and faulty winding configurations by appropriate definition of the machine connection matrix [21]. The connection matrices for investigated winding configurations including open- and short-circuit faults are shown in Appendix 2. Model results for electromagnetic torque were analysed in a bandwidth of 0–1 kHz, as this bandwidth is generally of interest when monitoring ac machine fault frequencies [13, 16, 18]. To achieve good frequency resolution 2^{13} fast Fourier transform (FFT) data points are used in the analysis.

Table 1 Electromagnetic torque frequencies

Windings	Torque frequencies
balanced	$\lvert 6k(1-s)\rvert f$
	$\lvert 2 \pm 6k(1-s)\rvert f$
unbalanced	$\lvert (k/p)(1-s)\rvert f$
	$\lvert 2 \pm k/p(1-s)\rvert f$

3.2 Model study

A model study was undertaken to explore the spectral content of the laboratory machine electromagnetic torque signal for the examined healthy and faulty winding conditions. A typical super-synchronous operating speed of 1590 rpm was investigated in the calculations for illustration purposes. The simulations assume a stator phase voltage root-mean-square unbalance of $\simeq 1\%$, equal to that measured in the laboratory grid supply during experiments. It is important to point out that the model calculations neglect any time-varying interactions between the electromagnetic torque and the machine mechanical system (inherent rotor and/or shaft unbalance, natural/resonant frequencies) and therefore any pulsating torque damping or excitation that may result from these [15].

Machine operation with balanced windings and unbalanced supply was first simulated and numerical model predictions for the steady-state torque signal spectrum are shown in Fig. 2. In addition to the dominant DC component, the spectrum is seen to contain a number of pulsating torques at harmonic and inter-harmonic frequencies. These originate from the interaction of healthy machine harmonic air-gap fields, as discussed in Section 2. To understand and evaluate the influence of inherent stator winding unbalance, machine operation with a 3% resistive unbalance between parallel winding groups in individual stator phase windings was simulated in the model and the resulting torque spectrum shown in Fig. 2. The considered resistive unbalance value was measured between individual phase winding groups on the laboratory machine. The inherent winding unbalance is seen to result in a number of additional low magnitude frequency components in the

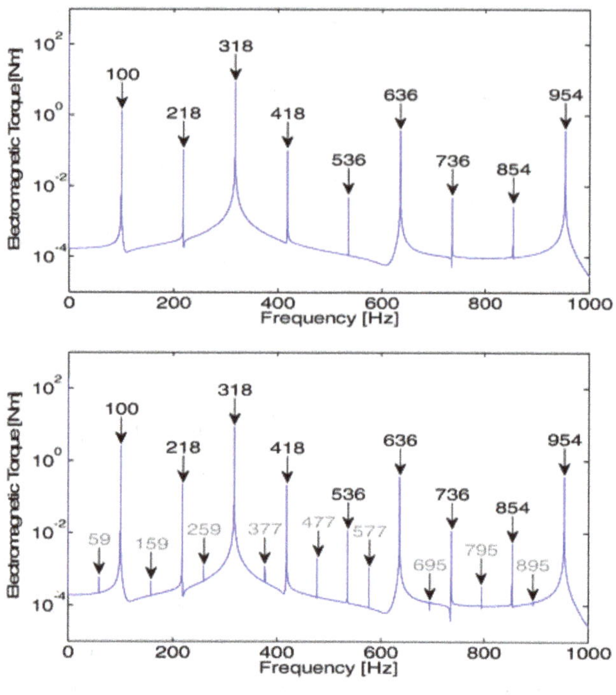

Fig. 2 *Electromagnetic torque spectrum: balanced windings (top) and inherent winding unbalance (bottom), unbalanced supply, 1590 rpm*

predicted torque spectrum. These arise because of interaction of the unbalance induced field harmonic components analysed in Section 2.2. The predicted torque spectra for machine operation with unbalanced supply and open-circuit and short-circuit faults shown in Figs. 1*b–c*, are presented in Figs. 3 and 4, respectively. The model study suggests that, even under perceived 'healthy' machine operation, the existence of inherent machine stator winding asymmetry can produce distinct small magnitude torque pulsations. However when an actual stator winding fault takes place, the resulting field distortion is significantly amplified and the magnitudes of these pulsating torques are much increased, becoming several orders of magnitude larger. Monitoring these changes may therefore provide useful information on the integrity of stator windings.

The pulsating torque frequencies identified in the model study results in Figs. 2–4 confirm the theoretical principles established in Section 2. The observed torque frequencies can be calculated for the corresponding operating conditions using the appropriate closed-form expressions from Table 1, coupled with the knowledge of the machine operating speed and the supply frequency. For the considered operating conditions, Table 2 lists the calculated pulsating torque frequency values in ascending order with the corresponding equations and parameters that yield the identified frequencies.

4 Torque and vibration measurements

4.1 Laboratory test-rig description

The laboratory test rig, shown in Fig. 5, comprises a four-pole 30 kW wound rotor induction machine coupled to a 40 kW DC motor. The DC motor torque and speed are controlled using a DC speed drive to operate the induction machine as a generator. The rig can be driven in WRIG or DFIG configuration; it contains a 40 kW back-to-back converter that can be used to inject/recover slip power into the rotor circuit to establish steady-state DFIG operation. The converter open loop controller enables the output magnitude and frequency to be separately set to a desired value. The converter switching frequency in this application was set to 8 kHz. For the purpose of this research, the WRIG rotor windings were either short circuited or interfaced to the grid via the back-to-back converter. The stator windings were connected to the grid via a three-phase variable transformer. To enable experimental emulation of stator winding faults, the WRIG stator was wound so that the individual coil connections are taken out to an external terminal box. The desired healthy and faulty winding configurations can be achieved by appropriate connection of the accessible coil ends. The available laboratory machine imposes the limit for the minimum amount of turns that can be shorted at seven. This enables the emulation of a practical short-circuit fault comprising 6% of the effective phase winding turns. Healthy WRIG winding configurations corresponding to manufacturer's specifications, along with the investigated short-circuit and open-circuit fault scenarios are shown in Fig. 1. To avoid irreparable damage to the stator windings during short-circuit experiments, a variable resistor was used in the short-circuit path to limit the current. Different levels of incipient fault severity were investigated by adjusting the variable resistance to manipulate the magnitude of short-circuit current. Three-phase power analysers were used in the WRIG stator and rotor circuits to monitor and record currents and voltages. The rotational speed was measured by a stub shaft mounted 1024 ppr incremental encoder. The electromagnetic air-gap torque signal for a balanced machine was estimated from recorded stator voltage and current measurements [14, 22]. To accurately measure the dynamic shaft torque, the IM was mounted on a Kistler 9281B force platform containing three-axis piezoelectric transducers [23]. The stator frame vibration signal was recorded by installing a Bruel & Kjaer accelerometer (DT4394) on the top of the drive end bearing in the radial direction. The accelerometer output signal was conditioned using the Bruel & Kjaer Pulse vibration analysis platform. The vibration signal FFT analysis was performed using the Pulse platform proprietary routine at 6400 line resolution for the investigated 0–1 kHz bandwidth. The air-gap and shaft torque FFT measurements were performed for the identical bandwidth with a resolution of ≃0.1 Hz.

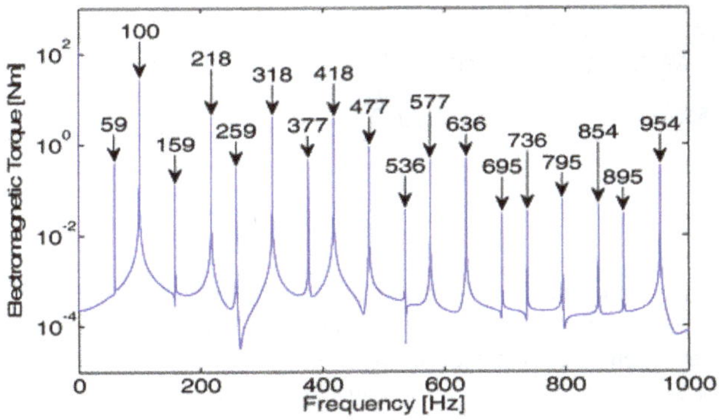

Fig. 3 *Electromagnetic torque spectrum for a single stator open-circuit fault and unbalanced supply, 1590 rpm*

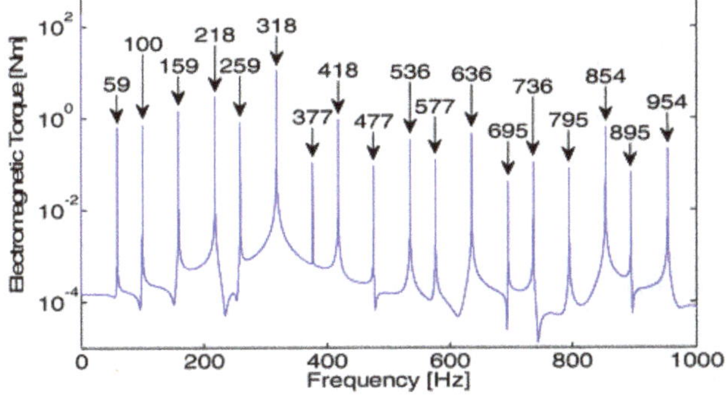

Fig. 4 *Electromagnetic torque spectrum for stator short-circuit fault and unbalanced supply, 1590 rpm*

Table 2 Pulsating torque signal frequencies and corresponding closed-form expression parameters (50 Hz, 1590 rpm)

#	Balanced windings		Winding unbalance/fault	
	Equation number/*k*	Frequency, Hz	Equation number/*k*	Frequency, Hz
1	(11)/0	100	(13)/6	59
2	(11)/1	218	(12)/6	159
3	(10)/1	318	(13)/6	259
4	(11)/1	418	(13)/18	377
5	(11)/2	536	(12)/18	477
6	(10)/2	636	(13)/18	577
7	(11)/2	736	(13)/30	695
8	(11)/3	854	(12)/30	795
9	(10)/3	954	(13)/30	895

4.2 Experimental results

The analysis in Sections 2 and 3 indicates that stator winding faults can increase torque pulsations at predictable frequencies that can be numerically determined. The identified electromagnetic torque pulsations will be transferred to the machine shaft, but can also be manifested mechanically on the machine frame as radial vibrations, because of the interaction of forces produced by torque pulsations with the machine's mechanical system [15, 16]. Not all frequencies

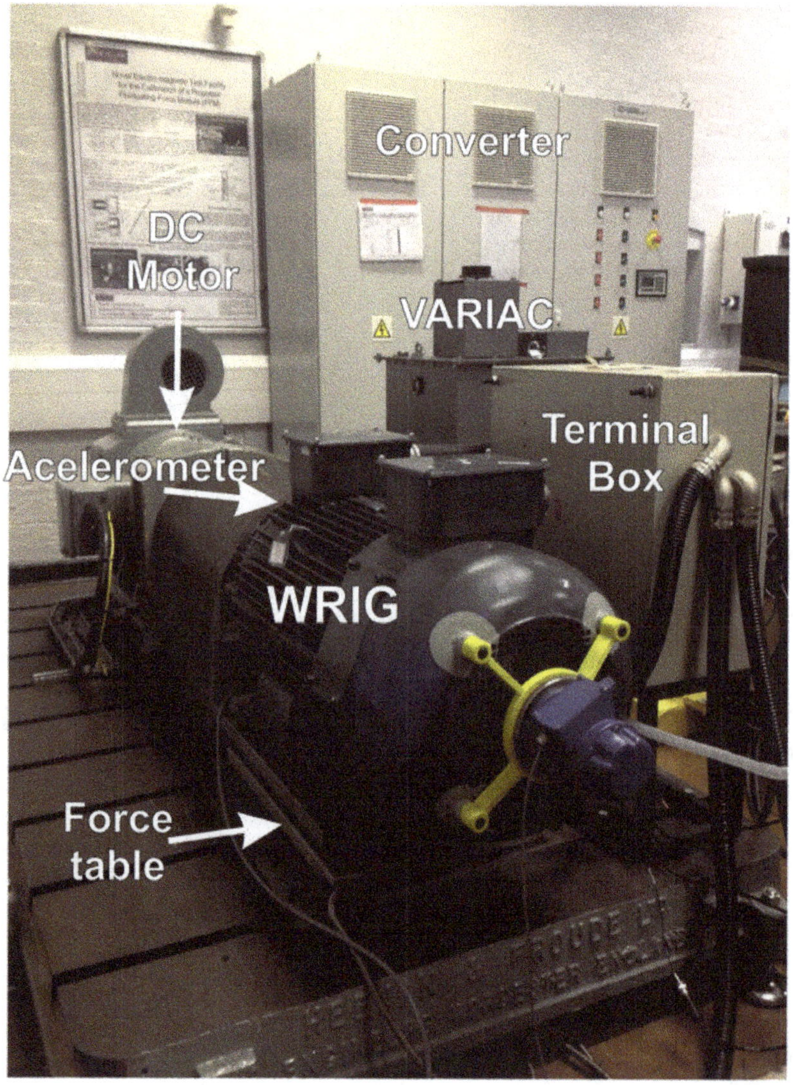

Fig. 5 *30 kW WRIG laboratory test rig*

observed in the torque signal will be equally present as vibration, since how the electrically excited vibration components are established largely depends on the considered machine's frame mechanical response. Excitation forces acting near natural frequencies of the machine's mechanical system are likely to result in a pronounced motion response [16].

To verify the spectral phenomena observed in numerical simulations, a series of experiments were undertaken on the laboratory test rig. The generator was driven at a typical operating speed of 1590 rpm and vibration, shaft torque, stator voltage and current signals recorded. Stator short-circuit fault was simulated by shorting seven turns in laboratory tests. The open-circuit fault was established by open circuiting one leg in a phase winding. To avoid damage to machine windings and because of laboratory supply limitations (≃60 A), the short-circuit current was limited to being approximately equal to the machine rated current (50 A) by adjusting the variable resistance in the short-circuit path to 0.3 Ω.

The estimated air-gap torque spectrum obtained from stator voltage and current measurements for the machine operating with balanced windings is shown in Fig. 6. The electromagnetic torque

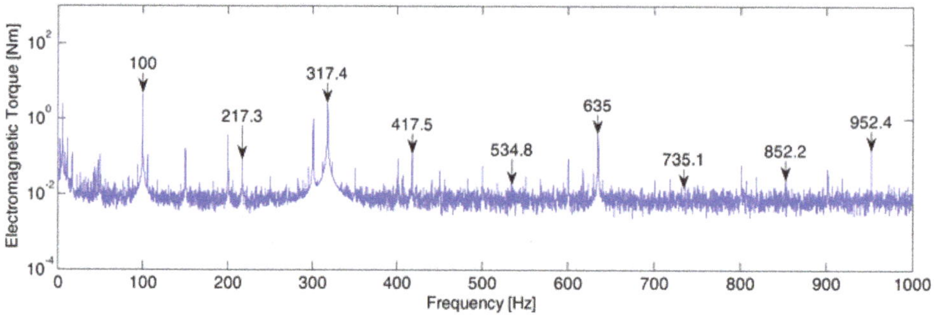

Fig. 6 *Estimated torque spectrum for balanced windings and unbalanced supply, 1590 rpm*

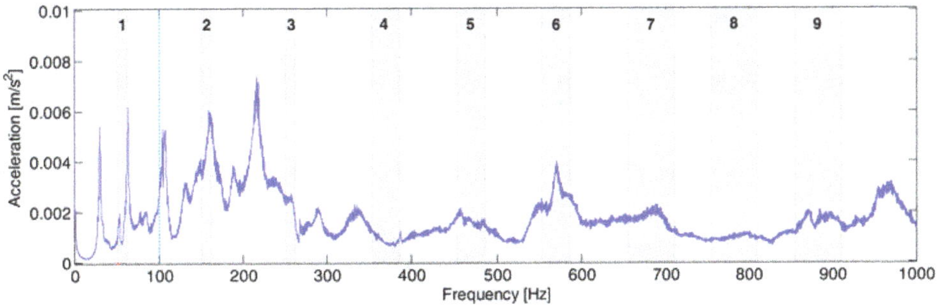

Fig. 7 *Test-rig impact test frequency response, vertical acceleration*

frequencies predicted by the model simulation in Fig. 2 for the same operating conditions and calculated using the equations in Table. 2 are clearly present in the experimental results. This observation confirms the validity of the developed frequency expressions and the efficacy of the numerical model. There exist additional components in the estimated torque signal, which are believed to mostly originate from the supply harmonic effects and unbalances that this paper neglects [24]. The estimated air-gap torque signal does not account for unbalanced operation or mechanical interactions in the system so the shaft torque signal measurements are used in further analysis in this paper.

To investigate the influence of the mechanical response of the laboratory test rig on the electromagnetically induced vibrations, a series of impact tests were performed recording the mechanical response of the machine frame to vertical axis excitation. Fig. 7 presents the measured frequency response obtained from impact testing for the vertical acceleration component; this corresponds to the frame vibration measurements on the vertical plane investigated

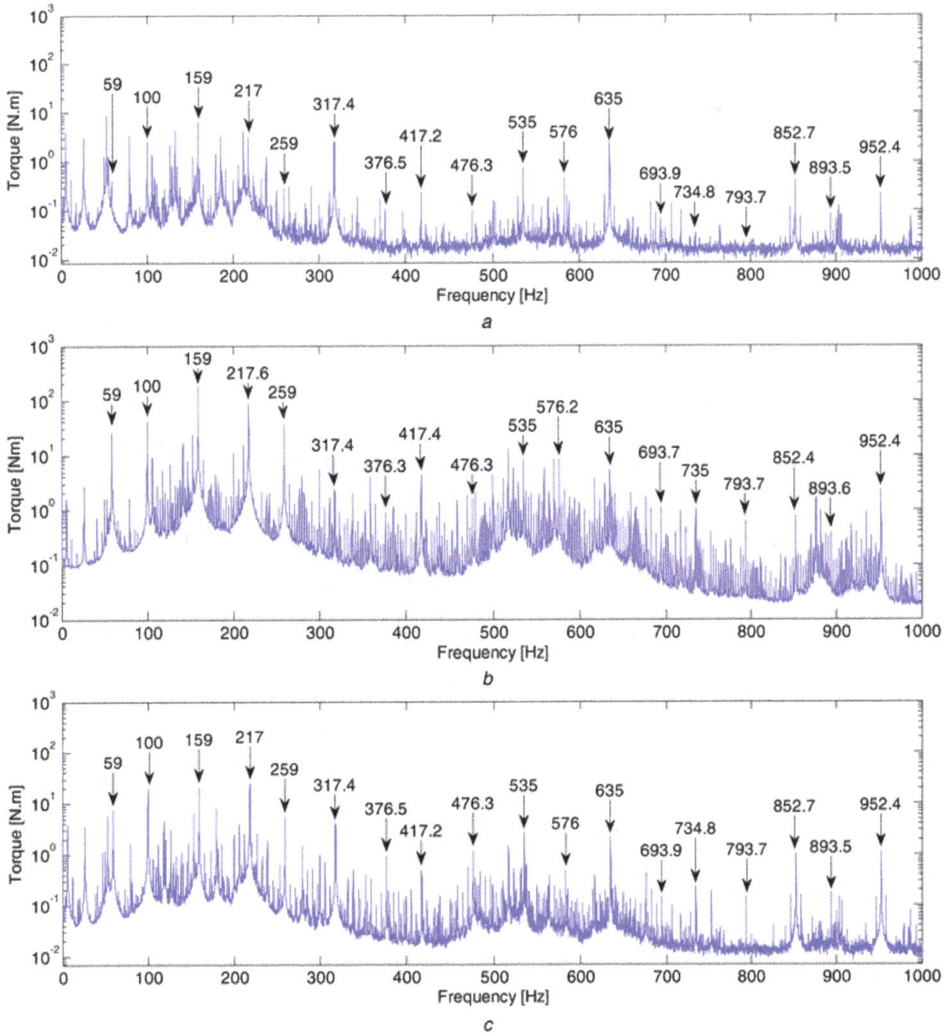

Fig. 8 *Measured shaft torque spectrum, 1590 rpm*
a Balanced windings
b Open-circuit fault
c Short-circuit fault

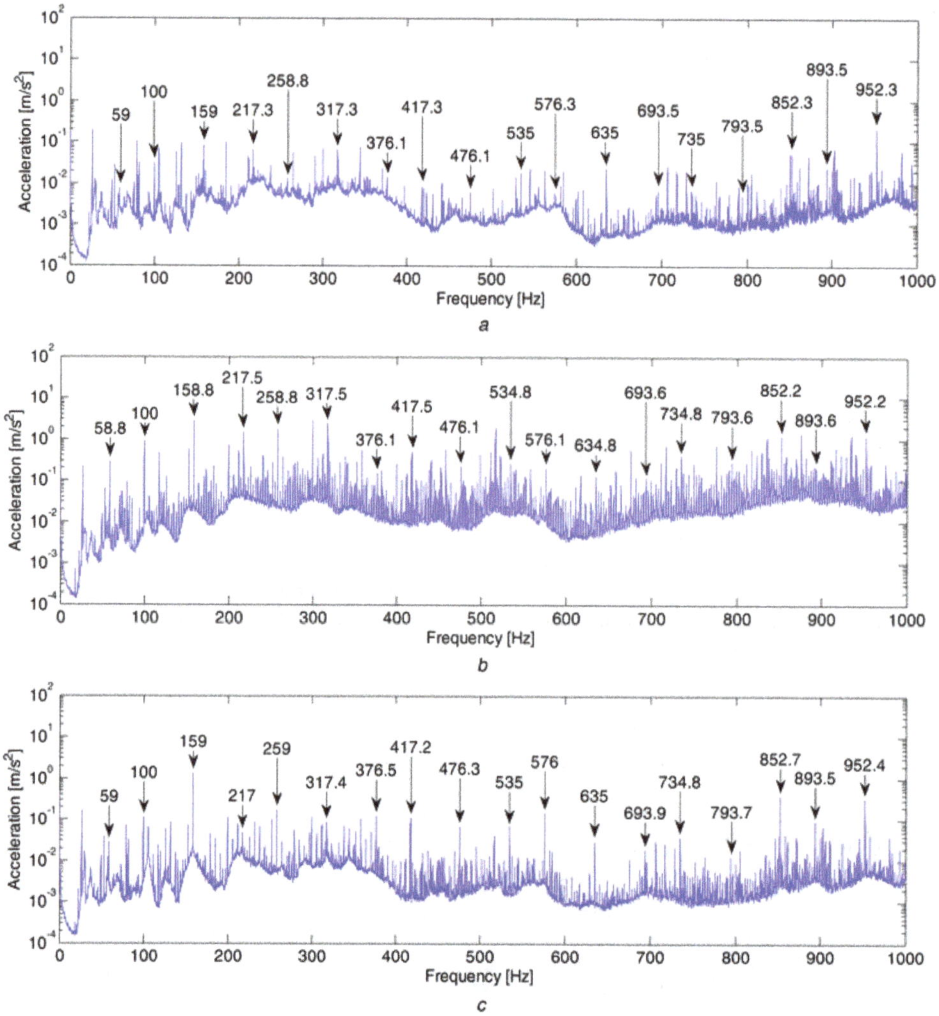

Fig. 9 *Measured vibration signal spectrum, 1590 rpm*
a Balanced windings
b Open-circuit fault
c Short-circuit fault

experimentally in this paper. The frequency bands containing each of the nine slip dependent fault frequencies identified in Table 2 are mapped in the figure for the machine rated operating speed range (≃1500–1620 rpm).

The recorded response indicates that frame excitation at the second unbalance induced frequency [(12), $k = 6$ from Table 2] can be expected to exhibit the strongest response in the vibration spectrum for the machine rated operating region. The 1st, 3rd and 6th fault frequency excitations can also be expected to exhibit a moderately strong response. It is interesting to note that, in addition to being present for machine operation with healthy windings (Fig. 6), the excitation at twice supply frequency $2f$ (100 Hz) that is commonly monitored for global electrical fault detection [17] is not expected to exhibit the strongest response in vibration as result of a stator winding failure. This implies that, depending on machine design, alternative frequency components may be better suited for electrical fault detection using vibration analysis.

The measured vibration and shaft torque signals spectra for machine operation with balanced windings, open-circuit and short-circuit faults are shown in Figs. 8 and 9, respectively. Frequencies corresponding to the electromagnetic torque signal components identified in Table 2 are labelled in the measurements. Minute discrepancies between the numerical data in Table 2 and frequencies measured in Figs. 8 and 9 are because of an assumed ideal 50 Hz supply and 1590 rpm speed; the measured values of these two quantities are slightly different and time-varying because of inherent supply frequency oscillations and limitations in velocity measurement accuracy. The measured vibration and shaft torque data account for the mechanical effects in the test-rig drive train, as well as the supply harmonic effects, and are therefore significantly noisier than the estimated air-gap torque signal. A number of components seen in the measured spectra originate from inherent unbalances in the

Table 3 Measured vibration signal winding unbalance-specific components magnitude

#	Frequency, Hz	Vibration, m/s^2	Vibration [normalised]	
		Healthy	Short-circuit	Open-circuit
1	59	0.004938	4.76	16.51
2	159	0.065290	15.57	43.40
3	259	0.011670	9.69	113.36
4	377	0.011100	7.43	13.48
5	477	0.009648	22.31	50.66
6	577	0.004445	22.78	23.46
7	695	0.002061	11.10	210.19
8	795	0.002318	11.13	254.53
9	895	0.024170	3.52	45.01

Table 4 Measured vibration signal balanced windings specific components magnitude

#	Frequency, Hz	Vibration, m/s^2	Vibration [normalised]	
		Healthy	Short-circuit	Open-circuit
1	100	0.017180	3.56	90.34
2	218	0.037640	5.74	34.25
3	318	0.107800	3.05	16.31
4	418	0.009524	10.42	94.86
5	536	0.016000	3.86	28.89
6	636	0.019090	2.95	23.54
7	736	0.010460	3.66	44.07
8	854	0.047630	7.56	39.74
9	954	0.119600	7.03	7.65

machine mechanical system and are found at frequencies equal to integer multiples of the rotor mechanical speed.

The measured shaft torque signals are seen to be in good agreement with the spectral content patterns observed in the model predictions of electromagnetic torque in Figs. 2–4 and the measured air-gap torque spectrum in Fig. 6. All relevant torque signal components predicted by the model are clearly manifested in the measured shaft torque spectrum for the considered healthy and faulty operating conditions. As expected, the mechanical response of the machine frame is seen to have a moderate effect on how the existing shaft torque pulsations are transferred to machine frame as vibration. All shaft torque signal spectral components identified as originating from the interaction of air-gap fields are clearly present in the corresponding measured vibration signals shown in Fig. 9.

The experimental results in Figs. 8 and 9 indicate a clear relationship between the magnitude of all electrically excited vibration signal spectral components and the presence of winding fault. The recorded increase in individual component magnitude averaged from three separate vibration measurements is summarised in Tables 3 and 4 for winding unbalance-specific components and those existing for healthy windings, respectively. The data in Fig. 9 and Tables 3 and 4 demonstrate a general vibration level increase in presence of winding fault. However, on examination of the individual components magnitude change with fault, it becomes clear that the winding unbalance-specific components exhibit a significantly higher rise in relative magnitude when compared with other electrically excited vibration components. The unbalance-specific components are seen to be manifested at different levels for short-circuit and open-circuit faults at the investigated load point, indicating different sensitivity to the existing unbalance intensity. This is especially evident for the 477 and 577 Hz components that exhibit the highest increase for a short-circuit fault. Similarly, the magnitude increase of 695 and 795 Hz components is dominant for an open-circuit fault. The latter pair of components is also seen to exhibit a much higher sensitivity to fault than the component at double supply frequency, which is a commonly used stator electrical unbalance indicator. The 418 Hz component is also seen to have a strong presence in the spectrum with existence of winding fault. All these high sensitivity frequency components were found to be in close proximity to local resonances of the mechanical systems.

To provide a more clear illustration of the winding fault induced effects, a detailed side-by-side comparison of the measured shaft torque and vibration signal spectra is shown in Fig. 10. To separately identify the influence and highlight the natural modes of the laboratory rig mechanical system, the figure also includes the WRIG vibration and shaft torque spectra measured with no excitation and the rotor driven at constant speed. For the presented narrowband spectra, a clear increase in the magnitude of ≃59 Hz torque and vibration component occurs with the presence of a winding fault. A much less significant fault induced magnitude variation is present in other spectral components.

To assess the impact of rotor converter noise on the observed spectral phenomena when WRIG is configured for DFIG operation, a series of experiments were conducted with the machine's rotor circuit interfaced to the grid via the back-to-back converter. Fig. 11 shows the shaft torque and vibration signals frequency spectra for DFIG operation at 1590 rpm and the test rig operating conditions consistent with those analysed in Fig. 10. The converter operation is seen to induce the appearance of additional low magnitude frequency components and general noise in the shaft torque and vibration spectra of the WRIG. These effects are clearly

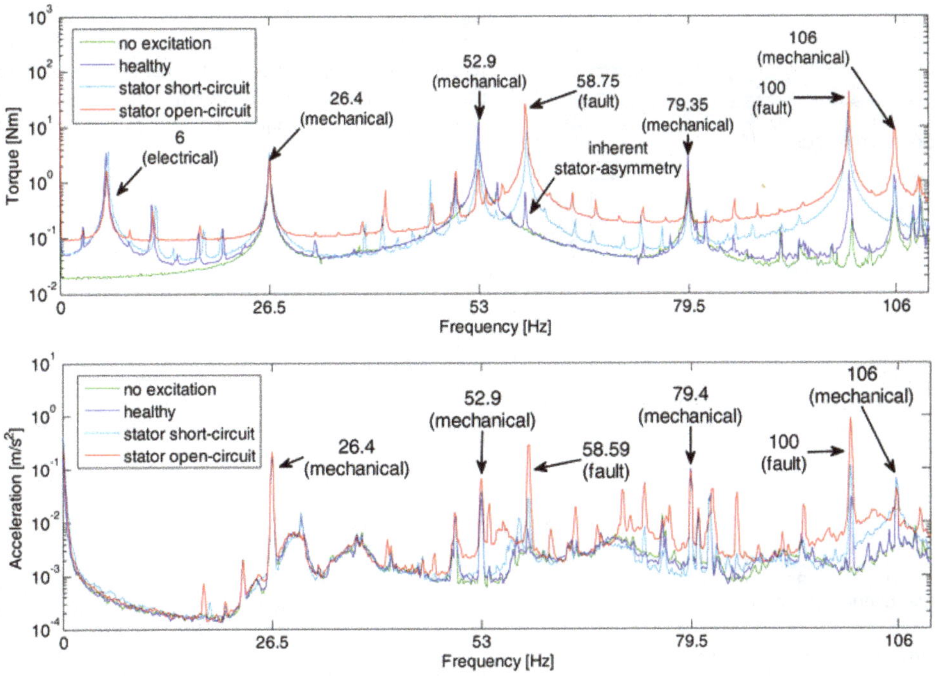

Fig. 10 *Measured torque (top) and vibration (bottom) signals spectra for machine operation at 1590 rpm*

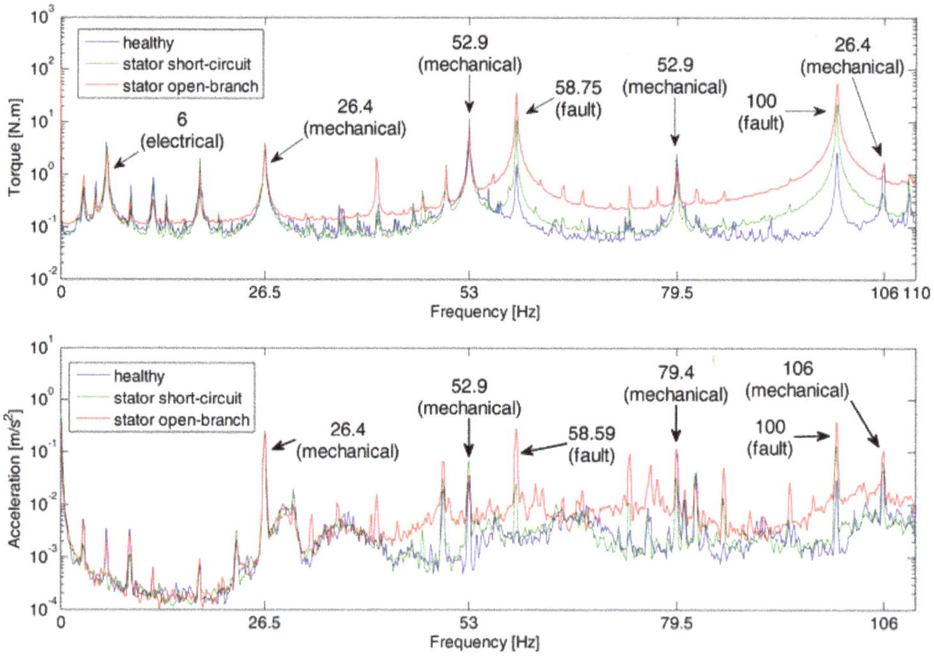

Fig. 11 *Measured torque (top) and vibration (bottom) signals spectra for DFIG machine operation at 1590 rpm*

illustrated in the frequency spectrum below the fundamental rotational frequency (26.4 Hz) for the measured bandwidth shown in Fig. 11. More importantly, the observed effects were not found to have a significant detrimental influence on the identified fault specific components, suggesting these could be used as fault indicators for DFIG operation.

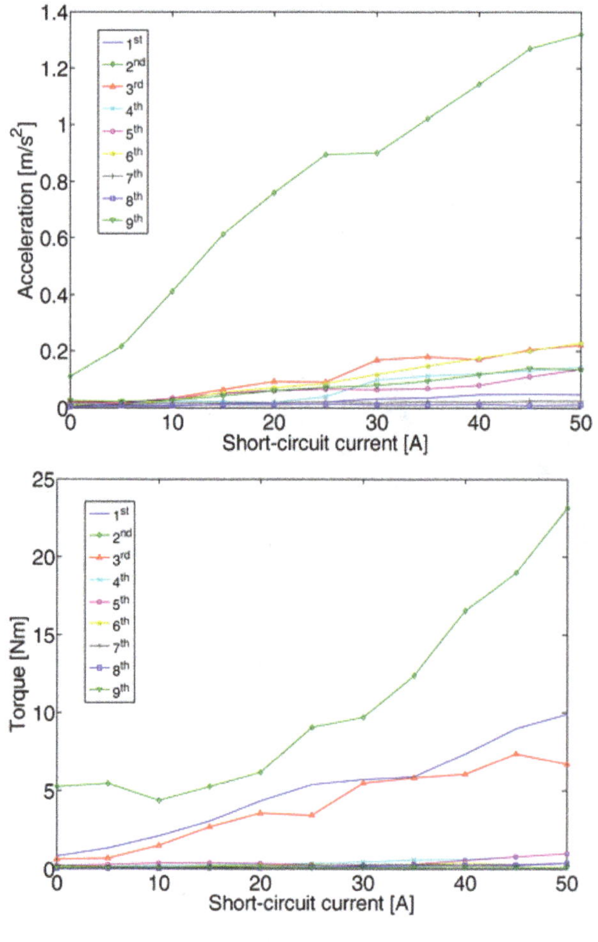

Fig. 12 *Measured fault sensitivity of the winding unbalance-specific vibration (top) and shaft torque (bottom) frequencies for different levels of short-circuit current, 1590 rpm*

4.3 *Sensitivity assessment*

The results of an experimental study investigating the magnitude sensitivity of the nine winding unbalance induced components identified in Table 2 to different short-circuit fault levels are shown in Fig. 12. The data clearly indicate that the 2nd frequency component is exhibited at the highest magnitude level suggesting that its monitoring would be most feasible for the investigated machine design. It is interesting to note in Fig. 12 that, although good correlation exists between the shaft torque and vibration signals, the relative magnitude differs considerably between some torque and vibration components at matching frequencies. This difference may be explained by the fact that the shaft torque measurement is established by considering the interaction between forces acting in both horizontal and vertical planes [23], whereas the

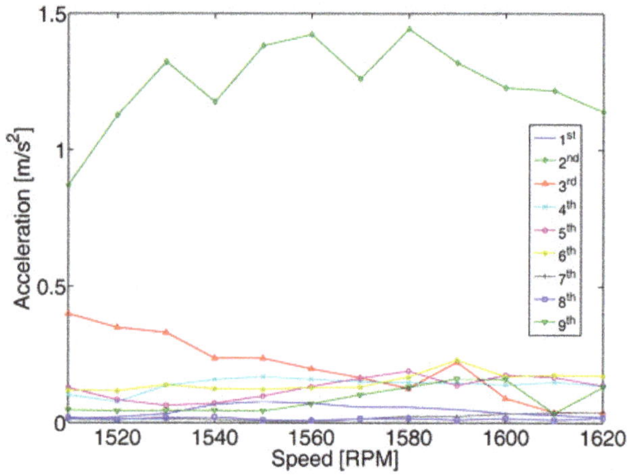

Fig. 13 *Measured load dependency of the winding unbalance-specific vibration frequencies for 50 A short-circuit current*

vibration signal is reported for the vertical component of acceleration only.

A load dependency study was conducted, where a series of signal measurements were taken in uniform steps within the machine's rated operating region, that is, from no load conditions until rated current (≃59 A stator) was achieved. For each different steady-state load levels, three separate measurements were taken for stator short-circuit fault and averaged in order to minimise the sensitivity to variations in the grid supply. The study results are presented in Fig. 13 for the nine winding unbalance induced components. The data show a predominantly rising trend in magnitude of components with increasing load. However, individual components exhibit different load dependencies and some are seen to be more pronounced than others. In concordance with Fig. 7, the results in Fig. 13 highlight the 2nd fault frequency component as having the highest magnitude throughout the operating range. In combination with the sensitivity study results this data confirm that, for the investigated machine design, the 2nd [(12), $k=6$] unbalance-specific vibration component would be the most suitable for providing winding fault information.

5 Conclusions

This paper investigates the electromagnetically induced shaft torque and vibration signal pulsations for wound rotor induction machine operation with and without stator winding unbalance or fault. These effects are first examined through developing a generalised theoretical analysis of the machine air gap fields' interaction and the resulting electromagnetic torque pulsations. A detailed harmonic model of the investigated industrial machine design is then employed to perform a numerical study of the winding fault and electrical unbalance effects in the electromagnetic torque signal. This paper then uses a purpose built fully instrumented 30 kW wound rotor induction machine laboratory test rig to undertake an in-depth experimental study of winding unbalance and fault effects. The measured shaft torque and stator frame vibration signals are examined for the rated operating range of the laboratory machine. The test rig can be configured to operate in steady-state DFIG regime and was used to investigate the influence of the rotor circuit converter on the observed spectral effects.

It is shown that a stator winding asymmetry gives rise to a range of theoretically and numerically predictable pulsating electromagnetic torques that are detectable in the machine shaft torque and stator frame vibration signals. These are demonstrated to be significantly amplified by the presence of stator short-circuit and open-circuit faults. Monitoring of these components may therefore provide useful information on the stator windings electrical condition. Further investigation of the observed effects on a wider range of industrial wound rotor machine designs would be required to confirm the generality of the reported phenomena.

6 Acknowledgment

This work was supported by the UK EPSRC as part of the Supergen Wind Consortium under grant EP/H018662/1.

7 References

[1] 'EU Energy Policy to 2050', The European Wind Energy Association Report, EWEA, March 2011

[2] 'UpWind: Design Limits and Solutions for Very Large Wind Turbines', The European Wind Energy Association Report, EWEA, March 2011

[3] Spinato F., Tavner P.J., Van Bussel G.J.W., Koutolakos E.: 'Reliability of wind turbine subassemblies', *IET Renew. Power Gener.*, 2009, **3**, (4), pp. 1–15

[4] Wilkinson M., Hendriks B., Spinato F., *et al.*: 'Methodology and results of the reliawind reliability field study'. Scientific Proc. European Wind Energy Conf. (EWEC), Warsaw, Poland, March 2010

[5] Alewine K., Chen W.: 'Wind turbine generator failure modes analysis and occurrence'. Windpower 2010, Dallas, Texas, May, 2010, pp. 1–6

[6] Alewine K., Chen W.: 'A review of electrical winding failures in wind turbine generators'. Proc. IEEE Electrical Insulation Conf. (EIC) 2011, June 2011, pp. 392–397

[7] Tavner P.J.: 'Review of condition monitoring of rotating machines', *IET Renew. Power Gener.*, 2008, **2**, (4), pp. 215–247

[8] Williamson S., Djurović S.: 'Origins of stator current spectra in dfigs with winding faults and excitation asymmetries'. Proc. IEEE IEMDC 2009, Miami, USA, 2009, pp. 563–570

[9] Yazidi A., Henao H., Capolino G.A., Betin F.: 'Rotor inter-turn short circuit fault detection in wound rotor induction machines'. Proc. IEEE ICEM 2010, Rome, Italy, 2010, pp. 1–6

[10] Shah D., Nandi S., Neti P.: 'Stator inter-turn fault detection of doubly-fed induction generators using rotor current and search coil voltage signature', *IEEE Trans. Ind. Appl.*, 2009, **45**, (5), pp. 1831–1842

[11] Crabtree C.J.: 'Survey of commercially available condition monitoring systems for wind turbines'. SuperGen Wind Energy Technologies Consortium, November 2010

[12] Yang W., Tavner P.J., Crabtree C.J., Feng Y., Qiu Y.: 'Wind turbine condition monitoring: technical and commercial challenges', Wind Energy, 2012

[13] Li W., Mechefske C.K.: 'Detection of induction motor faults: a comparison of stator current, vibration and acoustic methods', *J. Vib. Control*, 2006, **12**, (2), pp. 165–188

[14] Hsu J.S.: 'Monitoring of defects in induction motors through air gap torque observation', *IEEE Trans. Ind. Appl.*, 1995, **31**, (5), pp. 1016–1021

[15] Ding F., Trutt F.C.: 'Calculation of frequency spectra of electromagnetic vibration for wound-rotor induction machines with winding faults', *Electr. Mach. Power Syst.*, 1988, **14**, pp. 137–150

[16] Trutt F.C., Sottile J., Kohler J.L.: 'Detection of AC machine winding deterioration using electrically excited vibrations', *IEEE Trans. Ind. Appl.*, 2001, **37**, (1), pp. 10–14

[17] Rodriguez P.J., Belahcen A., Arkkio A.: 'Signatures of electrical faults in the force distribution and vibration pattern of induction motors', *Proc. Inst. Electr. Eng. – Electr. Power Appl.*, 2006, **153**, p. 523

[18] Djurovic S., Vilchis-Rodriguez D., Smith A.C.: 'Vibration monitoring for wound rotor induction machine winding fault detection'. 2012 XXth Int. Conf. Electrical Machines (ICEM), 2–5 September 2012, pp. 1906–1912

[19] Djurović S., Williamson S., Renfrew A.: 'Dynamic model for doubly-fed induction generators with unbalanced excitation, both with and without faults', *IET Electr. Power Appl.*, 2009, **3**, (3), pp. 171–177

[20] Tuohy P., Djurović S., Smith A.C.: 'Finite element analysis of winding fault effects in a wound rotor induction machine with experimental validation'. Proc. IET PEMD 2012, Bristol, UK, 2012, pp. 1–6

[21] Williamson S., Mirzoian K.: 'Analysis of cage induction motors with stator winding faults', *IEEE Trans. Power Appar. Syst.*, 1985, **PAS-104**, (7), pp. 1838–1842

[22] Hsu J.S., Woodson H.H., Weldon W.F.: 'Possible errors in measurement of air-gap torque pulsations of induction motors', *IEEE Trans. Energy Convers.*, 1992, **7**, pp. 202–208

[23] Healey R.C., Lesley S., Williamson S., Palmer P.R.: 'The measurement of transient electromagnetic torque in high performance electrical drives'. Proc. Sixth Int. Conf. Power Electronics and Variable Speed Drives, Nottingham, UK, 1996, pp. 226–229

[24] Djurović S., Williamson S.: 'Influence of supply harmonic voltages on dfig stator current and power spectrum'. Proc. IEEE ICEM 2010, Rome, Italy, 2010, pp. 1–6

8 Appendix

8.1 *Appendix 1: Machine parameters*

MarelliMotori E4F-225 wound rotor induction machine. About 240 V, 50 Hz, 30 kW, 3phase, 4 poles, 59 A stator rated current and 56 A rotor rated current.

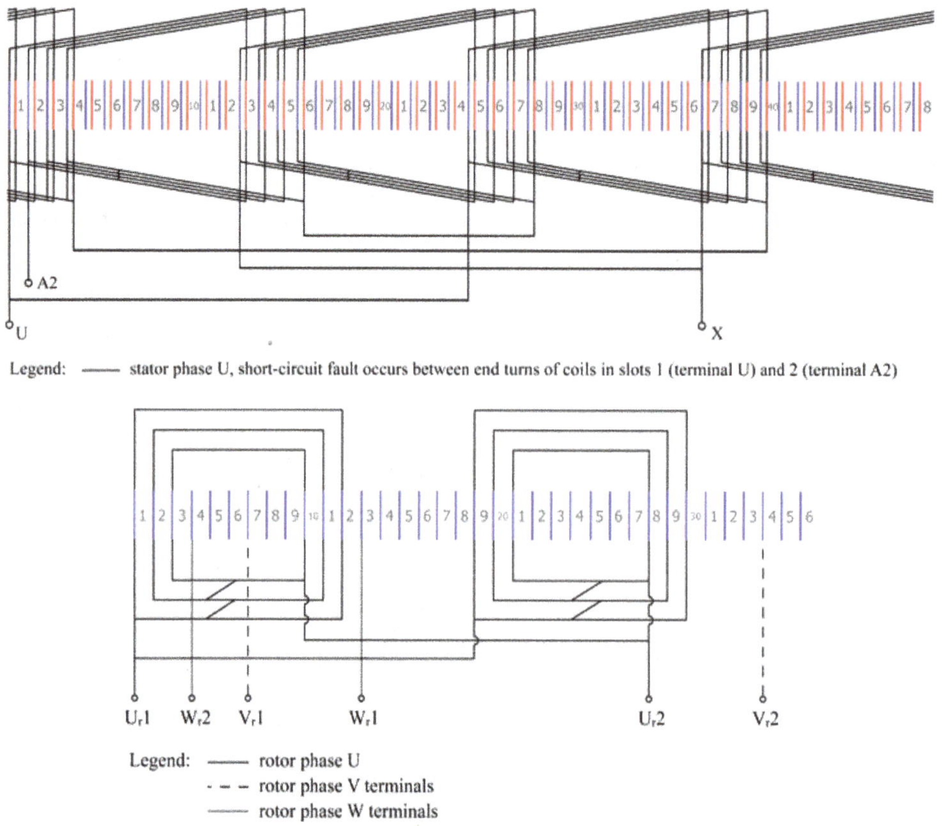

Fig. 14 *Connection diagrams of stator (top) and rotor (bottom) windings for the experimental machine*

8.2 Appendix 2: Machine model connection matrices

8.2.1 Healthy

$$
\boldsymbol{C} = \begin{bmatrix}
1 & 0 & -1 & 0 & 0 & 0 & 0 & 0 & 0 & 0 & 0 \\
0 & 0 & 1 & 0 & 0 & 0 & 0 & 0 & 0 & 0 & 0 \\
0 & 1 & 0 & 0 & -1 & 0 & 0 & 0 & 0 & 0 & 0 \\
-1 & 0 & 0 & 0 & 1 & 0 & 0 & 0 & 0 & 0 & 0 \\
0 & -1 & 0 & 1 & 0 & 0 & 0 & 0 & 0 & 0 & 0 \\
0 & 0 & 0 & -1 & 0 & 0 & 0 & 0 & 0 & 0 & 0 \\
0 & 0 & 0 & 0 & 0 & 1 & 0 & -1 & 0 & 0 & 0 \\
0 & 0 & 0 & 0 & 0 & 0 & 0 & 1 & 0 & 0 & 0 \\
0 & 0 & 0 & 0 & 0 & 0 & 1 & 0 & 0 & -1 & 0 \\
0 & 0 & 0 & 0 & 0 & -1 & 0 & 0 & 0 & 1 & 0 \\
0 & 0 & 0 & 0 & 0 & 0 & 0 & 0 & -1 & 0 & 0 \\
0 & 0 & 0 & 0 & 0 & 0 & -1 & 0 & 1 & 0 & 0 \\
0 & 0 & 0 & 0 & 0 & 0 & 0 & 0 & 0 & 0 & 1
\end{bmatrix}
$$

8.2.2 Short-circuit fault

$$
\boldsymbol{C} = \begin{bmatrix}
1 & 0 & -1 & 0 & 0 & 0 & 0 & 0 & 0 & 0 & 0 \\
0 & 0 & 1 & 0 & 0 & 0 & 0 & 0 & 0 & 0 & -1 \\
0 & 1 & 0 & 0 & -1 & 0 & 0 & 0 & 0 & 0 & 0 \\
-1 & 0 & 0 & 0 & 1 & 0 & 0 & 0 & 0 & 0 & 0 \\
0 & -1 & 0 & 1 & 0 & 0 & 0 & 0 & 0 & 0 & 0 \\
0 & 0 & 0 & -1 & 0 & 0 & 0 & 0 & 0 & 0 & 0 \\
0 & 0 & 0 & 0 & 0 & 1 & 0 & -1 & 0 & 0 & 0 \\
0 & 0 & 0 & 0 & 0 & 0 & 0 & 1 & 0 & 0 & 0 \\
0 & 0 & 0 & 0 & 0 & 0 & 1 & 0 & 0 & -1 & 0 \\
0 & 0 & 0 & 0 & 0 & -1 & 0 & 0 & 0 & 1 & 0 \\
0 & 0 & 0 & 0 & 0 & 0 & 0 & 0 & -1 & 0 & 0 \\
0 & 0 & 0 & 0 & 0 & 0 & -1 & 0 & 1 & 0 & 0 \\
0 & 0 & 0 & 0 & 0 & 0 & 0 & 0 & 0 & 0 & 1
\end{bmatrix}
$$

8.2.3 Open-circuit fault

$$
\boldsymbol{C} = \begin{bmatrix}
0 & 0 & -1 & 0 & 0 & 0 & 0 & 0 & 0 & 0 & 0 \\
1 & 0 & 0 & 0 & 0 & 0 & 0 & 0 & 0 & 0 & 0 \\
0 & 1 & 0 & 0 & -1 & 0 & 0 & 0 & 0 & 0 & 0 \\
-1 & 0 & 0 & 0 & 1 & 0 & 0 & 0 & 0 & 0 & 0 \\
0 & -1 & 0 & 1 & 0 & 0 & 0 & 0 & 0 & 0 & 0 \\
0 & 0 & 0 & -1 & 0 & 0 & 0 & 0 & 0 & 0 & 0 \\
0 & 0 & 0 & 0 & 0 & 1 & 0 & -1 & 0 & 0 & 0 \\
0 & 0 & 0 & 0 & 0 & 0 & 0 & 1 & 0 & 0 & 0 \\
0 & 0 & 0 & 0 & 0 & 0 & 1 & 0 & 0 & -1 & 0 \\
0 & 0 & 0 & 0 & 0 & -1 & 0 & 0 & 0 & 1 & 0 \\
0 & 0 & 0 & 0 & 0 & 0 & 0 & 0 & -1 & 0 & 0 \\
0 & 0 & 0 & 0 & 0 & 0 & -1 & 0 & 1 & 0 & 0 \\
0 & 0 & 0 & 0 & 0 & 0 & 0 & 0 & 0 & 0 & 1
\end{bmatrix}
$$

3

Sensorless speed control of a five-phase induction machine under open-phase condition

Ahmed S. Morsy[1], Ayman S. Abdelkhalik[1], Shehab Ahmed[2], Ahmed Mohamed Massoud[3]

[1]Electrical Power and Machines Department, Faculty of Engineering, Alexandria University, Alexandria, Egypt
[2]Electrical & Computer Engineering, Texas A&M University at Qatar, Doha 23874, Qatar
[3]Electrical Engineering Department, Qatar University, Doha, Qatar
E-mail: ahmed.salah.morsy@ieee.org

Abstract: Recently, multiphase machines have been promoted as competitors to their three-phase counterparts in high-power safety-critical drive applications. Among numerous advantages of multiphase induction machine (IM) drives, self-starting and operation under open phase(s) stand as the most salient features. With open phase(s), optimal current control provides disturbance- free operation given a set of objective functions. Although hysteresis current control was merely employed in the literature as it offers a simple controller structure to control the remaining healthy phases, it is not suitable for high-power applications. In the literature, multiple synchronous reference frame (dq) control can be an alternative; however, it requires back and forth transformations with several calculations and additional sophistication. In this paper, a simple technique employing adaptive proportional resonant (PR) current controllers is presented to control a five-phase IM under open-phase conditions. Results for both volt/hertz (V/f) and field oriented control (FOC) systems are presented. Moreover, sensorless operation under fault condition is also demonstrated by estimating the machine speed using a rotor flux-based model reference adaptive system (MRAS) speed estimator. The proposed controllers are experimentally verified and compared. Although FOC provides better dynamic performance, V/f control offers a simpler control structure and a lower number of PR controllers.

1 Introduction

In high power and safety-critical applications, multiphase induction motors (IMs) are strong competitors to their three-phase counterparts based on the numerous advantages offered by multiphase systems [1]. First, multiphase machines can be designed with a reduced per-phase current correspondingly reduced semiconductor devices' current rating. Second, multiphase systems offer additional degrees of freedom that improve system performance, increase system fault tolerant capability and enhance machine power density using harmonic current injection [1–3]. Among these vast features, fault tolerant capability is recognised as the most salient feature of multiphase systems. Theoretically, multiphase machines with 'n' phases can continue running with up to '$n-2$' disconnected phases [4]. Multiphase machine performance with open circuited phases has been addressed in the literature [4, 5] to a large extent, and control strategies to ensure disturbance-free operation with the same pre-fault magneto-motive force have been demonstrated.

Field oriented control is usually employed to control multiphase IMs. Several recent papers [6–11] introduce control schemes to ensure motor operation when one or more phases are open-circuited while satisfying specific optimisation criteria [6]. Generally, minimum torque ripples, equal phase currents and minimum copper losses are among the most common targeted optimisation functions [6]. Most of the proposed controllers are mainly based on rotor field oriented control (FOC) which requires accurate machine parameters to ensure proper orientation.

Another widely used control technique for IMs is the constant volt/hertz (V/f) method [12, 13]. Despite the fact that vector control gives better dynamic response; yet, scalar control is simpler and still widely used in industrial fields. In the literature, the control of multiphase IM using V/f is addressed for the healthy case [12, 13]; however under the open phase case, little work has been conducted [14]. In earlier work [14], a simple open-loop V/f controller is provided for open-circuit faults using proportional resonant (PR) controllers to ensure equal line currents and hence minimum torque ripples. Under open phase, the sequence components of stator currents as calculated by the control system may contain a significant negative sequence, that is, at angular frequency $-\omega_s$. Additionally, optimal current control requires unbalanced current components in the non-fundamental sequence planes which yield negative sequence components of frequency $-\omega_s$ in the reference currents. Hence, employing conventional synchronous reference frame proportional–integral (PI) regulators yield non-zero tracking errors. This problem has been tackled in [6], but with adding current regulators in negative sequence (dq) synchronous reference frame which sophisticates the controller structure. On the other hand, PR controllers [15] are advantageous over conventional PI controllers in the synchronous reference frame with their capability to track unbalanced reference currents without sophisticated transformations [14].

Sensorless operation of a three-phase IM is well recognised in the literature [16, 17]; however, little work has been conducted for multiphase IMs [18–22] and assuming unbalanced operation. In this paper, two sensorless closed-loop speed controllers based on V/f and FOC to control a five-phase IM under phase open conditions are presented and compared. The earlier work for open-loop V/f control of a five-phase IM presented in [14] is extended to obtain sensorless closed-loop operation. Moreover, a controller based on rotor FOC is presented where PR controllers are employed for each sequence plane to generate the stator voltage components corresponding to these planes. Although this proposed controller is functionally equivalent to that presented in [6], which used synchronous PI controllers, the total number of controllers is now reduced from six PI controllers to three PR controllers. Moreover, the required forward and backward transformations are dispensed. Sensorless operation is provided by estimating the machine speed using an model reference adaptive system (MRAS) observer [18] using the fundamental sequence components of the rotor flux (RF-MRAS). The proposed controllers are experimentally verified using a 1 hp prototype five-phase IM. Both steady-state and dynamic performances are presented.

2 Optimal current control

In multiphase systems, disturbance-free operation can be provided with some phases open by controlling the currents of the remaining healthy phases to ensure certain optimisation criterion [6]. For a five-phase machine with one phase open, the remaining healthy phases are usually controlled to ensure equal stator phase currents, minimum machine torque ripples and maximum torque production by nullifying the fundamental negative sequence [23].

For a five-phase system, the sequence currents can be obtained from the phase values using the transformation shown in (1) [6]

$$\left[\boldsymbol{i}_{s\alpha\beta}\right] = [\boldsymbol{T}]\left[\boldsymbol{i}_{ph}\right] \tag{1}$$

where $\left[\boldsymbol{i}_{s\alpha\beta}\right] = \left[\, i_{s\alpha1} \;\; i_{s\beta1} \;\; i_{s\alpha3} \;\; i_{s\beta3} \;\; i_{s0} \,\right]^{\mathrm{T}}$ and $\left[\boldsymbol{i}_{ph}\right] = \left[\, i_a \;\; i_b \;\; i_c \;\; i_d \;\; i_e \,\right]^{\mathrm{T}}$. The transformation matrix, $[\boldsymbol{T}]$, is defined by (2)

$$[T] = \frac{2}{5}\begin{bmatrix} 1 & \cos\gamma & \cos 2\gamma & \cos 3\gamma & \cos 4\gamma \\ 0 & -\sin\gamma & -\sin 2\gamma & -\sin 3\gamma & -\sin 4\gamma \\ 1 & \cos 3\gamma & \cos 6\gamma & \cos 9\gamma & \cos 12\gamma \\ 0 & -\sin 3\gamma & -\sin 6\gamma & -\sin 9\gamma & -\sin 12\gamma \\ 0.5 & 0.5 & 0.5 & 0.5 & 0.5 \end{bmatrix} \tag{2}$$

where $\gamma = 2\pi/5$.

For a star-connected five-phase machine with line 'a' assumed open, the remaining healthy phase currents that maintain same rated fundamental magneto-motive force (MMF) and nullify the fundamental negative sequence component are given in per-unit as [5]

$$\begin{aligned} i_a &= 0 \\ i_b &= 1.382\ \sin\left(\omega t - 36^{\circ}\right) = -i_d \\ i_c &= 1.382\ \sin\left(\omega t - 144^{\circ}\right) = -i_e \end{aligned} \tag{3}$$

Substituting in (1) by these optimum current in (3), the corresponding sequence current components in per-unit are as in (4)

$$\begin{aligned} i_{s\alpha1} &= \sin(\omega t) \\ i_{s\beta1} &= \cos(\omega t) \\ i_{s\alpha3} &= -\sin(\omega t) \\ i_{\beta s3} &= 0.236\ \cos(\omega t) \end{aligned} \tag{4}$$

This results in the identities shown in (5)

$$i_{s\alpha3} = -i_{s\alpha1}, \qquad i_{s\beta3} = 0.236\, i_{s\beta1} \tag{5}$$

Controlling the $\alpha\beta$ current components of the third sequence plane to fulfil the conditions given by (5) yields equal stator phase currents in the remaining healthy phases, maximum output torque and minimum torque ripples. Nevertheless, a corresponding increase in the machine copper loss by ~53% is expected. Practically, the machine should be deloaded to avoid excessive copper loss [5].

As a first glance, four current controllers are needed to control the four current components, which is the case for the available literature. However, for star connection with phase 'a' open and based on the inverse transformation given by (1), one can write

$$i_a = i_{s\alpha1} + i_{s\alpha3} = 0 \tag{6}$$

Hence, the first identity in (5) is always achieved, which reduces the system degrees of freedom and, hence, the required controllers to only three. Therefore, the second identity in (5) can be properly achieved by applying proper voltage component $v_{s\beta3}$, whereas the other voltage component $v_{s\alpha3}$ is redundant (can be simply set to zero). Reducing the total number of current controllers to three represents one of the main contributions of this paper when compared with available controllers in the literature.

3 Proposed controllers

In this section, two speed controllers are proposed based on V/f control and FOC to control the speed of a five-phase IM under one-phase open with MRAS speed estimation. PR controllers are employed to control the current components of the third sequence plane in accordance with the current components of the fundamental sequence plane to fulfil the relations given by (5).

3.1 V/f control

The complete system block diagram for the proposed V/f controller is shown in Fig. 1. In this controller, the fundamental stator voltage is decided as in conventional V/f control. The speed error is used to calculate the required slip frequency through a PI controller which is then added to the estimated rotor speed to determine the required synchronous speed. The measured currents are decomposed to their sequence components. The fundamental sequence is used to calculate the optimum reference of third sequence current component $i^*_{s\beta3}$, using (5), that ensures equal stator currents under phase open. Consequently, the third sequence stator voltage component $v_{s\beta3}$ that ensures equal remaining phase currents is obtained using a PR controller from the error in $i^*_{s\beta3}$, whereas $v_{s\alpha3}$ is set to zero since $i^*_{s\alpha3}$ is already achieved as shown in (6). Both fundamental voltage and current components are used to estimate the machine speed using a conventional MRAS observer.

3.2 Field oriented control

The complete system block diagram for the proposed FOC is shown in Fig. 2. In this controller, the fundamental flux and torque-producing current components i^*_{ds1} and i^*_{qs1}, respectively, are obtained as in conventional indirect rotor field orientation [3], then they are transformed to their stator frame current components $i^*_{\alpha s1}$ and $i^*_{\beta s1}$ using the machine estimated speed and the calculated machine slip. Based on (5), the corresponding third sequence β current component $i^*_{\beta s3}$ that ensures equal stator currents under phase open is determined, whereas $i^*_{\alpha s3}$ is naturally achieved as previously shown in (6). These reference current components are compared with their actual sequence current components and three PR controllers, two for the first sequence plane and one for the third sequence plane, are then used to obtain the corresponding first and third sequence voltage components. Both fundamental voltage and current components are used to estimate the machine speed using a conventional MRAS observer as will be shown in the next section.

3.3 PR controller

When compared with PI controllers in synchronous reference frame (*dq* control), the PR controller can effectively achieve zero tracking error for unbalanced reference currents without sophisticated transformations, using reduced calculations and without the need for the transformation angle [6]. To achieve zero tracking error at a variable fundamental frequency, its resonant poles in (7) must be adaptively tuned to the fundamental frequency of the tracked signal [15]. Figs. 1 and 2 show adaptive PR controller implemented (in V/f and FOC, respectively) to track the stator reference currents as

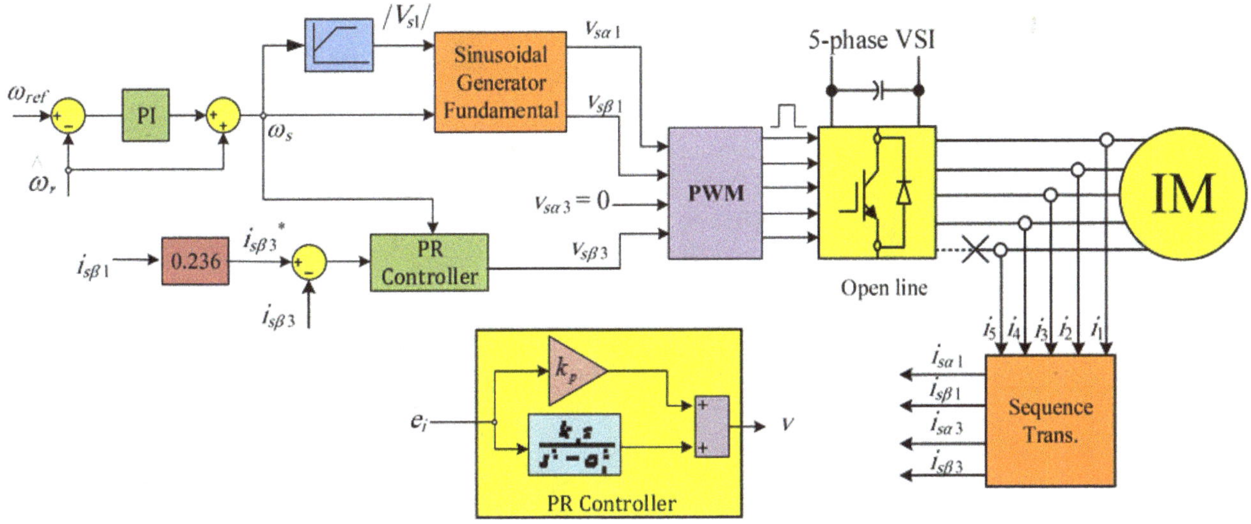

Fig. 1 *Block diagram of the proposed V/f sensorless speed controller*

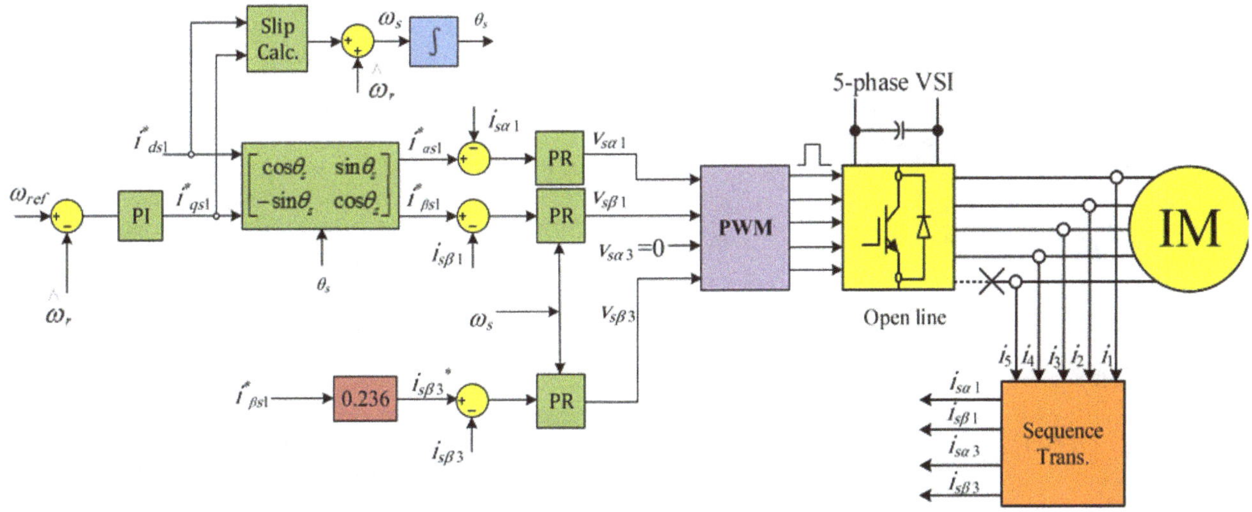

Fig. 2 *Block diagram of the proposed FOC sensorless speed controller*

discussed

$$\text{PR controller} \quad H(S) = K_p + K_i \frac{S}{S^2 + \omega_0^2} \tag{7}$$

For digital implementation on the digital signal processor (DSP), the PR controllers given in (7) are discretised at 5 kHz using a zero-order hold method [14].

The controller gains K_p, K_{i1} and K_{i3}, shown in Table 1, can be found as in [15]. The current control loop is required to be much faster than the speed control loop. The bandwidth of the current control loop (third sequence frame) is in the order of several hundreds of rad/s and it is set according to the machine time constant (L/r), whereas the speed loop bandwidth is in the order of tens of rad/s (depending on the inertia/friction time constant of the drive system).

Table 1 Current control parameter

Current control parameters	K_p	K_{i1}	K_{i3}
Value	7.5	75	50

4 Speed estimation using MRAS observer

MRAS schemes have been extensively employed for speed estimation in various control applications. Depending on the output states that form the error function, various MRAS observers have been introduced in the literature, where the most common are those based on RF-MRAS [24] and back EMF (BEMF-MRAS) [25].

The design of an MRAS estimator for speed estimation of IM drives requires the definition of two models having similar outputs. One model, termed the reference model, should be independent of the rotor speed, whereas the other, the adaptive model, is speed dependent on it. An adaptive mechanism, based on a PI controller, is employed to generate the value of the estimated speed in such a way as to minimise the error between the reference and estimated outputs [26]. A block diagram showing the MRAS speed estimator is given in Fig. 3. Since the fundamental sequence is the torque-producing sequence plane, this sequence plane can be effectively used to estimate the machine speed even under one-phase open.

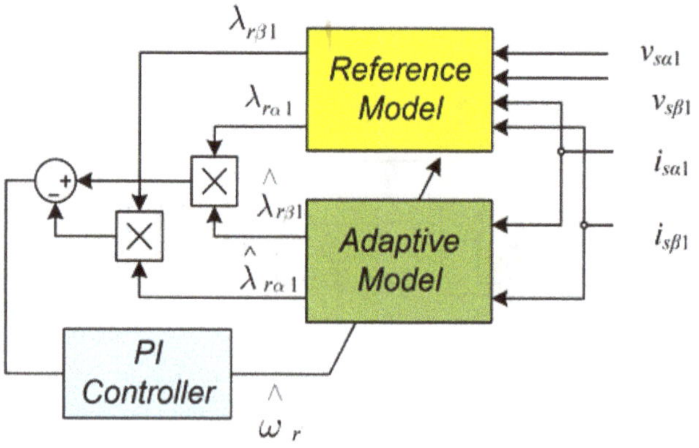

Fig. 3 *Block diagram of MRAS speed estimator*

In the RF-MRAS scheme, the reference model represents the stator equation (voltage model) of the fundamental sequence plane and can be given by (8) and (9) [24, 26]

$$p\lambda_{r\alpha} = \frac{L_{r1}}{L_{m1}}\left(v_{s\alpha 1} - R_s i_{s\alpha 1} - \sigma L_{s1} p i_{s\alpha 1}\right) \quad (8)$$

$$p\lambda_{r\beta} = \frac{L_{r1}}{L_{m1}}\left(v_{s\beta 1} - R_s i_{s\beta 1} - \sigma L_{s1} p i_{s\beta 1}\right) \quad (9)$$

where $\sigma = 1 - \left(L_{m1}^2 / L_{s1} L_{r1}\right)$ is the leakage factor.

The adaptive model represents the rotor equation (current model) and can be given by (10) and (11)

$$p\hat{\lambda}_{r\alpha} = \frac{R_{r1} L_{m1}}{L_{r1}} i_{s\alpha 1} - \frac{R_{r1}}{L_{r1}} \hat{\lambda}_{r\alpha} - \hat{\omega}_r \hat{\lambda}_{r\beta} \quad (10)$$

$$p\hat{\lambda}_{r\beta} = \frac{R_{r1} L_{m1}}{L_{r1}} i_{s\beta 1} - \frac{R_{r1}}{L_{r1}} \hat{\lambda}_{r\beta} + \hat{\omega}_r \hat{\lambda}_{r\alpha} \quad (11)$$

The estimated speed is then given by (12)

$$\hat{\omega}_r = \left(k_{po} + \frac{k_{io}}{p}\right)\left(\lambda_{r\beta}\hat{\lambda}_{r\alpha} - \lambda_{r\alpha}\hat{\lambda}_{r\beta}\right) \quad (12)$$

where the MRAS PI control parameters K_{po} and K_{io} are equal to (500, 10 000), respectively.

5 Experimental setup

A five-phase stator was built to fit an existing squirrel cage rotor of a three-phase machine to obtain the same power rating. The stator comprises a 4-pole five-phase single layer winding occupying 40 slots. The machine was fed from a custom made five-phase inverter operating at a 5 kHz switching frequency and fed from a 350 V DC-link. An eZdsp^{TM} DSP kit hosting a Texas Instrument floating point DSP (F28335) is used to provide the pulse width modulation (PWM) signals. The DSP is sending real-time measurement and control signals (voltages, currents and speed) on controller area network (CAN)-bus. The host personal computer is connected to the CAN-bus via CANcaseXL in order to display DSP signals and log them for post-analysis. Four Hall-effect transducers are used to measure the motor currents. The IM is coupled to a PM DC-generator of the same power rating which acts as a mechanical load. The IM output power can be estimated from the generator output after adding the estimated mechanical and generator copper losses. Fig. 4 shows the actual laboratory setup. The detailed ratings of the prototype machine are given in Table 2. The machine parameters for the fundamental sequence plane are given in Table 3 and estimated using conventional no-load, locked rotor and DC tests.

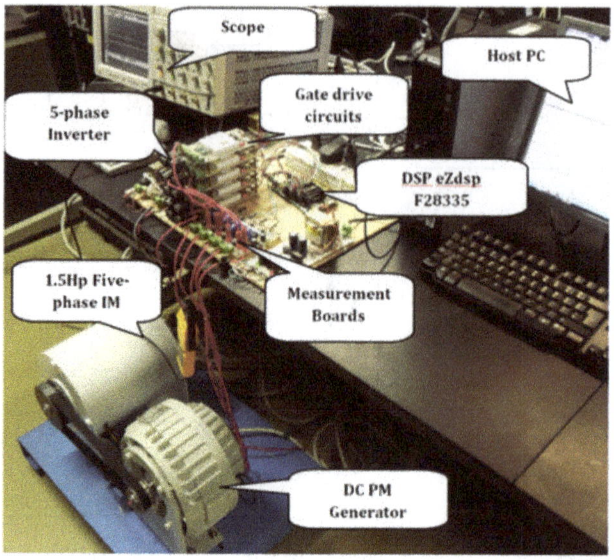

Fig. 4 *Experimental setup*

Under V/f control, the V/f constant is calculated based on rated peak voltage and frequency. With FOC, the rated direct current component is selected based on the machine magnetising current, which is calculated by dividing the V/f constant by the fundamental magnetising inductance, L_{m1}.

6 Experimental results

In this section, the prototype machine is tested using different controllers and under healthy as well as open phase cases. First, the machine characteristic curves are plotted with the machine reference speed set to the rated synchronous speed, 1500 rpm. Next, the machine dynamic response and current waveforms are shown for both controllers.

To estimate the losses efficiency and torque, the following procedure is carried out. First, the system is run by powering the DC machine (in motoring mode) up to the rated speed then measuring its current to estimate the copper losses, hence mechanical losses can be found as the difference between the DC input power and the copper losses of the DC machine (running as a motor). Then, the five-phase machine is controlled in closed loop to run at the rated speed and the load connected to the DC-generator is varied. Generator copper losses are estimated from the current and armature

Table 3 Prototype five-phase IM parameters

R_{s1}	l_{s1}	R_{r1}	l_{r1}	L_{m1}
2.6 Ω	13.4 mH	2.43 Ω	16.1 mH	157 mH

Table 2 Ratings of the prototype five-phase IM

Rated phase voltage, V	Rated DC-link voltage, V	Rated power, hp	Rated phase current, A	Rated frequency, Hz	Number of poles	Rated speed, rpm
90	250	1	3.2	50	4	1410

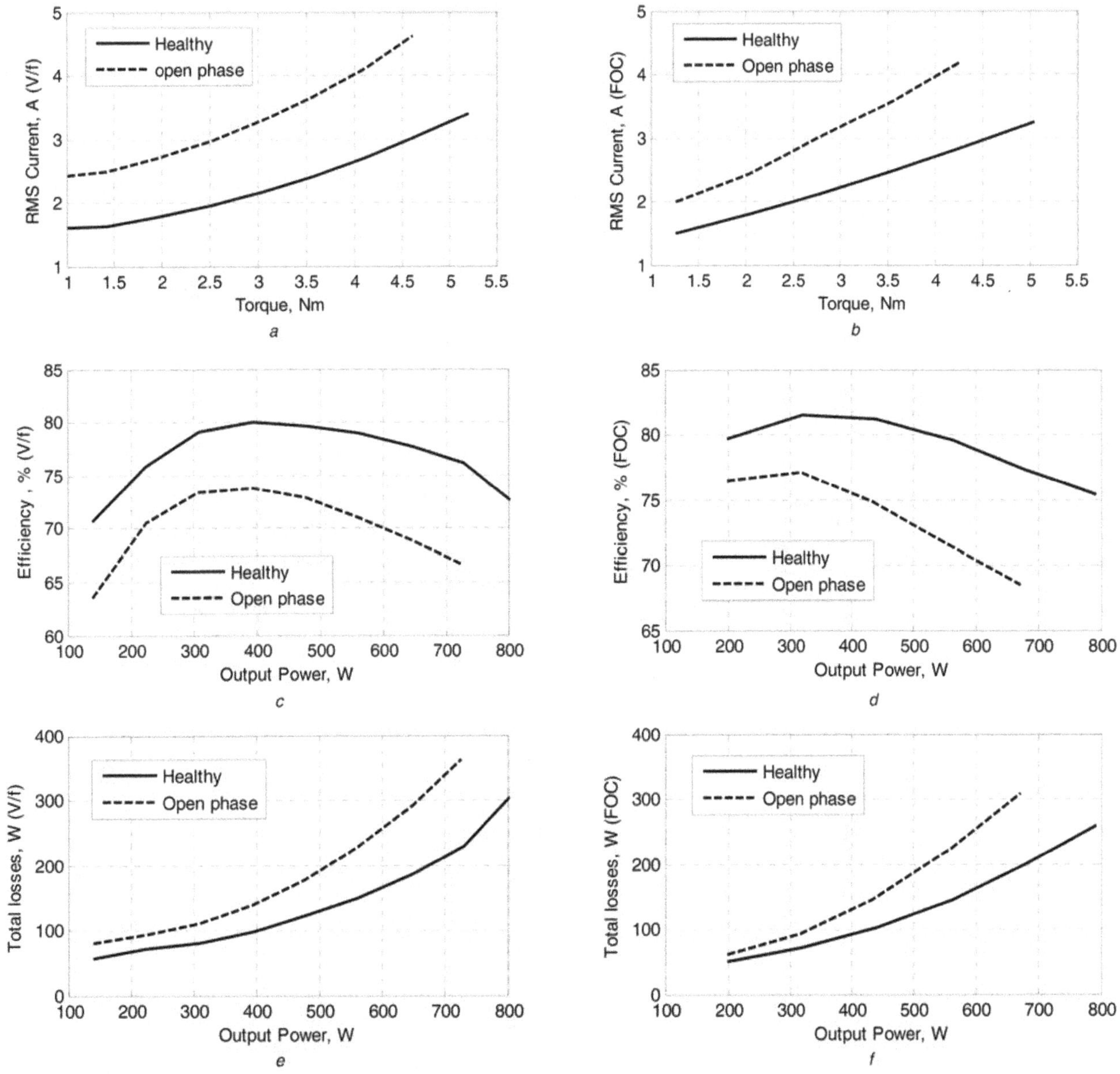

Fig. 5 *Machine characteristics under healthy and open-phase cases*
a, b RMS current against torque characteristic
c, d Total losses against output power characteristic
e, f Efficiency against output power characteristic

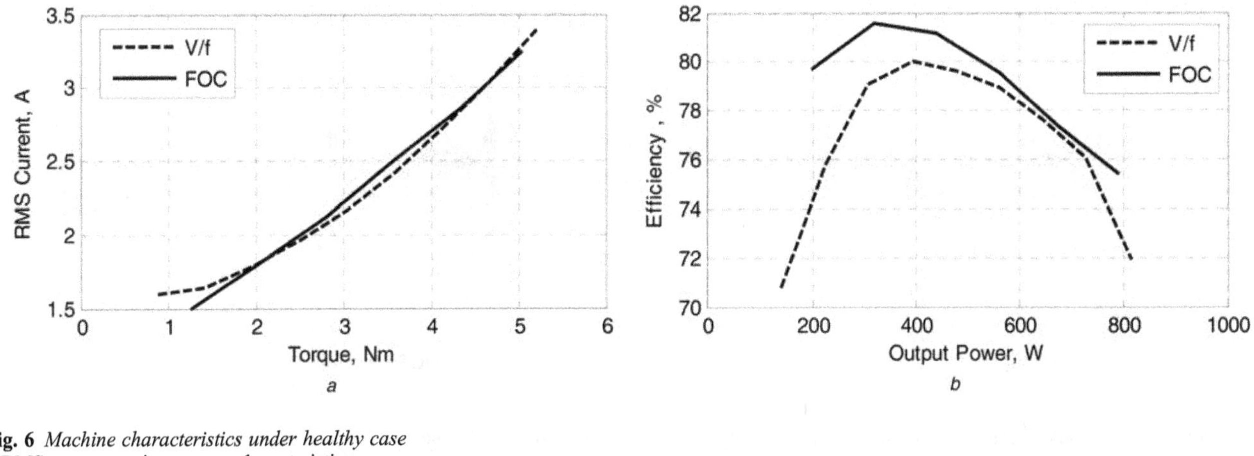

Fig. 6 *Machine characteristics under healthy case*
a RMS current against torque characteristic
b Efficiency against output power characteristic

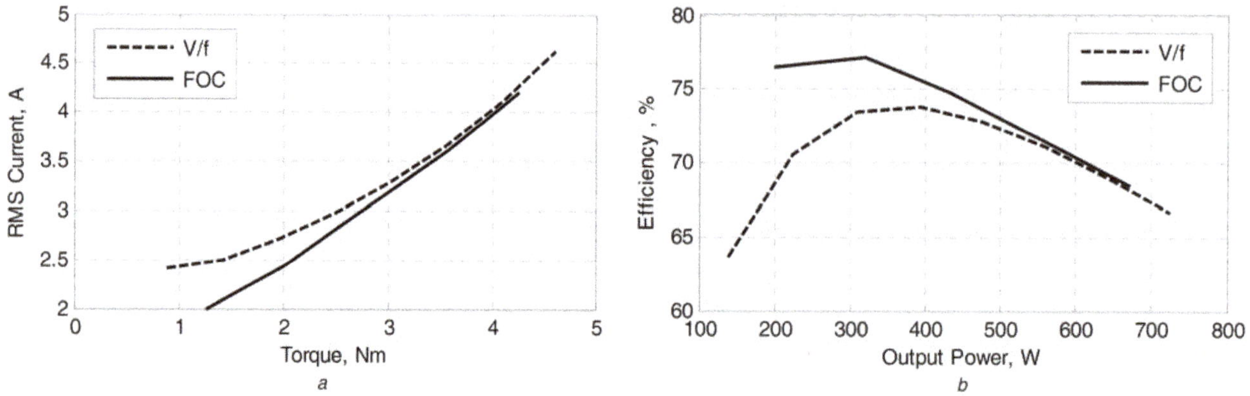

Fig. 7 *Machine characteristics with one-phase open*
a RMS current against torque characteristic
b Efficiency against output power characteristic

resistance. The five-phase machine input power and the generator output power are measured. The five-phase machine output power is found from subtracting the generator losses and mechanical losses from the input power to the five-phase machine. Torque is found from the computed output power five-phase machine and the shaft speed.

6.1 Steady-state characteristic curves

In this section, the machine steady-state characteristic curves under healthy as well as open-phase cases are compared for both controllers with machine reference speed set to 1500 rpm. Figs. 5*a* and *b* show the root-mean-square (RMS) stator phase current against torque characteristic under V/f control and FOC, respectively,

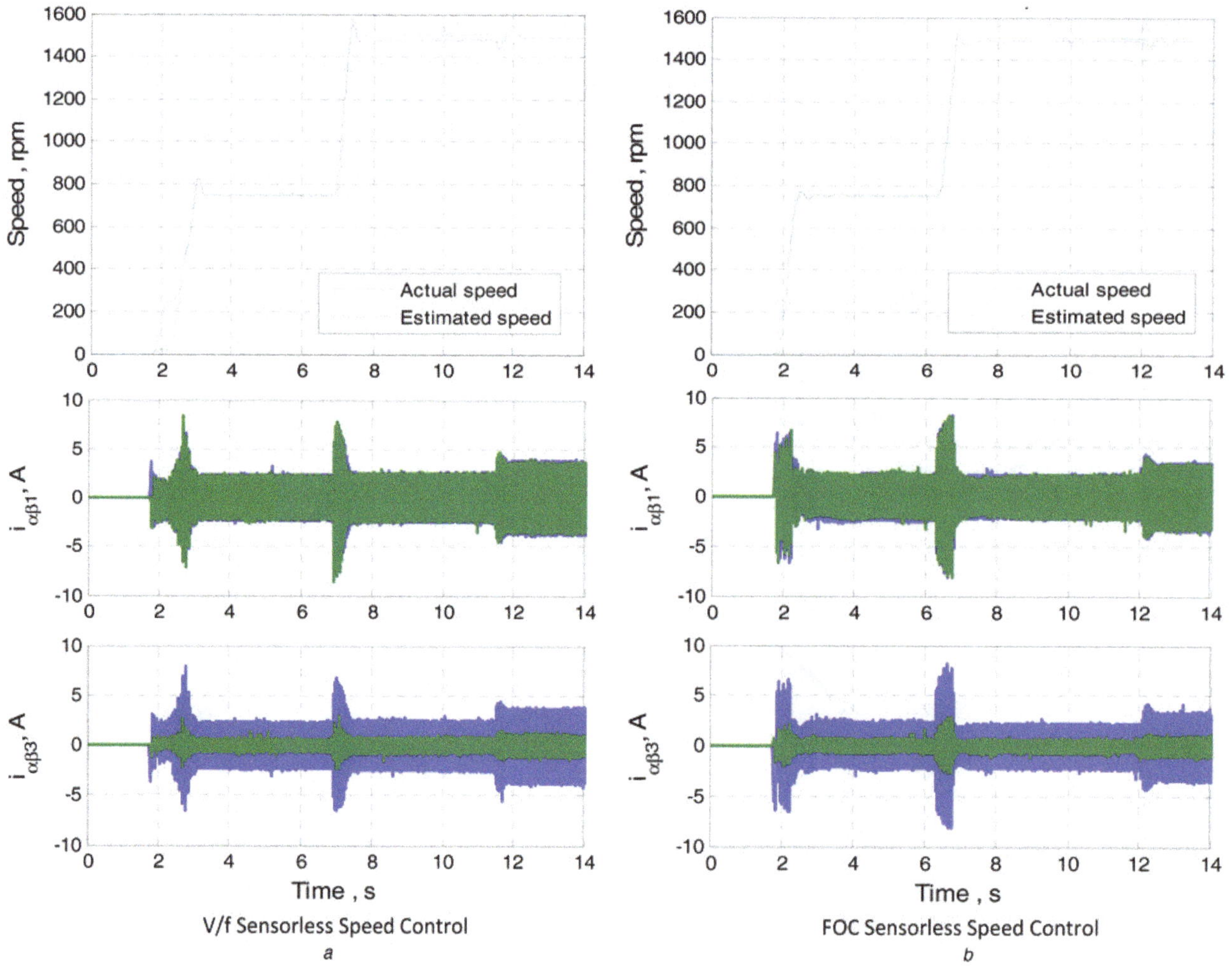

Fig. 8 *Actual and estimated speed, fundamental sequence current component (αβ1), and third sequence current component (αβ3)*
a V/f control
b FOC

whereas the machine efficiencies against output power are depicted in Figs. 5*c* and *d*. Figs. 5*e* and *f* show the relation between the output power and the machine total loss for V/f control and FOC, respectively. The latter characteristic curves can be used to find the maximum output load under open-phase condition to avoid excessive machine losses. It is clear that for the same load torque, the phase current magnitude for the open-phase case is increased by an approximate factor of 1.38 compared with healthy case, as depicted by (4). As given by Figs. 5*e* and *f*, the rated machine losses are ~250 W for both cases which correspond to an output power of 750 W, 1 hp. Hence, for the same total losses, the maximum machine load under open phase is limited to 592 W, 0.79 pu.

For sake of comparison between the two controllers, both healthy and open-phase cases are plotted in Figs. 6 and 7, respectively. It is shown that for the healthy case the FOC offers slightly higher efficiency than V/f control especially for low mechanical loads. Under open phase, FOC also offers lower stator current, lower copper loss and, hence, higher efficiency.

6.2 *Dynamic response*

In this section, the machine dynamic response is compared for both controllers with MRAS speed estimation and with one-phase open. It is assumed that a step reference speed of 750 rpm is applied at 2 s with the machine mechanically unloaded and then it is increased to 1500 rpm at 7 s. Then the machine is loaded with its full-load torque of 4.7 Nm. The actual and estimated speeds are shown in Fig. 8. It is clear that FOC results in better dynamic response. It is also shown that both controllers yield a steady-state speed error which increases with the reference speed and reduces with mechanical loading. For FOC the steady-state speed error is lower. The corresponding $\alpha\beta$ current components for fundamental and third sequence planes are also shown in Fig. 8. The relation between them is shown to follow (5), with improved dynamic response using FOC.

Fig. 9 shows the no-load and full-load currents with the reference speed set to 1500 rpm. It is evident that the proposed controller ensures equal remaining healthy phase currents under different loading cases. The wave distortion is mainly because of the effect of saturation and machine non-linearities and asymmetries. Unlike FOC, the V/f controller experiences some deviation in current magnitude for the remaining healthy phases as well as higher wave distortion which are mainly because of machine asymmetry and saturation effects [27]. To investigate this point, the $\alpha\beta$ current and voltage components for each period are shown in Figs. 10 and 11 for V/f control and FOC, respectively. The FOC shows more sinusoidal current components as both planes are completely controlled using three PR controllers. However for V/f control, the reference value of the third sequence plane is decided based on the measured fundamental $\alpha\beta$ current components. In V/f control, the fundamental $\alpha\beta$ voltage components, $v_{s\alpha\beta1}$, are perfectly balanced as these components are derived from a sinusoidal generator, whereas the corresponding $\alpha\beta$ current components, $i_{s\alpha\beta1}$, experience some degree of unbalance because the prototype machine has notable unbalance and winding asymmetry. On the other hand in FOC, the fundamental voltage components, $v_{s\alpha\beta1}$, are unbalanced as they are derived from two PR controllers to ensure balanced $i_{s\alpha\beta1}$ irrespective of

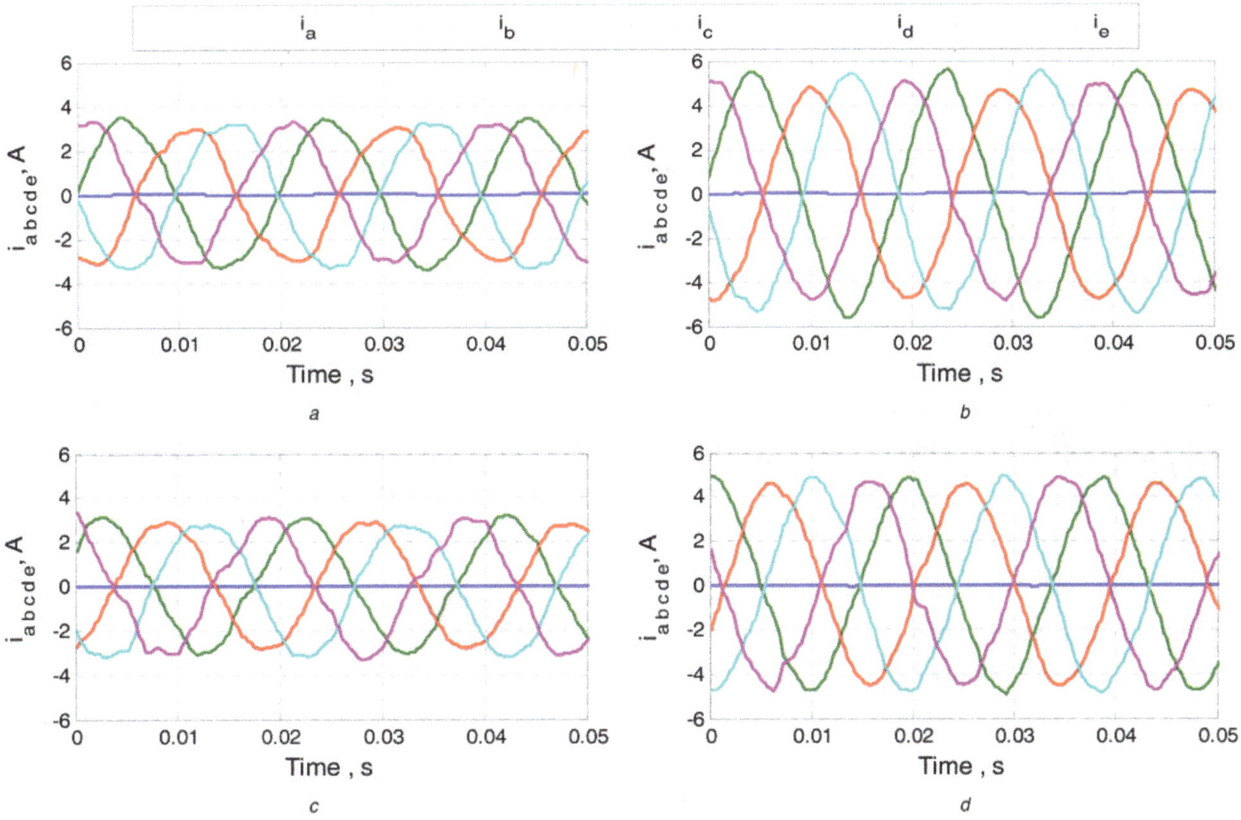

Fig. 9 *Machine phase currents at different cases*
a, b V/f sensorless speed control
c, d FOC sensorless speed control vertical axis scale (2.0 A/division), horizontal axis scale (5 ms/division) fault case with V/f control at no load
a Fault case with V/f control at no load
b Fault case with V/f control at full load
c Fault case with FOC control at no load
d Fault case with FOC control at full load

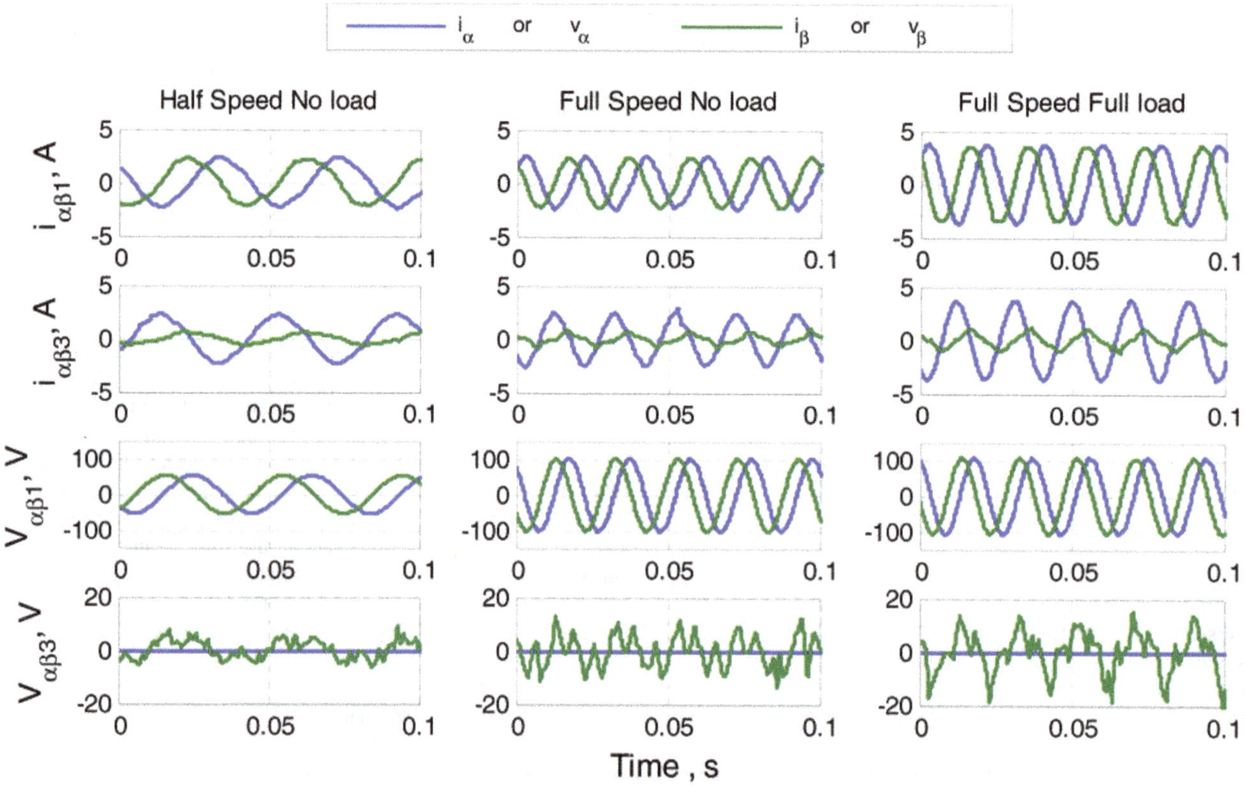

Fig. 10 *Experimental CAN results for V/f sensorless speed control*

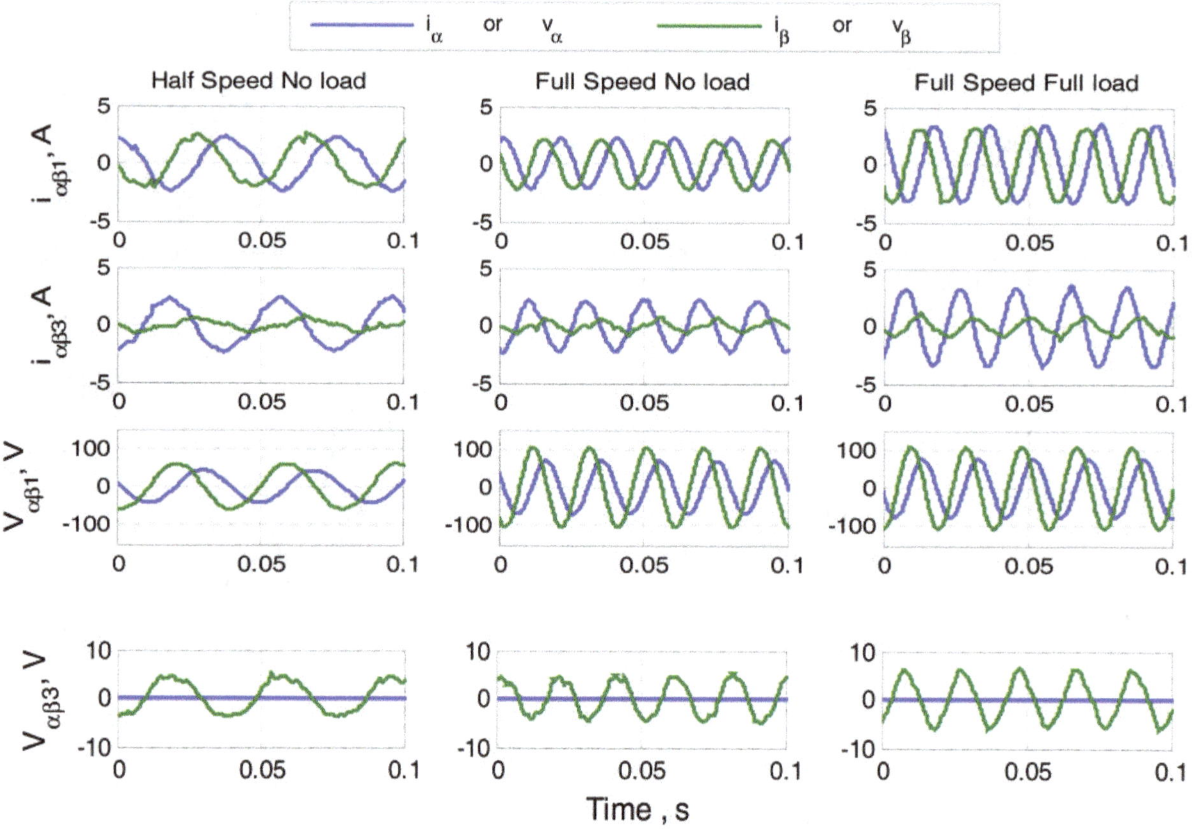

Fig. 11 *Experimental CAN results for FOC sensorless speed control*

winding asymmetry. It can be shown also that the voltage and current components of the third sequence plane with V/f control experience notable distortion because of the induced third harmonic caused by magnetic saturation [27]. On the other hand, FOC corresponds to a more sinusoidal waveform in both sequence planes.

The main advantage of V/f control over FOC control is that it does not require accurate parameter determination. On the other hand, FOC is highly affected by accurate machine parameter determination, which is still challenging for multiphase machines [28].

7 Conclusions

In this paper, two simple sensorless fault tolerant control schemes based on conventional V/f control and FOC control of a five-phase IM are presented. In the literature, FOC is usually employed with relatively sophisticated controllers that require several transformations to maintain certain optimisation objectives. On the contrary, the proposed control schemes allow for disturbance-free operation in cases of open-circuit faults using only one and three PR controllers for the V/f control and FOC. The PR controllers are used to control the third sequence current component in accordance with the fundamental sequence current to maintain equal current magnitude in the remaining healthy phases, which in turn maximises torque production, minimises torque ripples and improves overall efficiency. Sensorless operation is provided by estimating the machine speed using a RF-based MRAS speed observer based on fundamental sequence voltage and current components. The comparison between V/f control and FOC based on experimental investigation reflects the following conclusions:

- Both controllers can effectively maintain equal remaining phase currents under one-phase open. Deviations mainly occur because of winding asymmetry and saturation effects.
- Although FOC provides better dynamics as well as steady-state performance, V/f control requires less current controllers and has a simpler overall controller structure.
- The main disadvantage of FOC is that the performance of FOC is highly affected by accurate machine parameter determination, which is still challenging for multiphase machines. On the other hand, V/f control does not require accurate machine parameters.
- Sensorless operation of the multiphase machine under fault can be simply provided using a conventional technique used for the three-phase case by using the fundamental voltage and current components.

8 Acknowledgments

This work was supported by a National Priorities Research Program (NPRP) grant NPRP 4-941-2-356 from the Qatar National Research Fund (QNRF). The statements made herein are solely the responsibility of the authors.

9 References

[1] Levi E.: 'Multiphase electric machines for variable-speed applications', *IEEE Trans. Ind. Electron.*, 2008, **55**, (5), pp. 1893–1909

[2] Levi E., Bojoi R., Profumo F., Toliyat H.A., Williamson S.: 'Multiphase induction motor drives – a technology status review', *IET Proc., Electr. Power Appl.*, 2007, **1**, (4), pp. 489–516

[3] Abdel-Khalik A.S., Masoud M.I., Williams B.W.: 'Improved flux pattern with third harmonic injection for multiphase induction machines', *IEEE Trans. Power Electron.*, 2012, **27**, (3), pp. 1563–1578

[4] Apsley J., Wiiliamson S.: 'Analysis of multiphase induction machines with winding faults', *IEEE Trans. Ind. Appl.*, 2006, **42**, (2), pp. 465–472

[5] Abdel-Khalik A.S., Morsy A., Ahmed S., Massoud A.: 'Effect of stator winding connection on performance of five-phase induction machines', *IEEE Trans. Ind. Electron.*, 2014, **61**, (1), pp. 3–19

[6] Tani A., Mengoni M., Zarri L., Serra G., Casadei D.: 'Control of multi-phase induction motors with an odd number of phases under open circuit faults', *IEEE Trans. Power Electron.*, 2012, **27**, (2), pp. 565–577

[7] Betin F., Capolino G.: 'Shaft positioning for six-phase induction machines with open phases using variable structure control', *IEEE Trans. Ind. Electron.*, 2012, **59**, (6), pp. 2612–2620

[8] Fnaiech M., Betin F., Capolino G., Fnaiech F.: 'Fuzzy logic and sliding-mode controls applied to six-phase induction machine with open phases', *IEEE Trans. Ind. Electron.*, 2010, **57**, (1), pp. 354–364

[9] Guzman H., Duran M., Barrero F., Toral S.: 'Fault-tolerant current predictive control of five-phase induction motor drives with an open phase'. Conf. IECON'11, 2011, pp. 3680–3685

[10] Moghadasian M., Kianinezhad R., Betin F., Yazidi A., Lanfranchi V., Capolino G.: 'Position control of faulted six-phase induction machine using genetic algorithms'. Conf. SDEMPED,11, 2012, pp. 385–390

[11] Peng Z., Xiao-feng Z., Ming-zhong Q., Wei C., Jing-Hui L.: 'A novel control strategy for five-phase concentrated full-pitch windings induction motor under open-phase fault'. Conf. ICMA'11, 2011, pp. 950–955

[12] Suetake M., da Silva I.N., Goedtel A.: 'Embedded DSP-based compact fuzzy system and its application for induction-motor V/f speed Control', *IEEE Trans. Ind. Electron.*, 2011, **58**, (3), pp. 750–760

[13] Scharlau C., Pereira L., Pereira L., Haffner S.: 'Performance of a five-phase induction machine with optimized air gap field under open loop V/f control', *IEEE Trans. Energy Convers.*, 2008, **23**, (4), pp. 1046–1056

[14] Morsy A., Abdel-khalik A.S., Abbas A., Ahmed S., Massoud A.: 'Open loop V/f control of multiphase induction machine under open-circuit phase faults'. Conf. APEC 2013, 2013, pp. 1170–1176

[15] Holmes D., McGrath B., Parker S.: 'Current regulation strategies for vector-controlled induction motor drives', *IEEE Trans. Ind. Electron.*, 2012, **59**, (10), pp. 3680–2689

[16] Cirrincione M., Accetta A., Pucci M., Vitale G.: 'MRAS speed observer for high-performance linear induction motor drives based on linear neural networks', *IEEE Trans. Power Electron.*, 2013, **28**, (1), pp. 123–134

[17] Abdelsalam A., Masoud M., Hamad M., Williams B.: 'Improved Sensorless operation of a CSI-based induction motor drive: long feeder case', *IEEE Trans. Power Electron.*, 2013, **28**, (8), pp. 4001–4012

[18] Abu-Rub H., Khan M.R., Iqbal A., Ahmed S.M.: 'MRAS-based sensorless control of a five-phase induction motor drive with a predictive adaptive model'. Conf. ISIE, 2010, pp. 3089–3094

[19] Zheng L., Fletcher J.E., Williams B.W., Xiangning H.: 'A novel direct torque control scheme for a Sensorless five-phase induction motor drive', *IEEE Trans. Ind. Electron.*, 2011, **58**, (2), pp. 503–513

[20] Khaldi B., Abu-Rub H., Iqbal A., Kennel R., Mahmoudi M., Boukhetala D.: 'Sensorless direct torque control of five-phase induction motor drives'. Conf. IECON 2011, 2011, pp. 3501–3506

[21] Khan M.R., Iqbal A., Ahmad M.: 'MRAS-based sensorless control of a vector controlled five-phase induction motor drive', *Electr. Power Syst. Res.*, 2008, **78**, pp. 1311–1321

[22] Khan M.R., Iqbal A.: 'Extended Kalman filter based speeds estimation of series-connected five-phase two-motor drive system', *Simul. Model. Pract. Theory*, 2009, **17**, pp. 1346–1360

[23] Fu J.R., Lipo T.A.: 'Disturbance-free operation of a multiphase current-regulated motor drive with an opened phase', *IEEE Trans. Ind. Appl.*, 1994, **30**, (5), pp. 1267–1274

[24] Schauder C.: 'Adaptive speed identification for vector control of induction motors without rotational transducers', *IEEE Trans. Ind. Appl.*, 1992, **28**, (5), pp. 1054–1061

[25] Peng F., Fukao T.: 'Robust speed identification for speed-sensorless vector control of induction motors', *IEEE Trans. Ind. Appl.*, 1994, **30**, (5), pp. 1234–1240

[26] Vas P.: 'Sensorless vector and direct torque control' (Oxford University Press, New York, 1998)

[27] Abdel-Khalik A.S., Ahmed S., Elserougi A., Massoud A.: 'A voltage-behind-reactance model of five-phase induction machines considering the effect of magnetic saturation', *IEEE Trans. Energy Convers.*, 2013, **28**, (3), pp. 576–592

[28] Yepes A.G., Riveros J.A., Doval-Gandoy J., ET AL.: 'Parameter identification of multiphase induction machines with distributed windings – part 1: sinusoidal excitation methods', *IEEE Trans. Energy Convers.*, 2012, **27**, (4), pp. 1056–1066

4

One watt gallium arsenide class-E power amplifier with a thin-film bulk acoustic resonator filter embedded in the output network

Kyle Holzer, Jeffrey S. Walling

University of Utah PERFIC Laboratory, Salt Lake City, UT 84112, USA
E-mail: jeffrey.s.walling@utah.edu

Abstract: Integration of a class-E power amplifier (PA) and a thin-film bulk acoustic wave resonator (FBAR) filter is shown to provide high power added efficiency in addition to superior out-of-band spectrum suppression. A discrete gallium arsenide pseudomorphic high-electron-mobility transistor is implemented to operate as a class-E amplifier from 2496 to 2690 MHz. The ACPF7041 compact bandpass FBAR filter is incorporated to replace the resonant LC tank in a traditional class-E PA. To reduce drain voltage stress, the supply choke is replaced by a finite inductance. The fabricated PA provides up to 1 W of output power with a peak power added efficiency (PAE) of 58%. The improved out-of-band spectrum filtering is compared to a traditional class-E with discrete LC resonant filtering. Such PAs can be combined with linearisation techniques to reduce out-of-band emissions.

1 Introduction

Switched-mode power amplifiers (PAs) offer higher efficiency than their linear counterparts, however, natively generate a rich spectrum of harmonics and intermodulation products. This paper presents an efficient class-E PA with an embedded thin-film bulk acoustic wave resonator (FBAR, Fig. 1*a*) filter. The FBAR filter is incorporated into the class-E output circuit, replacing the traditional LC resonant tank (Fig. 1*b*). This provides superior in-band frequency selectivity and out-of-band spectrum suppression.

Piezoelectric materials have been used for microwave frequency filtering for decades. Thin film technologies including surface acoustic wave (SAW) and bulk acoustic wave (BAW) filters have proven advantages compared to traditional discrete inductor and capacitor derived counterparts because of the acoustic wavelength at a given frequency being several orders of magnitude shorter than the analogous electrical wavelength in a comparable electromagnetic filter medium. The FBAR builds on BAW filter technology with the advantages of improved performance and improved ease of integration with standard planar device technologies. FBAR filters do not require exotic materials typical of both SAW and BAW filters, such as quartz, lithium niobate, ZnO and others. FBAR [1] can be fabricated in the same process as a typical high performance amplifier, enabling the integration of a power efficient, and spectrally pure, switching amplifier design; an integrated extension of the discrete design presented in this paper.

2 Design details

The amplifier implementation is fabricated including a custom printed circuit board with chip-on-board (COB) construction as shown in Fig. 2. The small size of this implementation shows future potential for fully integrated PA dies with an FBAR either integrated into the process technology or bonded directly to the die in a system-in-package implementation.

2.1 Class-E amplifier design

A broadband TGF2080 gallium arsenide pseudomorphic high-electron-mobility transistor (GaAs pHEMT) device is chosen as the power transistor for this switching PA. This device offers low input capacitance and moderate optimal impedance load magnitude across the target design frequency.

A high pass dual element conjugate match is implemented at the input to minimise insertion loss and reduce component count. The optimal class-E output shunt susceptance is achieved with a combination of the device's drain capacitance, C_1 and L_1 as shown in Fig. 1 [2]. The device's output capacitance is larger than required; hence, finite L_1 is used to offset the additional shunt capacitance. Finite inductance in the drain has also been shown to reduce the voltage swing at the drain to as low as $2.5 \times V_{DD}$ [3]. Breakdown voltage for this 0.25 μm GaAs process is in excess of +20 V enabling continuous class-E operation.

Series inductance (L_2, Fig. 1) selection optimises transition from the class-E output into the load impedance as derived in [4]

$$\frac{\omega L}{R} = 1.1525 \tag{1}$$

The series inductance aligns the voltage and current incident to the load.

2.2 Class-E with embedded FBAR

Intrinsic to the class-E amplifier output tuning circuit is a series resonant circuit to filter undesired harmonic content that arises as a result of the periodic pulsed operation of the amplifier. In the classic implementation, this series resonant circuit can be designed with any quality factor (Q) and performs well even for relatively low Q. This series resonant circuit can be replaced by any bandpass filter or matching network, and can be comprised of more than two components. Hence, the implementation is a trade-off between the required out-of-band (OOB) filter rejection and the number of filter poles. A higher number of filter poles increases both the amplifier size and its insertion loss.

In this design, we replace these traditional LC tanks with an FBAR filter that increases the OOB rejection, significantly, while minimising the insertion loss associated with a high-order LC filter. The size is comparable to that of a single LC tank filter while providing significantly better OOB rejection. Owing to the operating voltage and desired output power, the 50 Ω FBAR filter provides the real part of the impedance termination for the PA; hence no additional matching circuits are required. Using the previous equation, L_2 is optimised using the FBARs input impedance, R = 50 Ω. Subsequently, in the case where a traditional LC tank is used, the selection of L_2 does not change for operation at the resonant frequency. Given the antenna input impedance closely matches 50 Ω.

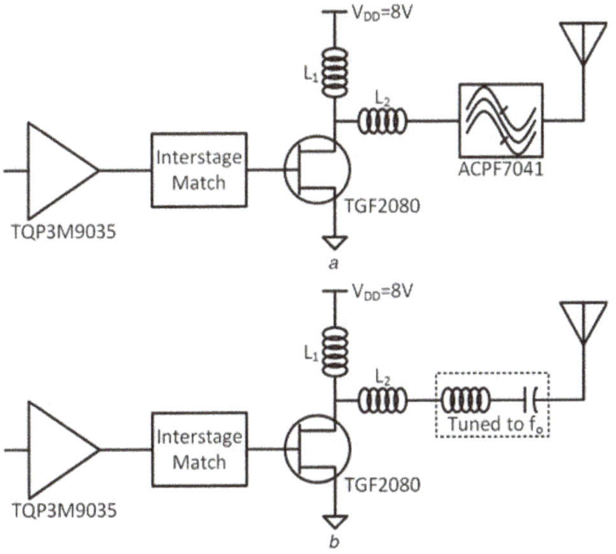

Fig. 1 *Schematic of*
a Class-E PA with embedded FBAR filter and
b Traditional class-E PA

2.3 Class-E with traditional LC tank

For comparison to the embedded FBAR resonator a single LC tank resonator is designed for the centre frequency of operation. High frequency multi-layer ceramic components are chosen to maximise the realisable network Q. To aid the comparison between the two amplifiers, no impedance transformation is used, hence, the Q of the LC tank is limited by the load resistance, which is 50 Ω, and yields a 20% fractional bandwidth.

2.4 Amplifier input match optimisation

Owing to amplifier reverse isolation, S_{12}, the input match can be affected with changes to the load impedance. This was verified through simulation of input impedance comparing a traditional LC tank and the FBAR filter resonant elements. The previously designed high pass dual element input amplifier match proves sufficient in both cases.

2.5 COB implementation

To minimise parasitics associated with packaging, the amplifier die is COB bonded directly to the PCB. This provides additional advantages of smaller size compared to packaged surface mount devices as well as a lower junction to board thermal conductivity.

3 Measurement results

The PA was fabricated and its performance characteristics measured. A TQP3M9035 driver amplifier is included on the printed circuit to lower the required power level for input test signals.

In order to validate the FBAR tuned class-E PA (Fig. 2*a*), measurements are included for a class-E with LC tank PA (Fig. 2*b*). Both amplifiers use the same TGF2080 device and PCB with only modifications to the output network components (e.g. FBAR, LC tank).

The peak output power of 29.95 dBm is achieved by the class-E FBAR PA at 2.64 GHz (Fig. 3). The passband ripple is a characteristic of the FBAR filter. The class-E LC tank PA has the highest average output power across the passband because of the small insertion of the single LC tank compared to 2.5 dB insertion loss for the FBAR.

The PAE of each amplifier configuration is plotted against frequency in Fig. 4. The low insertion loss of the single LC tank

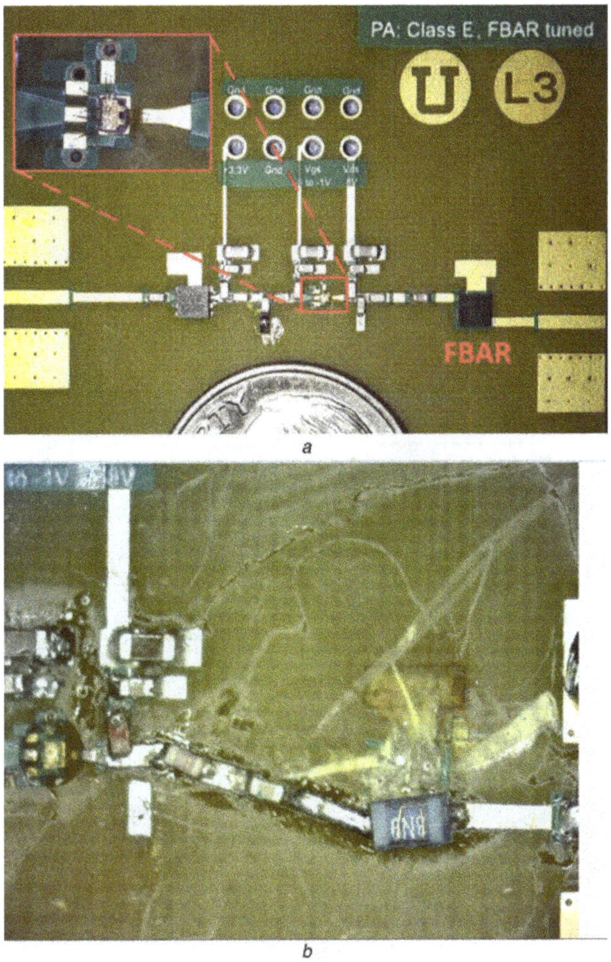

Fig. 2 *PCB layouts for*
a Class-E PA with FBAR filter and
b Class-E PA with LC tank

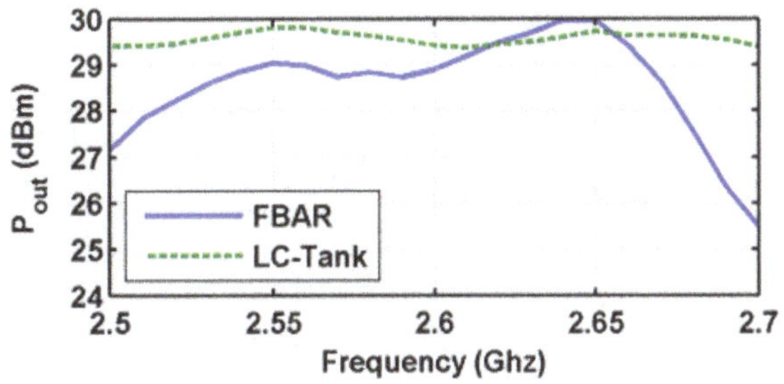

Fig. 3 *Measured output power against amplifier configuration*

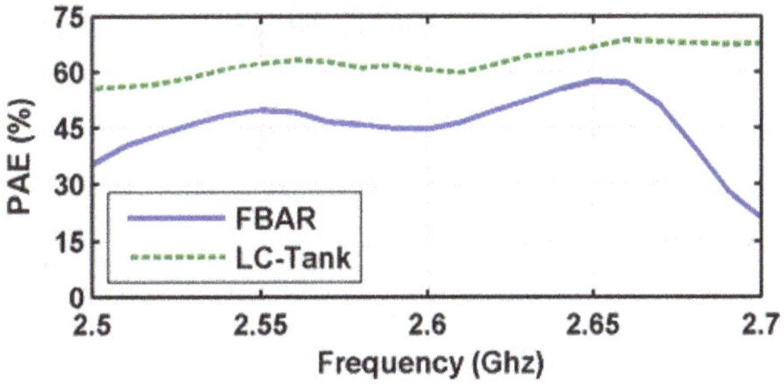

Fig. 4 *Measured PAE against amplifier configuration*

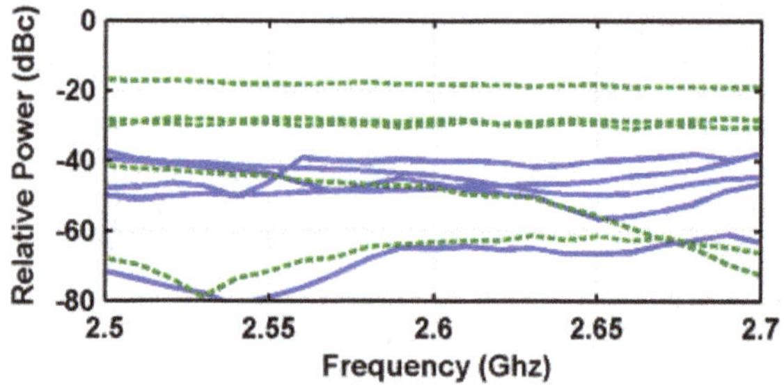

Fig. 5 *Measured harmonics against amplifier configuration: (horizontal bar line) = class-E FBAR harmonics 2–6 and (dashed line) = class-E LC tank harmonics 2–6*

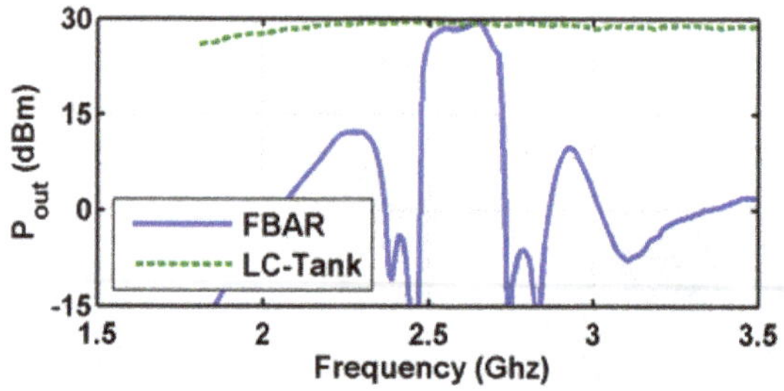

Fig. 6 *Measured class-E PA output spectrum comparison*

clearly yields the highest PAE. The peak PAE for the Class-E FBAR PA is only moderately lower at 58%. The PAE can be significantly improved with integration onto a single die because of a combination of lower insertion loss through the FBAR filter optimised for target filter rejection and minimised signal transition losses between component interfaces.

The distinct advantage of FBAR integration is the superior filtering capability in a small footprint. The two amplifier configurations harmonic content are measured and plotted against frequency in Fig. 5.

The class-E LC tank PA that offers a PAE advantage is also unable to adequately suppress harmonic content. The FBAR filter shows similarly strong harmonic rejection for both the linear mode and the class-E mode, as expected.

The PA output is plotted over one octave of spectrum in Fig. 6 and demonstrates the adjacent channel rejection afforded by the FBAR filter, in this case nulling among others the adjacent ISM 2.4 GHz band used by WLAN and Bluetooth. The LC tank offers virtually no spectrum shaping. This application of harmonic filtering has a performance against size advantage over alternate approaches, such as tuned distributed element filters [5]. Active harmonic suppression [6] requires considerable stability optimisation for mitigated harmonic improvement.

In order to provide linear operation, the class-E PA must be linearised via external means. One example is envelope elimination and restoration (EER); to accurately recreate the amplitude envelope modulation at the amplifier load [7]. The optimal amplifier load impedance will vary with the drain voltage. A drain voltage of 8 V was used for device design and optimisation. The performance over a

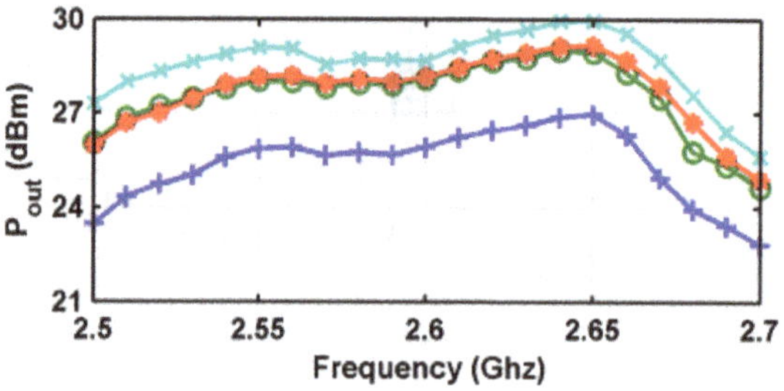

Fig. 7 *Measured class-E PA FBAR P_{out} against V_{DD} (plus symbol 5 V, circle symbol 6 V, asterisk symbol 7 V and multiplication sign 8 V)*

wider range of voltages is plotted in Fig. 7. Increasing the drain voltage beyond 8 V offers minimal effect on output power as the drain current is at saturation. Decreasing the drain voltage reduces the output signal swing and power as expected without significantly affecting load match performance.

4 Conclusions

An integration of a class-E PA and an FBAR filter is shown to provide high PAE in addition to superior out of band spectral content suppression. The PA is designed using a broadband discrete GaAs pHEMT device configured for class-E operation in concert with the ACPF7041 FBAR filter (2496–2690 MHz). The PA provides up to 1 W of output power with a peak PAE of 58%. Superior out of band spectrum filtering is shown in comparison to discrete LC resonant filtering.

5 Acknowledgments

The authors wish to acknowledge Triquint Semiconductor for their donation of the PA driver and transistor dies and Avago Technologies for their donation of the FBAR Filters. The authors also wish to thank L-3 Communications for their measurement support.

6 References

[1] Morkner H., Ruby R., Frank M., Figueredo D.: 'An integrated FBAR filter and PHEMT switched-amp for wireless applications'. MTT-S Int. Microwave Symp. Digest, 1999, pp. 1393–6

[2] Sokal N.O., Sokal A.D.: 'Class E-A new class of high-efficiency tuned single-ended switching power amplifiers', *IEEE J Solid-State Circuits*, 1975, **10**, (3), pp. 168–76

[3] Yoo C., Huang Q.: 'A common-gate switched 0.9-W class-E power amplifier with 41% PAE in 0.25 μm CMOS', *IEEE J Solid-State Circuits*, 2001, **36**, (5), pp. 823–30

[4] Grebennikov A.: 'RF and microwave power amplifier design' (McGraw-Hill Professional, 2005)

[5] Lim J.-S., Jeong Y.-C., Ahn D., Lee Y.-T., Cho H., Nam S.: 'Size-reduction and harmonic-rejection of microwave amplifiers using spiral-defected ground structure'. 33rd European Microwave Conf., 2003, pp. 1421–4

[6] Park J.W., Razavi B.: 'A harmonic-rejecting CMOS LNA for broadband radios', *IEEE J Solid-State Circuits*, 2013, **48**, (4), pp. 1072–84

[7] Walling J.S., Allstot D.J.: 'Linearizing CMOS switching power amplifiers using supply regulators', *IEEE Trans. Circuits Syst. II Express Briefs*, 2010, **57**, (7), pp. 497–501

5

Robust leakage-based distributed precoder for cooperative multicell systems

Daniel Castanheira, Adão Silva, Atílio Gameiro

DETI, Instituto de Telecomunicações, University of Aveiro, Aveiro, Portugal
E-mail: dcastanheira@av.it.pt

Abstract: Coordinated multipoint (CoMP) from long term evolution (LTE)-advanced is a promising technique to enhance the system spectral efficiency. Among the CoMP techniques, joint transmission has high communication requirements, because of the data sharing phase through the backhaul network, and coordinated scheduling and beamforming reduces the backhaul requirements, since no data sharing is necessary. Most of the available CoMP techniques consider perfect channel knowledge at the transmitters. Nevertheless for practical systems this is unrealistic. Therefore in this study the authors address this limitation by proposing a robust precoder for a multicell-based systems, where each base station (BS) has only access to an imperfect local channel estimate. They consider both the case with and without data sharing. The proposed precoder is designed in a distributed manner at each BS by maximising the signal-to-leakage-and-noise ratio of all jointly processed users. By considering the channel estimation error in the design of the precoder, they are able to reduce considerably the impact of these errors in the system's performance. The results show that the proposed scheme has improved performance especially for the high signal-to-noise ratio regime, where the impact of the channel estimation error may be more pronounced.

1 Introduction

Multicell cooperation is a promising solution for cellular wireless systems to mitigate intercell interference, improving system fairness and increasing capacity in the years to come [1–3], and thus is already under study in LTE-advanced under the coordinated multipoint concept [4]. There are several cooperative multicell approaches depending on the amount of information shared by the transmitters through the backhaul network and where the processing takes place, that is, centralised if the processing takes place at the central unit (CU) [5] or distributed [2, 6] if it takes place at different transmitters. Coordinated centralised approaches promise larger spectral efficiency gains than distributed interference coordination techniques, but typically at the price of larger backhaul and more severe synchronisation requirements [3].

Some sub-optimal centralised precoding schemes have been discussed in [5]. The interference is eliminated by joint and coherent coordination of the transmission from the base stations (BSs) in the network, assuming that they share all downlink signals. In [7], inner bounds on capacity regions for downlink transmission were derived with or without BS cooperation and under per-antenna power or sum-power constraint. Two centralised multicell precoding schemes based on the waterfilling technique have been proposed in [8]. It was shown that these techniques achieve close to optimal weighted sum rate performance. Based on the statistical knowledge of the channels, the CU performs a centralised power allocation and jointly minimises the outage probability of the user terminals (UTs) [9]. In [10], a clustered BS coordination is enabled through a multicell block diagonalisation (BD) strategy to mitigate the effects of interference in multicell multiple-input–multiple-output (MIMO) systems. A BD cooperative multicell scheme was proposed in [11] where the weighted UTs sum-rate achievable is maximised. Non-linear centralised multicell precoding was considered in [12].

Distributed precoding approaches, where the precoder vectors are computed at each BS in a distributed fashion, have been proposed in [13] for the particular case of two UTs and generalised for K UTs in [14]. It is assumed that each BS has only the knowledge of local channel state information (CSI) and based on that a parameterisation of the beamforming vectors used to achieve the outer boundary of the achievable rate region was derived. Distributed precoding schemes based on zero-forcing (ZF) criterion with several centralised power allocation approaches, which minimise the average bit error-ratio (BER) and sum of inverse of signal-to-noise ratio was proposed in [15]. In [13–15], it was considered that the BSs share the entire data of the all jointly processed users, whereas in [16] the distributed precoding was designed so that the transmitters do not share the data which fall into the interference channel framework. One of the considered criteria to design the precoders was the signal-to-leakage-and-noise (SLNR) ratio maximisation, introduced first in the context of multiuser MIMO [17]. This technique balances the received signal power of the target user against the interference power imposed on the remaining users. Basically, it combines the benefits of both the egoistic distributed maximum ratio transmission and the altruistic ZF techniques [18]. In the previous distributed approaches, the precoders were designed by assuming perfect knowledge of local CSI. In [13–18], the authors assume that perfect channel knowledge is available. Nevertheless, this is not a realistic assumption for practical scenarios. In this paper, we tackle this limitation. More specifically, the main contributions of this paper are the following:

- Design of a new SLNR-based precoder, where the channel errors are explicitly taken into account.
- In the precoder design, we tackle both the case where the BSs share their users data (extension of the paper presented in [13–15]) and where there is no data sharing (extension of the paper presented in [16]).
- By using the SLNR metric, we are able to design each user's precoder independently of the others, which enable the derivation of a closed-form solution for the proposed robust precoder, unlike the signal-to-interference-and-noise metric.

The remainder of this paper is organised as follows: Section 2 presents the multicell system model for both scenarios with and without data sharing. In Section 3, we derive the proposed robust distributed precoder for these two multicell-based approaches. Section 4 presents the main performance results. The conclusions will be drawn in Section 5.

Notations: Throughout this paper, we will use the following notations. Lowercase letters, boldface lowercase letters and boldface uppercase letters are used for scalars, vectors and matrices, respectively. $(.)^{H}$ represents the conjugate transpose operator, $\mathbb{E}[.]$ represents the expectation operator, $\boldsymbol{I}_N$ is the identity matrix of size $N \times N$, $\mathcal{CN}(.,.)$ denotes a circular symmetric complex Gaussian vector and $\|\boldsymbol{h}\|$ denotes the norm of vector $\boldsymbol{h}$.

2 System model

We consider two downlink multicell multiple-input–single-output (MISO)-based systems: in the first approach we consider that the BSs know the data symbols of all joint processing users which are shared by the backhaul network, and in the second one the BSs only know its own data symbols and therefore the backhaul network is not needed. It is assumed, for both approaches, that each BS has only access to an imperfect local channel estimate, that is, the channels between a given BS and all the joint processing users.

2.1 Multicell system with data sharing

We consider B BSs, each equipped with N_{t_b} antennas, transmitting to K single antenna UTs sharing the same physical channel, that is, the information for all UTs is transmitted at the same frequency band. The data symbols of all joint processing users are shared by the backhaul network as shown in Fig. 1. Under the assumption of linear precoding, the signal transmitted by the BS b is given by

$$\boldsymbol{x}_b = \sum_{k=1}^{K} \sqrt{p_{b,k}} \boldsymbol{w}_{b,k} s_k \tag{1}$$

where $p_{b,k}$ represents the power allocated to UT k at the BS b, $\boldsymbol{w}_{b,k} \in \mathbb{C}^{N_{t_b} \times 1}$ is the distributed precoder of user k at BS b with unit norm, that is, $\|\boldsymbol{w}_{b,k}\| = 1, \quad b = 1, \ldots, B, \ k = 1, \ldots, K$. The data symbol s_k, with $\mathbb{E}\left[|s_k|^2\right] = 1, \forall k$, is intended for UT k and is assumed to be available at all BSs.

The received signal at the UT k can be expressed as

$$y_k = \sum_{b=1}^{B} \boldsymbol{h}_{b,k}^{\mathrm{H}} \boldsymbol{x}_b + n_k \tag{2}$$

where $\boldsymbol{h}_{b,k} \in \mathbb{C}^{N_{t_b} \times 1}$ represents the channel between the BS b and user k and $n_k \sim \mathcal{CN}(0, \sigma^2)$ is the Gaussian noise.

From (1) and (2) the received signal at UT k can be decomposed in

$$y_k = \underbrace{\sum_{b=1}^{B} \sqrt{p_{b,k}} \boldsymbol{h}_{b,k}^{\mathrm{H}} \boldsymbol{w}_{b,k} s_k}_{\text{desired signal}} + \underbrace{\sum_{b=1}^{B} \boldsymbol{h}_{b,k}^{\mathrm{H}} \sum_{j=1, j \neq k}^{K} \sqrt{p_{b,j}} \boldsymbol{w}_{b,j} s_j}_{\text{multiuser multicell interference}} + \underbrace{n_k}_{\text{noise}} \tag{3}$$

2.2 Multicell system without data sharing

Here we also consider a downlink multicell MISO-based system, where B BSs, each equipped with N_{t_b} antennas, transmit data to K single antenna UTs sharing the same physical channel. For this scenario, each BS only serves one user and has only access to its data symbols. Therefore for this case the backhaul network is not needed, since there is no data sharing between BSs, see Fig. 2. However, the BSs have access to an imperfect version of channels between themselves and all the joint processing users such that coordinated precoding can be performed. In the following, we consider that UT k is server by BS b_k. Under the assumption of linear precoding, the signal transmitted by BS b_k is given by

$$\boldsymbol{x}_{b_k} = \sqrt{P_{b_k}} \boldsymbol{w}_{b_k} s_{b_k} \tag{4}$$

where $\boldsymbol{w}_{b_k} \in \mathbb{C}^{N_{t_b} \times 1}$, with $\|\boldsymbol{w}_{b_k}\|^2 = 1$, is the precoder at BS b_k, s_{b_k} with $\mathbb{E}\left[|s_{b_k}|^2\right] = 1$, denotes the BS b_k data symbol intended for UT k and P_{b_k} is the transmit power of the b_kth BS. The received signal at UT k, *served by BS* b_k is now given by

$$y_k = \underbrace{\sqrt{P_{b_k}} \boldsymbol{h}_{b_k,k}^{\mathrm{H}} \boldsymbol{w}_{b_k} s_{b_k}}_{\text{desired signal}} + \underbrace{\sum_{i=1, i \neq b_k}^{B} \sqrt{P_i} \boldsymbol{h}_{i,k}^{\mathrm{H}} \boldsymbol{w}_i s_i}_{\text{multiuser multicell interference}} + \underbrace{n_k}_{\text{noise}} \tag{5}$$

where $\boldsymbol{h}_{i,k} \in \mathbb{C}^{N_{t_b} \times 1}$ represents the channel between the BS i and user k.

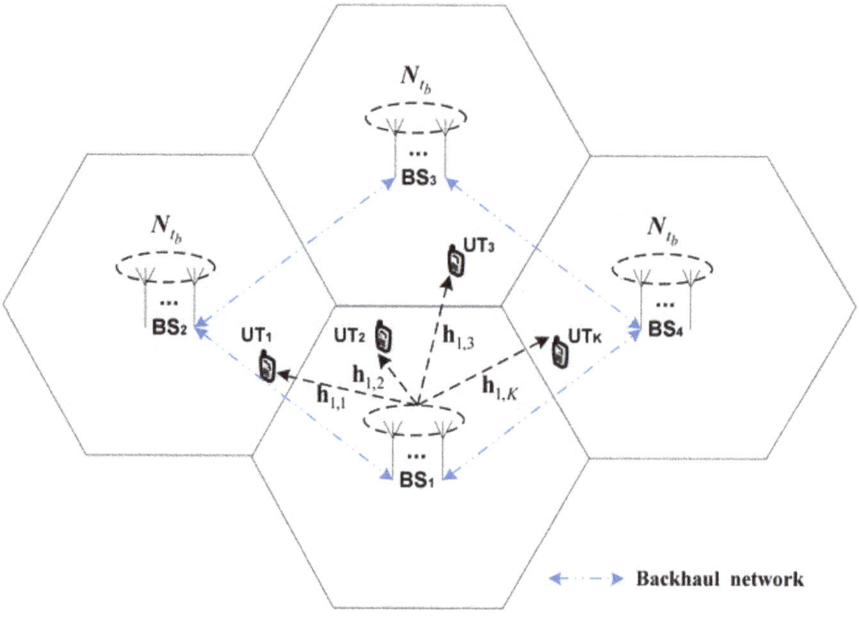

Fig. 1 *Considered data sharing scenario with K UTs (illustrated for B = 4 BSs equipped with N_{t_b} antennas)*

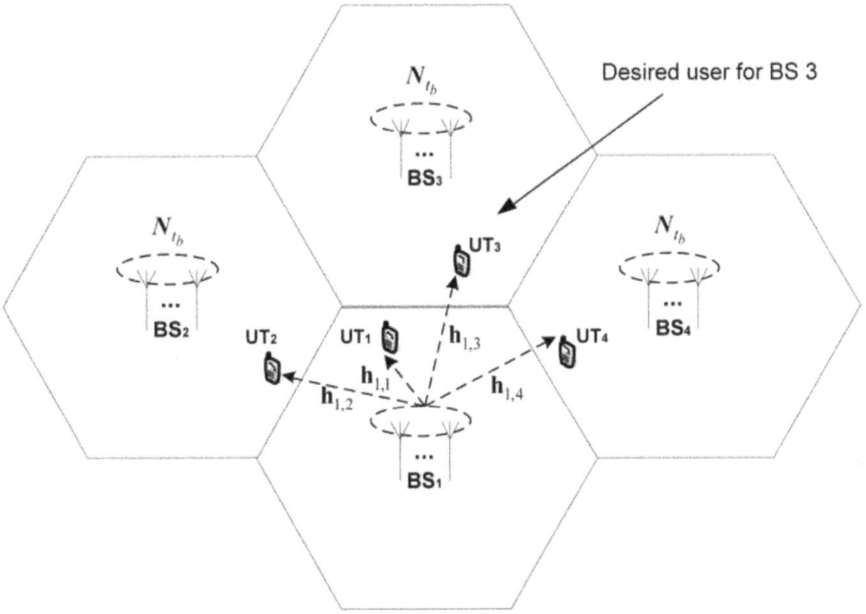

Fig. 2 *Considered scenario without data sharing (illustrated for B = 4 BSs equipped with N_{t_b} antennas each one served a single user)*

2.3 Channel estimation error model

In both scenarios, we assume that BS b has only knowledge of an estimate of its own channels $\boldsymbol{h}_{b,j}^{\mathrm{H}}$, $j = 1, \ldots, K$ and has no access to the channels from the other BSs. The channel estimate at the transmitter b will be modelled as

$$\hat{\boldsymbol{h}}_{b,j}^{\mathrm{H}} = \boldsymbol{h}_{b,j}^{\mathrm{H}} - \boldsymbol{e}_{b,j}^{\mathrm{H}} \tag{6}$$

where $\boldsymbol{e}_{b,j} \in \mathbb{C}^{N_{t_b} \times 1}$represents the overall channel estimation error and its elements are assumed to be independent identically distributed zero-mean complex Gaussian distributed with variance σ_h^2 and spatially white. The variance of the channel estimation error σ_h^2 is assumed to be known at all BSs [19].

3 Robust leakage precoder

3.1 Multicell system with data sharing

From (3), we verify that the power of the desired signal component coming from BS b at user k is

$$I_{b,k} = p_{b,k}\left|\boldsymbol{h}_{b,k}^{\mathrm{H}}\boldsymbol{w}_{b,k}\right|^2 \tag{7}$$

Likewise the interference power induced by BS b on UT k is given by

$$L_{b,k} = p_{b,k}\sum_{\substack{j=1\\ j\neq k}}^{K}\left|\boldsymbol{h}_{b,j}^{\mathrm{H}}\boldsymbol{w}_{b,k}\right|^2 \tag{8}$$

BS b has only an imperfect estimate of its channel vectors. Let us define $\boldsymbol{A}_{b,k} = \hat{\boldsymbol{h}}_{b,k}\hat{\boldsymbol{h}}_{b,k}^{\mathrm{H}} + \sigma_h^2\boldsymbol{I}$ and $\boldsymbol{D}_{b,k} = \sum_{j\neq k}\hat{\boldsymbol{h}}_{b,j}\hat{\boldsymbol{h}}_{b,j}^{\mathrm{H}} + (K-1)\sigma_h^2$, then by averaging over the channel errors the desired signal component and channel leakage power are as follows

$$\bar{I}_{b,k} = p_{b,k}\boldsymbol{w}_{b,k}^{\mathrm{H}}\left(\hat{\boldsymbol{h}}_{b,k}\hat{\boldsymbol{h}}_{b,k}^{\mathrm{H}} + \sigma_h^2\boldsymbol{I}\right)\boldsymbol{w}_{b,k} = p_{b,k}\boldsymbol{w}_{b,k}^{\mathrm{H}}\boldsymbol{A}_{b,k}\boldsymbol{w}_{b,k} \tag{9}$$

$$\begin{aligned}\bar{L}_{b,k} &= p_{b,k}\boldsymbol{w}_{b,k}^{\mathrm{H}}\left(\sum_{\substack{j=1\\ j\neq k}}^{K}\hat{\boldsymbol{h}}_{b,j}\hat{\boldsymbol{h}}_{b,j}^{\mathrm{H}} + (K-1)\sigma_h^2\boldsymbol{I}\right)\boldsymbol{w}_{b,k}\\ &= p_{b,k}\boldsymbol{w}_{b,k}^{\mathrm{H}}\boldsymbol{D}_{b,k}\boldsymbol{w}_{b,k}\end{aligned} \tag{10}$$

The equality in (9) and (10) follows from

$$\mathbb{E}\left[\left|\boldsymbol{h}_{b,k}^{\mathrm{H}}\boldsymbol{w}_{b,k}\right|^2\right] = \boldsymbol{w}_{b,k}^{\mathrm{H}}\left(\hat{\boldsymbol{h}}_{b,k}\hat{\boldsymbol{h}}_{b,k}^{\mathrm{H}} + \sigma_h^2\right)\boldsymbol{w}_{b,k}, \quad k = 1, \ldots, B$$

In the following, we use the SLNR [17] as a figure of merit for the design of precoders $\boldsymbol{w}_{b,k}$. For this scenario, the SLNR is given by

$$\mathrm{SLNR}(\boldsymbol{w}_{b,k}) = \frac{\bar{I}_{b,k}}{\sigma^2 + \bar{L}_{b,k}} = \frac{\boldsymbol{w}_{b,k}^{\mathrm{H}}\boldsymbol{A}_{b,k}\boldsymbol{w}_{b,k}}{\boldsymbol{w}_{b,k}^{\mathrm{H}}\boldsymbol{\Delta}_{b,k}\boldsymbol{w}_{b,k}} \tag{11}$$

where $\boldsymbol{\Delta}_{b,k} = (\sigma^2/p_{b,k})\boldsymbol{I} + \boldsymbol{D}_{b,k}$. The aim of BS b is to maximise SLNR($\boldsymbol{w}_{b,k}$). The BS b optimisation problem can now be mathematically described by

$$\boldsymbol{w}_{b,k} = \arg\max_{\|\boldsymbol{w}_{b,k}\|^2 = p_{b,k}} \frac{\boldsymbol{w}_{b,k}^{\mathrm{H}}\boldsymbol{A}_{b,k}\boldsymbol{w}_{b,k}}{\boldsymbol{w}_{b,k}^{\mathrm{H}}\boldsymbol{\Delta}_{b,k}\boldsymbol{w}_{b,k}} \tag{12}$$

The merit function of optimisation problem (12) is a Rayleigh quotient [20] and thus the optimum precoder for BS b is given by

$$\begin{aligned}\boldsymbol{w}_{b,k} &= \frac{\bar{\boldsymbol{w}}_{b,k}}{\|\bar{\boldsymbol{w}}_{b,k}\|}, \quad \bar{\boldsymbol{w}}_{b,k} = \frac{\boldsymbol{C}_{b,k}^{-1}\boldsymbol{v}_{b,k}}{\hat{\boldsymbol{h}}_{b,k}^{\mathrm{H}}\boldsymbol{C}_{b,k}^{-1}\boldsymbol{v}_{b,k}}\\ \boldsymbol{v}_{b,k} &= \max\ \mathrm{eigenvector}\left(\left(\boldsymbol{C}_{b,k}^{-1}\right)^{\mathrm{H}}\boldsymbol{A}_{b,k}\boldsymbol{C}_{b,k}^{-1}\right)\end{aligned} \tag{13}$$

where $\boldsymbol{C}_{b,k}$ denotes the Cholesky decomposition of matrix $\Delta_{b,k}$. The solution given by (13) ensures that $\hat{\boldsymbol{h}}_{bk}^{\mathrm{H}}\boldsymbol{w}_{b,k}$ is real valued and positive. This facilitates the decoding process at the UTs, as a scalar multiplication will not affect the value of the merit function given by (11).

3.2 Multicell system without data sharing: For this case, we obtain for the desired and leakage powers, at BS b_k

$$I_{b_k} = \left|\boldsymbol{h}_{b_k,k}^{\mathrm{H}}\boldsymbol{w}_{b_k}\right|^2 \tag{14}$$

$$L_{b_k} = \sum_{j=1, j\neq k}^{K}\left|\boldsymbol{h}_{b_k,j}^{\mathrm{H}}\boldsymbol{w}_{b_k}\right|^2 \tag{15}$$

Averaging over the channel estimation errors, the corresponding SLNR at BS b_k is

$$\mathrm{SLNR}(\boldsymbol{w}_{b_k}) = \frac{I_{b_k}}{\sigma^2 + L_{b_k}} = \frac{\boldsymbol{w}_{b_k}^{\mathrm{H}} \boldsymbol{A}_{b_k} \boldsymbol{w}_{b_k}}{\boldsymbol{w}_{b_k}^{\mathrm{H}} \boldsymbol{\Delta}_{b_k} \boldsymbol{w}_{b_k}} \tag{16}$$

where $\boldsymbol{A}_{b_k} = \hat{\boldsymbol{h}}_{b_k,k} \hat{\boldsymbol{h}}_{b_k,k}^{\mathrm{H}} + \sigma_h^2 \boldsymbol{I}$, $\boldsymbol{\Delta}_{b_k} = \left(\sigma^2 / P_{b_k}\right)\boldsymbol{I} + \boldsymbol{D}_{b_k}$ and $\boldsymbol{D}_{b_k} = \sum_{j=1, j\neq k}^{K} \hat{\boldsymbol{h}}_{b_k,j} \hat{\boldsymbol{h}}_{b_k,j}^{\mathrm{H}} + (B-1)\sigma_h^2$. The aim of BS b_k is to maximise $\mathrm{SLNR}(\boldsymbol{w}_{b_k})$ like in (12). The solution of this optimisation problem is

$$\boldsymbol{w}_{b_k} = \frac{\bar{\boldsymbol{w}}_{b_k}}{\left\| \bar{\boldsymbol{w}}_{b_k} \right\|}, \quad \bar{\boldsymbol{w}}_{b} = \frac{\boldsymbol{C}_{b_k}^{-1} \boldsymbol{v}_{b_k}}{\hat{\boldsymbol{h}}_{b_k,jk}^{H} \boldsymbol{C}_{b_k}^{-1} \boldsymbol{v}_{b_k}}$$

$$\boldsymbol{v}_{b_k} = \max \ \mathrm{eigenvector}\left(\left(\boldsymbol{C}_{b_k}^{-1}\right)^{\mathrm{H}} \boldsymbol{A}_{b_k} \boldsymbol{C}_{b_k}^{-1} \right) \tag{17}$$

where $\boldsymbol{C}_{b_k}$ denotes the Cholesky decomposition of matrix $\boldsymbol{\Delta}_{b_k}$. The solution given by (17) ensures that $\hat{\boldsymbol{h}}_{b_k,k}^{\mathrm{H}} \boldsymbol{w}_{b_k}$ is real valued and positive.

4 Performance results

In this section, we assess the performance of the proposed robust leakage-based precoder both for the scenario with and without data sharing. We compare the proposed method against the non-robust approach. We consider a scenario with $B=4$, $K=4$ and $N_{t_b} = 4$. In addition, we consider the same power constraint for all BSs, that is, that $P_b = P_k$, $\forall k \neq b$ for the non-data sharing scenario, and $p_{b,k} = P_b/K$ for the data sharing one. The components of the channels $\boldsymbol{h}_{b,j}$ are assumed to be complex Gaussian, that is, the envelope Rayleigh distributed. The results are presented in terms of average BER as a function of the signal-to-noise ratio $(\mathrm{SNR}) = P_b/\sigma^2$.

In Figs. 3 and 4, we show the impact of the channel estimation on the BER performance, by considering different values for the variance of the estimation error, for the scenario with and without data sharing, respectively. As the robust method takes into account both the contributions of the additive noise and channel estimation error, it achieves improved performance. For example, for the case where $\sigma_h^2 = -5$ dB, and considering the data sharing scenario, the BER of the non-robust approach is lower bounded by 3×10^{-2} and degrades for higher SNR values. For the robust method this bound is reduced to 10^{-3}, a reduction of about 66%. For the scenario without data sharing and for $\sigma_h^2 = -20$ and $\sigma_h^2 = -30$ dB, the improvement is about 40 and 50%, respectively. For this case, the channel estimation error has a higher impact on the BER. This occurs as for the scenario with data sharing each user is served by more than one BS and therefore benefits from a diversity advantage. However, the system complexity is higher as the data symbols of all joint processed users must be shared by the backhaul network. As verified from Figs. 3 and 4, the improvement is higher for lower values of the estimation error.

For the least-squares (LS) channel estimation-based method, the estimation error is expected to be equal to the system SNR, that is, $\sigma_h^2 = \sigma^2$. In Figs. 5 and 6, we show the performance of the proposed method when the LS estimator is used, for the scenario with and without data sharing, respectively. For this case, we verify, that there is a gap of about 1.5 dB (3 dB) from the robust to the non-robust approach and of about 8 dB (8 dB) from the robust to the case where $\sigma_h^2 = 0$, at a target BER of 10^{-3}, for the scenario with (without) data sharing. The gap between the robust and the $\sigma_h^2 = 0$ case is constant over the SNR, contrarily to the case of the non-robust method, which indicates that by using the non-robust method the noise is considerably enhanced.

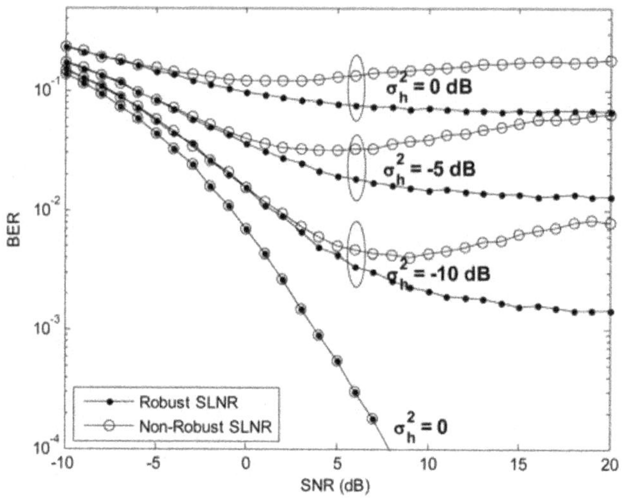

Fig. 3 *Performance evaluation of the robust and non-robust precoding schemes for the multicell system with data sharing and* $\sigma_h^2 = \{-5, \ -10, \ -15\}$ *dB,* $B=4$ *and* $N_{t_b} = 4$

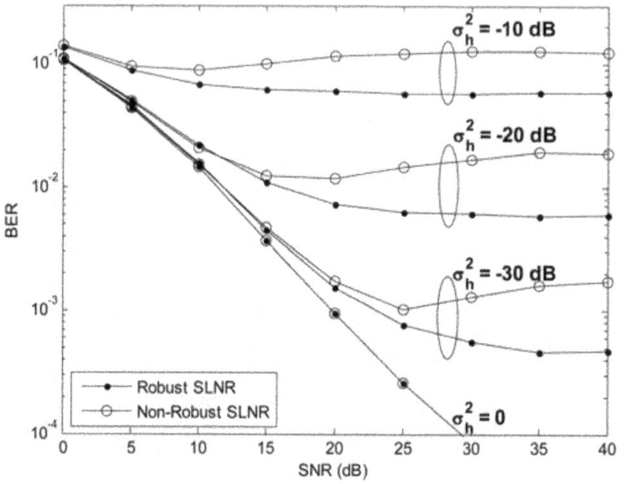

Fig. 4 *Performance evaluation of the robust and non-robust precoding schemes for the multicell system without data sharing and* $\sigma_h^2 = \{-10, \ -20, \ -30\}$ *dB,* $B=4$ *and* $N_{t_b} = 4$

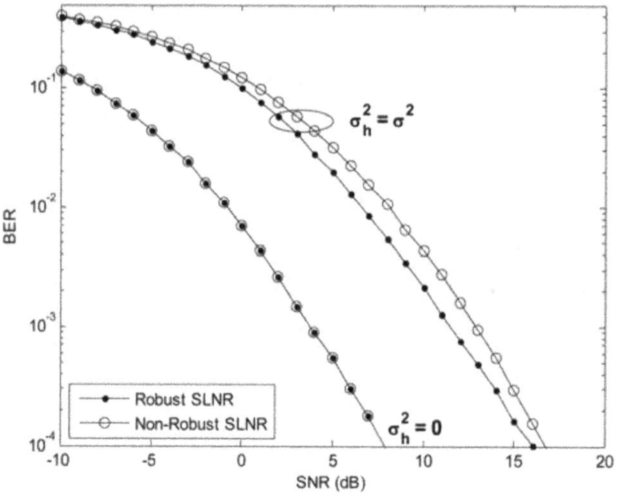

Fig. 5 *Performance evaluation of the robust and non-robust precoding schemes, for the multicell system with data sharing and* $\sigma_h^2 = \{0, \ \sigma^2\}$*,* $B=4$ *and* $N_{t_b} = 4$

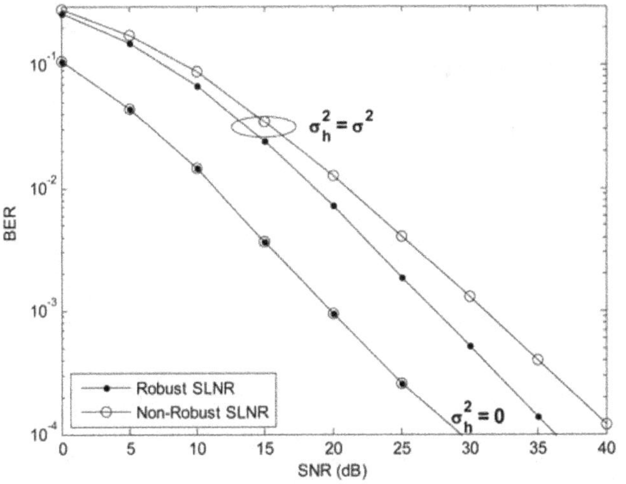

Fig. 6 *Performance evaluation of the robust and non-robust precoding schemes, for the multicell system without data sharing and* $\sigma_h^2 = \{0, \sigma^2\}$, $B = 4$ *and* $N_{t_b} = 4$

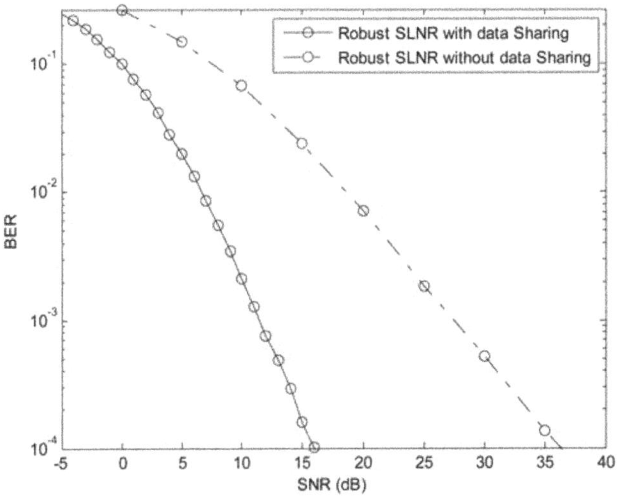

Fig. 7 *Performance evaluation of the robust precoding schemes, with and without data sharing*

Finally, in Fig. 7 we compare the performance of the proposed robust method with and without that sharing, that is, for both scenarios considered. In the data sharing scenario, all BSs transmit data to all users, contrarily to the case without data sharing, where a user only receives its data from one of the BSs. As a consequence, in the data sharing scenario the received data are made available through multiple independent paths increasing the receive signal diversity. This phenomenon is easy to verify from Fig. 7 where the BER curve for the scenario with data sharing, the solid one, has a high diversity order than for the case without data sharing.

5 Conclusions

In this paper, we have proposed a robust distributed precoding method by taking into account the channel estimation errors. A leakage-based approach was considered leading to a closed-form precoder with low complexity. Two approaches were considered: the BSs know the data symbols of all users, shared by the backhaul network and the BSs only know its own data symbols.

The performance of the proposed precoder is considerably better than the non-robust approach for both multi-cell-based systems. By considering the estimation error in the precoder design, we were able to address the shortcomings of the non-robust method by removing the noise enhancement, inherent to such a precoder. The presented results show that the proposed precoder is of special interest for the next generation networks as it deals effectively with the channel errors, which are always present in practical wireless systems.

6 Acknowledgments

This work was supported by the Portuguese FCT (Fundação para a Ciência e Tecnologia) Projects ADIN (PTDC/EEI-TEL/2990/2012), COPWIN (PTDC/EEI-TEL/1417/2012) and HETCOP (PEst-OE/EEI/LA0008/2013).

7 References

[1] Karakayali M., Foschini G., Valenzuela R.: 'Network coordination for spectral efficient communications in cellular systems', *IEEE Commun. Mag.*, 2006, **13**, (4), pp. 56–61

[2] Mudumbai T.R., Brown D.R., Madhow U.: 'Distributed transmit beamforming: challenges and recent progress', *IEEE Commun. Mag.*, 2009, **47**, (2), pp. 102–110

[3] Gesbert D., Hanly S., Huang H., Shamai S., Simeone O., Yu W.: 'Multi-cell MIMO cooperation networks: a new look at interference', *IEEE J. Sel. Areas Commun.*, 2010, **28**, (9), pp. 1380–1408

[4] 3GPP: 'Coordinated multi-point operation for LTE physical layer aspects'. Technical Report, 3GPP TR 36.819, 2011

[5] Jing S., Tse D.N.C., Sorianga J.B., Hou J., Smee J.E., Padovani R.: 'Multicell downlink capacity with coordinated processing', *EURASIP J. Wirel. Commun. Netw.*, 2008, **2008**, (18), pp. 1–19

[6] Castanheira D., Gameiro A.: 'Distributed antenna system capacity scaling', *IEEE Commun. Mag.*, 2010, **17**, (3), pp. 78–75

[7] Marsch P., Fettweis G.: 'On downlink network MIMO under a constrained backhaul and imperfect channel knowledge'. Proc. IEEE GLOBECOM'09, 2009

[8] Armada A.G., Fernándes M.S., Corvaja R.: 'Waterfilling schemes for zero-forcing coordinated base station transmissions'. Proc. IEEE GLOBECOM'09, 2009

[9] Kobayashi M., Debbah M., Belfiore J.: 'Outage efficient strategies in network MIMO with partial CSIT'. Proc. ISIT'09, 2009

[10] Zhang J., Chen R., Andrews J.G., Ghosh A., Heath R.W.Jr.: 'Networked MIMO with clustered linear precoding', *IEEE Trans. Wirel. Commun.*, 2009, **8**, (4), pp. 1910–1921

[11] Zhang R.: 'Cooperative multi-cell block diagonalization with per-base-station power constraints'. Proc. IEEE WCNC'10, 2010

[12] Castanheira D., Silva A., Gameiro A.: 'Linear and nonlinear precoding schemes for centralized multicell MIMO-OFDM systems', *Wirel. Pers. Commun.*, 2013, **72**, (1), pp. 759–777

[13] Zakhour R., Gesbert D.: 'Distributed multicell-MISO precoding using the layered virtual SINR framework', *IEEE Trans. Wirel. Commun.*, 2010, **9**, (8), pp. 2444–2448

[14] Bjornson E., Zakhour R., Gesbert D., Ottersten B.: 'Cooperative multicell precoding: rate region characterization and distributed strategies with instantaneous and statistical CSI', *IEEE Trans. Signal Process.*, 2010, **58**, (8), pp. 4298–4310

[15] Silva A., Holakouei R., Gameiro A.: 'A novel distributed power allocation scheme for coordinated multicell systems', *EURASIP J. Wirel. Commun. Netw.*, 2013, **30**, pp. 1–11

[16] Zakhour R., Gesbert D.: 'Coordination on the MISO interference channel using the virtual SINR framework'. Proc. of ITG/IEEE Workshop on Smart Antennas, 2009

[17] Sadek M., Tarighat A., Sayed A.H.: 'A leakage-based precoding scheme for downlink multi-user MIMO channels', *IEEE Trans. Wirel. Commun.*, 2007, **6**, (5), pp. 1711–1721

[18] Larsson E., Jorswieck E.: 'Competition versus cooperation on the MISO interference channel', *IEEE J. Sel. Areas Commun.*, 2008, **26**, (7), pp. 1059–1069

[19] Yoo T., Goldsmith A.: 'Capacity and power allocation for fading MIMO channels with channel estimation error', *IEEE Trans. Inf. Theory*, 2006, **52**, (5), pp. 2203–2214

[20] Goluband G., Loan C.V.: 'Matrix computations' (The Johns Hopkins University Press, Baltimore, MD, 1996, 3rd edn.)

23 μW 8.9-effective number of bit 1.1 MS/s successive approximation register analog-to-digital converter with an energy-efficient digital-to-analog converter switching scheme

Lei Sun[1], *Chi Tung Ko*[1], *Marco Ho*[1], *Wai Tung Ng*[2], *Ka Nang Leung*[1], *Chiu Sing Choy*[1], *Kong Pang Pun*[1]

[1]*Department of Electronic Engineering, Chinese University of Hong Kong, Shatin 852, Hong Kong, People's Republic of China*
[2]*Department of Electrical & Computer Engineering, University of Toronto, 10 King's College Road, Toronto, Ontario, Canada M5S 3G4*
E-mail: sunleiy@gmail.com

Abstract: This study presents a successive approximation register analog-to-digital converter with an energy-efficient switching scheme. A split-most significant bit capacitor array is used with a least significant bit-down switching scheme. Compared with the conventional binary-weighted capacitor array, it reduces the area and average switching energy by 50 and 87% under the same unit capacitor. Moreover, capacitor matching requirement is relaxed by 75%. A prototype design was fabricated in a 0.13 μm complementary metal oxide semiconductor process. It consumes 23.2 μW under 1 V analog supply and 0.5 V digital supply. Measured results show a peak signal-to-distortion-and-noise ratio of 55.2 dB and an effective resolution bandwidth up to 1.1 MHz when it operates at 1.1 MS/s. Its figure-of-merit is 44.1 fJ/conversion-step.

1 Introduction

With the permeation of wireless sensor networks and handheld or wearable devices with built-in sensors, compact and low-power analog-to-digital converters (ADCs) with medium bandwidth are highly demanded [1–5]. The successive approximation register (SAR) ADC earned its dominance in this field of applications because of its area and power efficiency resulting from its nature of binary search algorithm [6]. An SAR ADC normally consists of a digital controller, a comparator and a capacitive digital-to-analog converter (CDAC). It requires the least active circuities and avoids the use of amplifiers, which increasingly become a bottleneck in scaled complementary metal oxide semiconductor (CMOS) technologies because of the decline of intrinsic gain in short channel length transistors. When a dynamic comparator (without pre-amplification) is used, the power dissipation of the SAR ADC is partly because of switching activities and partly because of leakages (which is more important in lower speed applications) [7]

$$P_{ADC} \simeq \alpha C V^2 f_{clk} + P_{leakage} \tag{1}$$

Intuitively, reducing the capacitance lowers the power dissipation. Digital circuits naturally benefit from smaller device and wire parasitic capacitances in scaled CMOS. The comparator's power does not scale as in digital circuits [8] because it has to drive a large load capacitance for noise immunity, unless special techniques such as data-driven noise-reduction method [9] are applied. The switching power in the CDAC can be lowered by using a smaller unit capacitor [9]. However, the minimum value of the unit capacitor is commonly set by matching [10], noise requirement or process limitation. Decreasing the supply voltage saves some power at the cost of reduced analog signal swings. Lastly, the CDAC structure and switching scheme have a great impact on its switching power consumption. The split-capacitor [11], energy saving [12], set-to-down [13] and V_{cm}-based [14] CDACs reduce the average switching energy by 37, 56, 81 and 87%, respectively, compared with the CDAC using the conventional binary-weighted capacitor array (BWA) with the same unit capacitor. However, the set-to-down encounters the problem of common-mode (CM) voltage variation at the comparator's inputs; the V_{cm}-based CDAC needs an extra reference voltage V_{cm} and the buffer driving it. To overcome these minor problems, a CDAC switching concept, namely 'split-most significant bit (MSB) with least significant bit (LSB) down' is reported in [15, 16] [The group of researchers of [16] developed a similar technique. Our work was done independently before the publication of [16].], which in theory saves the CDAC switching power and area by 87 and 50%, respectively, compared with the BWA CDAC with no side effects. In this work, an SAR ADC adopting this concept is designed in a 0.13 μm process. It measures an 8.9-effective number of bits (ENOBs) at 1.1 MS/s, consuming 23.2 μW.

2 SAR ADC architecture

2.1 Architecture

Fig. 1 shows the SAR ADC architecture, which is fully differential in order to suppress substrate, supply and other CM noise. It consists of four blocks, namely, CDAC, comparator, digital controller (operating from 0.5 V) and level shifter between the controller and the rest of the ADC (operating from 1 V). The capacitances of the N-bit ADC are given in the following equation

$$C_i = C_{2,i} = 2^{N-i} C_0, \quad i \in \{3, \ldots, N\} \tag{2}$$

where C_0 (and $C_{2,0}$) is the unit capacitor. The total capacitance in a split-MSB array is $2^{N-2}C_0$. The subscript '2' indicates those capacitors are the split ones of the largest capacitor C_2 in the conventional BWA.

The operation of the ADC is described as follows. Consider the positive half circuit. During the sampling phase switches S_a close, the input (V_{IP}) is sampled on the top plate of all the CDAC capacitors in the half circuit. The bottom plates of the split-MSB capacitors are connected to V_{ref}, whereas the others are connected to ground. Top plate sampling is employed here for better energy efficiency because the MSB can be resolved right after the sampling without switching any capacitors. The other half circuit works on V_{IN} in the same manner.

After sampling, the inputs are held by opening switches S_a. The comparator outputs $b_1 = 1$ (b_1 stands for MSB here) if $V_{DACP} > V_{DACN}$; otherwise $b_1 = 0$. In the next clock phase, if $b_1 = 1$, the

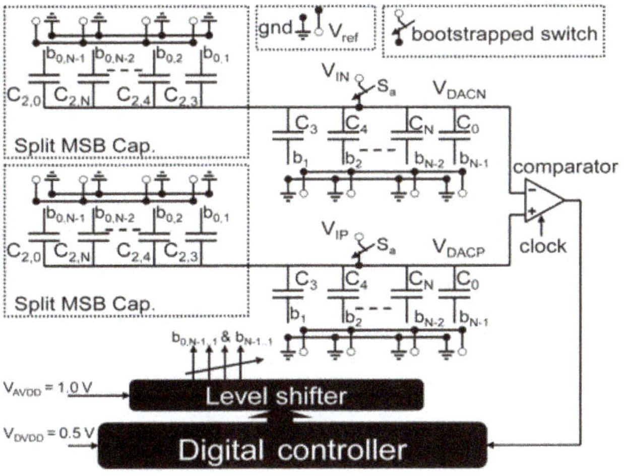

Fig. 1 *SAR ADC architecture*

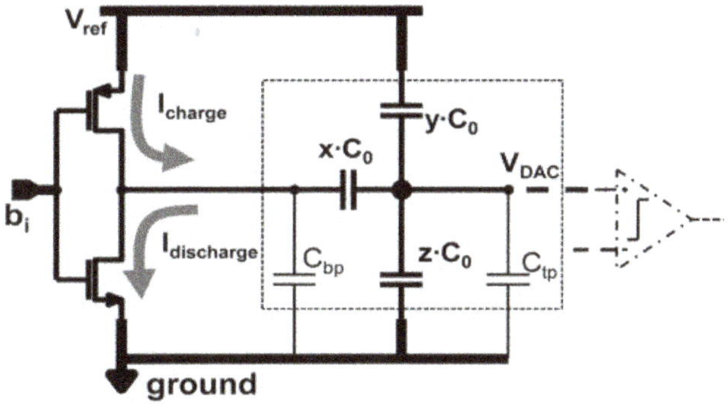

Fig. 3 *Charging and discharging the CDAC*

largest split-MSB capacitor $C_{2,3}(=2^{N-3}C_0)$ in the positive array switches from V_{ref} to ground to decrease V_{DACP} by $V_{\text{ref}}/4$. Simultaneously, $C_3(\text{also} = 2^{N-3}C_0)$ in the main array of the negative half circuit switches from ground to V_{ref} to increase V_{DACN} by $V_{\text{ref}}/4$. The opposite takes place if $b_1 = 0$. Then V_{DACP} is compared with V_{DACN} to resolve the second MSB, b_2. The process repeats until the second LSB (b_{N-1}) is resolved as illustrated in an example in Fig. 2.

To resolve the LSB (b_N), only one unit capacitor ($C_{2,N}$) in the split-MSB array, in either positive or negative half circuit (depending on b_{N-1}), switches from V_{ref} to ground to change ($V_{\text{DACP}} - V_{\text{DACN}}$) by a positive or negative amount of $V_{\text{ref}}/2^{N-1}$. This single-ended LSB switching reduces the required total capacitance by half, compared to the conventional split-MSB capacitor array approach [11]. The only problem it brings is a CM voltage variation at the inputs of the comparator during the LSB switching. However, this CM voltage variation, with amplitude of LSB/2, is insignificant. In addition, unlike the V_{cm}-based SAR ADCs, this architecture does not require an extra dc voltage V_{cm}, and the buffer driving it.

Lastly, a prior work [7] showed that different supplies for the analog and digital parts lower the overall power consumption without hurting the performance of an SAR ADC that operates at 2 kS/s. The dual supply approach is adopted in this ADC which operates at a higher rate, up to 1.1 MS/s.

2.2 *CDAC switching energy*

The switches for connection to V_{ref} or ground are realised by P-type metal-oxide-semiconductor (PMOS) or N-type metal-oxide-semiconductor (NMOS). The switch configuration is practically an inverter as illustrated in Fig. 3. The inverter is controlled by its corresponding bit (b_i) during the binary searching process. Both charging and discharging target capacitor xC_0 draw energy from V_{ref}. The capacitors zC_0 and yC_0 represent the rest of the capacitors in the CDAC that are connected to ground and V_{ref}. Let the parasitic capacitor C_{tp} at the top plate be λC_0, which is often much smaller than the total capacitance of the CDAC [10]. The energies delivered from V_{ref} for charging (E_C) and discharging (E_D) xC_0 are derived as

$$E_C = \frac{x \cdot (z + \lambda)}{x + y + z\lambda} C_0 V_{\text{ref}}^2 \tag{3}$$

$$E_D = \frac{x \cdot y}{x + y + z\lambda} C_0 V_{\text{ref}}^2 \tag{4}$$

For the parasitic capacitance C_{bp} at the bottom plate, it presents a direct loading to the inverter, which should be minimised with careful layout. Note here that the energies dissipated by C_{bp} are not covered in (3) and (4).

Based on (3) and (4), the switching energy for different output codes (different values of x, y and z) is evaluated as shown in Fig. 4, which ignores the parasitic capacitances. The averaged energy consumption, assuming an even distribution of the output codes, is also shown in Fig. 4.

For the LSB-down scheme, its switching energy is about the same as that of the current best one, namely, the V_{cm}-based

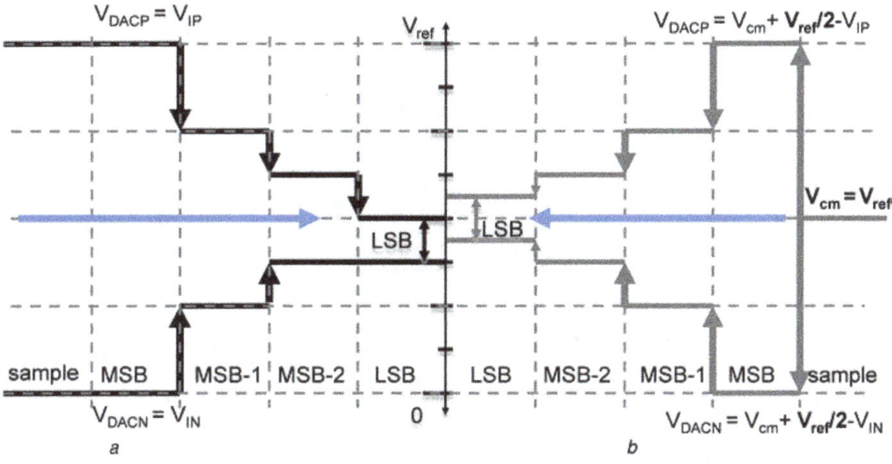

Fig. 2 *Voltage waveforms at the CDAC outputs in*
a Split-MSB with LSB-down scheme
b Conventional BWA scheme
$V_{\text{IP}} = V_{\text{ref}}$ and $V_{\text{IN}} = 0$ V are assumed in this example

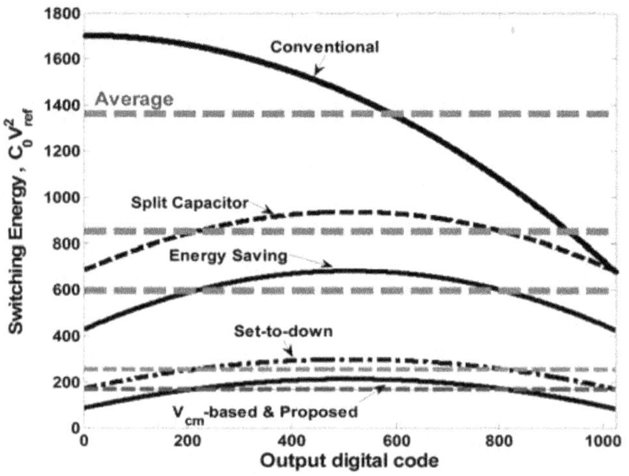

Fig. 4 *CDAC switching energy in different switching schemes*

approach. The peak switching energy happens at the mid-scale because it invokes the largest charge redistribution. The idea of energy saving is to reduce the capacitance by half for the same unit capacitor, while that behind the V_{cm}-based approach is to reduce the step-size of voltage to be switched by 1/2 during the transition with an extra reference voltage V_{cm}.

A CDAC architectural coefficient β is defined here, which measures the average switching energy of a CDAC normalised to that of the BWA CDAC. The values of β for different CDACs are compared in Table 1.

2.3 *CDAC linearity*

The ADC linearity is mainly affected by the capacitor mismatch in the CDAC. Model the actual value of a capacitor as the sum of its nominal capacitance and an error term δ_i

$$C_i = 2^{N-i}C_0 + \delta_i, \quad \sigma_i^2 = E[\delta_i^2] = 2^{N-i}\sigma_0^2 \tag{5}$$

where the error term δ_i is a random variable with a zero mean and a variance of σ_i^2. The σ_0 is the standard deviation of the unit capacitance. It is reasonable to assume that the capacitors in the positive half circuits have the same mismatch properties as those in the negative half. The largest accumulated capacitor mismatch, and thus the worst differential non-linearity (DNL) and integral non-linearity (INL), occurs at $V_{in} = 1/4V_{ref}$ and $3/4V_{ref}$. The standard deviations of the worst DNL and INL (end-point fit) of this SAR ADC, because of capacitor mismatch, are found as

$$\sigma_{\text{DNL, max}} = \frac{1}{2}2^{N/2}\frac{\sigma_0}{C_0}\text{LSB} \tag{6}$$

Table 1 Comparison of different CDAC switching schemes

Switching procedures	Normalised switching energy (β)	Normalised total capacitance	Normalised σ_{DNL}	Normalised σ_{INL}
BWA	1	1	1	1
split-capacitor	0.63	1	0.71	1
energy saving	0.44	1	0.5	0.25
set-to-down	0.19	0.5	0.71	0.35
V_{cm}-based	0.13	0.5	0.71	0.5
split-MSB w/LSB down	0.13	0.5	0.5	0.35

$$\sigma_{\text{INL, max}} = \frac{\sqrt{2}}{4}2^{N/2}\frac{\sigma_0}{C_0}\text{LSB} \tag{7}$$

The $\sigma_{DNL,max}$, being $\sqrt{2}$ times larger than $\sigma_{INL,max}$, determines the allowed minimum size of the unit capacitor given that the resulted value fulfils the thermal noise requirement. Table 1 also compares the linearity and total capacitance with other switching procedures.

The metal–insulate–metal (MIM) capacitor available in the adopted 0.13 μm CMOS process is used, which has 2.01 fF/μm^2 and a matching coefficient about 2.6%-μm. To maintain $\sigma_{DNL,max} < 1/2$ LSB for 10 bit resolution, $C_0 \geq 11$ fF. The smallest MIM capacitor allowed in the technology is 59.6 fF (5.24 μm × 5.24 μm). Two 59.6 fF capacitors connected in series are used to realise an LSB capacitor (29.8 fF), resulting in $\sigma_0/C_0 = 0.7\%$. From (6) and (7), the theoretical DNL and INL are 0.1 and 0.08 LSB, respectively. Lastly, customised capacitors [17] can be used for smaller area and power.

3 Building block circuits

3.1 *Passive sample-and-hold (S/H) circuit*

The passive S/H circuit in the SAR ADC must be fast enough to meet the settling requirements. For a sampling error less than 1/2 LSB for N-bit resolution, the −3 dB bandwidth of the S/H circuit must be at least $0.69(N+1)(1/T)$, where T is the duration of the available sampling time. In a synchronous SAR ADC, it takes $(N+1)$ clock periods to convert one sample. As a result, the minimum −3 dB bandwidth of the S/H is $0.69(N+1)^2 f_{clk}$, where f_{clk} is the clock rate. The sampling time is assumed to occupy one clock period.

In this design, a switch on-resistance less than 690 Ω is required for $N = 10$, $C_S \cong 15$ pF and $1/f_{clk} = 80$ ns. To avoid using large size switches, bootstrapped switches [17] are used. Monte Carlo runs under the worst corner (ss, hot and low V_{DD}) gives a signal-to-distortion-and-noise ratio (SNDR) with a mean of 66 dB and a standard deviation of 1 dB, which is within the design target.

3.2 *Dynamic comparator*

Fig. 5 shows the dynamic comparator [18] designed for this ADC. A number of measures have been taken. First, NMOS input pairs are used for higher speed and smaller size. Second, a pair of balanced buffers is inserted before set-reset (SR) latch for smaller offset. The offset of comparator hurts the input swing and degrades the signal-to-noise ratio (SNR) although it does not affect the linearity of the ADC [13]. Monte Carlo simulations show that the comparator's offset voltage is less than 25 mV with a 99.7% confidence level. This amount of offset degrades the SNR by

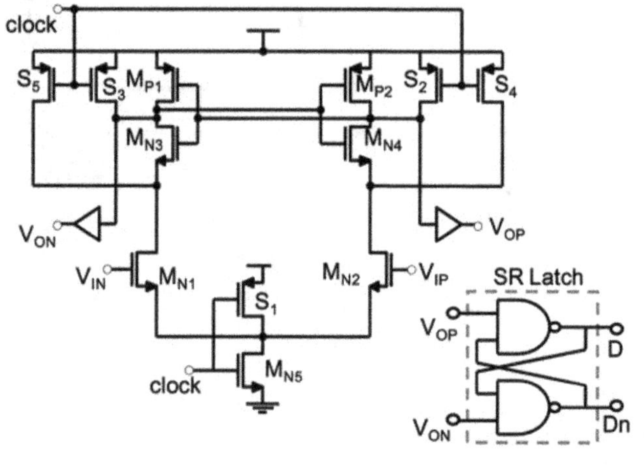

Fig. 5 *Dynamic comparator and its SR latch*

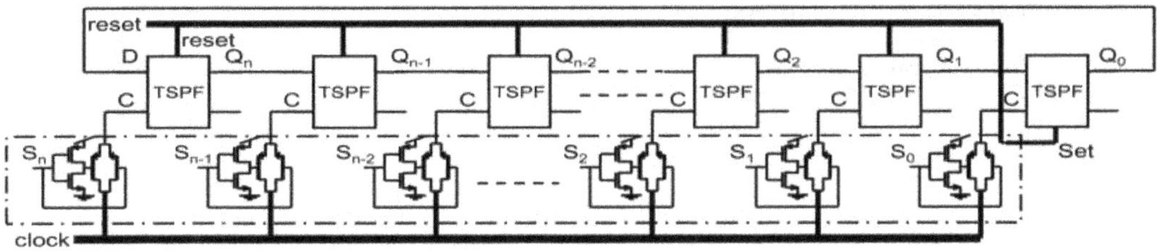

Fig. 6 *Clock-gated shifter register based on TSPF*

0.22 dB because of the reduction on the dynamic range. Third, the thermal noise of the comparator is tailored into the level of quantisation noises by setting the total load capacitor (including parasitic) at the comparator output to be larger than 10 fF [7].

3.3 Digital controller

A low-power digital controller is designed using clock gating technique, as shown in Fig. 6. True single-phase flip-flops (TSPF) [19] are used for its fewer transistor numbers and low-power consumption. In each bit-resolving clock cycle, at most two registers alter their outputs. A transmission gate, controlled by proper logics, is thus inserted to modulate the clock to save power. Besides, enabled by low-V_t transistors, 0.5 V supply is used in the digital controller to reduce its dynamic power dissipation. A level shifter [20] is used for interface between the analog (1.0 V) and digital parts (0.5 V). A limitation of the TSPF, like other dynamic

Fig. 7 *Micrograph of the ADC*

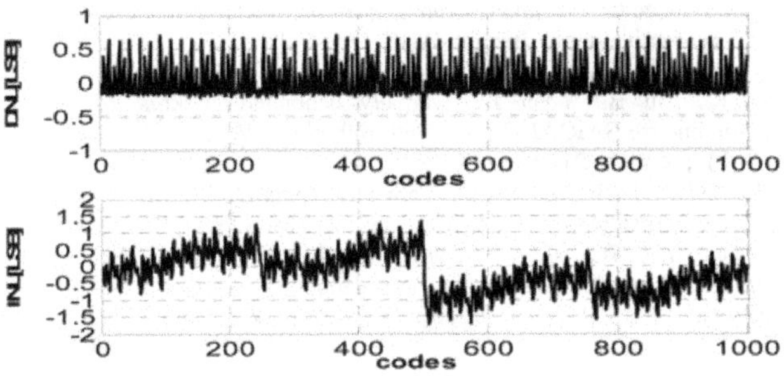

Fig. 8 *Measured DNL/INL plot*

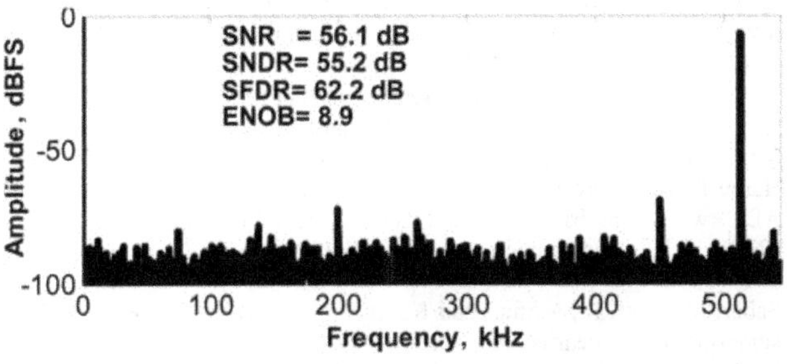

Fig. 9 *4 k-point FFT for an input frequency near $f_s/2$ and $f_s = 1.1$ MS/s*

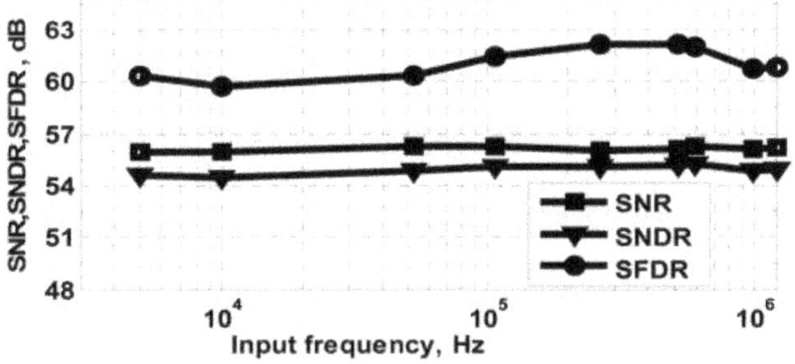

Fig. 10 *Measured dynamic performance against input frequency at 1.1 MS/s*

Table 2 Comparison of different switching procedures

technology	0.13 μm 1P6M CMOS			
sampling rate	1.1 MS/s			
core area (w/I/O buffer)	821 × 665 μm^2			
DNL [LSB]	+0.6/ −0.7			
INL [LSB]	+1.3/ −1.6			
SNDR/SFDR (at $f_{in} - f_s/2$)	55.2 dB/62.2 dB			
ADC power breakdown at 1.1 MS/s				
supply voltage (V) in comparator and level shifter	1.2	1.2	1.0	1.0
DAC	1.2	1.0	1.2	1.0
logics	0.5	0.5	0.5	0.5
power consumption (μW) in comparator and level shifter	8.4	8.5	5.0	5.0
DAC	24.4	16.3	28.3	17
logics	1.2	1.2	1.2	1.2
ADC FOM				
FOM [fJ/conv.-step]	64.7	49.4	65.5	44.1

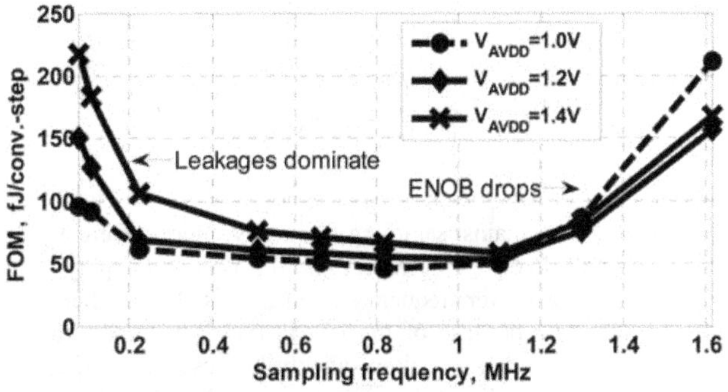

Fig. 11 *Measured FOM against sampling frequency*

Table 3 SAR ADC comparison with different switching procedures

	Switching procedures	Process, nm	Supplies, V	Unit cap., fF	f_s, MS/s	ENOB, bit	Power, μW	FOM, fJ/ conv.-step	FOM$_a$, fJ/ conv.-step-fF
[1] Hong'07	single-ended BWA	180	0.9	24	0.4	7.31	6.15	97	4
[11] Ginsburg'07	split capacitor	65	1.2	—	500	4.04	5900	750	—
[12] Chang'07	energy saving	180	1.0	20	0.5	75	7.75	86	4.3
[22] Craninckx'07	charge sharing	90	1.0	64	20	7.8	290	65.1	1.0
[6] Elzakkler'08	adiabatic charging	65	1.0	~0.6	1	8.75	1.9	4.4	7.3
[13] Liu'10	set-to-down	130	1.2	4.8	50	9.18	826	29	6.0
[14] Zhu'10	V_{cm}-based	90	1.2	50	100	9.1	3000	55	1.1
[16] Tripathi'13	split-MSB w/LSB down	65	1.2	0.75	450	7.56	6480	76	101.3
[17] Harpe'11	custom capacitor	90	1.0	0.4932	10.24	7.77	26.3	12	24.3
[2] Xu'12	cap. with calibration	180	3.3/1.8	20	0.768	9.8	58	74	3.7
[23] Liu'10	windowed switching	180	1	5	10	9.83	98	11	2.2
[24] Liou'13	charge average	90	0.4	5	0.5	8.72	0.5	2.37	0.5
[25] Harpe'13	custom capacitor	65	0.6	0.25	0.04	9.4	0.072	2.7	10.8
this work	split-MSB w/LSB down	130	1.0/0.5	29.8	1.1	8.9	23.2	44.1	1.5

memory circuits, is that it must operate above a certain clock rate to avoid losing its internal memory status because of charge leakages.

4 Measurement results

A 10 bit SAR ADC was designed and fabricated in a 0.13 μm CMOS process. A micro-photograph of the die is shown in Fig. 7. The ADC core area is 821 × 665 μm^2, including the I/O buffers. The capacitor array occupies about 55% of the active chip area because we used the standard MIM capacitors, the size of one of which is 5.24 μm × 5.24 μm or larger.

A differential 5 kHz sinusoidal input signal with 0 dB FS amplitude was applied to the 1.1 MS/s, 10 bit ADC for the static linearity measurement of ADC [21]. Fig. 8 shows the measured DNL and INL. The peak DNL error is +0.6/−0.7 LSB and the peak INL error is +1.3/−1.6 LSB, which are larger than the theoretical values from (6) and (7). The deviation is attributed to wide separation between some capacitors and systematic mismatch errors, including routing mismatch, second-order oxide thickness variation and parasitic effects at the internal nodes of LSB capacitors.

Fig. 9 shows a measured 4*k*-point fast Fourier transform (FFT) of the output of the ADC when it operates at 1.1 MS/s. The amplitude of the test stimulus was set to −0.2 dB FS, where FS = V_{ref} = 1 **V**. The measured SNR, SNDR (ENOB = 8.9 bits) and spurious-free dynamic range (SFDR) are 56.1, 55.2 and 62.2 dB, respectively. When the input frequency varies from 5 kHz to 1.1 MHz, the measured dynamic performance in Fig. 10 shows that the SNDR holds virtually constant over the entire effective resolution bandwidth of 1.1 MHz.

The total power consumption of the SAR ADC is 23.2 μW at 1.1 MS/s. It breaks down into 1.2, 5.0 and 17 μW from the 1 V analog supply, 1 V reference and 0.5 V digital supply, respectively. Table 2 summarises the measured performance. The dynamic power dissipation, as in (1), scales with the sampling frequency and the supply voltage. However, the figure-of-merit (FOM), defined below does not scale

$$\mathrm{FOM} = \frac{\mathrm{Power}}{2^{\mathrm{ENOB}}\min\{2\mathrm{ERBW}, f_s\}} \quad (8)$$

Fig. 11 displays FOM against sampling frequency under different analog supplies with fixed 1 V reference and 0.5 V digital supply. The FOM degrades at lower frequency regime (<200 kHz) when the leakages dominate because of the low-V_t transistors used in the design. The ENOB starts to drop quickly because of limited bandwidth when the sampling frequency is greater than 1.3 MS/s. Thus the FOM degrades.

5 Conclusions

An SAR ADC with an energy-efficient CDAC, namely, the split-MSB with LSB down, has been presented. Theoretically, given the same unit capacitor, this switching scheme achieves the best energy efficiency under the same DNL and INL performance. With an optimised clock-gated digital controller using TSPF and a lower digital supply, the 0.13 μm CMOS ADC achieves 8.9-ENOB at 1.1 MS/s, and an FOM of 44.1 fJ/conversion-step. Table 3 compares the measured result to state-of-the art SAR ADCs with different switching schemes. Normalising the FOM to the unit capacitance, denoted as FOM$_a$ representing the architectural FOM, this switching scheme is among the most energy efficient.

6 Acknowledgment

The authors would like to express their gratitude to the CMC Microsystems for implementing the test chips.

7 References

[1] Hong H., Lee G.: 'A 65fJ/conversion-step 0.9-V 200kS/s rail-to-rail 8-bit successive approximation ADC', *IEEE JSSC*, 2007, **42**, (10), pp. 2161–2168

[2] Xu R., Liu B., Yuan J.: 'Digitally calibrated 768kS/s 10-bit minimum-size SAR ADC array with dithering', *IEEE JSSC*, 2012, **47**, (9), pp. 2129–2140

[3] Chen D.G., Tang F., Bermak A.: 'A low-power pilot-DAC based column parallel 8b SAR ADC with forward error correction for CMOS image sensors', *IEEE TCAS-I, Regul. Pap.*, 2013, **60**, (10), pp. 2572–2583

[4] Zhu Z.M., Xiao Y., Song X.: 'Vcm-based monotonic capacitor switching scheme for SAR ADC', *Electron. Lett.*, 2013, **49**, (5), pp. 327–329

[5] Zhu Z.M., Xiao Y., Wang W., Guan Y., Liu L., Yang Y.: 'A 1.33uW 10-bit 200KS/s SAR ADC with a tri-level based capacitor switching procedure', *Microelectron. J.*, 2013, **44**, (12), pp. 1132–1137

[6] Elzakker M.V., Tuijl E.V., Geraedts P., *ET AL.*: 'A 1.9 μW 4.4fJ/conversion-step 10 b 1MS/s charge redistribution ADC'. IEEE ISSCC, February 2008, pp. 244–610

[7] Zhang D., Bhide A., Alvandpour A.: 'A 53-nW 9.1-ENOB 1-kS/s SAR ADC in 0.13-μm CMOS for medical implant devices', *IEEE JSSC*, 2012, **47**, (7), pp. 1585–1593

[8] Nuzzon P., de Bernardinis F., Terreni P., ven der Plas G.: 'Noise analysis of regenerative comparators for reconfigurable ADC architecture', *IEEE TCAS-II*, 2008, **55**, (6), pp. 1441–1454

[9] Harpe P., Cantatore E., Roermund A.V.: 'A 10 b/12 b 40 kS/s SAR ADC with data-driven noise reduction achieving up to 10.1b ENOB at 2.2 fJ/conversion- step', *IEEE JSSC*, 2013, **48**, (12), pp. 3011–3018

[10] Saberi M., Lotfi R., Mafinezhad K., *ET AL.*: 'Analysis of power consumption and linearity in capacitive digital-to-analog converters used in successive approximation ADCs', *IEEE TCAS-I, Regul. Pap.*, 2011, **58**, (8), pp. 1736–1748

[11] Ginsburg B.P., Chandrakasan A.P.: '500MS/s 5bit ADC in 65-nm CMOS with split capacitor array DAC', *IEEE JSSC*, 2007, **42**, (4), pp. 739–747

[12] Chang Y.K., Wang C.S., Wang C.K.: 'A 8-bit 500 kS/s low power SAR ADC for bio-medical application'. IEEE ASSC., November 2007, pp. 228–231

[13] Liu C.C., Chang S.J., Huang G.Y., Lin Y.Z.: 'A 10-bit 50-MS/s SAR ADC with a monotonic capacitor switching procedure', *IEEE JSSC*, 2010, **45**, (4), pp. 731–740

[14] Zhu Y., Chan C.H., Chio U.F., ET AL.: 'A 10-bit 100-MS/s reference-free SAR ADC in 90 nm CMOS', *IEEE JSSC*, 2010, **45**, (6), pp. 1111–1121

[15] Sun L., Pun K.P., Ng W.T.: 'CDACs with LSB down in differential SAR ADCs', *IET J. Eng.*, 2014, doi:10.1049/joe.2013.0219

[16] Tripathi V., Murmann B.: 'An 8-bit 450-MS/s single-bit/cycle SAR ADC in 65-nm CMOS'. IEEE ESSCIRC, September 2013, pp. 117–120

[17] Harpe P., Zhou C., Bi Y., ET AL.: 'A 26 μW 8 bit 10 MS/s asynchronous SAR ADC for low energy radios', *IEEE JSSC*, 2011, **46**, (7), pp. 1585–1595

[18] van der Plas G., Verbruggen B.: 'A 150 MS/s 133 μW 7 b ADC in 90 nm digital CMOS using a comparator-based asynchronous binary search sub-ADC'. IEEE ISSCC, February 2008, pp. 242–243

[19] Yuan J., Svensson C.: 'New single-clock CMOS latches and flip-flops with improved speed and power savings', *IEEE JSSC*, 1997, **32**, (1), pp. 62–69

[20] Tran C.Q., Kawaguchi H., Sakurai T.: 'Low-power high-speed level shifter design for block-level dynamic voltage scaling environment'. Proc. ICICDT, May 2005, pp. 229–232

[21] Doernberg J., Lee H.S., Hodges D.A.: 'Full-speed testing of A/D converters', *IEEE JSSC*, 1984, **SC-19**, (6), pp. 820–827

[22] Craninckx J., van der Plas G.: 'A 65fJ/conversion-step 0-50-MS/s 0-to-0.7 mW 9b charge-sharing SAR ADC in 90 nm digital'. IEEE ISSCC Dig. of Tech. Papers, February 2007, pp. 246–248

[23] Liu C.C., ET AL.: 'A 1 V 11fJ/conversion-step 10bit 10Ms/s asynchronous SAR ADC in 0.18 μm CMOS'. IEEE Symp. on VLSI Circuits, June 2010, pp. 241–242

[24] Liou C.Y., Hsieh C.C.: 'A 2.4-to-5.2fJ/conversion-step 10b 0.5-to-4MS/s SAR ADC with charge-average switching DAC in 90 nm CMOS'. IEEE ISSCC, February 2013, pp. 280–282

[25] Harpe P., Cantatore E., Roermund A.: 'A 2.2/2.7fJ/conversion-step 10/12b 40kS/s SAR ADC with data-driven noise reduction'. IEEE ISSCC, February 2013, pp. 270–272

Complete permutation Gray code implemented by finite state machine

Li Peng[1], Pingliang Zeng[2], Hao Li[1], Xin Li[1]

[1]*Department of Electronics and Information Engineering, Huazhong University of Science and Technology, Wuhan 430074, People's Republic of China*
[2]*Power System Department, China Electric Power Research Institute, Beijing 100192, People's Republic of China*
E-mail: pengli@hust.edu.cn

Abstract: An enumerating method of complete permutation array is proposed. The list of $n!$ permutations based on Gray code defined over finite symbol set $Z_n = \{1, 2, \ldots, n\}$ is implemented by finite state machine, named as n-RPGCF. An RPGCF can be used to search permutation code and provide improved lower bounds on the maximum cardinality of a permutation code in some cases.

1 Introduction

In this Letter, we regard a permutation defined over finite symbol set $Z_n = \{1, 2, \ldots, n\}$ as a vector of state space with dimension n, which is denoted as $\pi = [\pi_1, \ldots, \pi_i, \ldots, \pi_n]$ for π_i, $i \in Z_n$. The set formed by all $n!$ permutations over Z_n is referred to as the complete permutation array (CPA) and is denoted as set $S_n = \{\pi_1, \pi_2, \ldots, \pi_{k-1}, \pi_k, \pi_{k+1}, \ldots, \pi_{n!-1}, \pi_{n!}\}$. A special permutation array (PA) (or permutation code), denoted as PA (n, M, d), is a subset of S_n, where n is an integer, M is the cardinality, $1 \leq M \leq n!$ and d is the minimum Hamming distance. Recently, PA (n, M, d) found applications in several fields such as multilevel flash memory [1] and power line communication [2, 3]. In addition, general algebraic methods of constructing PA have also received great attention [4].

Generation of the CPA with dimension n has been studied for many years. Numerous computer programmes have been developed to enumerate all $n!$ permutations. Sedgewick in [5] reviewed research and development in construction of CPA using computer programmes, and especially presented a permutation network diagram with exchange modules to represent the process of generating all permutations. However, algorithms in [5] are inefficient and computing resource intensive, and can only list those permutations for small values of n. In [1], two ways of listing $n! = 4! = 24$ permutations with dimension $n = 4$ were provided in the form of Gray code, the enumerating method is first-order recursive, that is, given the enumeration with dimension $n-1$, the enumeration with dimension n can be completed. This Letter attempts to establish a general mathematical model of enumerating $n!$ permutations using finite state machine (FSM), we called it the recursive complete permutation Gray code (PGC) based on FSM, RPGCF for short. A RPGCF uses the shifting module which is completely different from exchange module as described in [5]. The method not only can be realised on field programmable gate array (FPGA) devices but also is a fast algorithm for searching PA. We have implemented an RPGCF with $n = 8$ using an FPGA circuit which can enumerate $8! = 40\,320$ permutations in 605 ms under the condition that the FPGA machine period is 10 ns. In comparison, Sedgewick's work [5] showed that generating $8! = 40\,320$ permutations took ~40 ms. Let $P(n, d)$ be the maximal size of a PA (n, M, d), that is, $M_{\max} = P(n, d)$. It can be shown that RPGCF improves the lower bounds on $P(n, d)$ in some cases comparing with some lower bounds shown in [3, 4]. Therefore it can be speculated that RPGCF will be an important tool in permutation code design and interleave algorithm.

2 Several definitions

State transition of a permutation in S_n involves operational behaviour of all elements in a permutation. This behaviour is assumed to be able to continually and regularly change positions of at least two elements in a permutation with minimum cost, just like Boolean system which can trigger the state transition by means of flipping at least one bit of a binary sequence. A type of operation available for all elements in a permutation is shifting and swapping, rather than inserting an element into or deleting it from a permutation which damages the structure of a permutation with dimension n. From the viewpoint of computer implementation, shifting operation (which can directly be executed by circular shifting register) is easier than swapping operation (which involves programming at least three assignment statements). Therefore we focus on shifting operation in this Letter. The shifting operation defined below is similar to the 'push-to-the-top' operation referred to in [1].

Definition 1: [element forward operation]: Let $\pi = [\pi_1, \pi_2, \ldots, \pi_n] \in S_n$ and T_{right} denote a set of operation functions. If an operation function $t_i \in T_{\text{right}}$ ($i = 2, 3, \ldots, n$) satisfies $t_i\pi = t_i[\pi_1, \ldots, \pi_i, \ldots, \pi_n] = [\pi_i, \pi_1, \ldots, \pi_{i-1}, \pi_{i+1}, \ldots, \pi_n] \in S_n$, that is, the ith element π_i of π is placed to the first position, all elements before π_i are right-shifted in turn and all elements after it remain unchanged, then we call $t_i \in T_{\text{right}}$ element forward operation function of the ith element and $T_{\text{right}} = \{t_2, \ldots, t_n\}$ element forward operation function set.

Generally, t_n is called the right-circular shift (CS) operation function. The specific combination of different operation functions in $T_{\text{right}} = \{t_2, \ldots, t_n\}$ can generate a sequence of operation functions (SOF) like $f = (t_x \cdots (t_2(t_n)^{u_1})^{u_2} \cdots)^{u_x}$ which is characterised with a nested structure, where $t_n, t_2, \ldots, t_x \in T_{\text{right}}$, $u_1, u_2, \ldots, u_x$, $x \in Z_n$, $t_n \neq t_2 \neq \cdots \neq t_x$ and $(t)^u$ means that operation function t is used u times. If we make an SOF $f = (t_2(t_n)^{u_1})^{u_2}$ act on a permutation π, then the set $\{(t_2(t_n)^{u_1})^{u_2}\pi\}$ contains $(u_1+1)u_2+1$ permutations with dimension n. For example, $f = (t_n)^{n-1}$ act on a permutation π, then the set $\{(t_n)^{n-1}\pi\}$ contains n permutations which form a Latin square (LS). It is called in this Letter a CS-LS, where $u_1 = n-1$ and $u_2 = 0$.

The Gray code is a set of distinct binary vectors with dimension n in which every two adjacent words differ by exactly one bit flip. It has since been generalised in countless ways, and also defined as an ordered set of distinct states in [1] where every two adjacent states s_i and s_{i+1} differ by exactly one operation such that $s_{i+1} = t(s_i)$, where $t \in T$ is a transition function from a predetermined set T. Accordingly, we define the following Gray code for $n!$ permutations of the CPA.

Definition 2: [complete permutation Gray code]: For a positive integer n, let $\pi_k \in S_n$; $k \in [1, n!]$; $t_i \in T_{\text{right}} = \{t_2, \ldots, t_n\}$. If $n!$

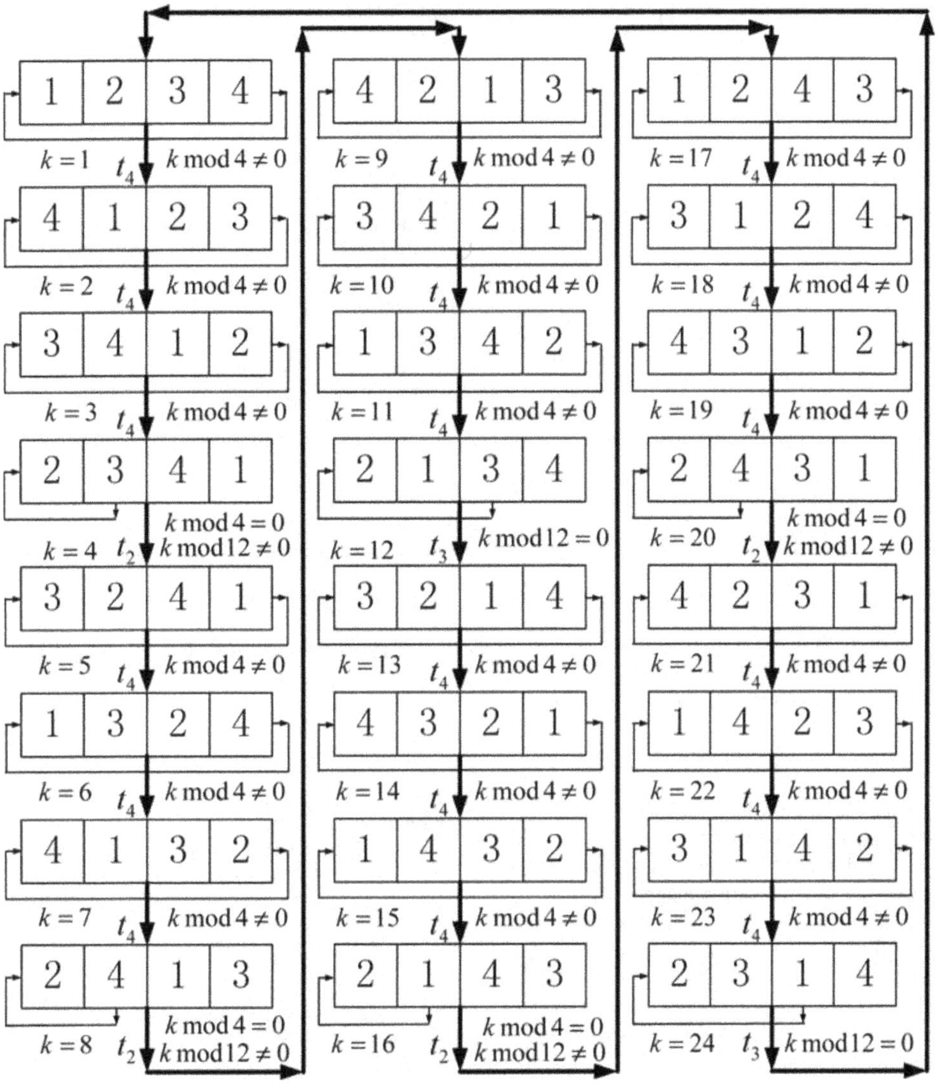

Fig. 1 *State transition diagram of the 4-RPGCF*

permutations of S_n are an ordered set where every permutation π_k can transit to its next adjacent permutation π_{k+1} by applying a single forward operation, that is, $\pi_{k+1} = t_i(\pi_k)$, then this ordered set is called complete permutation Gray code, denoted as n-PGC. If acted on by one operation function of $T_{\text{right}} = \{t_2, \ldots, t_n\}$, the last permutation in n-PGC returns to the initial permutation, then n-PGC is cyclic (recursive) and is denoted as n-RPGC.

Obviously, an n-RPGC contains $n! + 1$ permutations in which there are $n!$ distinct permutations, and the last permutation repeats the initial one. For the sake of finding a method of quickly enumerating all $n!$ permutations, we use FSM to implement an n-RPGC.

Definition 3: [FSM of n-RPGC]: Let $Z_n = \{1, 2, \ldots, n\}$ be the input symbol set, S_n be the state set, any permutation $\pi_k \in S_n$ $(k = 1, 2, \ldots, n!)$ be an initial permutation, $T_{\text{right}} = \{t_2, \ldots, t_n\}$ be the state transition function set with each element $t_i \in T_{\text{right}}$ being a function t_i: $S_n \times T_{\text{right}} \rightarrow S_n$; $S_{\text{out}} \subset S_n$ be the set of final states where S_{out} consists of $n - 1$ permutations whose structures are formed in such a way that the element π_1 is inserted into a permutation $[\pi_2, \pi_3, \ldots, \pi_{n-1}, \pi_n]$ with the size $n - 1$ from the second position to the last position. The quintuple $(Z_n, S_n, \pi_k, T_{\text{right}}, S_{\text{out}})$ defines a recursive FSM model of an n-RPGC which is denoted as n-RPGCF.

The following theorem provides a general mathematical model for constructing an n-RPGCF.

Theorem 1: : Let n be a positive integer, $k \in [1, n!]$ the index of $n!$ permutations and $i_k \in [2, n]$ the index of $n - 1$ 'element forward operation' functions. Let $t_{i_k} \in T_{\text{right}} = \{t_2, \ldots, t_n\}$ be an 'element forward operation' function and $\pi = \pi_k = [\pi_1, \pi_2, \ldots, \pi_n] \in S_n$ be an initial permutation. Then there exists a nested sequence formed by $n!$ 'element forward operation' functions in which each of $T_{\text{right}} = \{t_2, \ldots, t_n\}$ can be used several times, denoted as $f_{\text{SOF}}(n, n!)$, that exactly constructs an n-RPGCF by making $f_{\text{SOF}}(n, n!)$ act on the initial permutation π. $f_{\text{SOF}}(n, n!)$ consists of 'element forward operation' functions, $t_{i_k} \in T_{\text{right}} = \{t_2, \ldots, t_n\}$, in the order determined as follows (see equation (1) at the bottom of the next page)

where $t_a \in T_{\text{right}} = \{t_2, \ldots, t_n\}$ denotes the last operation function in the sequence of $f_{\text{SOF}}(n, n!)$; $t_{n-l}, t_{l+2} \in T_{\text{right}} = \{t_2, \ldots, t_n\}$, where l is an integer, $l = 0, 1, 2, \ldots, m$, $m = (n - 3)/2$, $a = (n + 1)/2$ when n is an odd number or $m = (n - 2)/2$, $a = (n + 2)/2$ when n is an even number; $(n)_{2l} = n(n - 1)\ (n - 2) \cdots (n - 2l + 1) = n!/(n - 2l)!$ denotes the $2l$th falling factorial power of n; $k(\text{mod}(n)) \neq 0$ means number k, $k \in [1, n!]$, is not a multiple of n, similarly, $k(\text{mod}(n)) = 0$ means number k, $k \in [1, n!]$, is a multiple of n.

We can easily complete the proof of Theorem 1 by explaining clearly four problems: (i) an n-RPGCF traverses all $n!$ permutations; (ii) an n-RPGCF does not contain any other repetitive permutation except for initial permutation π; (iii) any transition from a permutation to another only involves one operation function; and (iv) the last permutation in an n-RPGCF returns to the initial permutation π by using only one operation function in $T_{\text{right}} = \{t_2, \ldots, t_n\}$. Here we omit the proof because of limited space. However, we give the following nested recursive complete SOF $f_{\text{SOF}}(n, n!)$ of (2) and (3). They are the other alternative of representing (1).

For an even number n, $l = 0, 1, 2, \ldots, m-1, m$ and $m = (n-2)/2$, all t_{i_k}'s satisfying (1) construct a nested recursive complete SOF

$$f_{\text{SOF}}(n, n!) = (t_{(n+2)/2}(t_n)^{n-1}(t_{n/2}(t_n)^{n-1} \cdots (t_{l+2}(t_n)^{n-1}(t_{n-l}(t_n)^{n-1} \cdots (t_2(t_n)^{n-1})^{n-2\cdots})^{n-2l-1})^{n-2l-2\cdots})^2)^2 \tag{2}$$

For an odd number n, $l = 0, 1, 2, \ldots, m-1, m$ and $m = (n-3)/2$, all t_{i_k}'s satisfying (1) construct a nested recursive complete SOF

$$f_{\text{SOF}}(n, n!) = (t_{(n+1)/2}(t_n)^{n-1}(t_{(n+3)/2}(t_n)^{n-1} \cdots (t_{l+2}(t_n)^{n-1}(t_{n-l}(t_n)^{n-1} \cdots (t_2(t_n)^{n-1})^{n-2\cdots})^{n-2l-1})^{n-2l-2\cdots})^2)^2 \tag{3}$$

Example: : Let $n = 4$ and $\pi = [1234] \in S_4$ be the initial permutation. From Theorem 1, for $k \in [1, 24]$, we have (see equation at the bottom of the page).

According to the natural order of k values, the sequence t_{i_k} (from right to left) is listed as $t_3t_4t_4t_4t_2t_4t_4t_4t_2t_4t_4t_4t_3t_4t_4t_4t_2t_4t_4t_4t_2t_4t_4t_4$. It can be rearranged as a nested recursive SOF: $f_{\text{SOF}}(4, 24) = (t_3(t_4)^3(t_2(t_4)^3)^2)^2$. Fig. 1 shows the state transition diagram of this 4-RPGCF with the initial permutation $\pi = [1234]$, where each permutation in the set $\{(t_3(t_4)^3(t_2(t_4)^3)^2)^2\pi\}$ is placed inside a virtual circular shift register (VCSR). The $3 \times 3! = 18$ VCSRs corresponding to $k(\text{mod}(4)) \neq 0$ complete the CS operation to right; the rest six VCSRs corresponding $k(\text{mod}(4)) = 0$ and $k(\text{mod}(12)) = 0$ complete the CS operation from the first unit to the ith (here $i = 2, 3$) units to right. All VCSRs being controlled by the SOF $f_{\text{SOF}}(4, 24) = (t_3(t_4)^3(t_2(t_4)^3)^2)^2$ complete the recursive operation of $4! = 24$ permutations.

3 Application

An n-RPGCF serves as the searching algorithm of $P(n, d)$ in that we begin with an empty set S_{PA}, run through all permutations of n-RPGCF and add a permutation into S_{PA} if it has distance at least d from every member of the current S_{PA}. Using this method, we constructed $P(n, d)$ with $n = 11, 12$ and $d = 5, 6$. We obtained (11, 95 260, 5), (11, 16 148, 6) and (12, 129 864, 6) PAs, which provide the improved lower bounds on $P(n, d)$ in contrast to (11, 60 940, 5), (11, 9790, 6) and (12, 117 480, 6) PAs resulting from the work of Chu *et al.* in [3].

4 Conclusion

An algebraic method of enumerating the CPA is presented. It can be seen as a searching algorithm of PA (n, M, d) which can give improved lower bounds on $P(n, d)$ in some cases.

5 Acknowledgment

This work is supported by the National Natural Science Foundation of China under grant no. 61071069.

6 References

[1] Jiang A., Mateescu R., Schwartz M., Bruck J.: 'Rank modulation for flash memories', *IEEE Trans. Inf. Theory*, 2009, **55**, pp. 2659–2673

[2] Vinck A.J.H.: 'Coded modulation for powerline communications', *AEU Int. J. Electron. Commun.*, 2000, **54**, (1), pp. 45–49

[3] Chu W., Colbourn C.J., Dukes P.: 'Constructions for permutation codes in powerline communications', *Des. Codes Cryptography*, 2004, **32**, pp. 51–64, Kluwer Academic Publishers, Manufactured in the Netherlands

[4] Fu F.-W., Klove T.: 'Two constructions of permutation arrays', *IEEE Trans. Inf. Theory*, 2004, **50**, pp. 881–883

[5] Sedgewick R.: 'Permutation generation methods', *Comput. Surv.*, 1977, **9**, (2), pp. 137–164

$$t_{i_k} = \begin{cases} t_n, & k(\bmod (n)) \neq 0, & \text{for } l = 0 \\ t_2, & k(\bmod (n)) = 0 \text{ and } k(\bmod (n)_2) \neq 0, & \text{for } l = 0 \\ t_{n-1}, & k(\bmod (n)_2) = 0, \text{ and } k(\bmod (n)_3) \neq 0, & \text{for } l = 1 \\ t_3, & k(\bmod (n)_3) = 0 \text{ and } k(\bmod (n)_4) \neq 0, & \text{for } l = 1 \\ & \cdots & \\ t_{n-l}, & k(\bmod (n)_{2l}) = 0 \text{ and } k(\bmod (n)_{2l+1}) \neq 0, & \text{for } l = 2, \ldots, m-1 \\ t_{l+2}, & k(\bmod (n)_{2l+1}) = 0 \text{ and } k(\bmod (n)_{2l+2}) \neq 0, & \text{for } l = 2, \ldots, m-1 \\ & \cdots & \\ t_a, & k(\bmod (n)_{n-2}) = 0, & \text{when} \begin{cases} a = (n+1)/2, & \text{for } n \text{ being an odd} \\ a = (n+2)/2, & \text{for } n \text{ being an even} \end{cases} \end{cases} \tag{1}$$

$$t_{i_k} = \begin{cases} t_4, & \text{when } k = 1, 2, 3, 5, 6, 7, 9, 10, 11, 13, 14, 15, 17, 18, 19, 21, 22, 23, \text{ for } k(\bmod 4) \neq 0 \\ t_2, & \text{when } k = 4, 8, 16, 20, \text{ for } k(\bmod 4) = 0 \text{ and } k(\bmod (4)_2) \neq 0 \\ t_3 = t_{n-l} = t_{4-1}, & \text{when } k = 12, 24, \text{ for } k(\bmod (4)_2) = 0 \text{ and } k(\bmod (4)_3) \neq 0 \end{cases}$$

8

Distributed demand-side management optimisation for multi-residential users with energy production and storage strategies

Emmanuel Chifuel Manasseh[1], Shuichi Ohno[2], Toru Yamamoto[2], Aloys Mvuma[3]

[1]Department of Electronics and Telecommunications Engineering, School of Computational and Communication Science and Engineering, Nelson Mandela African Institution of Science and Technology, P.O. Box 447, Arusha, Tanzania
[2]Department of System Cybernetics, Graduate school of Engineering, Hiroshima University, 1-4-1 Kagamiyama, Higashi-Hiroshima, 739-8527, Japan
[3]Department of Telecommunications and Communications Networks, School of Informatics, College of Informatics and Virtual Education, University of Dodoma, P.O. Box 490, Dodoma, Tanzania
E-mail: manasejc@ieee.org

Abstract: This study considers load control in a multi-residential setup where energy scheduler (ES) devices installed in smart meters are employed for demand-side management (DSM). Several residential end-users share the same energy source and each residential user has non-adjustable loads and adjustable loads. In addition, residential users may have storage devices and renewable energy sources such as wind turbines or solar as well as dispatchable generators. The ES devices exchange information automatically by executing an iterative distributed algorithm to locate the optimal energy schedule for each end-user. This will reduce the total energy cost and the peak-to-average ratio (PAR) in energy demand in the electric power distribution. Users possessing storage devices and dispatchable generators strategically utilise their resources to minimise the total energy cost together with the PAR. Simulation results are provided to evaluate the performance of the proposed game theoretic-based distributed DSM technique.

1 Introduction

The smart grid presents several opportunities for end-users to save energy and for the utility company to operate the grid in a more efficient, effective and reliable way. In recent years, much attention has been paid to optimisation of energy consumption in smart grid. Demand-side management (DSM) is one of the notable functions in a smart grid that enables end-users to modify their demand for energy through various methods, such as financial incentives and education [1–6]. The deployment of DSM will motivate end-users to utilise less energy during peak hours, or to shift the time of energy use to off-peak times [1, 7–9], which will help the utility company to reduce the peak load demand and reshape the load profile. Consequently, end-users will save money on electricity and the society will conserve electricity [5, 7, 10, 11].

Basically, the outcomes of DSM programmes depend on a portion of the total load that can be controllable [1, 8, 12]. End-users consisting of adjustable loads, such as plug-in hybrid electric vehicles (PHEVs) and dishwashers, offer significant benefit to this end [5, 7, 11]. Moreover, end-users with storage devices and dispatchable energy generators offer an exceptional opportunity to increase the percentage of controllable load. The control of end-users demand and supply of energy can be done through various methods such as financial incentives, new tariff schemes and education. The end-users agree to involvement, if they may be charged less for consuming electricity during off-peak hours and paid for supplying electricity during peak hours. Suppose the utility company pays more for user-generated energy during peak hours of energy demand, and pay less for off-peak power, and then end-users will be motivated to generate more energy and consume minimum energy during peak hours, which in turn achieves the main goal of DSM [3, 4, 8, 11].

Practically, to meet all energy demands from the end-users, the grid capacity should be designed such that it satisfies the peak power demand instead of just the average power demand [6, 13–15]. However, the utility company supplying energy to the grid prefers to use the least expensive sources of energy to generate electricity (which might not be enough to meet the required grid capacity) and use expensive energy sources only when the demand increases [3, 4, 16]. When costly energy sources are employed by the utility company, end-users will also pay high prices for the energy. Thus, by strategically engaging end-users in energy production, storage and shifting the energy consumptions of their adjustable load appliances, the utility company will alleviate the use of expensive base load generators and both end-users and the utility company will benefit from the strategy [1, 8, 10, 12].

In addition, renewable energy sources (RESs) such as solar and wind turbines play an important role in reducing the total load. Typically solar and wind turbines (without some added component for storage) are non-dispatchable because the sunlight or wind is periodic and cannot be predicted and controlled [10, 12, 17]. With a rapid advancement of battery technology, it is likely that storage devices will become an integral part of the means by which energy generated from renewable sources can be stored and utilised when needed. Storage devices can be used to store some of the energy generated by renewable sources and discharge them during peak hours.

Home automation systems play an important role in determining the success of the proposed energy strategies. Through the use of automation, the decision about the energy schedule, amount of power consumption, charging and discharging as well as the running of dispatchable generators can be facilitated. An automatic scheme that requires minimum effort from the end-users is desired since most end-users do not have knowledge and/or interest to respond to the energy costs [8].

In this study, we present an energy scheduling strategy for the future smart grid network. A distributed game-theoretic cost minimisation demand-side optimisation and energy scheduling scheme that takes into account load uncertainty are presented. We consider a scenario where the main source of energy is shared by several end-users; some of the users are equipped with the RESs, some with dispatchable distribution generators (DGs) and/or storage devices. Each end-user is utilising an energy scheduler (ES) deployed inside the smart meters for the adjustable loads, charging and discharging of their storage devices as well as generating energy from the dispatchable generators.

The smart meters are linked to the electric power line and communication network [5, 8, 11, 18]. The smart meters with ES scheme communicate automatically by executing a distributed algorithm to obtain the optimised energy schedule for each end-user [5, 7, 8, 11]. The DSM optimisation and scheduling problem is formulated as a non-cooperative cost minimisation game among the end-users and an iterative algorithm that optimises user energy costs is proposed.

Unlike the methods in [5, 8, 11], where ES is considered for a system with adjustable energy appliances without energy storage and DGs, the proposed scheme considers a setup with both storage devices, RESs as well as dispatchable generators. The scheme in [12] includes the storage devices and dispatchable energy sources as well as non-dispatchable energy sources; however, it does not include energy shifting of adjustable appliances, which are crucial in reducing peak-to-average ratio (PAR) energy costs of the end-users as well as the cost incurred by the utility company. The proposed algorithm includes optimisation of energy consumption of adjustable appliances and that of the energy generators and storers.

The proposed iterative algorithm demands each end-user to give some information related to his/her daily total load. Residential users can exchange limited information related to their daily total load (without giving detailed information regarding the energy consumption of their appliances or their storage and production strategies [8, 11]). A simple pricing mechanism can give motivation for the end-users to cooperate (i.e. self-enforcing mechanism) [5, 11]. Thus, it is in the end-user's personal interest to disclose local information accurately to improve the overall system performance and minimise his/her daily energy costs.

Simulation results demonstrate that, with our proposed distributed scheme, the ES can substantially minimise the PAR as well as the daily energy cost of each end-user. Furthermore, by utilising energy producers and storers, end-users without energy strategies (dispatchable generators and storage devices) can also benefit and pay less for the same daily total energy load.

2 System model and problem statement

Fig. 1 depicts the smart power system model considered, where each residential user has non-adjustable load appliances and adjustable load appliances. Non-adjustable load appliances include electric bulbs, TVs, refrigerators etc., and these are loads whose instantaneous power or starting time cannot be adjusted. Adjustable load refers to the loads whose instantaneous power, starting time or both can be adjusted. Adjustable load appliances include PHEVs, dish washers, washing machines etc. In addition, some residential users possess storage devices, and/or RESs, DGs or both (see Fig. 1).

We consider N multiple end-users connected to the electrical grid from the same energy source. Each end-user has a smart meter that communicates with several appliances per end-user as well as the utility company via the advanced metering infrastructure. It is generally assumed that the cost of supplied energy from the utility company is dictated in advance within a specified period of time 1, …, T. Each time slot of the scheduling horizon can stand for, for example, 1 h, with $T=24$ representing one day. For an end-user $n \in N$, let $\mathcal{K}_n$ denote a set of adjustable load appliances. For each device $k \in \mathcal{K}_n$, we define energy consumption scheduling vector $\boldsymbol{x}_{n,k} \triangleq [x^1_{n,k}, \ldots, x^T_{n,k}]$. Here, $x^t_{n,k}$ corresponds to the 1 h energy consumption scheduled for device k of user n, whereas the energy consumption for non-adjustable load at slot t is denoted as $y^t_{n,0}$.

2.1 Energy storage

The residential end-user n may have a storage device (such as a battery). Let $p^t_n \geq 0$, $t=1, \ldots, T$, be the available energy in a battery at the end of slot t; and $p^{\max}_n$ is the capacity of the battery. The energy available at the beginning of the horizon can be represented as p^0_n. The battery can be either charged or discharged throughout the time slot t. Let b^t_n be the energy discharged from or charged to the battery at slot t. Here $b^t_n < 0$ represents that the battery is discharging while $b^t_n > 0$ represents that the battery is charging. The charge/discharge variables and the accumulated

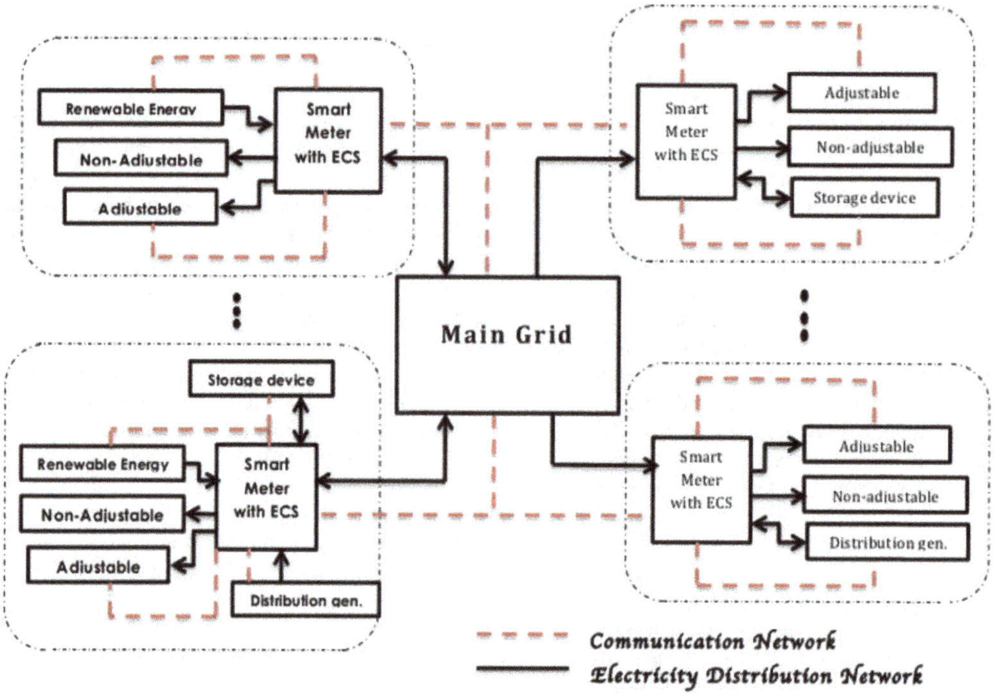

Fig. 1 *Links between connecting end-users and the smart grid*

energy in the battery at time slot t are given by the equation [9]

$$p_n^t = p_n^{t-1} + b_n^t, \quad t = 1, \ldots, T \tag{1}$$

The variable b_n^t is constrained by the maximum charge and discharge as $b_n^{\text{dis}} < b_n^t < b_n^{\text{ch}}$, also we assume that the batteries have limited capacity, thus the energy supplied by the battery is no more than the current consumed energy, that is, $b_n^t + \sum_k x_{n,k}^t + p_{n,0}^t \geq 0$. Every single battery has an efficiency $\eta_n \in (0, 1)$, implying that if p_n^{t-1} is accumulated at the end of the time slot $t-1$, then the discharge at time slot t is constrained by $b_n^t \geq -\eta_n p_n^{t-1}$. With ES, residential users can utilise energy storage devices to store energy during off-peak hours and discharge them during peak hours [9].

2.2 Energy production

Apart from storage devices, some end-users own RES and/or distributed generators (DGs). Users deploy DG and/or RES to produce energy rather than just consuming energy supplied by the utility company. By integrating DG and/or RES, end-users reduce their energy cost since they can produce energy to power their own appliances, to sell it to the utility company or to charge their batteries during peak hours. These energy sources are classified as dispatchable or non-dispatchable energy producers [12]. Typically, RESs such as solar and wind turbines (without some added component for storage) are non-dispatchable, since the supply of sunlight or wind is periodic and cannot be predicted and controlled. Possessing only fixed (initial plus maintenance) costs, they produce electricity at their maximum available power, which indicates no optimal scheme regarding the production of energy [12]. Conversely, dispatchable energy sources are those sources that can be turned on or off or can adjust their energy productions on demand; these include fuel generators, gas turbines or internal combustion engines [12]. DGs are categorised as dispatchable energy sources, thus end-users possessing DGs are concerned with the optimisation of their energy production strategies.

We denote the non-dispatchable energy generated by user n per-time slot as $g_{n,r}^t$. The dispatchable energy generated by user n per-time slot is denoted as $g_{n,d}^t$. We introduce the production cost function $\mathcal{W}(g_{n,d}^t)$, which provides the variable costs for producing a certain quantity of energy $g_{n,d}^t$ at a time slot t, where $\mathcal{W}(0) = 0$.

Let $g_n^{\max}$ denote the maximum energy production capability for end-user n during a time slot t. Then, energy production profile per-time slot is bounded as

$$0 \leq g_{n,d} \leq g_n^{\max} \tag{2}$$

Here, $g_n^{\max}$ represents the quantity of energy produced when user n's energy source is operated throughout the time slot t. Besides, the cumulative energy production has to satisfy the constraint

$$\sum_{t=1}^{T} g_{n,d}^t \leq \lambda_n^{\max} \tag{3}$$

where $0 \leq \lambda_n^{\max} \leq T \times g_{n,d}^{\max}$. We define the non-dispatchable energy production vector and the dispatchable energy production scheduling vector as $\boldsymbol{g}_{n,r} = (g_{n,r}^t)_t^T$ and $\boldsymbol{g}_{n,d} = (g_{n,d})_t^T$, respectively.

The total hourly energy profile for user $n \in N$ can be defined as

$$l_n^t \triangleq \sum_{k \in \mathcal{K}_n} x_{n,k}^t + y_{n,0}^t + b_n^t - g_{n,r}^t - g_{n,d}^t \tag{4}$$

and the daily total load for user n is defined as $\mathcal{I}_n = [l_n^1, l_n^2, \ldots, l_n^T]$. Here, $l_n^t \geq 0$ if the energy flows from the utility company to the end-user n, else $l_n^t \leq 0$.

2.3 Energy scheduler

The aim of utilising ESs is to minimise the operational costs and PAR as well as the energy costs of the end-users. The involvement of the end-users is a response to aspects such as incentive pricing, tariff schemes etc. End-users participation may involve either active behavioural changes or passive responses, using automation techniques. For example, if someone chooses not to charge his/her PHEV battery with regard to a period of high demand, he/she is only rescheduling (deferring) that use and will still charge his/her PHEV battery later, thus the energy is consumed at a different time instead of being minimised [9].

The total hourly energy consumption for $n \in N$ is given by

$$e_n^t = \sum_{k \in \mathcal{K}_n} x_{n,k}^t + y_{n,0}^t + b_n^t \tag{5}$$

Hourly energy consumption of each user includes the energy consumed by the devices as well as that consumed to charge the battery. Since the net energy of the battery charge is zero at the time of observation, the total energy used to charge the battery is equivalent to the discharged energy.

Thus we have

$$l_n^t = e_n^t - g_{n,d}^t - g_{n,r}^t \tag{6}$$

For users without generators, either $g_{n,r}^t$ or $g_{n,d}^t$ can be set to zero based on what they possess. Although RESs are intermittent and uncertain, the use of storage devices can absorb this variability. Thus, for end-users possessing both RES and storage device, RES can improve the active participation of users with storage devices.

Operation of the adjustable loads can be shifted to a different time, so each residence selects the time interval $[\alpha_{n,k}, \beta_{n,k}]$ (i.e. beginning time $\alpha_{n,k}$ and end time $\beta_{n,k}$) that the energy consumption for device k can practically be scheduled. We define the total energy consumption for device k from user n as [8, 11]

$$E_{n,k} = \sum_{t=\alpha_{n,k}}^{\beta_{n,k}} x_{n,k}^t \tag{7}$$

and $x_{n,k}^t = 0, \forall t \notin \{\alpha_{n,k}, \ldots, \beta_{n,k}\}$.

Similar to [5, 8, 11], the proposed scheme does not intend to change the amount of power consumed by appliances, but systematically control and adjust it to minimise the energy cost of the end-users as well as to reduce the PAR that likely happen during the peak hours. Thus, ES is about optimising energy consumption over a number of factors, rather than just a simple 'do not consume at these hours' command.

Many devices may have some maximum power levels $\gamma_{n,k}^{\max}$ as well as the minimum power level $\gamma_{n,k}^{\min}$; this sets the upper bound and lower bound constraints on the ES vector $\boldsymbol{x}_{n,k}$ for each device [1], that is

$$\gamma_{n,k}^{\min} \leq x_{n,k}^t \leq \gamma_{n,k}^{\max}, \quad \forall t \in \{\alpha_{n,k}, \ldots, \beta_{n,k}\} \tag{8}$$

The total load for N residences at each hour of the day is given by

$$L_t = \sum_{n \in N} l_n^t \tag{9}$$

Let $C^t(L_t)$ denote the energy cost over a time slot t. This is the cost that the utility company incurs to provide energy to the end-users or the cost that the utility company pays to buy electricity from the end-users. We also consider that the price of the same load may differ at different times of the day [5, 7, 8, 11]. The cost paid by end-user n to buy electricity l_n^t from the utility company (if $l_n^t > 0$) or the cost paid to the end-user for selling the same energy load to the utility company (if $l_n^t < 0$) is given by

$$C_t(L_t)\frac{l_n^t}{L_t} \tag{10}$$

The dispatchable energy production cost function is given by $\mathcal{W}(y) = vy$, where v is a constant. The cumulative expenses incurred by user n over the period of analysis T is given by

$$\mathcal{F}(\boldsymbol{x}_n, \boldsymbol{g}_{n,d}, \boldsymbol{b}_n) = \sum_{t=1}^{T}\left(C^t(L_t)\frac{l_n^t}{L_t} + \mathcal{W}(g_{n,d}^t)\right) \tag{11}$$

We define the total cumulative expenses of all users as

$$\Gamma(\mathcal{I}) = \sum_{n\in N}\mathcal{F}(\boldsymbol{x}_n, \boldsymbol{g}_{n,d}, \boldsymbol{b}_n) \tag{12}$$

where $\mathcal{I}$ is an $N \times T$ matrix representing the daily total load of all users, that is, $\mathcal{I} = \mathcal{I}_1, \ldots, \mathcal{I}_N{}^T$.

The multi-residential load control task amounts to minimising the cumulative expense of electricity, that is

$$\begin{aligned}
\min_{\substack{x_1,\ldots,x_N\\ g_{1,d},\ldots,g_{N,d}\\ b_1,\ldots,b_N}} \quad & \Gamma(\mathcal{I}_1, \ldots, \mathcal{I}_N)\\
\text{subject to} \quad & \sum_{t=\alpha_{n,k}}^{\beta_{n,k}} x_{n,k}^t = E_{n,k}\\
& \gamma_{n,k}^{\min} \le x_{n,k}^t \le \gamma_{n,k}^{\max}, \quad \forall t \in \{\alpha_{n,k}, \ldots, \beta_{n,k}\}\\
& x_{n,k}^t = 0, \quad \forall t \notin \{\alpha_{n,k}, \ldots, \beta_{n,k}\}\\
& b_n^{\text{dis}} < b_n^t < b_n^{\text{ch}}, \quad t = 1, \ldots, T\\
& p_n^t = p_n^{t-1} + b_n^t, \quad t = 1, \ldots, T\\
& b_n^t \ge -\eta_n p_n^{t-1}\\
& 0 \le g_{n,d} \le g_n^{\max}\\
& \sum_{t=1}^{T} g_{n,d}^t \le \lambda_n^{\max}\\
& 0 \le \lambda_n^{\max} \le T \times g_{n,d}^{\max}
\end{aligned} \tag{13}$$

The constraints of the optimisation problem in (13) are linear, thus if the objective function is convex then the problem can be solved using convex optimisation techniques. The problem above is in a centralised fashion, thus some modifications are required to solve the problem distributively. A distributed approach is desirable in order to address possible concerns regarding data privacy and integrity [7].

3 Energy consumption game

We assume that the price that each user pays or receives is proportional to his/her daily energy load. For each end-user $n \in N$, let d_n represent the daily price in dollars to be charged to the end-user n by the utility company or the amount of money that user $n \in N$ receives from the utility company for generating energy. Thus

$$d_n \propto \sum_{t}^{T} l_n^t, \quad \forall n \in N \tag{14}$$

Using the proportionality constant, we can equate users' energy consumption and their bill as

$$\frac{d_n}{d_m} = \frac{\sum_{t=1}^{T} l_n^t}{\sum_{t=1}^{T} l_m^t}, \quad \forall n, m \in N \tag{15}$$

From (15), we have

$$d_m = d_n\frac{\sum_{t=1}^{T} l_m^t}{\sum_{t=1}^{T} l_n^t} \tag{16}$$

The total monetary expenses for all end-users can be expressed as

$$\sum_{m\in N} d_m = \sum_{m\in N}\left(d_n\frac{\sum_{t=1}^{T} l_m^t}{\sum_{t=1}^{T} l_n^t}\right) = d_n\frac{\sum_{m\in N}\sum_{t=1}^{T} l_m^t}{\sum_{t=1}^{T} l_n^t} \tag{17}$$

From (17), we can express d_n as

$$\begin{aligned}
d_n &= \frac{\sum_{t=1}^{T} l_n^t}{\sum_{m\in N}\sum_{t=1}^{T} l_m^t}\left(\sum_{m\in N} d_m\right)\\
&= \Phi_n\sum_{m\in N} d_m
\end{aligned} \tag{18}$$

where

$$\Phi_n = \frac{\sum_{t=1}^{T} l_n^t}{\sum_{m\in N}\sum_{t=1}^{T} l_m^t} = \frac{\sum_{t=1}^{T} l_n^t}{\sum_{t=1}^{T} l_n^t + \sum_{m\in N\setminus n}\sum_{t=1}^{T} l_m^t} \tag{19}$$

Φ_n is not constant for daily energy because of the uncertainty of the RESs such as solar or wind turbine in producing energy. At some hours in a day, users with such generators might produce more power than their own energy demand and thereby feed their excess energy to the grid, which may lead to zero or negative aggregate load l_n^t for such a user n.

From (18), it can be seen that the charge on each user depends on his/her energy strategy and the strategies of other users. This leads to the game theory among the users. In this game, users are players and their strategies are their daily energy schedule. Next, we investigate different approaches of end-users in responding to the price values.

The cumulative cost of user n $\mathcal{F}(\boldsymbol{x}_n, \boldsymbol{g}_{n,d}, \boldsymbol{b}_n)$ is proportional to his/her daily load, that is

$$\mathcal{F}(\boldsymbol{x}_n, \boldsymbol{g}_{n,d}, \boldsymbol{b}_n) \propto \sum_{t=1}^{T} l_n^t \tag{20}$$

For the utility company to generate profit, it is expected that the cost of electricity for the end-users d_n to be equal or slightly higher than the cumulative cost [8, 11], that is

$$d_n \ge \mathcal{F}(\boldsymbol{x}_n, \boldsymbol{g}_{n,d}, \boldsymbol{b}_n) \tag{21}$$

where the left-hand side represents the total daily charge to the end-users, whereas the right-hand side indicates the daily cumulative cost. Following the inequality in (21), we can define

$$\mu = \frac{d_n}{\mathcal{F}(\boldsymbol{x}_n, \boldsymbol{g}_{n,d}, \boldsymbol{b}_n)} \ge 1 \tag{22}$$

For $\mu = 1$, the billing system is budget balanced and the energy supplier pays/charges the end-users equivalent amount corresponding to their cumulative costs. From (18) and (22), it can be shown that regardless of the value of μ

$$\mathcal{F}(\boldsymbol{x}_n, \boldsymbol{g}_{n,d}, \boldsymbol{b}_n) = \Phi_n\sum_{m\in N}\mathcal{F}(\boldsymbol{x}_m, \boldsymbol{g}_{m,d}, \boldsymbol{b}_m) \tag{23}$$

3.1 Equilibrium among users

Given the daily total load for user n as $\mathcal{I}_n$, we define the daily total load of other users as $\mathcal{I}_{-n}$ such that $\mathcal{I}_{-n} = \mathcal{I}\setminus\mathcal{I}_n$. The problem can be formulated as a non-cooperative energy cost minimisation game. In game theory, a non-cooperative game is one in which players make decisions independently [19]. The game consists of

- *Player*: a set of end-users $n = 1, \ldots, N$.
- *Strategies*: energy scheduling vectors $\boldsymbol{x}_n, \boldsymbol{g}_{n,d}, \boldsymbol{b}_n$ for all end-users with adjustable load devices, dispatchable generators, battery or both.
- Payoff functions $P_n(\mathcal{I}_n, \mathcal{I}_{-n}) = -\mathcal{F}(\boldsymbol{x}_n, \boldsymbol{g}_{n,d}, \boldsymbol{b}_n)$, define the user payoffs for the joint strategies.

To maximise payoff, the goal of the users is to minimise the expected overall cost of energy. From (23), the payoff can be expressed as

$$P_n(\mathcal{I}_n, \mathcal{I}_{-n}) = -\Phi_n \sum_{m \in N} \mathcal{F}(\boldsymbol{x}_m, \boldsymbol{g}_{m,d}, \boldsymbol{b}_m) \tag{24}$$

Using (12), the payoff of user n can be expressed as

$$P_n(\mathcal{I}_n, \mathcal{I}_{-n}) = -\Phi_n \Gamma(\mathcal{I}) \tag{25}$$

End-users attempt to determine their energy strategies to minimise the cost paid to the utility company or maximise their profit. Using Nash equilibrium, we can characterise how players play a game [8, 11]. The optimal performance with regard to energy cost minimisation achieves at Nash equilibrium of power consumption game. The Nash equilibrium of this game always exists. The energy consumption variable ($\mathcal{I}_n^*$, $\forall n \in N$) is in a Nash equilibrium of the game if for every user $n \in N$

$$P_n(\mathcal{I}_n^*, \mathcal{I}_{-n}^*) \geq P_n(\mathcal{I}_n; \mathcal{I}_{-n}^*) \tag{26}$$

Once the energy scheduling game is at unique Nash equilibrium, none of the end-users would attempt to diverge from the schedule ($\mathcal{I}_n^*$, $\forall n \in N$). Moreover, the user cannot influence the value of Φ_n with the choice of their strategies.

3.2 *Distributed algorithm*

Suppose all other end-users fix their corresponding energy schedule $\mathcal{I}_{-n}$, then the end-user n can maximise his/her own payoff by solving the local optimisation problem

$$\begin{aligned} \max_{\boldsymbol{x}_n, \boldsymbol{g}_{n,d}, \boldsymbol{b}_n} \quad & P_n(\mathcal{I}_n; \mathcal{I}_{-n}) \\ \text{subject to} \quad & \sum_{t=\alpha_{n,k}}^{\beta_{n,k}} x_{n,k}^t = E_{n,k} \\ & \gamma_{n,k}^{\min} \leq x_{n,k}^t \leq \gamma_{n,k}^{\max}, \quad \forall t \in \{\alpha_{n,k}, \ldots, \beta_{n,k}\} \\ & x_{n,k}^t = 0, \quad \forall t \notin \{\alpha_{n,k}, \ldots, \beta_{n,k}\} \\ & b_n^{\text{dis}} < b_n^t < b_n^{\text{ch}}, \quad t = 1, \ldots, T \\ & p_n^t = p_n^{t-1} + b_n^t, \quad t = 1, \ldots, T \\ & b_n^t \geq -\eta_n p_n^{t-1} \\ & 0 \leq g_{n,d}^t \leq g_n^{\max} \\ & \sum_{t=1}^{T} g_{n,d}^t \leq \lambda_n^{\max} \\ & 0 \leq \lambda_n^{\max} \leq T \times g_{n,d}^{\max} \end{aligned} \tag{27}$$

This is equivalent to the minimisation of the cost function

$$\begin{aligned} \min_{\boldsymbol{x}_n, \boldsymbol{g}_{n,d}, \boldsymbol{b}_n} \quad & \Phi_n\left(\mathcal{F}(\boldsymbol{x}_n, \boldsymbol{g}_{n,d}, \boldsymbol{b}_n) + \sum_{m \in N\backslash n} \mathcal{F}(\boldsymbol{x}_m, \boldsymbol{g}_{m,d}, \boldsymbol{b}_m)\right) \\ \text{subject to} \quad & \sum_{t=\alpha_{n,k}}^{\beta_{n,k}} x_{n,k}^t = E_{n,k} \\ & \gamma_{n,k}^{\min} \leq x_{n,k}^t \leq \gamma_{n,k}^{\max}, \quad \forall t \in \{\alpha_{n,k}, \ldots, \beta_{n,k}\} \\ & x_{n,k}^t = 0, \quad \forall t \notin \{\alpha_{n,k}, \ldots, \beta_{n,k}\} \\ & b_n^{\text{dis}} < b_n^t < b_n^{\text{ch}}, \quad t = 1, \ldots, T \\ & p_n^t = p_n^{t-1} + b_n^t, \quad t = 1, \ldots, T \\ & b_n^t \geq -\eta_n p_n^{t-1} \\ & 0 \leq g_{n,d}^t \leq g_n^{\max} \\ & \sum_{t=1}^{T} g_{n,d}^t \leq \lambda_n^{\max} \\ & 0 \leq \lambda_n^{\max} \leq T \times g_{n,d}^{\max} \end{aligned} \tag{28}$$

We assume that each end-user has a predetermined amount of energy consumption and active end-users have some limit in the capacity of generating power for each particular day, thus even with the uncertainty of the RESs in generating power, user influence on the value of Φ_n is minimum. Consequently, the value of Φ_n can be assumed to be constant for daily consumption. Under this assumption, (28), can be written as

$$\begin{aligned} \min_{\boldsymbol{x}_n, \boldsymbol{g}_{n,d}, \boldsymbol{b}_n} \quad & \mathcal{F}(\boldsymbol{x}_n, \boldsymbol{g}_{n,d}, \boldsymbol{b}_n) + \sum_{m \in N\backslash n} \mathcal{F}(\boldsymbol{x}_m, \boldsymbol{g}_{m,d}, \boldsymbol{b}_m) = \Gamma(\mathcal{I}) \\ \text{subject to} \quad & \sum_{t=\alpha_{n,k}}^{\beta_{n,k}} x_{n,k}^t = E_{n,k} \\ & \gamma_{n,k}^{\min} \leq x_{n,k}^t \leq \gamma_{n,k}^{\max}, \quad \forall t \in \{\alpha_{n,k}, \ldots, \beta_{n,k}\} \\ & x_{n,k}^t = 0, \quad \forall t \notin \{\alpha_{n,k}, \ldots, \beta_{n,k}\} \\ & b_n^{\text{dis}} < b_n^t < b_n^{\text{ch}}, \quad t = 1, \ldots, T \\ & p_n^t = p_n^{t-1} + b_n^t, \quad t = 1, \ldots, T \\ & b_n^t \geq -\eta_n p_n^{t-1} \\ & 0 \leq g_{n,d}^t \leq g_n^{\max} \\ & \sum_{t=1}^{T} g_{n,d}^t \leq \lambda_n^{\max} \\ & 0 \leq \lambda_n^{\max} \leq T \times g_{n,d}^{\max} \end{aligned} \tag{29}$$

The optimisation problem in (29) is equivalent to the optimisation problem in (13). The problem in (29) can be solved distributively. The following is the proposed algorithm to solve the optimisation problem in (29) distributively.

The proposed algorithm requires only some limited information exchange between end-users when each of them attempts to maximise his/her own benefit. From Algorithm 1 (see Fig. 2), end-users minimise the optimisation problem in (29) based on the random order sequence $\mathcal{S}$ by optimising their load scheduling for adjustable loads $\boldsymbol{x}_n$, storage devices, dispatchable generators or both, such that the objective function $\Gamma(\mathcal{I})$ is strictly decreasing, that is, $\Gamma(\mathcal{I}^c) < \Gamma(\mathcal{I}^{c-1})$.

Each user minimises the cost function with respect to l_n^t, $\forall t \in T$, whereas the load of the other users (i.e. $\sum_{m \in N\backslash n} l_m^t$, $\forall t \in T$) is fixed. User n broadcasts its new load l_n^t, $\forall t \in T$ (without giving detailed information about his/her storage strategies, generators strategies or energy consumption of his/her appliances) provided that the objective function is decreasing. The energy consumption schedule for users $\mathcal{I}$ is updated and the next user in the generated sequence minimises the objective function in (29) with respect to its local load. This process is repeated until none of the users can improve their payoff by scheduling his/her load. The parameter ϵ_0 is a small fraction value for adjusting the cost function at the beginning of the algorithm to make the first two cost functions $\Gamma(\mathcal{I}^0)$ and $\Gamma(\mathcal{I}^1)$ different.

Algorithm 1:

Initialise $\mathcal{I}^0$ randomly, and calculate the corresponding cost function $\Gamma(\mathcal{I}^0)$;
Initialise counter $c = 1$ and set $\Gamma(\mathcal{I}^c) = \Gamma(\mathcal{I}^0) - \epsilon_0$;
while $\Gamma(\mathcal{I}^c) < \Gamma(\mathcal{I}^{c-1})$ **do**
  Generate a random sequence $\mathcal{S}$ for $\mathcal{N}$ users ;
  for $n \leftarrow 1$ **to** $\mathcal{N}$ **do**
    $\mathcal{I}_n = \mathcal{I}(\mathcal{S}(n), :)$ and $\mathcal{I}_{-n} = \mathcal{I} \backslash \mathcal{I}_n$;
    Optimise $\Gamma(\mathcal{I}_n, \mathcal{I}_{-n})$ and update $\mathcal{I}_n$;
    Convey a control message to make $\mathcal{I}_n$ known to all ES units and update $\mathcal{I}$;
  end
  Increment the counter, $c \leftarrow c + 1$;
  Update the cost function $\Gamma(\mathcal{I}^c) = \Gamma(\mathcal{I}_n, \mathcal{I}_{-n})$
end

Fig. 2 *Performed by every user $n \in \mathcal{N}$*

The fact that each user broadcasts its load schedule implies that each user reveals his/her strategy to all other users. The Nash equilibrium exists if no users change their strategy, despite knowing the actions of the other users [19]. From the algorithm it is clear that the termination will be reached when there is no change in cost function. Users are in Nash equilibrium since each user is making the best decision, taking into account the decisions of the others.

4 Simulation results

Simulation results are presented to access the performance of the proposed ES algorithm. We evaluate our distributed demand-side optimisation in a scenario consisting of $N = 1000$ end-users each having random devices from 15 to 25 non-adjustable loads and 15 to 25 adjustable loads. Non-adjustable loads have a fixed schedule and consume energy continuously; examples of these devices are electric stove, electric bulbs, refrigerators and TV. The adjustable loads include electrical appliances with flexible schedule such as PHEVs, dish washers, washing machines, clothes driers etc.

Out of N users, 10% of them possess either storage devices, RESs, DGs or both. These users are termed as active users because they can utilise ES together with the energy storage devices and/or their energy generators to optimise their interests.

The daily usage of adjustable and non-adjustable devices is set to be similar to those in [8, 9]. The average daily consumption of each user ranges from 12 to 16 kWh, and for users with storage devices the capacity of each user is between 2 and 5 kWh. We also considered that the RESs can produce a maximum of 6 kWh in a day. Users with dispatchable generators can produce up to 8 kWh in a day.

We consider that the higher energy demand (peak) occurs during day time, from 8:00 to 00:00, and the low energy peak occurs at night-time, from 00:00 to 8:00. This implies that the cost for day time covers the first 16 h of simulation and night-time cost covers the last 8 h of simulation. The selected cost function is quadratic given by $C^t(L_t) = \Phi_{\text{Day}} \times L_t^2$ for day time and $C^t(L_t) = \Phi_{\text{night}} \times L_t^2$ for night-time. We select $\Phi_{\text{night}} = \frac{1}{2}\Phi_{\text{Day}}$.

First, we examine a scenario where RESs together with ES are deployed compared with one without ES and RESs. Fig. 3 depicts hourly energy consumption and cost for users with RESs (i.e. 10% of the users possess RESs) compared with those without RESs. When ES is not deployed, the cost of electricity as well as the energy drawn from the utility company is reduced for users possessing RESs compared with the scenario where all users rely only on the utility company. However, RESs alone, without any optimisation scheme to effectively manage the hourly energy consumption of the users appliances, do not necessarily reduce the PAR. Using smart meters running the ES, users can optimally minimise the amount they pay to the utility company as well as PAR [2]. As it can be seen, the users without ES pay much more during the peak hour and they pay much less during the off-peak because they cannot shift their loads to the off-peak hours. Fig. 3 shows that the utilisation of the ES can help to shave off the peak load. The grid cost per-time slot depicted in Fig. 3*b* suggests that the ES helps in reducing the cost by shifting the adjustable loads to the valley of the energy cost. The results also show that the use of ES and RESs minimises the PAR as well as the daily energy cost. This is because, for each time slot, the deployed ES controls shiftable or adjustable load devices to operate at a certain power within the specified operational time of the appliances while taking the advantages of the energy generated by the RESs.

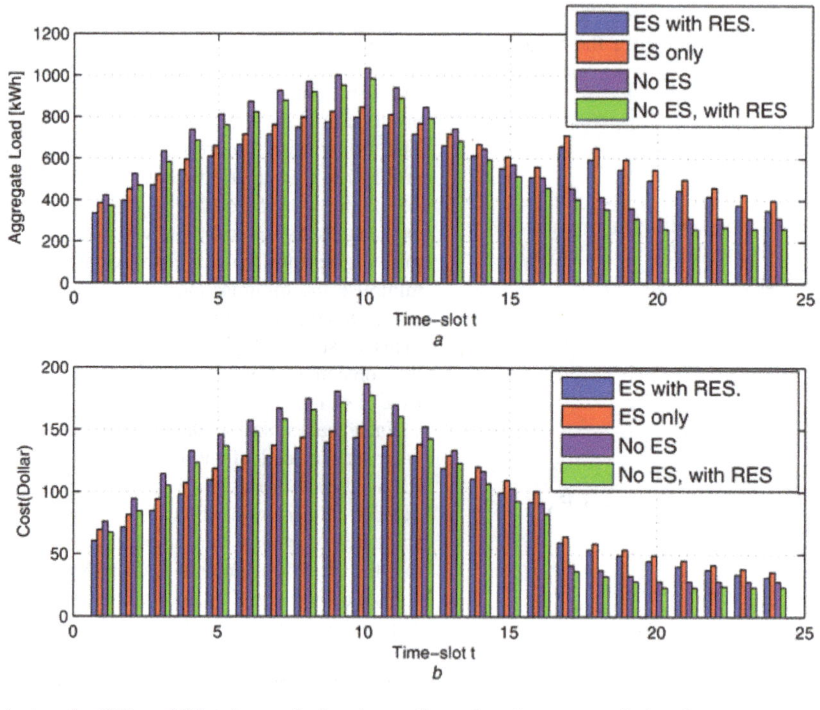

Fig. 3 *Scheduled energy load when the RESs and ES units are deployed as well as when they are not deployed*
a Comparison of the aggregate load of N users for different energy strategies
b Hourly cumulative energy cost of the aggregate load

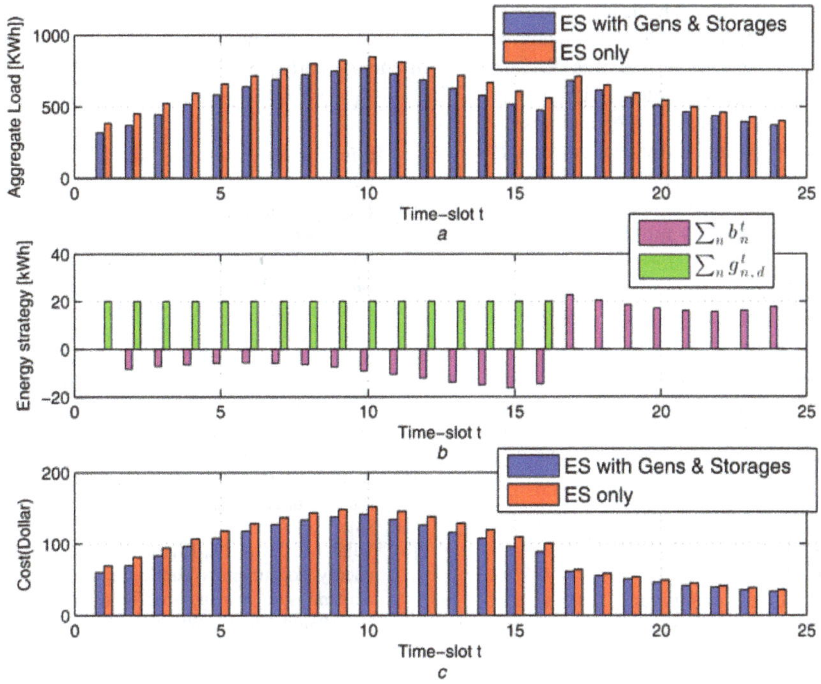

Fig. 4 *Effects of deploying ES for users with/without energy generators and storage strategies*
a Hourly aggregate load of *N* users with ES and storage and production devices compared with those with ES only
b Energy storage and production strategies
c Cumulative cost of the aggregate load

Next, we compare a scenario where some users possess RESs and/or storage devices and/or dispatchable generators or both. Again for $N = 1000$, only 10% of them possess storage and/or generators. For a fair comparison, we considered that at the end of the day, each storage device remains with its initial charging state. Fig. 4*a* depicts the aggregate energy consumption per-time

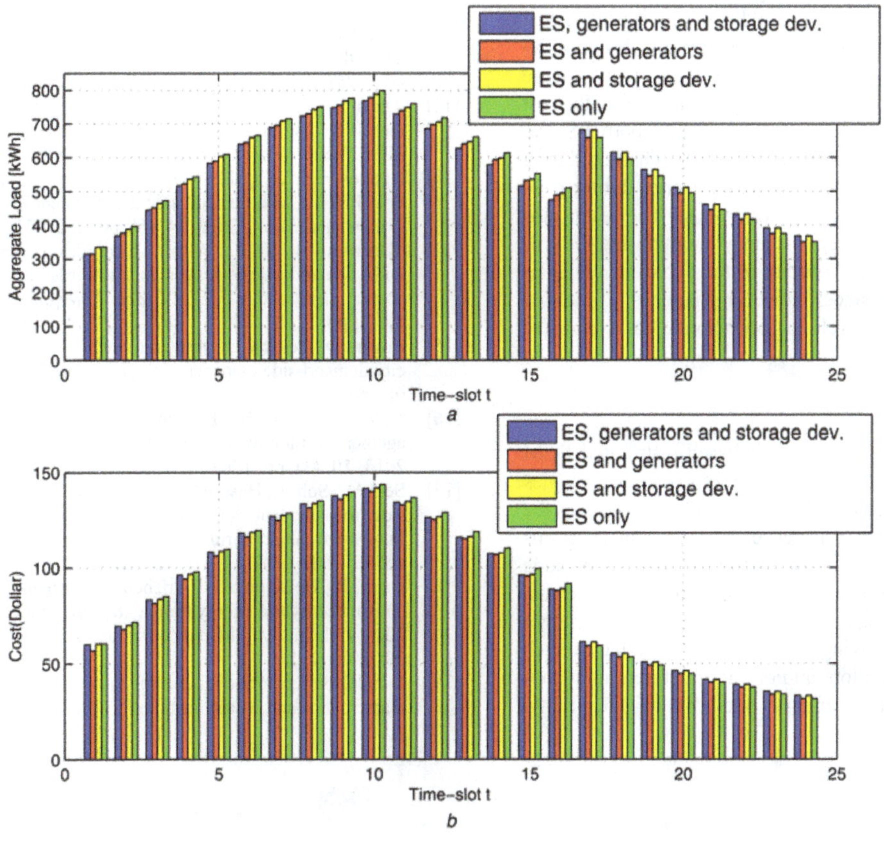

Fig. 5 *Comparison of ES together with influence of various energy strategies*
a Comparison of the aggregate load of *N* users for different energy strategies
b Hourly cumulative energy cost of aggregate load

slot t when ES is deployed. From the results, it is clear that there is no peak load, which implies that the utilisation of the ES can help to shave off the peak load. Moreover, from the results, it is obvious that deploying ES and energy generators together with energy storage devices significantly shaves the peak power and results into a stable grid. Both utility company and individual consumers benefit from the strategy as they can all maximise their payoffs.

Users with RES cannot directly optimise their energy production as they produce energy based on their capacity and the weather. On the contrary, users with dispatchable generators and/or storage devices strategically produce electricity or store electricity to optimise their payoffs. Fig. 4*b* shows that users with dispatchable generators produce power only when the cost of buying energy from the utility company is higher than that of generating energy with their dispatchable energy sources. These users can produce energy to power their own appliances or to sell it to the utility company. Similarly, users with storage devices (see Fig. 4*b*) strategically charges their devices during the low rate time and discharge their storage devices during the peak hours where the cost from the utility company is higher. In addition, the use of energy storage devices has a potential in increasing revenue earned by RESs by storing the excess produced energy and discharge it during the peak hours. It should be noted that charging and discharging are mutually exclusive (cannot occur at the same time) operations within the same time slot.

Most of the utility companies charge their customer based on the total hourly load they supply to the customers. Thus, users without energy generators or storage devices can benefit from the strategies as long as there are some active users who strategically generate and store energy. The grid cost per-time slot is depicted in Fig. 4*c*. From the results, it is clear that the ES helps in reducing the cost by shifting the adjustable loads to the valley of the energy cost. The use of energy storers and generators can further reduce the total cost of energy.

To realise the effect of deploying various energy strategies, we compare hourly load and the hourly energy cost. Fig. 5 shows the aggregate load for different strategies. From this figure, it is clear that ES minimises the PAR as well as the daily energy cost. By employing ES together with other energy strategies such as energy storage and/or dispatchable energy generators, the PAR and the daily cost of energy can further be minimised. Users utilising both ES and storage devices schedule their storage devices to be charged during low-price off-peak hours and discharge stored energy during peak hours to further minimise their consumption cost. The result also illustrates the influence of each energy strategy in optimising the utilisation of energy. Increasing the capacity of the deployed strategies may lead to better performance; however, there is a tradeoff between the initial costs as well as operational costs and the performance.

5 Conclusion

This paper demonstrated a game theoretic-based distributed DSM scheme, where end-users utilise ES and various energy strategies to minimise daily energy costs of the end-users as well as the utility costs. We verified that difference in pricing mechanisms employed by utility companies gives incentive for users to trade energy. Furthermore, increasing the hourly load results in increasing unit costs since more expensive energy sources are brought online. Thus, end-users deploying ES with energy strategies such as dispatchable energy generators and/or energy storers may substantially minimise the daily energy price of the electricity. Simulation results show that there is a significant reduction in the cost by just shifting the adjustable loads and strategic utilisation of energy storers and/or energy generators.

6 References

[1] Gatsis N., Giannakis G.: 'Residential load control: distributed scheduling and convergence with lost AMI messages', *IEEE Trans. Smart Grid*, 2012, **3**, (2), pp. 770–786

[2] Sioshansi F.P. (Ed.): 'Smart grid: integrating renewable, distributed & efficient energy' (Elsevier, Waltham, MA, USA, 2012)

[3] Baharlouei Z., Hashemi M.: 'Efficiency-fairness trade-off in privacy-preserving autonomous demand side management', *IEEE Trans. Smart Grid*, 2014, **5**, (2), pp. 799–808

[4] Chavali P., Yang P., Nehorai A.: 'A distributed algorithm of appliance scheduling for home energy management system', *IEEE Trans. Smart Grid*, 2014, **5**, (1), pp. 282–290

[5] Samadi P., Mohsenian-Rad H., Wong V.W.S, Schober R.: 'Tackling the load uncertainty challenges for energy consumption scheduling in smart grid', *IEEE Trans. Smart Grid*, 2013, **4**, (2), pp. 1007–1016

[6] Ullah M., Javaid N., Khan I., Mahmood A., Farooq M.: 'Residential energy consumption controlling techniques to enable autonomous demand side management in future smart grid communications'. Broadband and Wireless Computing, Communication and Applications (BWCCA), October 2013, pp. 545–550

[7] Logenthiran T., Srinivasan D., Tan Z.S.: 'Demand side management in smart grid using heuristic optimization', *IEEE Trans. Smart Grid*, 2012, **3**, (3), pp. 1244–1252

[8] Samadi P., Mohsenian-Rad H., Schober R., Wong V.: 'Advanced demand side management for the future smart grid using mechanism design', *IEEE Trans. Smart Grid*, 2012, **3**, (3), pp. 1170–1180

[9] Manasseh E., Ohno S., Yamamoto T.: 'Distributed demand-side management with load uncertainty'. ITU Kaleidoscope Academic Conf., 2014, pp. 1–6

[10] Atzeni I., Ordonez L., Scutari G., Palomar D.: 'Noncooperative and cooperative optimization of distributed energy generation and storage in the demand-side of the smart grid', *IEEE Trans. Signal Process.*, 2013, **61**, (10), pp. 2454–2472

[11] Mohsenian-Rad A.-H., Wong V., Jatskevich J., Schober R., Leon-Garcia A.: 'Autonomous demand-side management based on game-theoretic energy consumption scheduling for the future smart grid', *IEEE Trans. Smart Grid*, 2010, **1**, (3), pp. 320–331

[12] Atzeni I., Ordonez L., Scutari G., Palomar D.: 'Demand-side management via distributed energy generation and storage optimization', *IEEE Trans. Smart Grid*, 2013, **4**, (2), pp. 866–876

[13] Nguyen H.K., Song J.B.: 'Optimal charging and discharging for multiple PHEVs with demand side management in vehicle-to-building', *J. Commun. Netw.*, 2012, **14**, (6), pp. 662–671

[14] Shengrong B., Yu F.: 'A game-theoretical scheme in the smart grid with demand-side management: towards a smart cyber-physical power infrastructure', *IEEE Trans. Emerging Top. Comput.*, 2013, **1**, (1), pp. 22–32

[15] Rogers D., Polak G.: 'Optimal clustering of time periods for electricity demand-side management', *IEEE Trans. Power Syst.*, 2013, **28**, (4), pp. 3842–3851

[16] Nunna K., Doolla S.: 'Responsive end-user-based demand side management in multimicrogrid environment', *IEEE Trans. Ind. Inform.*, 2014, **10**, (2), pp. 1262–1272

[17] Seifi M., Soh A., Hassan M., Abd Wahab N.: 'An innovative demand side management for vulnerable hybrid microgrid'. 2014 IEEE Innovative Smart Grid Technologies – Asia (ISGT Asia), 2014, pp. 797–802

[18] Kallel R., Boukettaya G., Krichen L.: 'Demand side management of an isolated hybrid energy production unit supplying domestic loads'. 2014 11th Int. Multi-Conf. on Systems, Signals & Devices (SSD), 2014, pp. 1–5

[19] Mendelson E.: 'Introducing game theory and its applications (discrete mathematics and its applications)' (Chapman and Hall/CRC, 2004)

9

Dynamic crosstalk analysis of mixed multi-walled carbon nanotube bundle interconnects

Pankaj Kumar Das[1], *Manoj Kumar Majumder*[2], *Brajesh Kumar Kaushik*[2]

[1]*Department of Electronics and Communication Engineering, Sant Longowal Institute of Engineering and Technology, Longowal-148106, India*
[2]*Department of Electronics and Communication Engineering, Indian Institute of Technology Roorkee, Roorkee – 247667, India*
E-mail: manojbesu@gmail.com

Abstract: Multi-walled carbon nanotube (MWCNT) bundles have potentially provided attractive solutions in current nanoscale VLSI interconnects. From fabrication point of view, it is difficult to control the growth of a densely packed bundle having MWCNTs with similar diameters. A realistic bundle is combination of MWCNTs with different number of shells. Thus, this research work focuses on the analytical model of a bundle having the MWCNTs with different number of shells or in turn different diameters [mixed MWCNT bundle (MMB)]. Based on the multi-conductor transmission line theory, an equivalent single conductor (ESC) model is employed for the proposed MMB arrangements. The ESC model of MMB is used to compare the dynamic crosstalk delay with conventionally arranged bundle containing MWCNTs with similar diameters [MWCNT bundle (MB)] under different input transition time and spacing conditions. It is observed that a realistic MMB correctly estimates the crosstalk delay for the different transition time that overestimates the delay of a conventionally arranged MB by 1.35 times. Moreover, the MMB arrangement reduces the overall crosstalk delay by 47.26% compared with the conventional MB arrangements for an inter-bundle spacing ranging from 5 to 30 nm.

1 Introduction

During the recent past, researchers have considered carbon nanotubes (CNTs) as a potential interconnect material in high-speed electronics because of their unique physical [1], mechanical [2], electrical [3], chemical and thermal properties [4]. The unique atomic structure, formed by rolling of graphene sheet, provides a long mean free path (mfp) in the order of several micrometres [4]. The sp^2 bonding of carbon atoms in a graphene sheet makes CNTs a strong material that exhibits a variety of low-power interconnect applications in the areas of microelectronics/nanoelectronics [5]. The unique physical properties are mainly because of the structure of CNTs that primarily depends on the chirality, that is, rolling up direction of graphene sheets. Depending on the number of concentrically rolled up graphene sheets, CNTs can be categorised as single- (SWCNTs) and multi-walled CNTs (MWCNTs) [6]. Although the current carrying capability of a metallic SWCNT is similar to that of an MWCNT [1], but an MWCNT exhibits easier control on growth process compared with SWCNTs [7]. MWCNT consists of several coaxial cylindrical shells with different chirality that primarily depends on the rolled up direction of graphene sheets. Thus, MWCNTs have a large diameter ranging from a few to hundreds of nanometres [8]. Depending on the shell diameter, MWCNTs have a higher number of conducting channels and larger mfp in comparison to the SWCNTs. To explore these advantages, majority of the recent research [7–10] has targeted towards the fabrication and modelling of MWCNT interconnects.

From the fabrication point of view, it is quite difficult to control the growth of MWCNTs with similar diameters in a densely packed bundle. Yilmazoglu *et al.* [10] reported about the fabrication of a bundle with MWCNTs having similar diameter wherein a lot of free space was left out, thereby inefficiently using the space. It has become increasingly difficult to fabricate a densely packed bundle accommodating MWCNTs of similar diameters. Thus, in the current research scenario, it is preferred to fabricate and model the bundles where the MWCNTs of different diameters can be taken care of to achieve a densely packed bundle. To meet such requirements, this research paper presents two different types of CNT bundles: (i) bundles having MWCNTs with equal diameters [MWCNT bundle (MB)] and (2) bundles having MWCNTs with different diameters [mixed MWCNT bundle (MMB)].

Although, several electrical equivalent models for MWCNT have been proposed previously also, but they are inflicted by certain inaccuracies; Li *et al.* [11] proposed an equivalent resistor inductor capacitor (RLC) model of MWCNT by incorrectly assuming an mfp (λ_{mfp} = 1 µm) independent of the MWCNT shell diameter. In a similar way, Sarto and Tamburrano [8] and Majumder *et al.* [12] incorrectly considered the mfp and number of conducting channels independent of the shell diameter. This research paper presents an equivalent single conductor (ESC) model of MWCNT bundle by correctly modelling the diameter dependent mfp and number of conducting channels. Using a capacitively coupled interconnect line, the ESC model is used to analyse the crosstalk delay for different transition time and spacing conditions. Interestingly, it is observed that at global interconnect lengths; the crosstalk delay performance is significantly improved for realistic mixed MWCNT bundle (MMB) configuration compared with the conventionally modelled MB arrangements.

The organisation of this paper is as follows: Section 1 introduces the potential application and characteristics of a realistic MWCNT bundle in the context of current research scenario. Section 2 presents an ESC model for the bundles having MWCNTs with equal and different diameters. Section 3 provides a detailed description of capacitively coupled interconnect lines that is used to analyse the dynamic crosstalk delay for different bundle arrangements as demonstrated in Section 4. Finally, Section 5 draws a brief summary of the paper.

2 Geometry and ESC model

This section presents different specified arrangements of MWCNTs in a bundle that primarily depends on the basic geometry of an MWCNT. Initially, a multi-conductor transmission line (MTL) model of MWCNT bundle is employed to obtain different interconnect parasitics such as resistance, capacitance and inductance.

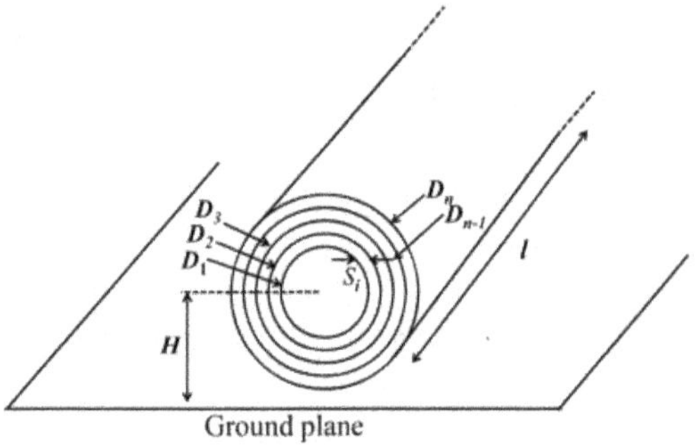

Fig. 1 *Geometry of MWCNT above ground plane*

Finally, the MTL model is simplified to an ESC for different MB and MMB configurations.

2.1 *Arrangements of bundled MWCNT having similar and different number of shells*

The geometry of MWCNT above the ground plane is shown in Fig. 1 that consists of several concentric rolled up graphene sheets with diameters D_1, D_2, ….., D_n. In current fabrication technology, the intershell spacing is approximately equivalent to the Van der Waal's gap (δ) between neighbouring carbon atoms [13] and can be expressed as

$$\delta = \frac{D_n - D_{n-1}}{2} \simeq 0.34\,\text{nm} \tag{1}$$

where innershell and outershell diameters are represented as D_1 and D_n, respectively, and n denotes the total number of shells. The distance between the centre of MWCNT and the ground plane is equivalent to $H = D_n/2 + h_t$, where h_t represents the distance of outermost shell from the ground plane. The outershell diameter of MWCNT primarily depends on the number of shells (n) and can be expressed as

$$D_n = D_1 + 2 \times \delta \times (n - 1) \tag{2}$$

Depending on the geometry of Fig. 1, this sub-section presents three different conventional bundle arrangements (Fig. 2) that consists of MWCNTs of uniform diameters (equal number of shells). Figs. 2*a–c* exhibits the bundle arrangements of MB-I, MB-II and MB-III, wherein MWCNTs with 4-, 8- and 12-shells are placed, respectively. The total numbers of MWCNTs [11] in terms of bundle height (h) and width (w) can be expressed as

$$N_{\text{MWCNT}} = \left[N_x N_y - \text{integer}\left[\frac{N_y}{2}\right]\right] \tag{3}$$

where

$$N_x = \frac{w - D_n}{D_n + \delta} + 1 \text{ and } N_y = \frac{h - D_n}{(D_n + \delta)\sqrt{3}/2} + 1 \tag{4}$$

where N_x and N_y represent the numbers of MWCNTs in horizontal and vertical directions, respectively.

Fig. 3 presents the arrangements of realistic mixed bundles wherein MWCNTs of different numbers of shells are placed. The mixed bundle follows the arrangements of MWCNTs with a higher number of shells at the periphery, whereas the MWCNTs of smaller diameters are placed at centre of the bundle. The primary advantage behind this type of arrangement indicates minimum unutilised areas that serve densely packed MWCNT bundles. Fig. 3*a* introduces the arrangement of MMB-I, wherein

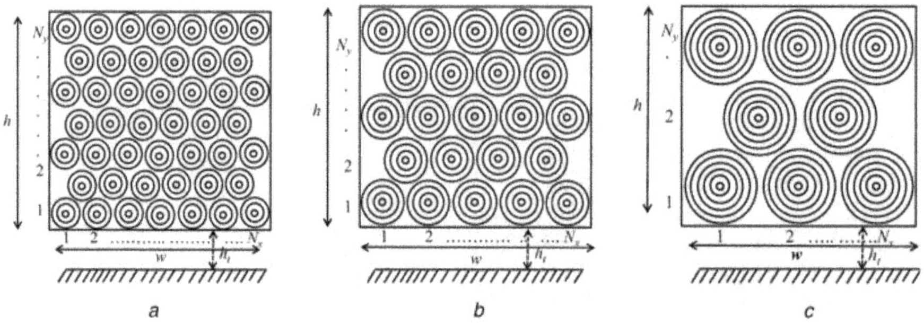

Fig. 2 *Arrangements of MWCNTs with 4-, 8- and 12-shells in*
a MB-I
b MB-II
c MB-III

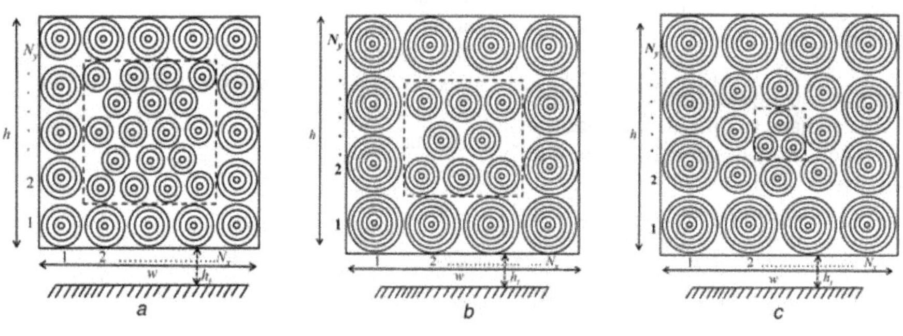

Fig. 3 *Arrangements of mixed MWCNT bundles of*
a MMB-I
b MMB-II
c MMB-III

Table 1 Number of MWCNTs in different bundle configurations

Bundle arrangements	MWCNTs with different number of shells	N_{MWCNT}	N_x	N_y
MB-I	$(MWCNT)_{4\text{-shell}}$	230	9	27
	$(MWCNT)_{8\text{-shell}}$	0	0	0
	$(MWCNT)_{12\text{-shell}}$	0	0	0
MB-II	$(MWCNT)_{4\text{-shell}}$	0	0	0
	$(MWCNT)_{8\text{-shell}}$	68	5	15
	$(MWCNT)_{12\text{-shell}}$	0	0	0
MB-III	$(MWCNT)_{4\text{-shell}}$	0	0	0
	$(MWCNT)_{8\text{-shell}}$	0	0	0
	$(MWCNT)_{12\text{-shell}}$	35	4	10
MMB-I	$(MWCNT)_{4\text{-shell}}$	127	0	0
	$(MWCNT)_{8\text{-shell}}$	32	5	13
	$(MWCNT)_{12\text{-shell}}$	0	0	0
MMB-II	$(MWCNT)_{4\text{-shell}}$	0	0	0
	$(MWCNT)_{8\text{-shell}}$	28	0	0
	$(MWCNT)_{12\text{-shell}}$	22	4	9
MMB-III	$(MWCNT)_{4\text{-shell}}$	27	0	0
	$(MWCNT)_{8\text{-shell}}$	22	0	0
	$(MWCNT)_{12\text{-shell}}$	22	4	9

MWCNTs with 8-shells are placed at periphery to the centrally located MWCNTs having 4-shells. The arrangement of MMB-II is similar to MMB-I, except that the peripheral and central MWCNTs contain 12-shells and 8-shells, respectively, as depicted in Fig. 3*b*. Similarly, Fig. 3*c* shows the arrangement of MMB-III, wherein MWCNTs having 12-, 8- and 4-shells are placed from periphery to centre. Therefore, the total number of MWCNTs in a bundle can be obtained using the following two approaches: (i) the MWCNTs placed at periphery ($N_{MWCNT(periphery)}$) and (ii) MWCNTs placed at centre ($N_{MWCNT(centre)}$) of the bundle. Thus, the total number of the MWCNTs in an MMB can be expressed as

$$N_{MWCNT} = N_{MWCNT(periphery)} + N_{MWCNT(center)} \tag{5}$$

where N_{xp} and N_{yp} represent the number of CNTs placed in periphery at horizontal and vertical directions, respectively. $N_{MWCNT(centre)}$ is calculated using (3) and (4) by replacing the number of CNTs at horizontal and vertical orientations with N_{xc} and N_{yc}, respectively. The quantitative values of N_{xc} and N_{yc} are similar to the N_x and N_y as obtained from (4). For different MB and MMB arrangements, the quantitative values of N_x and N_y along with the N_{MWCNT} are summarised in Table 1.

2.2 ESC model

This section provides a detailed description of ESC model of bundled MWCNT interconnects that takes into account each transmission line of MWCNTs in parallel. The interconnect parasitics of each transmission line is modelled using the total number of conducting channels associated with MWCNTs in a bundle. The number of conducting channels primarily takes into account the spin and sub-lattice degeneracy of carbon atoms. Naeemi and Meindl [13] modelled the number of conducting channels of each shell in MWCNT as

$$\begin{aligned} N_i(D_i) &\simeq k_1 T D_i + k_2, \quad D_i > d_T/T \\ &\simeq 2/3 \quad D_i \le d_T/T \end{aligned} \tag{6}$$

where D_i represents the diameter of *i*th shell, constants k_1 and k_2 have the values of $3.87 \times 10^{-4}\,\text{nm}^{-1}\,\text{K}^{-1}$ and 0.2, respectively. The thermal energy of electrons and gap between two sub-bands determines the quantitative value of d_T equivalent to 1300 nm. K at room temperature ($T = 300$ K) [13]. For $D_i > 4.3$ nm, the average number of conducting channels is proportional to its shell diameter. Thus, the total number of conducting channels in an MWCNT can be obtained using the summation of conducting channels (N_i) of each shell as

$$N^{channel} = \sum_{i=1}^{n} N_i \tag{7}$$

where *n* denotes the total number of shells in MWCNT. In a similar way, the total number of conducting channels in an MWCNT bundle can be obtained as

$$N_{total}^{bundle} = \sum_{j=1}^{N_{MWCNT}} N_j^{channel} \tag{8}$$

The conduction mechanism of CNTs is ballistic or dissipative because of large mfps in the range of micrometres. The mfp primarily depends on D_i and can be expressed as [14]

$$\lambda_{mfp,i} = \frac{10^3 D_i}{(T/T_0) - 2}, \quad T_0 = 100\,\text{K} \tag{9}$$

Depending on the MTL theory [11, 15], an equivalent RLC model of MWCNT bundle is presented in Fig. 4, which is further simplified to an ESC as shown in Fig. 5. The ESC model of Fig. 5 assumes that all the shells in MWCNT and all the MWCNTs in a bundle are parallel and participate in conduction. The tunnelling

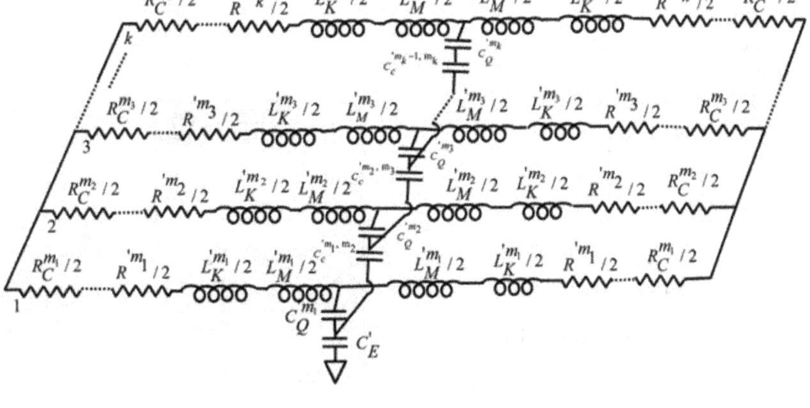

Fig. 4 *Equivalent RLC model of MWCNT bundle*

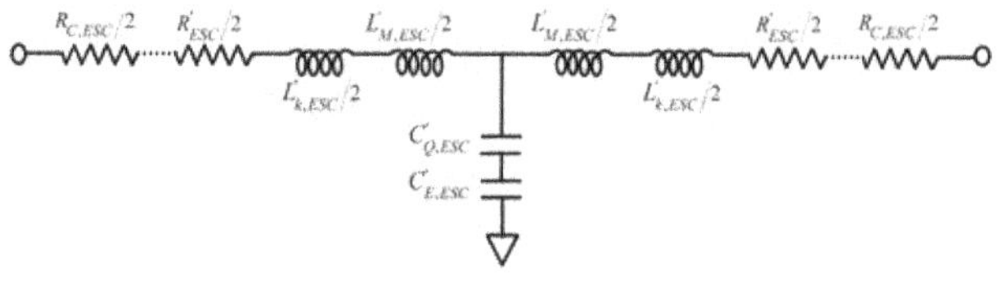

Fig. 5 *ESC model of MWCNT bundle*

conductance (G_t) that primarily represents the intershell and inter-CNT electron tunnelling effects has been neglected in the model because of a negligible interaction between two shells in MWCNT [8, 16–18]. For a defect free global interconnect length, Yoon *et al.* [19] also neglected the electron tunnelling between adjacent shells in MWCNT.

Each MWCNT in bundle demonstrates three types of resistances: (i) quantum or intrinsic resistance (R_q) that is due to the quantum confinement of electrons in a nano-wire [20], (ii) imperfect metal-nanotube contact resistance (R_{mc}) that exhibits a typical value of 3.2 kΩ depending on the fabrication process [21] and (iii) scattering resistance (R') that appears because of the static impurity scattering, defects, line edge roughness scattering, acoustic phonon scattering etc. The equivalent resistance of the ESC can be expressed as

$$R_{\mathrm{ESC}} = R_{C,\mathrm{ESC}} + R'_{\mathrm{ESC}}\, l \tag{10}$$

where

$$R_{C,\mathrm{ESC}} = \left[\left(\frac{1}{(N_{\mathrm{MWCNT}})m_1}\left(\frac{R_q}{2\times\sum_{i=1}^{n} N_i} + R_{\mathrm{mc}}\right)\right)^{-1} + \cdots + \left(\frac{1}{(N_{\mathrm{MWCNT}})m_k}\left(\frac{R_q}{2\times\sum_{i=1}^{n} N_i} + R_{\mathrm{mc}}\right)\right)^{-1}\right]^{-1} \tag{11}$$

and

$$R'_{\mathrm{ESC}} = \left[\left(\frac{1}{(N_{\mathrm{MWCNT}})m_1}\left(\frac{R_q}{2\times\sum_{i=1}^{n} N_i \lambda_{\mathrm{mfp},i}} + R_{\mathrm{mc}}\right)\right)^{-1} + \cdots + \left(\frac{1}{(N_{\mathrm{MWCNT}})m_k}\left(\frac{R_q}{2\times\sum_{i=1}^{n} N_i \lambda_{\mathrm{mfp},i}} + R_{\mathrm{mc}}\right)\right)^{-1}\right]^{-1} \tag{12}$$

where $m_1, m_2, \ldots, m_k$ are the different types of MWCNTs in the bundle.

The equivalent capacitance of the ESC model in Fig. 5 is categorised as (i) electrostatic capacitance ($C'_{E,\mathrm{ESC}}$) that is mainly because of the potential difference between the bundle and the ground plane and (ii) quantum capacitance ($C'_{Q,\mathrm{ESC}}$) that represents the finite density of electronic states in a quantum wire (i.e. CNT). The per unit length (p.u.l.) $C'_{Q,\mathrm{ESC}}$ and $C'_{E,\mathrm{ESC}}$ can be approximated as

$$C'_{Q,\mathrm{ESC}} = 2C'_{Q0} \times N_{\mathrm{total}}^{\mathrm{bundle}}, \quad \text{where} \quad C'_{Q0} = \frac{2e^2}{hv_{\mathrm{F}}} \tag{13}$$

$$C'_{E,\mathrm{ESC}} = \frac{2\pi\varepsilon_0\varepsilon_r}{\cosh^{-1}[((D_n + h_t)/D_n)]} \times N_x \tag{14}$$

where $\varepsilon_r = 2.2$ and $v_F = 8 \times 10^5$ m/s are the relative permittivity and the Fermi velocity of CNT, respectively [22, 23]. N_x represents the number of MWCNTs facing the ground plane as shown in Figs. 2 and 3. The quantitative values of N_x for different MB and MMB configurations are presented in Table 1. In addition, each shell in MWCNT and each MWCNT in the bundle experiences an innershell and an inter-CNT coupling capacitance, C_s and C_c, that primarily depends on the diameter of adjacent shells and centre-to-centre distance between neighbouring CNTs (d_{c-c}), respectively [8, 23]. The p.u.l. C'_s and C'_c can be expressed as

$$C_s'^{i+1,i} = \frac{2\pi\varepsilon_0\varepsilon_r}{\ln(D_{i+1}/D_i)} \tag{15}$$

$$C'_c = \frac{\pi\varepsilon_0\varepsilon_r}{\ln(d_{c-c}/2r) + \left(\sqrt{(d_{c-c}/2r)^2 + 1}\right)} \tag{16}$$

where r represents the mean radius of any two CNTs in bundle. The ESC model in Fig. 5 comprises of (i) kinetic inductance ($L'_{K,\mathrm{ESC}}$) that originates from the kinetic energy of electrons and (2) magnetic inductance ($L'_{M,\mathrm{ESC}}$) that arises because of the magnetic field induced by current flowing through a nanotube. The p.u.l. $L'_{K,\mathrm{ESC}}$ and $L'_{M,\mathrm{ESC}}$ can be expressed as

$$L'_{K,\mathrm{ESC}} = \frac{L'_{K0}}{2N_{\mathrm{total}}^{\mathrm{bundle}}}; \quad \text{where} \quad L'_{K0} = \frac{h}{2e^2 v_{\mathrm{F}}} \tag{17}$$

$$L'_{M,\mathrm{ESC}} = \frac{\mu_0}{2\pi}\cosh^{-1}\left(\frac{D_n + 2h_t}{D_n}\right) \tag{18}$$

3 Capacitively coupled interconnect lines

Crosstalk delays are analysed for different MB and MMB configurations using capacitively coupled interconnect lines as shown in Fig. 6. The interconnect lines in Fig. 6 are modelled by using the ESC of different MB and MMB configurations. The coupled interconnect lines, terminated by a load capacitance $C_L = 10$ aF, are connected with a supply voltage $V_{dd} = 1$ V. A CMOS driver is used for accurate estimation of crosstalk delay. It can be realised by noting the fact that a transistor in CMOS gate operates partially in linear region and partially in saturation region during switching. However, a transistor can be accurately approximated by a resistor only in the linear region. In the saturation region, the transistor is more accurately modelled as a current source with a parallel high resistance [24, 25]. Using the driver interconnect load (DIL) setup and the ESC model, performance is analysed for different

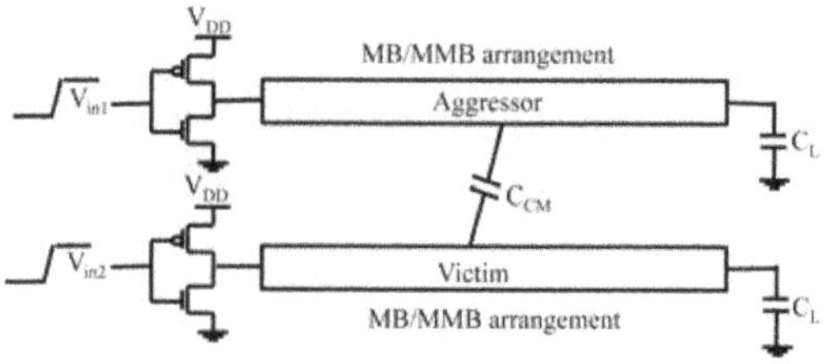

Fig. 6 *Capacitively coupled interconnect lines*

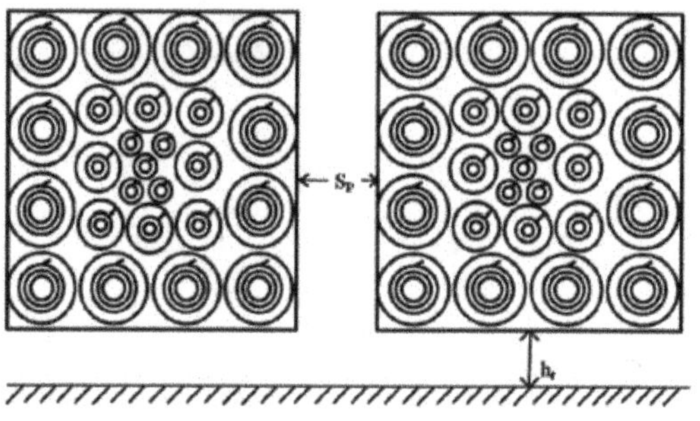

Fig. 7 *Placing of MWCNT bundle above ground plane*

Table 2 Equivalent interconnect parasitics for different MB and MMB configurations

Bundle arrangements	Equivalent resistance		Equivalent inductance		Equivalent capacitance		
	$R_{C,\mathrm{ESC}}$, kΩ	R'_{ESC}, Ω/μm	$L'_{K,\mathrm{ESC}}$, pH/μm	$L'_{M,\mathrm{ESC}}$, pH/μm	$C'_{Q,\mathrm{ESC}}$, fF/μm	$C'_{E,\mathrm{ESC}}$, fF/μm	C'_{CM}, fF/μm
MB-I	3.2	5.21	13.13	0.04	118.74	0.24	0.53
MB-II	3.2	4.96	24.43	0.07	63.79	0.15	0.36
MB-III	3.2	4.11	27.48	0.08	56.72	0.13	0.27
MMB-I	3.2	4.69	16.30	0.07	95.58	0.15	0.31
MMB-II	3.2	4.19	25.16	0.08	61.92	0.13	0.24
MMB-III	3.2	4.03	22.19	0.08	50.23	0.13	0.21

bundle configurations at global interconnect lengths ranging from 400 to 2000 μm.

The inter-bundle coupling capacitance (C'_{CM}) has a significant effect on crosstalk delay that primarily depends on the spacing between aggressor and victim (S_p) [26] and can be expressed as

$$C'_{\mathrm{CM}} = \frac{\pi\varepsilon_0\varepsilon_r}{\cosh^{-1}(S_p/D_n)} \times N_y \tag{19}$$

where D_n is the outershell diameter of MWCNTs facing each other (N_y in Table 1) as shown in Fig. 7. Table 2 summarises the quantitative values of interconnect parasitics associated with different bundle configurations. The quantitative values of these parasitics primarily depend on the bundle arrangement and the number of conducting channels associated with MWCNTs of different numbers of shells.

4 Crosstalk analysis

This section analyses propagation delay under the influence of dynamic crosstalk at different input transition times and spacing between coupled interconnect lines. Crosstalk is an important design concern in current nanoscale very large scale integration (VLSI) interconnects that primarily depends on the mutual capacitance, spacing between interconnect lines, relative transition-time skew etc. [26]. Crosstalk (an undesirable noise) is broadly classified into two categories: (i) functional and (ii) dynamic crosstalks. Under the functional crosstalk category, victim line experiences a voltage spike when an aggressor line switches. On the other hand, dynamic crosstalk is observed when aggressor and victim lines switch simultaneously. A change in propagation delay is experienced under dynamic crosstalk when adjacent line (aggressor and victim) switches either in-phase or out-of-phase.

4.1 Crosstalk delay for different transition times

For different transition times and MWCNT bundle configurations, Figs. 8–10 present the out-phase crosstalk delay at global interconnect lengths of 400, 1200 and 2000 μms, respectively. It is observed

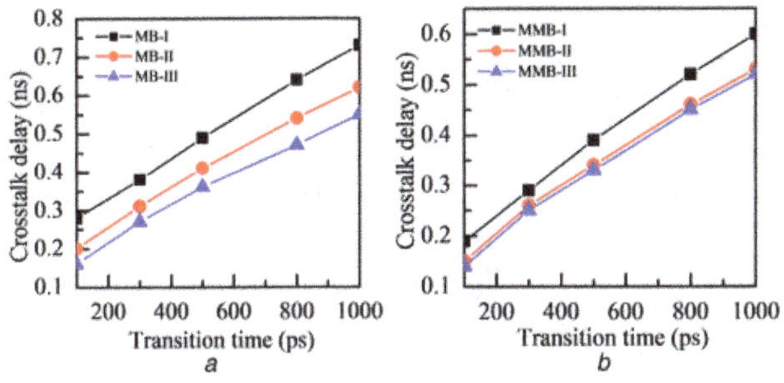

Fig. 8 *Crosstalk delay of*
a MB
b MMB configurations for different transition times at $l = 400$ μm

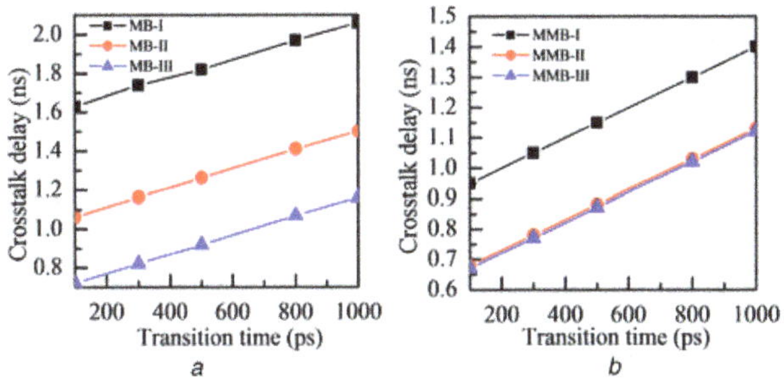

Fig. 9 *Crosstalk delay of*
a MB
b MMB configurations for different transition times at $l = 1200$ μm

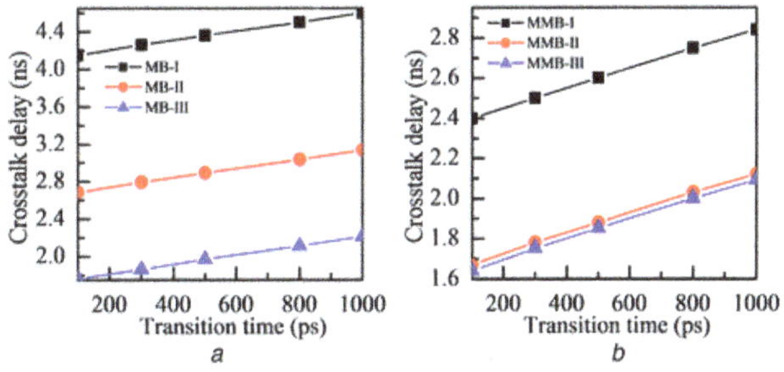

Fig. 10 *Crosstalk delay of*
a MB
b MMB configurations for different transition times at $l = 2000$ μm

that the crosstalk delay increases for large transition time that primarily depends on the signal rise and fall time. An increase in rise and fall times also increases the signal propagation delay under the influence of dynamic crosstalk. Apart from this, the percentage reduction in crosstalk delay for different bundle configurations are shown in Tables 3 and 4 at interconnect lengths of 400 and 2000 μm, respectively. It is observed that on using novel MMB-III arrangement, the overall crosstalk delay is reduced by 1.35 times compared with the conventional MB-I. The primary

Table 3 Percentage reduction in crosstalk delay of MMB-III at $l = 400$ μm

Transition time (ps)	Percentage reduction in crosstalk delay for MMB-III compared with				
	MB-I	MB-II	MMB-I	MB-III	MMB-II
100	100.00	42.86	35.71	14.28	7.14
300	52.00	24.24	18.18	9.09	4.00
500	48.48	24.00	16.00	8.00	3.03
800	42.22	20.00	15.55	5.76	2.22
1000	40.38	19.23	15.38	4.44	1.92

Table 4 Percentage reduction in crosstalk delay of MMB-III at $l = 2000$ µm

Transition time (ps)	Percentage reduction in crosstalk delay for MMB-III compared with				
	MB-I	MB-II	MMB-I	MB-III	MMB-II
100	153.05	63.41	46.34	7.32	1.83
300	143.42	59.42	42.85	6.48	1.71
500	135.67	56.22	40.54	6.28	1.62
800	123.88	50.75	36.81	5.74	1.43
1000	120.09	49.76	35.88	4.97	0.99

reason behind this reduction is the lower resistive and capacitive parasitics of novel MMB-III as presented in Table 2. The quantitative values of interconnect parasitics are primarily dependent on the total number of conducting channels that is substantially affected by the diameters of MWCNTs. Interconnect parasitics are considerably reduced for higher MWCNT diameters. Thus, the least parasitics associated with MMB-III results in a lesser crosstalk delay compared with other conventional bundle arrangements.

4.2 Crosstalk delay for different spacings

The spacing between aggressor and victim lines has a significant impact on crosstalk delay as presented in Figs. 11–13 at global interconnect lengths of 400, 1200 and 2000 µms, respectively. It is observed that the crosstalk delay is substantially reduced for an increase in spacing between the aggressor and victim lines. Crosstalk delay, analysed for opposite signal transitions in aggressor and victim lines, is primarily influenced by the coupling capacitance C'_{CM}. The C'_{CM} is mostly affected by the spacing between aggressor and victim as expressed in (19). Irrespective of interconnect lengths, the higher spacing between the coupled lines substantially reduces the C'_{CM} that in turn results in a reduced crosstalk delay. Moreover, it is observed that the overall crosstalk delay of novel MMB-III is reduced by 47.26% compared with the conventional MB-I arrangement presented in Tables 5 and 6 for 400 and 2000 µm lengths, respectively.

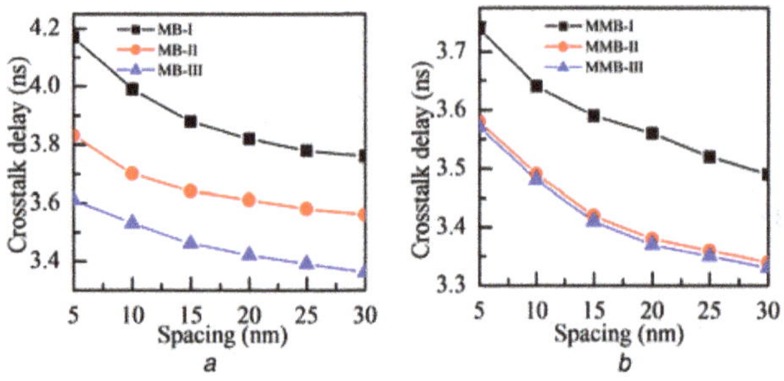

Fig. 11 *Crosstalk delay of*
a MB
b MMB configurations for different spacings at $l = 400$ µm

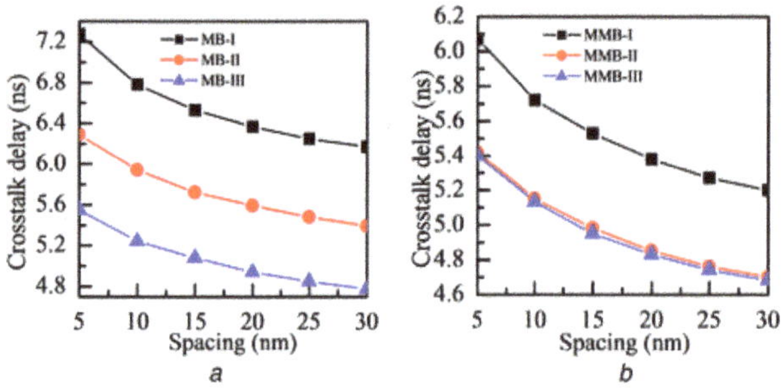

Fig. 12 *Crosstalk delay of*
a MB
b MMB configurations for different spacings at $l = 1200$ µm

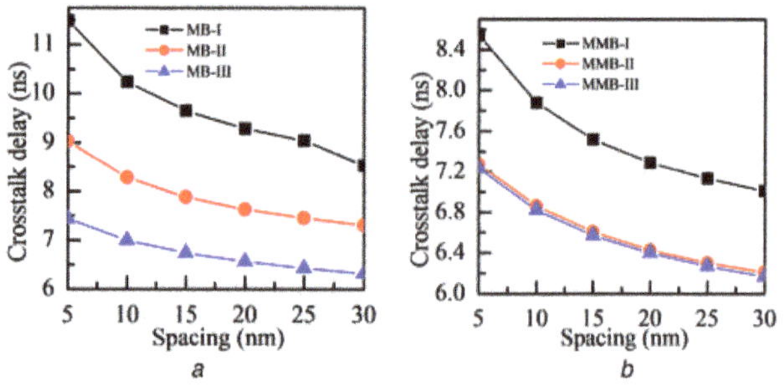

Fig. 13 *Crosstalk delay of*
a MB
b MMB configurations for different spacings at $l = 2000$ µm

Table 5 Percentage reduction in the crosstalk delay of MMB-III at $l = 400$ µm

Spacing, nm	Percentage reduction in crosstalk delay for MMB-III as compared with				
	MB-I	MB-II	MB-III	MMB-I	MMB-II
5	16.80	7.28	1.80	5.63	0.30
10	14.65	6.32	1.48	5.27	0.29
15	13.78	6.74	1.46	5.07	0.29
20	13.35	7.12	1.43	4.80	0.28
25	12.91	6.86	1.19	4.76	0.28
30	12.83	6.90	1.12	4.59	0.27

Table 6 Percentage reduction in crosstalk delay of MMB-III at $l = 2000$ µm

Spacing, nm	Percentage reduction in crosstalk delay for MMB-III as compared with				
	MB-I	MB-II	MB-III	MMB-I	MMB-II
5	58.92	24.75	2.77	18.12	0.64
10	50.15	21.26	2.50	15.54	0.60
15	46.88	19.78	2.49	14.46	0.58
20	45.16	19.06	2.43	13.90	0.55
25	44.18	18.66	2.39	13.71	0.47
30	38.25	18.31	2.26	13.61	0.47

5 Conclusion

This research paper presented different bundle configurations where MWCNTs with similar and different numbers of shells are specifically arranged. Based on the MTL theory, an ESC model is employed to analyse the dynamic crosstalk delay for different transition times and spacing between aggressor and victim. A capacitively coupled interconnect line, driven by a CMOS driver, is used to analyse the crosstalk delay. For different spacings between aggressor and victim lines, the overall crosstalk delay of MMB-III is reduced by 47.26% compared with the conventional MB arrangements. Therefore, a realistic mixed MWCNT bundle can be preferred over conventionally arranged MWCNT bundles for future global VLSI interconnects.

6 References

[1] Collins P.G., Hersam M., Arnold M., Martel R., Avouris P.: 'Current saturation and electrical breakdown in multiwalled carbon nanotubes', *Phys. Rev. Lett.*, 2011, **86**, pp. 3128–3131

[2] Thuau D., Koutsos V., Cheung R.: 'Electrical and mechanical properties of carbon nanotube', *J. Vac. Sci. Technol. B*, 2009, **27**, pp. 3139–3144

[3] Wei B.Q., Vajtai R., Ajayan P.M.: 'Reliability and current carrying capacity of carbon nanotubes', *Appl. Phys. Lett.*, 2001, **79**, pp. 1172–1174

[4] Javey A., Kong J.: 'Carbon nanotube electronics' (Springer-Verlag, Berlin, Germany, 2009)

[5] Avorious P., Chen Z., Perebeions V.: 'Carbon-based electronics', *Nat. Nanotechnol.*, 2007, **2**, pp. 605–613

[6] Li H., Xu C., Srivastava N., Banerjee K.: 'Carbon nanomaterials for next-generation interconnects and passive, physics, status and prospects', *IEEE Trans. Electron Devices*, 2009, **56**, pp. 1799–1821

[7] Sato S., Nihei M., Mimura A., *ET AL.*: 'Novel approach to fabricating carbon nanotube via interconnects using size-controlled catalyst nanoparticles'. Proc. IEEE Interconnect Technology Conf., Burlingame, CA, 2006, pp. 230–232

[8] Sarto M.S., Tamburrano A.: 'Single-conductor transmission-line model of multiwall carbon nanotubes', *IEEE Trans. Nanotechnol.*, 2010, **9**, pp. 82–92

[9] Liu Z., Bajwa N., Ci L., Lee S.H., Kar S., Ajayan P.M., Lu J.Q.: 'Densification of carbon nanotubes bundles for interconnect application'. Proc. IEEE Interconnect Technology Conf., Burlingame, CA, 2007, pp. 201–203

[10] Yilmazoglu O., Joshi R., Popp A., Pavlidis D., Schneider J.J.: 'Pronounced field emission from vertically aligned carbon nanotubes blocks and bundles', *J. Vac. Sci. Technol.*, 2011, **29**, pp. 02B106-1–02B106-5

[11] Li H., Yin W.-Y., Banerjee K., Mao J.-F.: 'Circuit modeling and performance analysis of multi-walled carbon nanotube interconnects', *IEEE Trans. Electron Devices*, 2008, **55**, pp. 1328–1337

[12] Majumder M.K., Pandya N.D., Kaushik B.K., Manhas S.K.: 'Dynamic crosstalk effect in mixed CNT bundle interconnects', *IET Electron. Lett.*, 2012, **48**, pp. 384–385

[13] Naeemi A., Meindl J.D.: 'Performance modeling for single- and multiwall carbon nanotubes as signal and power interconnects in gigascale systems', *IEEE Trans. Electron Devices*, 2008, **55**, pp. 2574–2582

[14] Naeemi A., Meindl J.D.: 'Physical modeling of temperature coefficient of resistance for single- and multi-wall carbon-nanotube interconnects', *IEEE Electron Device Lett.*, 2007, **28**, pp. 135–138

[15] Rabaey J.M., Chandrakasan A., Nikolic B.: 'Digital integrated circuits' (Prentice-Hall, NJ, USA, 2011), pp. 159–173

[16] Collins P.G., Avouris P.: 'Multishell conduction in multiwalled carbon nanotubes', *Appl. Phys. A*, 2002, **74**, pp. 329–332

[17] Bourlon B., Miko C., Forro L., Glattli D.C., Bachtold A.: 'Determination of the intershell conductance in multiwalled carbon nanotubes', *Phys. Rev. Lett.*, 2004, **93**, pp. 176806-1–176806-4

[18] Das D., Rahaman H.: 'Analysis of crosstalk in single- and multiwall carbon nanotubes interconnects and its impact on gate oxide reliability', *IEEE Trans. Nanotechnol.*, 2011, **10**, pp. 1362–1370

[19] Yoon Y.G., Delaney P., Louie S.G.: 'Quantum conductance of multiwall carbon nanotubes', *Phys. Rev. B*, 2002, **66**, pp. 073407-1–073407-4

[20] Burke P.J.: 'Lüttinger liquid theory as a model of the gigahertz electrical properties of carbon nanotubes', *IEEE Trans. Nanotechnol.*, 2002, **1**, pp. 129–144

[21] Srivastava A., Xu Y., Sharma A.K.: 'Carbon nanotubes for next generation very large scale interconnects', *J. Nanophotonics*, 2010, **4**, pp. 1–26

[22] Lamberti P., Tucci V.: 'Impact of variability of the process parameters on CNT-based nanointerconnects performances: a comparison between SWCNTs bundles and MWCNT', *IEEE Trans. Nanotechnol.*, 2012, **11**, pp. 924–933

[23] Subash S., Chowdhury M.H.: 'Mixed carbon nanotube bundles for interconnect applications', *Int. J. Electron., Taylor & Francis*, 2009, **96**, pp. 657–671

[24] Kaushik B.K., Sarkar S.: 'Crosstalk analysis for a CMOS-gate-driven coupled interconnects', *IEEE Trans. Comput. Aided Des. Integr. Circuits Syst.*, 2008, **27**, pp. 1150–1154

[25] Kaushik B.K., Sarkar S., Agarwal R.P., Joshi R.C.: 'An analytical approach to dynamic crosstalk in coupled interconnects', *Microelectron. J.*, 2010, **41**, pp. 85–92

[26] Rossi D., Cazeaux J.M., Metra C., Lombardi F.: 'Modeling crosstalk effects in CNT bus architecture', *IEEE Trans. Nanotechnol.*, 2007, **6**, pp. 133–145

10

Device and circuit performance analysis of double gate junctionless transistors at $L_g = 18$ nm

Chitrakant Sahu, Jawar Singh

Department of Electronics and Communication Engineering, PDPM Indian Institute of Information Technology, Design and Manufacturing Jabalpur, Madhya Pradesh, India
E-mail: chitrakant@iiitdmj.ac.in

Abstract: The design and characteristics of double-gate (DG) junctionless (JL) devices are compared with the DG inversion-mode (IM) field effect transistors (FETs) at 45 nm technology node with effective channel length of 18 nm. The comparison are performed at iso-V_{th} for both n- and p-type of devices. The JL device shows lower drain-induced barrier lowering, steep subthreshold slope and lower OFF state current. For the first time, the authors demonstrate a pass gate (PG) logic, inverter circuit and static random access memory (SRAM) stability analysis using JL devices, rather than a complementary metal-oxide semiconductor (CMOS) configuration. They observed that transient response of JL PG configuration is similar to that of conventional CMOS PGs. JL inverter also shows similar transient characteristics with 25% reduction in delay and 12% improvement in 6 T SRAM cell stability compared with IMFETs, which shows large potential in digital circuit applications. The simulations were performed using coupled device-circuit methodology in ATLAS technology aided computer design (TCAD) mixed-mode simulator.

1 Introduction

The aggressive scaling of metal-oxide semiconductor field effect transistor (MOSFET) has led to short-channel effects (SCEs) because of reduced gate controllability over the channel. In the nanoscale devices, the influence of SCE on the characteristics of conventional MOSFETs cannot be ignored. To reduce this influence, multi-gate structures such as double-gate (DG), surrounding-gate and FinFETs, which can suppress the SCEs and improve the capacity of control of the current, have been proposed [1–3]. However, aggressive scaling of semiconductor devices, the industry faces severe challenges while formation of steep source and drain junctions in short-channel devices. Therefore ultrafast annealing methods and the development of novel doping techniques are investigated which are complex and expensive. To address these issues, junctionless transistor (JLT) which contains a single doping species at the same level in its source, drain and channel has been proposed and investigated [4–6]. The JL devices are promising candidate for next generation high-speed and low-power integrated circuits owing to their excellent control of SCEs, ideal subthreshold swing (SS), low leakage current and good carrier transport efficiency [4–12]. Many reports on JLTs are available in the literature based on Lilienfeld's first transistor architecture [13] also Colinge *et al.* [4], Park *et al.* [5] and Jeon *et al.* [6] have successfully fabricated the multigate junctionless (JL) nanowire transistors. Different JLT architectures include double-gate architecture [7–9], bulk planar architecture [10], tri-gate nanowire architectures with silicon on insulator (SOI) as well as bulk substrate [11, 12, 14] and gate all-around architecture [15, 16].

The JL FET is very different from conventional MOSFET in terms of operating principle. The current in JLFET is because of majority carrier instead of minority and it flows in the volume instead of semiconductor-dielectric interface. In JLFET, most of the studies focused on the device performance estimation of JL transistors, but few addressed the circuit applications of such devices. Recently, Han *et al.* [14] has reported inverter based on tri-gate JL bulk FinFET and shown improved static noise margin (SNM) and reduced delay compared with inversion-mode (IM) bulk FinFET. As the double-gate/multigate devices are also promising candidates for the replacement of conventional MOSFETs because of their better electrostatic integrity, therefore one would be interested to know its circuit performance. A popular and widely used alternative to the conventional complementary metal-oxide semiconductor (CMOS) logic configuration is the pass gate (PG) configuration, which can significantly reduce the number of transistors required to implement a logic circuit. Also pass transistors are generally used as switches to pass logic levels between the nodes of a circuit, rather than as switches connected directly to the power supply. To transform the device level advantages of JLT into circuit level, we have investigated the circuit behaviour of JL devices. Mixed-mode simulation is typically used to simulate circuits that contain semiconductor devices for which accurate compact models do not exist, since JLT is an emerging device and from best of our knowledge there are no compact SPICE models available for circuit level simulation.

In this paper, we first optimised the JLFET for high-performance (HP) 45 nm technology node as defined by the Semiconductor Industry Association International Technology Roadmap for Semiconductors (ITRSs) [17], with effective channel length $L_g = 18$ nm, and compared the device and circuit performances with DG MOSFETs provided in [18]. The working and operation of n- and p-type DG JLT on PG configuration is described and its transient behaviour are observed using ATLAS mixed-mode simulation [19, 20]. Additionally, an inverter is designed with JL devices, where n and p channels were simply connected similar to existing CMOS inverter. We study how the increased gate capacitance and reduced drive capability affect the overall circuit performance and power dissipation. We also explored 6T static random access memory (SRAM) cell design making insightful comparisons with the DG MOSFET.

This paper is structured as follows. In Section 2, the simulation method and optimised parameters for studying the characteristics of devices and circuits are introduced. Section 3 consists of working and transient response of PG configuration JLT. In Section 4, the inverter and SRAM performances of IM and JL are compared. Finally, Section 5 summarises the conclusion.

2 Device and circuit simulation methodology

Fig. 1 show schematic views of (a) n-type and (b) p-type JL DGFETs. The JL device have silicon thickness ($T_{si} = 12$ nm), channel length ($L_g = 18$ nm) width ($W = 1$ μm) and source/drain (S/D) extensions are ($L_{ext} = 18$ nm), analogues to device design proposed in [18] for DG MOSFET. Gate oxide thickness (T_{ox}) of 1.2

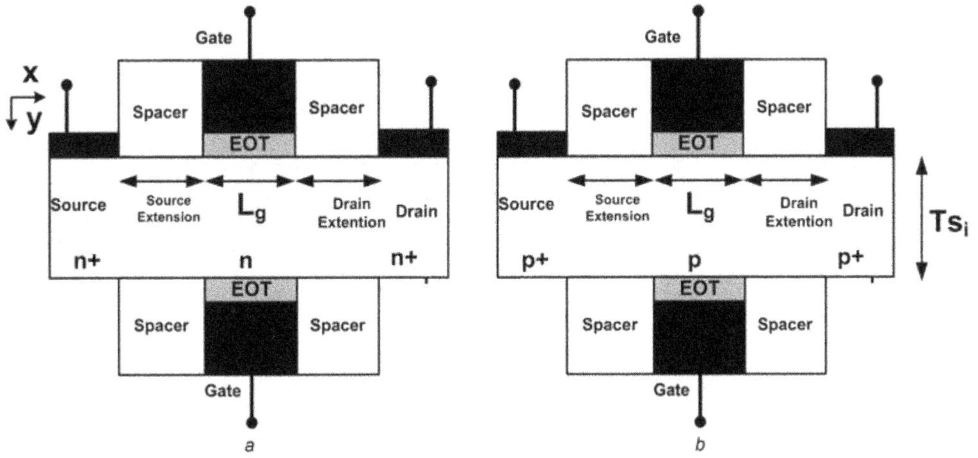

Fig. 1 *Cross-sectional view of*
a n-Type
b p-Type DG JLFET

nm have been used which is thick enough to avoid excessive gate tunnelling current in the HP device, as specified in the ITRS [17]. For a fair comparison between JL and IMFETs, linear threshold voltages V_{th} are adjusted to about ±0.3 V by tuning gate work-function and detailed parameters are provided in Table 1. The parasitic S/D series resistance R_S/R_D has two main components: the extension resistance R_{ext} and the contact resistance R_{con}. We estimated R_{ext} by numerically solving the following expression

$$R_{\text{ext}} = \int_0^{L_{\text{ext}}} \frac{1}{qN_{\text{SD}}(x)\mu(N_{\text{SD}})t_{\text{si}}} \, dx \tag{1}$$

where the carrier mobility depends on N_{SD} and R_{ext} is estimated as 46 and 77 Ωμm for n- and p-type DG MOSFET, respectively. Continued scaling of channel length decreases the channel resistance and increases the impact of parasitic S/D resistance on saturation current. The significant part of S/D resistance also lies on contact resistance since it is not scalable. Based on the transmission-line model, the contact resistance can be expressed as

$$R_{\text{co}} = \sqrt{\frac{\rho_{\text{sd}}\rho_{\text{c}}}{W}} \coth\left(l_{\text{c}}\sqrt{\frac{\rho_{\text{sd}}}{\rho_{\text{c}}}}\right) \tag{2}$$

where l_c is the width of the contact window, W is the transistor width, ρ_{sd} is the sheet resistivity of the source–drain diffusion and ρ_c is the interfacial contact resistivity of the ohmic contact between the metal and the silicon. R_{co} includes the resistance of the current crowding region in silicon underneath the contact. Equation (2) has two limiting cases: short contact and long contact. For the short contact limit $l_c \ll \sqrt{\rho_{\text{sd}}/\rho_c}$, we have

$$R_{\text{co}} = \frac{\rho_c}{Wl_c} \tag{3}$$

Equation (3) indicates that under short contact condition (which is the case for deep sub-micron devices) the contact resistance is dominated by the interfacial contact resistance and is determined by the contact window width. With the continued scaling of feature dimensions, l_c reduces and R_{co} increases. The channel resistance, however, decreases with the scaling of channel length. Therefore for deep sub-micron devices, without careful contact engineering, contact resistance can become dominant and the device current can be limited by S/D resistance instead of being controlled by gate, which results the transistors may cease to function properly. Thus, in order to capture the fundamental properties of double-gate fully depleted devices, in this study, we selected contact and S/D external resistance as specified in the ITRS [17]. The R_{ext} is estimated as 77 Ωμm, since the ITRS at 45 nm targets R_S/R_D = 125 Ωμm [17], then we assume a reasonable R_{con} = 48 Ωμm. Similarly, for p-type devices (1) yields R_{ext} = 46 Ωμm. Then, again assuming R_{con} = 110 Ωμm, we let R_S/R_D = 156 Ωμm. The JL and IM devices are simulated with same R_S/R_D and to match identical parasitic resistances, and to meet ITRS requirement for 45 nm technology node (with effective channel length 18 nm).

The simulation involves CVT model along with Shockley–Read–Hall (SRH) and Auger recombination models for minority carrier recombination. We have also included concentration dependent mobility (CONMOB) model for low field mobility related to doping density and field dependent mobility (FLDMOB) model for high field velocity saturation depending on parallel electric field in the direction of current flow. The inclusion of quantum effects in MOSFET circuit simulators is essential to accurately describe the electrical behaviours of today's nanoscale devices and to assess their performance limits. The necessary self-consistent solution of the two-dimensional (2D) Poisson and Schrodinger equation in a cross-section of the JL DGFET structure was incorporated by Silvaco ATLAS TCAD [19] with energy valley degeneration and bandgap narrowing taken into account in all the simulations. The keyword 2DXY.SCHRO is used to involve the 2D Poisson–Schrodinger solver. The keywords NUM.DIRECT and SP.DIR of models statement are used to select the valleys with appropriate effective mass. To refine small and noisy carrier concentration because of rapidly diminishing wave functions from the confining

Table 1 Device parameters of JL and IMFET

Parameters	JL FET	IMFET
channel doping concentration	N:1 × 10^{19} cm^{-3} P:1 × 10^{19} cm^{-3}	N:1 × 10^{15} cm^{-3} P:1 × 10^{15} cm^{-3}
source/drain doping concentration	N:1 × 10^{20} cm^{-3} P:1 × 10^{20} cm^{-3}	N:1 × 10^{20} cm^{-3} P:1 × 10^{20} cm^{-3}
gate work-function	N: 5.25 eV P: 4.17 eV	N: 4.8 eV P: 4.6 eV
S/D contact resistance	N: 48 Ωμm P: 110 Ωμm	N: 79 Ωμm P: 79 Ωμm
S/D extension resistance	N: 77 Ωμm P: 46 Ωμm	N: 46 Ωμm P: 77 Ωμm
S/D total resistance	N: 125 Ωμm P: 156 Ωμm	N: 125 Ωμm P: 156 Ωμm

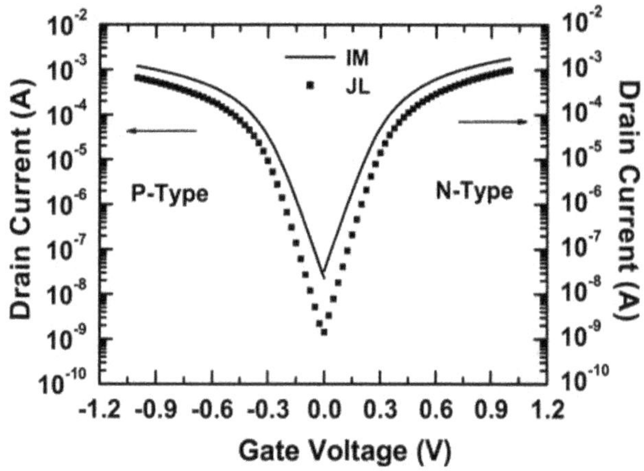

Fig. 2 *Transfer characteristics in logarithmic scales for the DG JLT and DG MOSFET at* $V_{ds} = \pm 1$ *V*

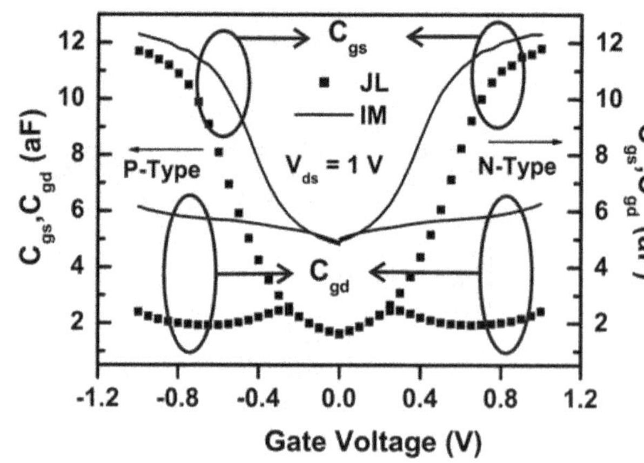

Fig. 3 *Gate to source* (C_{gs}) *and gate to drain capacitance* (C_{gd}) *as a function of gate voltage for the n-type JL and IM DGFET at* $V_{ds} = 1$ *V*

Table 2 Extracted parameters of DG JL and IMFET

Devices	I_{on}/I_{off}	DIBL, mV/V	SS, mv/dec
nJLT	7.0×10^5	44.7	68.13
nFET	0.5×10^5	101.3	79
pJLT	9.0×10^5	43.2	67.4
pFET	1.0×10^5	100.2	78.2

potential barriers, quantum minimum carrier concentration (QMINCONC) is used. For the calculation of μ_{eff} in this simulation, a relationship between the effective channel mobility is used as

$$\frac{1}{\mu_{eff}} = \frac{1}{\mu_{phon}} + \frac{1}{\mu_{rough}} + \frac{1}{\mu_{bulk}} \tag{4}$$

The mobility consists of three parts: surface acoustic phonon scattering μ_{phon}, surface roughness scattering μ_{rough} and bulk mobility with doping-dependent modification μ_{bulk}; the details are given in [12]. The effective mobility value extracted from the simulations is 100 cm^2/V for devices with an impurity concentration of 10^{19} cm^{-3} and Qss is assumed to be 3×10^{12} cm^{-2}.

One of the major challenge in downscaling of nanoscale MOSFETs for low power consumption and high-speed nanoscale digital applications is the control of drain current in subthreshold regime with the reduction in the channel length. The subthreshold current is the leakage current that affects dynamics of the circuits and determines the CMOS standby power. A steep SS is also an important parameter that is a measure of conversion speed of transistor from ON to OFF that is given by SS = d Log(I_d)/dV_g. In nanoscale MOSFET it is also seen that the influence of higher drain voltage leads to decreases in threshold voltage, which is also known as drain-induced barrier lowering (DIBL) effect. This effect is because of slope of surface potential in the drain side implying the effectiveness in reduction of electric field near the drain and presenting immunity to SCE because of increased drain voltages.

Fig. 2 shows $I_d - V_{gs}$ characteristics of *p*-channel (left) and *n*-channel (right) for both JL and IMFET, respectively. For a fair comparison, linear threshold voltages V_{th} are adjusted to about ±0.3 V by tuning gate work-function. The IM devices shows higher driving current but at the cost of higher subthreshold leakage current. JL device structure shows improved ON to OFF current ratio that can be observed from data provided in Table 2 because of reduced SCEs. JL devices also show lower SS and DIBL to those of the IM devices at gate lengths L_g of 18 nm. The

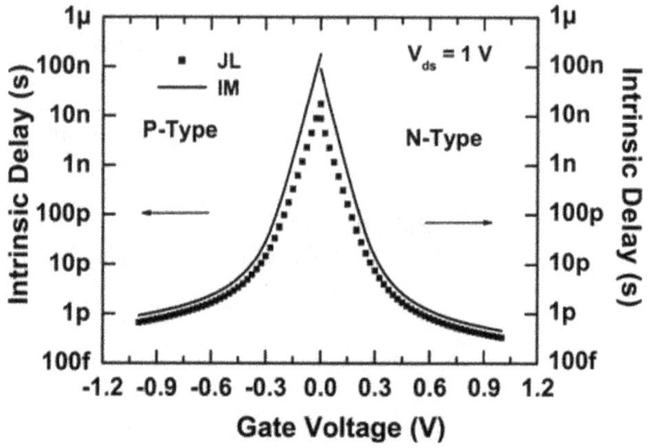

Fig. 4 *Intrinsic delay for both JL and IM devices as a function of gate voltage at* $V_{ds} = \pm 1$ *V*

reason for these differences is that, in conventional IM devices, some of the depletion charge is balanced by the source and drain at end of the channel. The severity of this phenomenon, which is known as the SCE, increases when the channel length is short. In contrast, JL devices do not have a concentration gradient between the source/channel and channel/drain to produce a junction. Therefore the charge is controlled by the gate alone, which substantially reduces SCEs. The steep SS and small DIBL values are key factors for future HP nanoscaled devices at low operating voltages.

Fig. 3 shows plot of C_{gs} and C_{gd} for JL and IM DGFET as a function of gate voltages. All the capacitances are extracted from the small-signal ac device simulations at a frequency of 1 MHz and drain voltage, V_{ds} = 1 V. The intrinsic capacitances depend on the operating region of the device, hence C_{gs} changes from subthreshold region to saturation region for both devices. However, C_{gd} is constant throughout because of effective voltage at drain end V_{gd} is not enough to make this end fully conducting. The larger values of C_{gs} and C_{gd} in the case of the IM devices result from lower channel doping as compared with JL devices. The gate capacitance $C_{gg} = C_{gs} + C_{gd}$ of JL and IM devices are observed 14aF and18aF, respectively, at $V_{gs} = V_{gs} = 1$ V. Fig. 4 shows intrinsic gate delay for JL and IM DGFET as a function of gate voltages. The intrinsic gate delay is important as it represents the frequency limit of the transistor operation. It is given by $C_{gg}V_{dd}/I_d$ where C_{gg} is the gate capacitance, V_{dd} is supply voltage and I_d is drain current. We observed that intrinsic gate delay is less in JL devices compared with the simulated IM devices. This is because of the lower gate capacitance of JL devices.

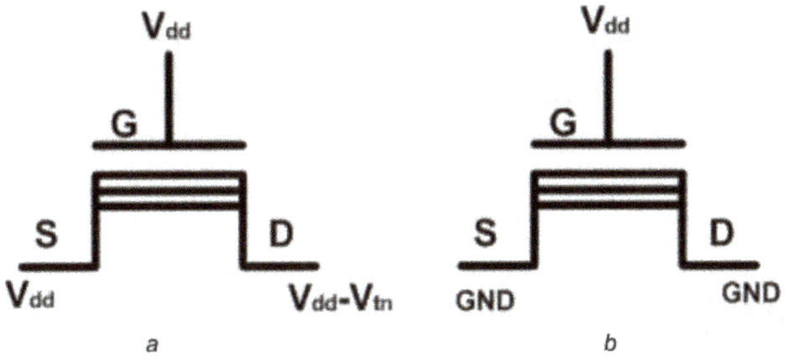

Fig. 5 *Pass gate configuration of p-type JLT*
a Passing '1'
b Passing '0'

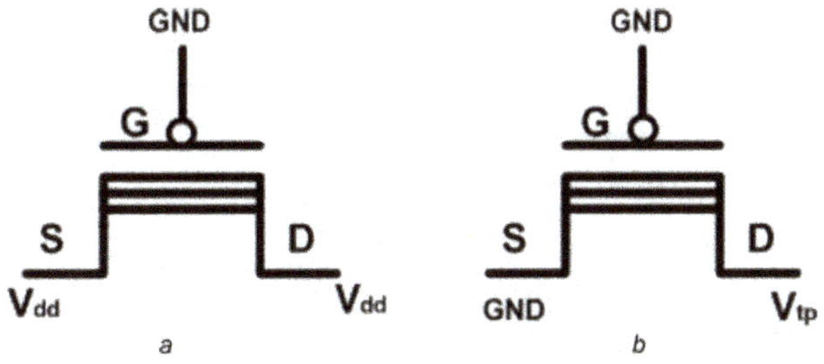

Fig. 7 *PG configuration of p-type JLT*
a Passing '1'
b Passing '0'

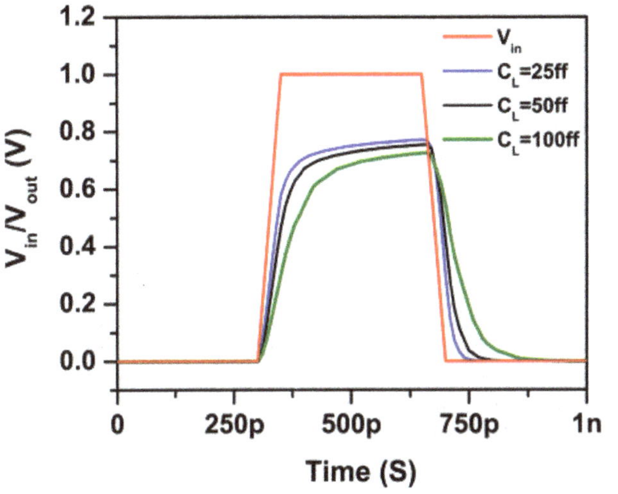

Fig. 6 *Transient behaviour of PG configuration n-type JLT at different capacitive loads*

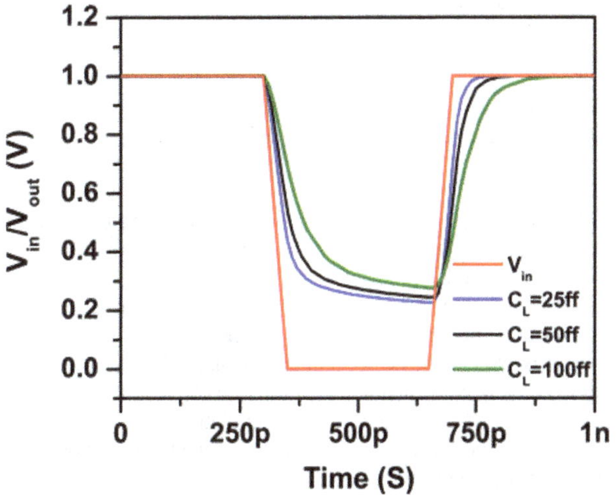

Fig. 8 *Transient behaviour of PG configuration p-type JLT at different capacitive loads*

3 JLT pass transistor configuration

A popular and widely used alternative to the conventional CMOS logic configuration is the PG configuration, which can significantly reduce the number of transistors required to implement a logic circuit. Also pass transistors are generally used as switches to pass logic levels between the nodes of a circuit, rather than as switches connected directly to the power supply. In conventional MOSFET, an n-type PG produces a 'strong zero' or ground and a 'weak one' by lowering the output below $V_{dd} - V_{tn}$, where V_{tn} is the threshold voltage of the nFET. In contrast, a p-type pass transistor produces a 'strong one', but a 'weak zero' by raising the output above V_{tp} when the input is zero, where V_{tp} is the threshold voltage of the pFET. The operating principle and simulation results for the PG configuration using both p-type and n-type JLT are described below.

3.1 DG nJLT PG

The configurations shown in Figs. 5*a* and *b* are n-type JL PG transistors. In new symbol of JLT parallel lines represent bulk conduction from source to drain. It passes the logic level on its source (input) to its drain (output), when its gate is driven to V_{dd} (when the PG is enabled). When the PG is disabled (its control gate is driven to ground), the resistance between the source (input) and drain (output) of the device, R_{ds}, is on the order of 100 kΩ. The PG can be useful when sharing a bus or a logic circuit and its inputs and outputs can be swapped that is, logic flow through the PG can be bi-directional. Fig. 6 shows how the output of a PG at various capacitive load conditions. The operation is similar to that of MOSFET PG and to keep the nJLT on, minimum voltage required is V_{tn}. It can be said that an nJLT device is not good at passing a '1' similar to MOSFET PG.

3.2 DG pJLT PG

Figs. 7*a* and *b* show PG configuration and Fig. 8 shows transient characteristics of pJLT at various capacitive load conditions. As expected the operation of the pJLT device is complementary to the nJLT's operation. The pJLT device turns on when its gate is driven to ground. If its gate is pulled to V_{dd}, the device is off (and the output is in the high resistance state). In Fig. 7*a*, the PG is passing a '1' to the output and (b) '0' is passed to the output. However, noting that the output only gets pulled down to V_{tp}. It can be said that a pJLT PG is good at passing a '1' and bad at passing a '0' similar to MOSFET PG.

4 Methodology of circuit simulation

Fig. 9*a* shows the JL inverter used as test circuits to estimate circuit characteristics. Fig. 9*b* shows the flowchart for coupled device-circuit approach for mixed-mode simulation. In mixed-mode simulation, operation is mainly divided in two parts. The first part includes the device description and simulation model parameters. The device descriptions provide information about device geometry, doping distribution and meshes. Simulation model parameters includes setting up models like mobility, recombination, impact ionisation etc. for simulation. The second part describes the circuit netlist and analysis. The circuit description includes the circuit topology (called 'netlist') and the electrical models and parameters of the circuit components. After that, bias condition and type of analysis like dc, ac and transient are described. First simulation

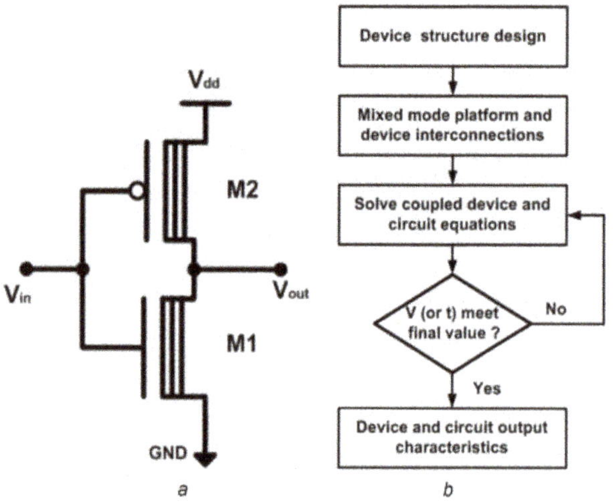

Fig. 9 *Test circuit and simulation methodology*
a JL FET inverter circuit
b Flowchart for the coupled device-circuit approach

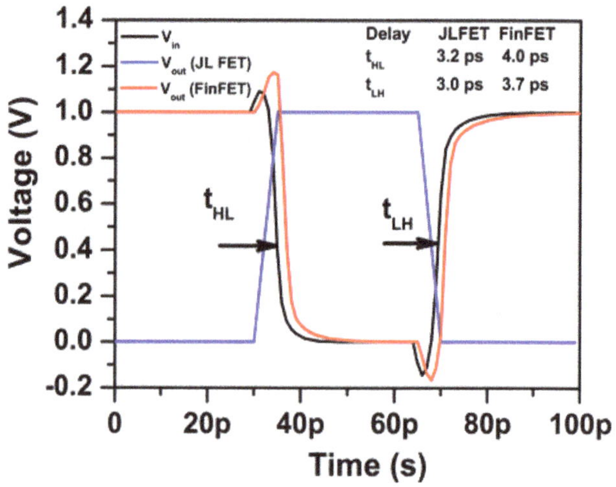

Fig. 10 *Transient characteristics of JL and IMFET inverter circuits, in which the definitions of high-to-low delay time (t_{HL}) and low-to-high delay time (t_{LH}) are indicated*

runs on the initial guess for the coupled device-circuit and then for nodal equations of the tested circuits. The nodal equation is formulated according to the Kirchhoff current law. Coupling the formulated circuit equations to the device transport equations obtains a large matrix containing both circuit and device equations. Solving the large matrix obtains the device and circuit characteristics simultaneously. The coupled simulation solves iteratively until it reaches its specified final voltage or time. When a simulation has finished it gives information about I-V data that is, voltages in all circuit nodes and currents in all circuit branches.

5 JL CMOS inverter performance

The CMOS inverter is a basic building block for digital circuit design and performs the logic operation of A to $\bar{A}$. The JL inverter, shown in Fig. 9*a* is a series combination of a *p*-channel and an *n*-channel JL DGFET. The gates of the two JL FETs connected together to form the input and the two drains are connected together to form the output similar to conventional CMOS inverter. When the input to the inverter is connected to ground, the output is pulled to V_{dd} through the pJLT device M_2 (M_1 shuts off). When the input terminal is connected to V_{dd}, the output is pulled to ground through the nJLT device M_1 (M_2 shuts off) similar to CMOS Inverter. As shown in Fig. 2 drain current of pJLT is approximately half of nJLT because of difference between mobility of electrons and holes. In inverter circuit both the devices are connected in series, in order to maintain same current, width of pJLT is taken twice of nJLT similar to conventional CMOS technology. The flowchart of coupled device-circuit approach using mixed-mode simulation of DG JL and MOSFET inverter circuits are shown in Fig. 9*b*. Results of transient and static characteristics are described below.

5.1 Propagation delay

Fig. 10 shows the transient characteristics of an inverter circuit based on JL and IMFETs and the high-to-low delay time (t_{HL}) and low-to-high delay time (t_{LH}) between and input output signals are also shown. When input signal switches from low to high voltage, the nJLT starts to turn on and pJLT is turned off. Similarly, when input signal switches from high to low voltage pJLT will turn on and nJLT will turn off. However, using the parameter setting in this paper, the simulated JL CMOS inverter circuit shows strikingly similar transient characteristics. The t_{HL} and t_{LH} values in the JL CMOS inverter are smaller than those observed in the IM CMOS inverter. Propagation delay t_p is defined as average of t_{HL} and t_{LH}. Reducing delay in digital circuit allows them to process data at a faster rate and improve overall performance. From transient response extracted values of t_{HL} and t_{LH} are 3.2 and 3.0 ps (JL inverter) and 4.0 and 3.7 ps (IM inverter), respectively. One can observe that the propagation delay of JL inverter is decreased by ~25%. Also the voltage overshoot in the transient response of inverter circuit is mainly attributed to the C_{gd} of the inverter transistors, which directly couples the steep voltage step at the input node to the output even before the response of transistor. Since in case of JL devices C_{gd} is low so the voltage overshoot is low as compared with IM inverter.

5.2 Static noise margin

The DC and transient behaviour exhibited by the proposed DG JL devices and inverter are very promising; therefore, we are extending this approach one more level higher for SRAM cell stability analysis. As SRAM is considered to be another benchmark circuit for evaluation of any emerging technology and its success absolutely depends of the proper realisation of SRAM cell. The stability of an inverter cell is often related to the SNM, which is defined as the highest dc noise voltage for which the cell state does not flip during its access. Generally, an SRAM cell operates in three different states: (i) standby; (ii) write; and (iii) read. As SRAM cell being most vulnerable to noise during read operations [21–23], therefore we estimated the read SNM. Fig. 11 shows the

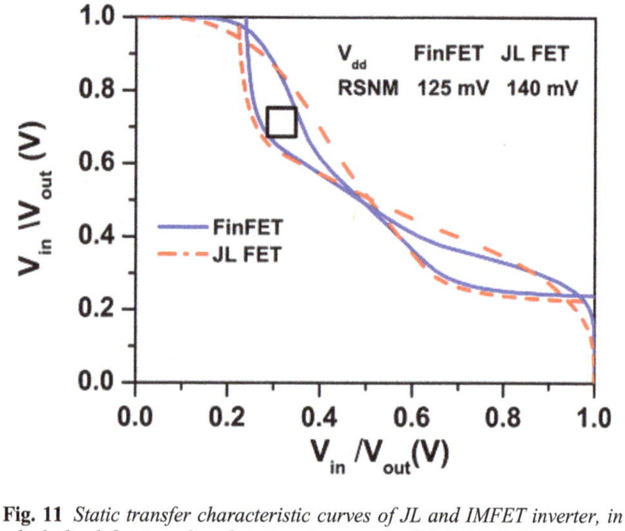

Fig. 11 *Static transfer characteristic curves of JL and IMFET inverter, in which the definition of read static noise margin (RSNM) is indicated*

plot for read SNM of standard 6 T SRAM cell. The JL FET-based SRAM's static transfer characteristics are very similar to those of the IMFET-based SRAM cell. The read SNM from both transfer characteristics are extracted by square fitting method that is the largest square to be fitted in between overlapped plot of inverter transfer characteristics and its inverse characteristics. The extracted values of read SNMs for JL- and IM DGFET-based SRAM cells are 140 and 125 mV, respectively. Hence, in terms of stability of SRAM cell, we can say that JL-based SRAM cell offers 12% improved stability as compared with DG MOSFET-based SRAM cell.

6 Conclusion

Simulations results show that the DG JLFET had a lower OFF current and better short-channel characteristics compared with its counterpart DG IMFET. Transient characteristics of PG configuration JL devices are similar to those of conventional MOSFET. The circuit performance comparisons show that the inverter and SRAM designed by JLT had similar transient and static transfer characteristics, comparable with those of DG MOSFETs. Also lowers gate capacitance in JL inverter devices leads to smaller miller capacitance, which speed up performance by 25% and 6 T SRAM cell stability is improved by 12%.

7 References

[1] Lee C.W., Afzalian A., Akhavan N.D., Yan R., Ferain I., Colinge J.P.: 'Junctionless multigate field-effect transistor', *Appl. Phys. Lett.*, 2009, **94**, pp. 053511-1–053511-2

[2] Colinge J.P., Gao M.H., Romano A., Maes H., Claeys C.: 'Silicon-on-insulator gate-all-around device'. IEDM Technical Digest, 1990, pp. 595–598

[3] Balestra F., Cristoloveanu S., Benachir M., Brini J., Elewa T.: 'Double-gate silicon-on-insulator transistor with volume inversion: a new device with greatly performance', *IEEE Electron Device Lett.*, 1987, **EDL-8**, (9), pp. 410–412

[4] Colinge J.P., Lee C.W., Afzalian A., ET AL.: 'Nanowire transistors without junctions', *Nat. Nanotechnol.*, 2010, **5**, (3), pp. 225–229

[5] Park C.-H., Ko M.-D., Kim K.-H., ET AL.: 'Electrical characteristics of 20-nm junctionless Si nanowire transistors', *Solid State Electron.*, 2012, **73**, (7), p. 710

[6] Jeon D.-Y., Park S.J., Mouis M., Barraud S., Kim G.-T., Ghibaudo G.: 'A new method for the extraction of flat-band voltage and doping concentration in tri-gate Junctionless transistors', *J. Solid-State Electron.*, **81**, (2013), pp. 113–118

[7] Chen Z., Xiao Y., Tang M., ET AL.: 'Surface-potential-based drain current model for long-channel junctionless double-gate mosfets', *IEEE Trans. Electron Devices*, 2012, **59**, (12), 3292–3298

[8] Duarte J.P., Choi S.J., Moon D.I., Choi Y.K.: 'Simple analytical bulk current model for long-channel double-gate junctionless transistors', *Trans. Electron Devices*, 2011, **32**, pp. 704–706

[9] Duarte J.P., Choi S.-J., Choi Y.-K.: 'A full-range drain current model for double-gate junctionless transistors', *IEEE Trans. Electron Devices*, 2011, **32**, (6), 704–706

[10] Bulk Planar Junctionless Transistor (BPJLT): 'An attractive device alternative for scaling', *IEEE Electron Device Lett.*, 2011, **32**, (3), 261–263

[11] Lee C.W., Afzalian A., Akhavan N.D., Yan R., Ferain I., Colinge J. P.: 'Junctionless multigate field-effect transistor', *Appl. Phys. Lett.*, 2009, **94**, pp. 1053511–1053511-2

[12] Trevisoli R., Doria R., de Souza M., Das S., Ferain I., Pavanello M.: 'Surface-potential-based drain current analytical model for triple-gate junctionless nanowire transistors', *IEEE Trans. Electron Devices*, 2012, **59**, (12), pp. 3510–3518

[13] Lilienfeld J.E.: 'Method and apparatus for controlling electric cur-rent'. U.S. Patent 1 745 175, 1925

[14] Han M.-H., Chang C.-Y., Chen H.-B., Cheng Y.-C., Wu Y.-C.: 'Device and circuit performance estimation of junctionless bulk finFETs', *IEEE Trans. Electron Devices*, 2013, **60**, (6), 1807–1813

[15] Han M.-H., Chang C.-Y., Jhan Y.-R., ET AL.: 'Characteristic of *p*-type junctionless gate-all-around nanowire transistor and sensitivity analysis', *IEEE Electron Device Lett.*, 2013, **34**, (2), pp. 157–159

[16] Su C.J., Tsai T.I., Liou Y.L., Lin Z.M., Lin H.C., Chao T.S.: 'Gate-all-around junctionless transistors with heavily doped polysilicon nanowire channels', *Electron Device Lett.*, 2011, **32**, pp. 521–523

[17] International Technology Roadmaps for Semiconductor, ITRS, London, UK, 2008 ed., 2008

[18] Agrawal S., Fossum J.G.: 'On the suitability of a high-k gate dielectric in nanoscale finFET CMOS technology', *IEEE Trans. Electron Devices*, 2008, **55**, (7), pp. 1714–1719

[19] A.U. Manual. Device simulation software. In Silvaco Int., Santa Clara, CA, 2008

[20] Grasser T., Selberherr S.: 'Mixed-mode device simulation', *Microelectron. J.*, 2000, **31**, (11-12), pp. 873–881

[21] Seevinck E., List F.J., Lohstroh J.: 'Static-noise margin analysis of MOS SRAM cells', *IEEE J. Solid-State Circuits*, 1987, **22**, (5), pp. 748–754

[22] Bhavnagarwala A.J., Tang X., Meindl J.D.: 'The impact of intrinsic device fluctuations on CMOS SRAM cell stability', *IEEE J. Solid-State Circuits*, 2001, **36**, (4), pp. 658–665

[23] Li Y., Cheng H.-W., Han M.-H.: 'Statistical simulation of static noise margin variability in static random access memory', *IEEE Trans. Semicond. Manuf.*, 2010, **23**, (4), pp. 509–516

[24] Razavi P., Ferain I., Das S., Yu R., Akhavan N.D., Colinge J.: 'Intrinsic gate delay and energy-delay product in junctionless nanowire transistors'. 2012 13th Int. Conf. Ultimate Integration on Silicon (ULIS), 6–7 March 2012, pp. 125–128

11

CMOS time-to-digital converters for mixed-mode signal processing

Fei Yuan

Department of Electrical and Computer Engineering, Ryerson University, Toronto, ON, Canada
E-mail: fyuan@ryerson.ca

Abstract: This study provides an in-depth review of the principles, architectures and design techniques of CMOS time-to-digital converters (TDCs). The classification of TDCs is introduced. It is followed by the examination of the parameters quantifying the performance of TDCs. Sampling TDCs including direct-counter TDCs, tapped delay-line TDCs, pulse-shrinking delay-line TDCs, cyclic pulse-shrinking TDCs, direct-counter TDCs with interpolation, vernier TDCs, flash TDCs, successive approximation TDCs and pipelined TDCs are studied and their pros and cons are compared. Noise-shaping TDCs that reduce in-band noise below technology limit are investigated. These TDCs include gated ring oscillator TDCs, switched ring oscillator TDCs, relaxation oscillator TDCs, ΔΣ TDCs and MASH TDCs. The performance of sampling and noise-shaping TDCs is compared. The direction of future research on TDCs is explored.

1 Introduction

The advance of CMOS technology has resulted in a sharply increasing time resolution and a rapidly deteriorating voltage resolution. As a result, time-mode circuits where information to be processed is represented by time variables, that is, the difference between the time of the occurrence of two digital events, rather than the nodal voltages or branch currents of electric networks offer a technology-friendly means to combat challenges such as deteriorating voltage accuracies and shrinking dynamic ranges encountered in design of mixed analogue–digital circuits. Time-to-digital converters (TDCs) that map a time variable to a digital code are the most important building blocks of time-mode circuits. Although the deployment of TDCs in particles and high-energy physics for time-of-flight measurement in nuclear science dates back to 1970s [1, 2], their applications in digital storage oscillators [3, 4], laser range finders [5], analogue-to-digital converters (ADCs) [6–8], audio signal processing [9], medical imaging [10], positron emission tomography [10], instrumentation [11], infinite impulse response (IIR) and finite impulse response (FIR) filters [12, 13], anti-imaging filters [14], all digital frequency synthesisers [15–19], multi-Gbps serial links [20], programmable band/channel select filters for software-defined radio [21], laser-scanner-based perception systems [22], to name a few, emerged only recently. Various architectures and design techniques have appeared to improve the performance of TDCs, an in-depth examination of the principles and design techniques of CMOS TDCs, however, is not available. The goal of this review paper is to provide readers with a comprehensive treatment of the principles, architectures and design techniques of TDCs, and a critical assessment of the pros and cons of each class of TDCs. The remainder of this paper is organised as follows: Section 2 provides a loose classification of TDCs. The key performance indicators of TDCs are depicted in Section 3. Section 4 investigates sampling TDCs. Noise-shaping TDCs are dealt with in Section 5. Section 6 explores the direction of future research on TDCs. This paper is concluded in Section 7.

2 Classification of TDCs

TDCs can be loosely classified into sampling TDCs and noise-shaping TDCs. The former digitise a time variable using a high-frequency reference clock or delay lines, whereas the latter suppress the quantisation noise of TDCs using system-level techniques such as ΔΣ modulation and multi-stage-noise-shaping (MASH). Sampling TDCs can be further classified into single-shot TDCs and averaging TDCs. The former digitise a time interval in a single measurement, whereas the latter digitise a time interval using the average of the results of a set of successive measurements to minimise the effect of random error in measurement so as to achieve a better accuracy. It can be shown that the precision of averaging TDCs is inversely proportional to the square root of the averaged results of measurement [23]. A large number of TDCs fall into the category of sampling TDCs. These TDCs include direct-counter TDCs, direct-counter TDCs with interpolation, tapped delay-line TDCs, pulse-shrinking delay-line TDCs, cyclic pulse-shrinking delay-line TDCs, vernier TDCs, pulse-stretching TDCs, flash TDCs, successive approximation TDCs (SA-TDCs) and pipelined TDCs. Sampling TDCs have the common characteristic that they digitise time variables directly in an open-loop manner with no suppression of quantisation noise. For each time variable, there is a corresponding digital code. Thus, a one-to-one mapping between a time input variable and a corresponding output digital code exists. The resolution of these TDCs is lower-bound by quantisation errors. Noise-shaping TDCs, on the other hand, suppress in-band noise. A number of noise-shaping TDCs such as gated ring oscillator (GRO) TDCs, switched ring oscillator (SRO) TDCs, relaxation oscillator (RO) TDCs and MASH TDCs emerged. Although the in-band noise of noise-shaping TDCs is lower than quantisation noise, there is no one-to-one mapping between input time variables and digital output codes.

3 Characterisation of TDCs

The performance of TDCs is characterised by a number of parameters, among them, resolution, precision, linearity, voltage and temperature sensitivities, conversion time and range are the most important [24].

The resolution of a TDC is the minimum time interval that the TDC can quantise ideally. In reality, the imperfections of TDCs such as non-linearity and clock jitter lower the resolution.

The precision of a TDC, often known as single-shot precision, is quantified by the standard deviation of measurement errors Δ_1 and Δ_2 when measuring a constant time interval. When a time variable specified by START and STOP pulses is measured using a reference clock and a counter, as shown in Fig. 1, since the reference clock is asynchronous with START and STOP, single-shot measurement errors Δ_1 and Δ_2 exist and their value is uniformly distributed in [0, T_c]. Their standard deviation thus provides an effective measure of the measurement error in a statistical sense. It can be shown that the precision of averaging TDCs is inversely proportional to the square root of the averaged results of measurement [23].

The non-linearity of a TDC is the deviation of the time-to-digital transfer characteristic of the TDC from that of an ideal TDC. For

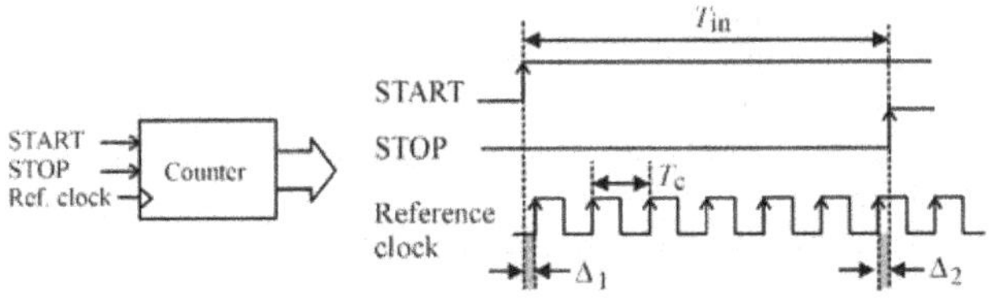

Fig. 1 *Direct-counter TDCs. The counter is enabled by START = 1, synchronised with the reference clock, and disabled by STOP = 1. Input time variable:* $T_{in} = NT_c + \Delta_1 + \Delta_2$ *where N is the number of the cycles of the reference clock in* T_{in}.Δ_1, $\Delta_2 \in [0, T_c]$ *are quantisation errors that are uniformly distributed in* $[0, T_c]$

delay-line TDCs, it is caused by the difference between the delay of delay stages arising from process variation. The non-linearity of the delay-line TDC is usually quantified by differential non-linearity (DNL) quantifying the mismatch of the delay of adjacent delay stages and integral non-linearity (INL) quantifying the accumulative effect of the mismatch of the delay of the delay stages over the entire delay line. Since the resolution of the TDC is measured using least significant bit (LSB), both DNL and INL are typically quantified in LSB. Alternatively, the effect of the non-linearity of the TDC can be depicted using signal-to-noise-plus-distortion ratio (SNDR) over a specific frequency range. SNDR is obtained from the frequency response of the TDC by computing the ratio of the power of the signal to that of the noise and harmonic tones over a specific frequency range. SNDR is widely favoured over signal-to-noise ratio when quantifying the performance of TDCs as the lower bound of the dynamic range of TDCs is often dictated by in-band harmonic tones rather than in-band noise.

The voltage and temperature sensitivities of TDCs quantify the effect of supply voltage fluctuation and temperature variation on the time accuracy of the TDCs with units ps/V and ps/°C, respectively. They are typically obtained by varying the supply voltage by ±10% of its nominal value, temperature over the range −40 ~ 100° C and measuring the resultant change of the time delay. Typical voltage sensitivity is 10–100 ps/V and temperature sensitivity is 0.1–0.01 ps/°C.

The conversion time of a TDC is the amount of the time that the TDC needs to complete the digitisation of a time variable. For a delay-locked loop (DLL)-stabilised delay-line TDC, the conversion time consists of one period of the reference clock, the lock time of the DLL, the propagation of D flip-flop (DFF) samplers and the time of thermometer-to-binary conversion. An alternative measure of the conversion time is throughput measured using the number of the samples that the TDC digitises per second. For applications such as laser range measurement, conversion time is typically not of a concern. However, for applications such as an all digital phase-locked loop (ADPLL), conversion time directly affects the time constant of TDC-based phase-frequency detector subsequently the loop stability of the phase-locked loop (PLL).

The range of a TDC is lower bound by the resolution and upper bound by the maximum time interval that it can digitise. For a tapped delay-line TDC, the lower bound is the time delay of one delay stage τ and the upper bound is given by $N\tau$ where N is the number of the delay stages of the delay line. Since INL deteriorates with the increase in the number of the delay stages, the value of N is set by the maximum allowable INL.

Although power consumption is not an explicit performance indicator of TDCs, it is often a determining factor that affects the architecture, resolution, conversion time and dynamic range of TDCs. Quite often, trade-offs between performance and power consumption are made.

4 Sampling TDCs

Sampling TDCs digitise time variables with no suppression of quantisation noise. The resolution of these TDCs is lower bound by quantisation noise. Noise-shaping TDCs whose in-band noise is lower than quantisation noise are investigated in Section 5.

4.1 Direct counter TDCs

A direct-counter TDC quantises a time variable T_{in} by counting the number of the cycles of a high-frequency reference clock within the duration of T_{in}, as shown in Fig. 1. The counter is started by START, synchronised with the reference clock and stopped by STOP. The random assertion of START and STOP results in quantisation errors Δ_1 and Δ_2 where $\Delta_1, \Delta_2 \in [0, T_c]$ are uniformly distributed in $[0, T_c]$. Direct-counter TDCs feature a large dynamic range lower bound by T_c and upper bound by the size of the counter. In addition, they enjoy a superior linearity as the linearity is only determined by the stability of the frequency of the reference clock [2]. The resolution of direct-counter TDCs can be increased if Δ_1 and Δ_2 are further quantised using interpolation, as to be seen shortly. Since the quantisation errors Δ_1, Δ_2 are caused by the asynchronisation of the reference clock with START and STOP, the quantisation errors can be reduced using delay-line TDCs to be studied in the next section. It should also be noted that when T_{in} is small, the timing jitter of the reference clock cannot be neglected. Both Δ_1 and Δ_2 will be affected by the jitter of the reference clock [25].

4.2 Tapped delay-line TDCs

Tapped delay-line TDCs shown in Fig. 2*a* where each delay stage has the same propagation delay improve the resolution from T_c of direct-counter TDCs to one buffer delay τ [26–28]. The operation of tapped delay-line TDCs is briefly depicted as follows: the START signal propagates through the delay line while the STOP signal enables the D-flipflops to sample the output of the delay stages at the rising edge of STOP. In the example shown in Fig. 2*b*, $D_N \ldots D_2D_1 = 0 \ldots 0111111$. Since $D_N \ldots D_2D_1$ is thermometer-coded, a thermometer-to-binary converter is often needed to convert $D_N \ldots D_2D_1$ to a binary code. The dynamic range of a tapped delay-line TDC is lower-bound by one buffer delay τ and upper-bound by the total time delay $N\tau$ of the delay line. The non-linearity of the delay-line TDC is determined by the mismatch of the delay of the

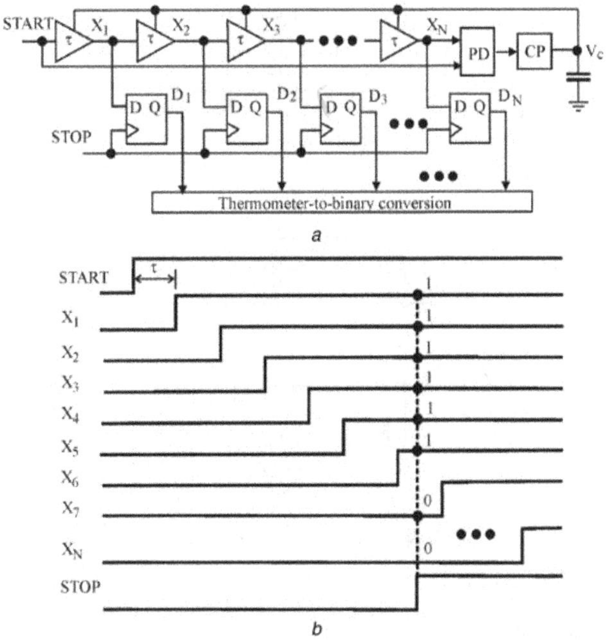

Fig. 2 *Delay-line TDCs [26, 28]*
a Configuration
b Operation. $X_N, \ldots, X_2$, and X_1 are sampled at the rising edge of STOP. The result is given by: $X_N \ldots X_2X_1 = 0 \ldots 0111111$

delay stages rising from the effect of process spread. Since INL of a tapped delay-line TDC deteriorates with the increase in the number of delay stages of the line, delay-line TDCs with a long delay line should be avoided. In addition to non-linearity, the delay of delay lines is also affected by supply voltage fluctuation and temperature variation. To minimise these effects, delay-locked loops used by Rahkonen and Kostamovaara [28] to stabilise the delay of delay stages two decades ago and shown in Fig. 2 are now a standard technique to minimise these effects. Since in this case START and X_N are phase-aligned in the lock state, we have $\tau = T_c/N$ where T_c is the period of START and N is the number of the delay stages. It becomes evident that τ is independent of the delay of the delay stages and is only determined by the period of the input and the number of the delay stages of the delay line in the lock state. TDCs with DLLs locked to a reference also emerged [29]. Since STOP is asynchronous with START, the sampling of the output of delay stages might occur in the meta state of the DFFs, resulting in a long propagation delay subsequently a long TDC conversion time [30].

Tapped delay-line TDCs are also known as flash TDCs because of their identical operation principle. A flash TDC digitises a time variable by comparing the edge of STOP to the time-displaced edge of START, as shown in Fig. 3*a* [31, 32]. In tapped delay-line TDCs, the displaced edges of START are connected to the data node of the arbiters, whereas in delay-line flash TDCs, they are routed to the clock node of the arbiters. The sampling of tapped delay-line TDCs takes place only at the rising edge of STOP, whereas delay-line flash TDCs samples STOP at each rising edge of the output of the delay stages. To improve resolution, vernier flash TDCs shown in Fig. 3*b*, similar to vernier TDCs, can be employed. The resolution of flash TDCs can be improved if the delay line is removed, as shown in Fig. 3*c* [33]. These TDCs operate based on the time offset of the arbiters caused by device mismatches and therefore termed sampling offset TDCs. Since the offset time of the arbiters differs, typically by 2–30 ps [31], these TDCs need to be calibrated prior to their operation [32]. The preceding flash TDCs employ balanced arbiters, that is, arbiters with a zero offset time between their two inputs. Alternatively, a flash TDC can be constructed with unbalanced arbiters with a gradually increased offset time, as shown in Fig. 3*d* [34, 35]. Flash TDCs with parallel delay elements shown in Fig. 3*e* also emerged recently [36, 37].

It was shown that the resolution of tapped delay-line TDCs is one buffer delay. Since the buffer is typically realised using two cascaded static inverters, one buffer delay is thus equal to two gate delays. The resolution of delay-line TDCs can be lowered to one gate delay using a pseudo-differential architecture [38]. To further improve the resolution to sub-buffer delay, interpolation between the rising edges of adjacent delay stages can be utilised. The active interpolation approach proposed in [39] (Fig. 4*a*) uses the weighted sum of the differential output voltages of v_1 and v_2. If $w = 1$, that is, M1 = ON and M2 = OFF, v_o is determined by v_1 only. Similarly, when $w = 0$, M1 = OFF and M2 = ON, v_o is determined by v_2. When both M1 and M2 are ON and their currents are determined by w. The preceding active interpolation consumes static power. Interpolation can also be implemented using passive networks such as resistor networks (Fig. 4*b*) with the drawback of non-negligible static power consumption as the resistance of the resistors needs to be small in order to meet speed requirements [40]. The phase interpolation method proposed in [41] (Fig. 4*c*) uses hierarchical trees to increase time resolution without static power consumption. Another technique to achieve a sub-gate resolution is to use an array of DLLs [42]. The time variable is measured by a primary delay-line TDC. The gate delay of each delay stage of the primary DLL is further measured by a secondary DLL to obtain a sub-gate resolution. This approach, however, is less attractive in terms of power and silicon consumption because of the need for multiple DLLs.

4.3 Pulse-shrinking delay-line TDCs

It was shown earlier that the resolution of tapped delay-line TDCs is one buffer delay unless inter-stage interpolation is used at the cost of additional silicon and power consumption. Rahkonen and

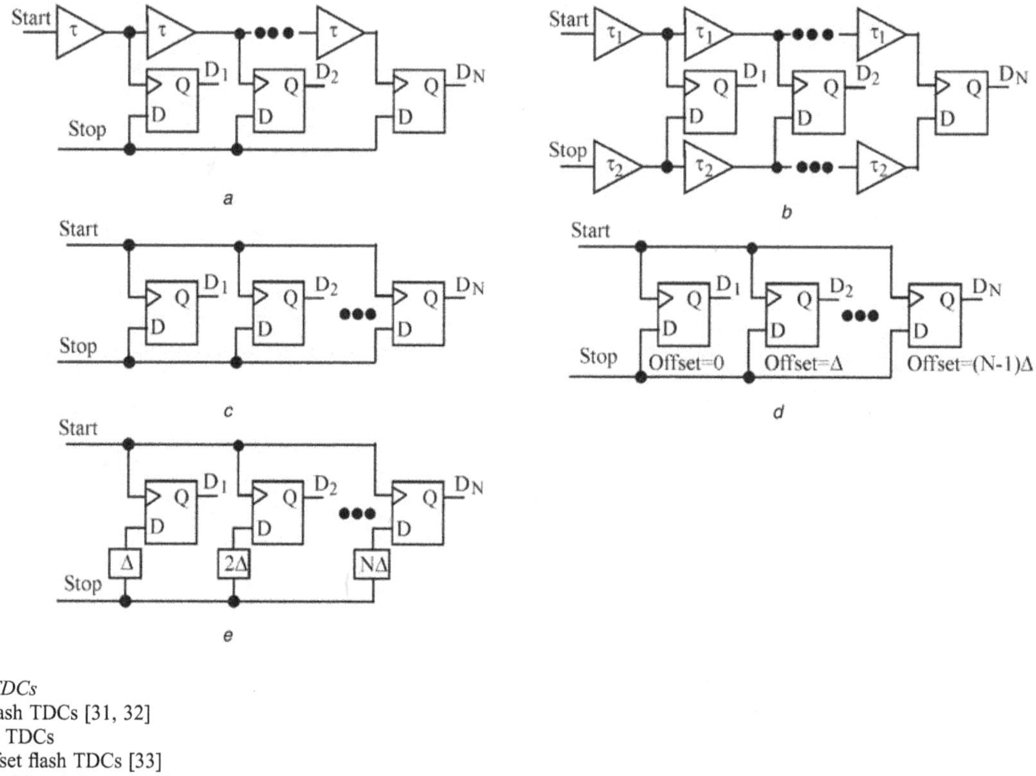

Fig. 3 *Flash TDCs*
a Delay-line flash TDCs [31, 32]
b Vernier flash TDCs
c Sampling offset flash TDCs [33]
d Flash TDCs with unbalanced arbiters [34, 35]
e Flash TDCs with parallel delay elements [36, 37]

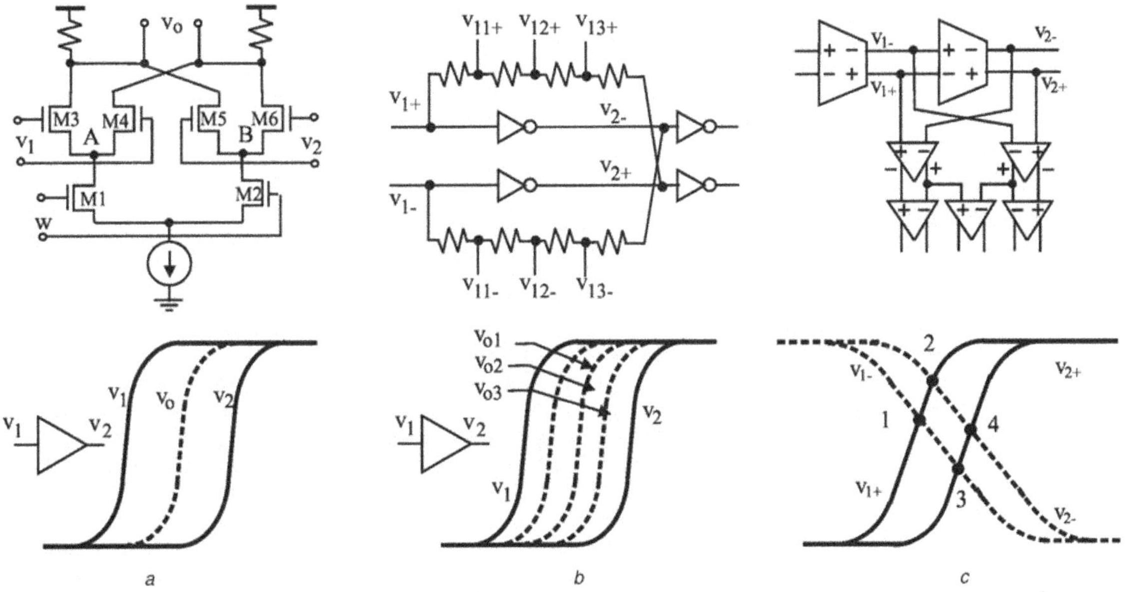

Fig. 4 *Inter-stage interpolation using*
a Weighted sum of the voltages of adjacent stages [39]
b Resistor networks [40]
c Hierarchical trees [41]

Kostamovaara showed that the resolution can be made below one buffer delay without interpolation if the tapped delay line is replaced with a pulse-shrinking delay line where the width of the propagating pulse decreases uniformly across the stages, as shown in Fig. 5 [28, 43]. Note that the outputs of the delay stages are sampled by RS-flipflips, rather than D-flipflops in delay-line TDCs studied earlier. In addition, these RS-flipflops are reset by RESET at the beginning of the measurement. Fig. 5 shows the simplified schematic of a pulse-shrinking delay stage consisting of a current-starving inverter and a generic static inverter. When a pulse of width T_{in} passes through the pulse-shrinking delay stage, the pulse width is reduced because of the controlled slow discharge of the load capacitor. The amount of time shrinkage can be adjusted by varying *J*. The per-stage shrinkage is determined as follows: a

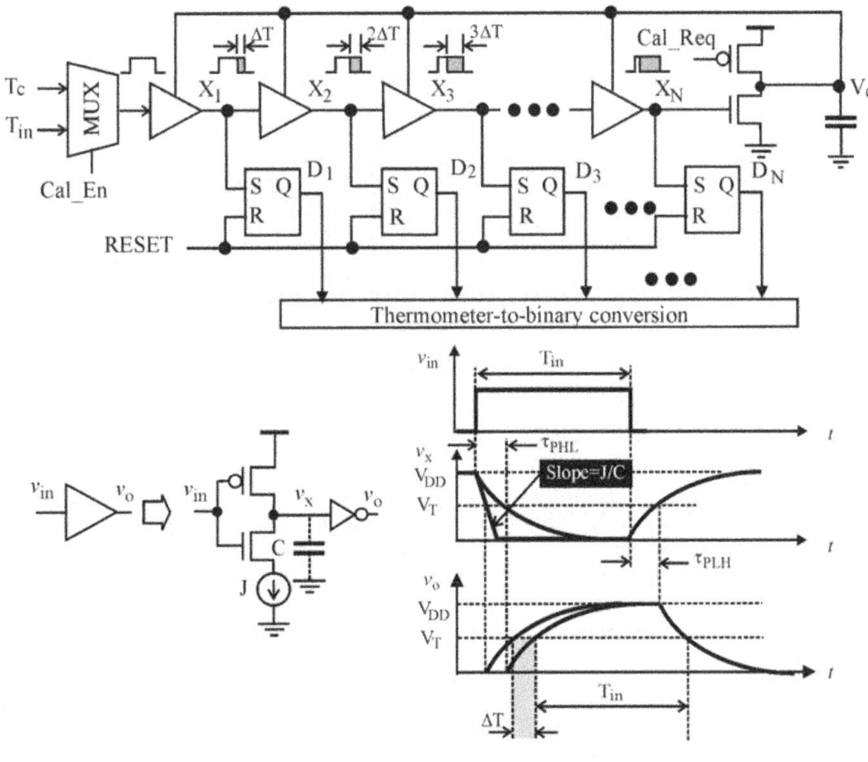

Fig. 5 *Pulse-shrinking delay-line TDCs [28, 43]*
Pulse width is reduced by ΔT in each stage. To calibrate the TDC, Cal_{Req} is asserted and the capacitor is fully charged. Calibration clock of known width T_c is routed to the delay line by the MUX on the arrival of Cal_{En}. If the width of the pulse X_n is not zero, the capacitor will be discharged via the NMOS transistor. This process ends when the width of X_n becomes zero

pulse of known pulse width T_c is applied to the TDC. The starving current J is tuned by a DLL in such a way that the pulse just vanishes when it reaches the output of the last delay stage. We therefore have $T_c = N\Delta T$ where ΔT is the shrinkage of the pulse width per stage and N is the number of the delay stages. If N is sufficiently large, ΔT can be made adequately small. Since the RS-flipflops will be triggered as long as the pulse width of X_j where $j = 1, 2, \ldots, N$ is not zero and the amount of the pulse-shrinkage per stage is ΔT, the resolution of the calibrated pulse-shrinking delay-line TDC is given by ΔT, much smaller than the average propagation delay $\tau = (1/2)(\tau_{PHL} + \tau_{PLH})$. The calibrated pulse-shrinking delay-line TDC can then be used to digitise an input time variable in a similar way as that of conventional delay-line TDCs with the digitised output from the DFF samplers. It should be noted that since a large N is needed for a better resolution, the mismatch between the delays of the pulse-shrinking delay stages will affect the non-linearity of the TDCs in a similar way as that of delay-line TDCs.

4.4 Cyclic pulse-shrinking TDCs

Cyclic pulse-shrinking TDCs can be considered as a special class of direct-counter TDCs and pulse-shrinking delay-line DTCs. They use skewed delay stages, that is, delay stages that have a larger propagation delay as compared with the delay of the remaining stages of a delay line, as shown in Fig. 6, to reduce the width of the propagating pulse [44–46]. In a cyclic pulse-shrinking delay-line TDC, a counter is used to record the number of the round trips that an input pulse T_{in} makes before it vanishes. The content of the counter when the pulse vanishes provides the digital code of the width of the input time variable. Since the amount of cycle-to-cycle time shrinkage remains unchanged, these TDCs exhibit a perfect linearity. One drawback of the cyclic pulse-shrinking delay-line TDC is that an input pulse can be applied only after the previous one diminishes completely [40]. Since only one skewed delay stage was used in the design in [44], the rate of pulse width shrinkage per around trip is rather small, resulting in a long conversion time. The effect of temperature on the delay could be as high as ±25% over 0–100°C and the shrinking pulse width prevents techniques such as DLLs to be used to minimise PVT effect [4], thermal compensation that minimises the effect of temperature on the delay is needed [5]. Pulse-shrinking delay-line TDCs is effective only if pulse width is sufficiently large as compared with pulse-shrinkage per around trip.

4.5 Direct counter TDCs with interpolation

The resolution of direct-counter TDCs can be increased if Δ_1 and Δ_2 are further quantised using interpolation. In this section, we examine interpolation techniques for digitising Δ_1 and Δ_2.

1) Direct-counter TDCs with tapped delay-line interpolation: It was shown earlier that the resolution of direct-counter TDCs is lower bound by quantisation errors Δ_1 and Δ_2 distributed uniformly over $[0, T_c]$. To increase the resolution of direct-counter TDCs, one needs to further digitise Δ_1 and Δ_2. This is known as interpolation. Δ_1 and Δ_2 can be digitised directly using tapped delay-line TDCs to improve the resolution from T_c of direct-counter TDCs to one buffer delay. To improve the resolution of interpolation, long delay lines are needed. The length of tapped delay-line TDCs, however, is upper bound by INL, which deteriorates with the increase in the number of delay stages. The achievable resolution of direct-counter TDCs with tapped delay-line interpolation is therefore rather limited.

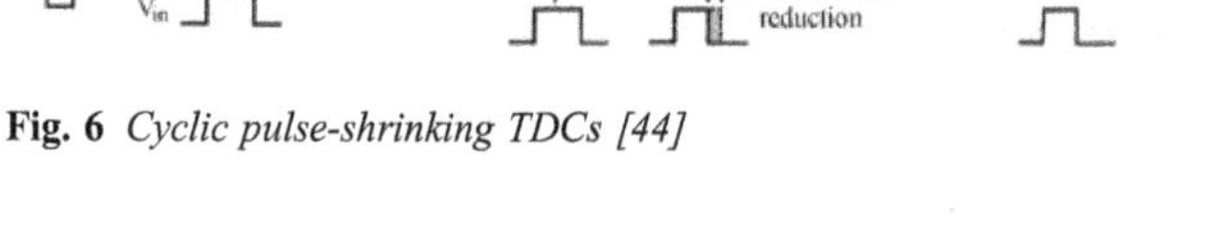

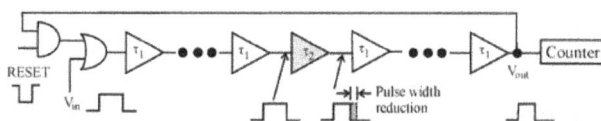

Fig. 6 *Cyclic pulse-shrinking TDCs [44]*

2) Direct-counter TDCs with 2-step tapped delay-line interpolation: The resolution of tapped delay-line interpolation can be improved using two-step interpolation without employing a long delay line. In the approach proposed in [47], Δ_1 and Δ_2 are first interpolated using two DLL-stabilised tapped delay lines of M stages that give resolution T_c/M where T_c is the period of the clock. Each two consecutive stages of the tapped delay lines are then interpolated using N-tap parallel interpolators to achieve a total of MN interpolation steps per clock period. The resolution is now $T_c/(MN)$ instead of T_c/M of one-step interpolation. In [30], the first-level interpolation is performed using a multi-phase sampling, whereas the second-level interpolation is carried out using vernier delay lines to achieve 24 ps resolution with a 160 MHz reference clock. A similar approach was used in [48].

3) Direct-counter TDCs with time-stretching and interpolation: The asynchronisation of START and STOP with the reference clock might result in small Δ_1 or Δ_2. In this case, tapped delay-line interpolation techniques will yield a poor result. To overcome this drawback, Δ_1 and Δ_2 can be first stretched using a time stretcher and then digitised. Both analogue interpolations that uses a time-to-amplitude converter (TAC) to convert Δ_1 and Δ_2 to a large voltage variation and then digitise the voltage variation using ADCs [23, 24, 49–51] and digital interpolation that first stretches Δ_1 and Δ_2 to $K\Delta_1$ and $K\Delta_2$ with $K \gg 1$ using the dual-slope approach of Nutt [52] and then digitises $K\Delta_1$ and $K\Delta_2$ using tapped delay-line TDCs [53, 54] exit. TAC is realised by discharging a pre-charged capacitor with a constant current from the start of T_{in} to the end of T_{in}. The voltage drop of the capacitor at the end of T_{in} is directly proportional to T_{in} and is digitised using an ADC, typically a flash ADC in order to meet time constraints [10, 50, 54, 55]. As pointed out in [50], analogue time stretching provides a good single-shot precision of a few picoseconds but suffers from poor stability typically 10–30 ps/C. This method also suffers from high power consumption because of the use of flash ADCs. The following methods have been proposed for pulse stretching:

1. *Charge-pump pulse stretching:* Raisanen-Ruotsalainen *et al.* showed that the quantisation errors Δ_1 and Δ_2 can be first stretched and then digitised, yielding a reduced quantisation error, as shown in Fig. 7 [55]. This approach was later used by many other researchers [4, 56, 57]. Pulses T_1 and T_2 are generated at the rising edge of START and STOP, respectively, with their falling edges aligned with the next rising edge of the reference clock. Pulse stretching starts with the assertion of the reset (RST) command that pre-charges C_1 and C_2 to V_{DD}. The discharge of C_1 and C_2 is controlled by J_1 and J_2, respectively. Since $J_1 = NJ_2$ and $C_2 = MC_1$ with $M, N > 1$, the discharge of C_1 is faster than that of C_2. Discharge is initiated by T_1. v_o is set to HIGH and will remain HIGH until $v_{c2} = v_{c1}$. Since v_{c2} drops slower, it will take k cycles of CLK to establish $v_{c1} = v_{c2}$. The number of the cycles is recorded by the counter. The content of the counter provides the digital code of the quantisation error Δ_1. The same process is followed when quantising Δ_2. To determine k, from $\Delta v_{c1} = \Delta v_{c2}$ where Δv_{c1} and Δv_{c2} are the voltage drop of C_1 and C_2 from V_{DD}, respectively, and noting $\Delta v_{c1} = (J_1/C_1)T_1$ and $\Delta v_{c2} = (J_2/C_2)T_c k$, we arrive at $k = MN(T_1/T_c)$ or equivalent $\hat{T}_1 = kT_c = MNT_1$ where $\hat{T}_1$ is the stretched version of T_1. It is evident that T_1 is stretched by MN times. A notable advantage of the dual-slope pulse stretching approach is the reduced effect of power, voltage and temperature (PVT). This approach suffers from a speed penalty because of the slow discharge of C_2. The need for a voltage comparator and two constant current sources also undermines its compatibility with technology scaling.
2. *Regeneration pulse stretching:* Abas *et al.* utilised the regenerative mechanism of SR-latches to stretch narrow pulses, as

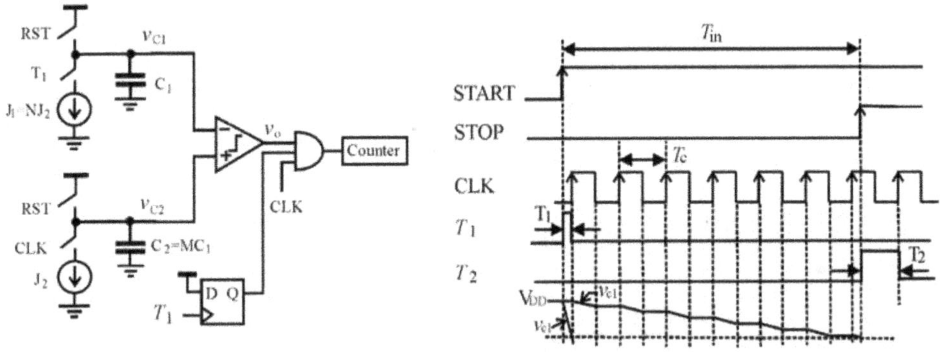

Fig. 7 *Charge pump time amplifier [4, 56, 57]*

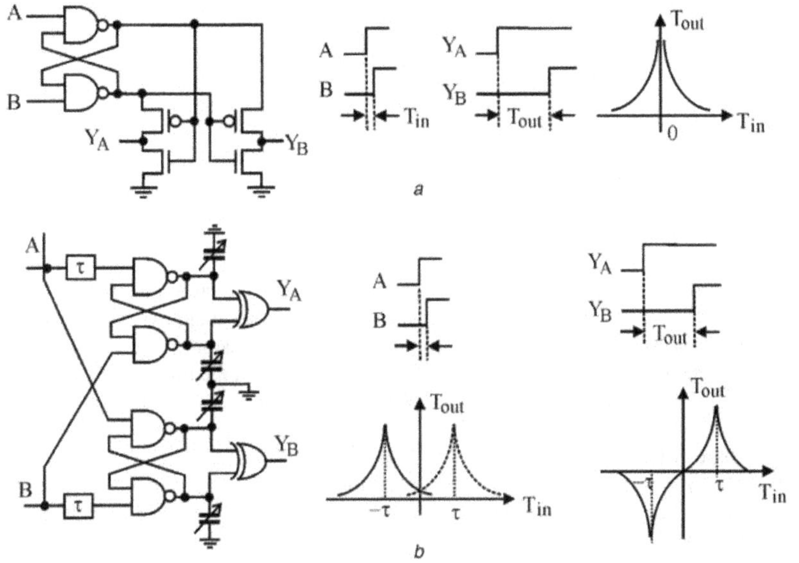

Fig. 8 *Regeneration pulse stretching*
a RS-latch time amplifier [58–61]
b RS-latch time amplifier with an improved input range [62]

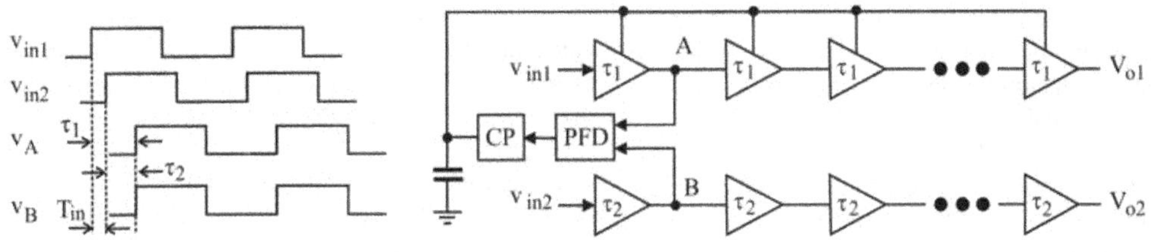

Fig. 9 *DLL-based time amplifier [64, 65]*

shown in Fig. 8*a* [58–61]. A key advantage of this approach is its fast response and ability to amplify a small time variable. Since the gain of the amplifiers is set by the characteristics of the latch, it is strongly subject to the effect of PVT. Other drawbacks include a small input range and the poor linearity of the gain. Lee and Abidi [62] improved the input range by inserting two delay units to generate an unbalanced re-generation mechanism, as shown in Fig. 8*b*. A drawback of Lee–Abidi time amplifier is that the delay mismatch of the buffers might be significant if the input time difference to be amplified is small, resulting in a non-negligible error. This drawback can be removed by using unbalanced active charge pump loads proposed in [63].

3. *Delay-lock loop pulse stretching:* The DLL-based time amplifier proposed by Rashidzadeh *et al.* and shown in Fig. 9 uses a closed-loop approach to amplify time while minimising the effect of PVT [64, 65]. Two inputs are fed to two delay lines of the same number of delay stages, but different stage delays. The waveforms at nodes *A* and *B* are phase-aligned. Since $\phi_A = \phi_{in1} + 2\pi(\tau_1/T)$ and $\phi_B = \phi_{in1} + 2\pi(\tau_2/T)$, we have from $\phi_A = \phi_B$ that $\phi_{in1} - \phi_{in2} = 2\pi/T(\tau_1 - \tau_2)$. Since the overall propagation delay of delay line 1 is $N\tau_1$ while that of delay line 2 is $N\tau_2$, we have $T_{out} = (N-1)(\tau_1 - \tau_2)$. The DLL must be locked in order to provide precision amplification, limiting the use of this technique for high-speed applications. In addition, although the rest of the delay cells are

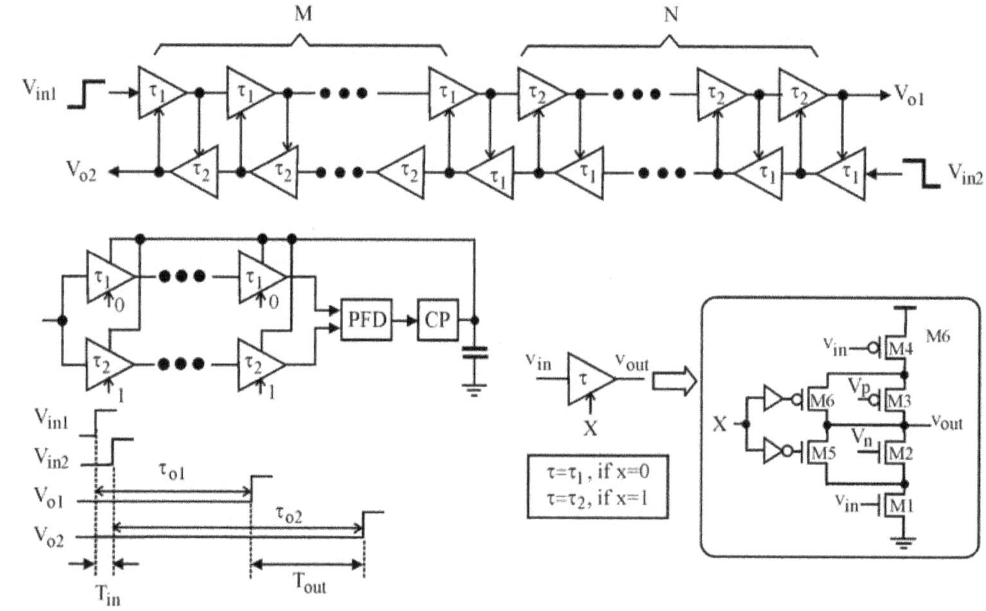

Fig. 10 *Time amplifier proposed by Nakura et al. [66–68]*
Top: configuration. Bottom left: wave forms. Middle left: DLL to precisely control the ratio of τ_2 to τ_1. The number of the delay cells in the two paths of the ratio control DLL is set such that the delay of the two delay lines is the same

the replicas of the first delay line, PVT does affect the delay of the delay cells.

4. *Nakura pulse stretching*: In [66–68], a closed-loop time amplification scheme shown in Fig. 10 was proposed. The two pulses whose time difference is to be amplified propagate in the opposite directions in two separate delay lines having the same number of delay cells. The delay of the delay cells can be toggled between τ_1 and τ_2 by control signal X, specifically, $\tau = \tau_2$ if $X = 1$ (M5 and M6 are OFF and the delay τ_2 is set by control voltages V_p and V_n) $\tau = \tau_1$ if $X = 0$ (M5 and M6 are ON) and with $\tau_2 = n\tau_1$ where n is an integer. The ratio of τ_1 to τ_2 is controlled by the DLL to minimise the effect of PVT. Before v_{in1} and v_{in2} meet, the delay of the cells ahead of v_{in1} and v_{in2} is τ_1. When v_{in1} and v_{in2} collide, the delay of the cells ahead of v_{in1} and v_{in2} becomes τ_2. It can be shown that $\tau_{o1} = M\tau_1 + N\tau_2$ and $\tau_{o2} = N\tau_1 + M\tau_2$. As a result, $T_{out} = \tau_{o1} - \tau_{o2} = (M - N)\tau_2 + (M - N)\tau_1$. The location where v_{in1} and v_{in2} meet is clearly determined by T_{in} and τ_1, specifically, $T_{in} = (M - N)\tau_1$. The amplifier completes time amplification when v_{o1} and v_{o2} reach the end of the delay lines. The minimum time interval between two time amplifications is given by $M\tau_1 + N\tau_2 + T_{in}$ or $N\tau_1 + M\tau_2$. In addition, the time gain is directly proportional to $\tau_2 - \tau_1$ and the length of the delay lines.

5. *Dual-slope pulse stretching:* Lee *et al.* [69] showed that a large time gain can be obtained by cascading multiple time amplifiers of gain of 2, as shown in Fig. 11. The capacitors at nodes 1 and 2 are pre-charged prior to the arrival of inputs A and B. When A arrives while $B = 0$, C_1 will be discharged by two pull-down paths, one provided by M1 and the other provided by M3 while no charge of C_2 is retained. When B arrives, since M4 is switched off because of the

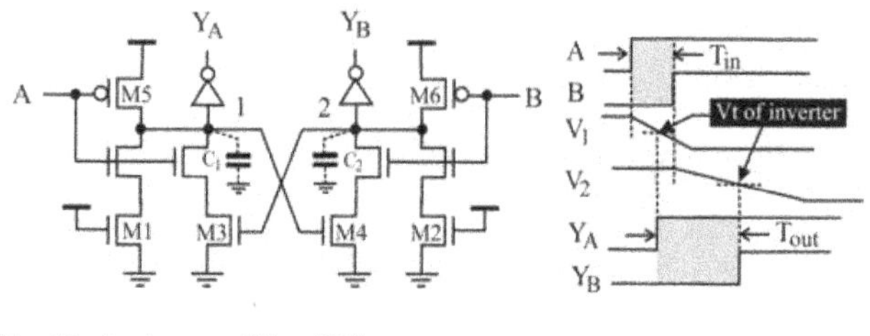

Fig. 11 *2x-time amplifier [69]*

drop of V_1, only one pull-down path provided by M2 exists to discharge C_2. V_2 drops at approximately half the rate of that of V_1 provided that all transistors have the same dimensions, as shown in Fig. 11. The gain of the time amplifier is approximately 2 if all transistors have the same dimensions. Since the proper operation of the time amplifier requires that V_1 drops below the threshold voltage of M4 so that M4 can be switched off prior to the arrival of B, the application of this amplifier for small T_{in} is rather difficult simply because C_1 cannot be drained completely. In addition, the gain of the amplifier is subject to PVT effect.

4.6 Vernier TDCs

TDCs with a sub-gate resolution can also be obtained using vernier delay lines where START and STOP propagate in two separate delay lines of the same number of delay stages but different per-stage delays, as shown in Fig. 12 [70–72]. Since $\tau_1 > \tau_2$, STOP signal in STOP-line will catch START signal in START-line provided that the lines are sufficiently long and $\Delta\tau = \tau_1 - \tau_2$ is not overly large. The time at which a catch-up takes place is determined from $T_{catch} = N\tau_1 = N\tau_2 + T_{in}$. For a given T_{in}, $N = \Delta T/(\tau_1 - \tau_2)$. Clearly, the dynamic range of vernier TDCs is upper-bound by the length of the lines and lower-bound by $\tau_1 - \tau_2$. To minimise the effect of PVT on the delay, DLL-stabilised vernier TDCs shown in Fig. 12 can be deployed.

To increase the dynamic range of vernier TDCs, two-level vernier TDCs consisting of a coarse vernier line and a fine vernier line with the former having a large delay difference $\Delta\tau_c$ and the latter having a small delay $\Delta\tau_f$ with $\tau_f \ll \tau_c$ can be utilised [73]. Alternatively, cyclic vernier TDCs shown in Fig. 13 can be deployed [29]. Two delay lines having N_1 and N_2 stages are employed with their delays τ_1 and τ_2 set by the two DLLs: $\tau_1 = 1/(M_1 N_1 f_o)$ and $\tau_2 = 1/(M_2 N_2 f_o)$, respectively, where f_o is the frequency of the reference clock and M_1 and N_1 are integers. START and STOP are fed to two cyclic loops; each consists of a delay stage and a NAND2 gate. Note that the delay of START cyclic loop is τ_1 and that for STOP is τ_2. The loop delays are $\tau_1 + \tau_{AND2}$ and $\tau_2 + \tau_{AND2}$. The loops are enabled on the arrival of START and STOP, and Y_f samples X_f at each cycle. A phase coincidence is detected when the sampled value changes from 1 to

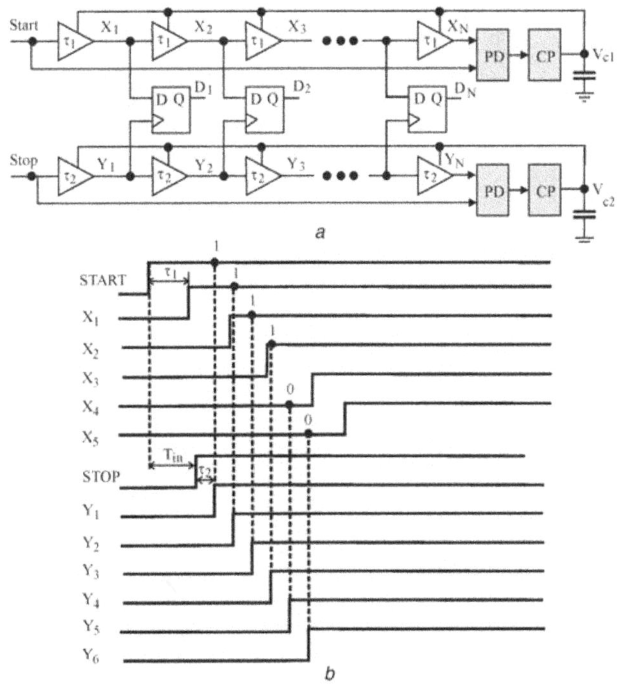

Fig. 12 *Vernier TDCs [71, 72]*
a Configuration
b Operation. *Xk* is sampled by *Yk*, $k = 1, 2, \ldots, N$. In this example, the output is given by $D_N \ldots D_2D_1 = 0 \ldots 01111$

0. Once this occurs, Y_f catches X_f. Clearly the cyclic operation of the two delay loops removes the need for two long delay lines. In addition, although τ_{NAND2} is comparable with τ_1 and τ_2, since we are only concerned with the difference of the delay of START and STOP loops, τ_{NAND2} plays no role in the accuracy of the TDC.

4.7 Successive approximation TDCs

Successive approximation is an effective means to perform analogue-to-digital conversion with low power consumption. SA-TDCs typically consists of a time comparator, a successive approximation register (SAR) and a digital-to-time converter (DTC), as shown in Fig. 14*a* [74]. The DTC maps a digital code D_{out} to a pulse of width T_f with the pulse width proportional to the value of D_{out} ideally. The widths of T_{in} and that of T_f are then compared by the time comparator that compares with the widths of T_{in} and T_f. When their difference ΔT is sufficiently small, time-to-digital conversion is completed and the output of the TDC is given by the SAR. The key component of SA-TDCs is the DTC that can be implemented using either a capacitor array (Fig. 14*b*) [74–77] or a delay line (Fig. 14*c*) [78, 79]. In delay-line DTCs, the input code selects the location of the output using a multiplexer, whereas in capacitor-array DTCs, the load capacitor of the current-starving inverter is digitally tuned by the output of the SAR. Delay-line DTCs enjoy a good linearity and low power consumption, but suffer from a small range. Capacitor-array DTCs, on the other hand, are silicon and power greedy especially when the number of bits is large.

4.8 Pipelined TDCs

Pipelined TDCs that provide a better throughput have also been proposed [80–82]. As compared with other TDCs, pipelined TDCs typically require more hardware simply because of the pipelined operation of these TDCs. Since in applications such as bio-electronics where power consumption is of a critical concern or wireless communications where a stringent constraint is typically imposed on SNDR, pipelined TDCs are less attractive for these applications.

4.9 Performance comparison of sampling TDCs

Table 1 tabulates the key performance indicators of sampling TDCs. It is seen that direct-counter TDCs enjoy a large dynamic range, but suffer from high quantisation noise and a low conversion speed. Direct-counter TDCs with interpolation have a much better resolution as compared with direct-counter TDCs, but suffer from a low conversion speed and high power consumption. Delay-line TDCs provide a better resolution and enjoy low power consumption, but suffer from a small dynamic range. Pulse-stretching TDCs enjoy low power consumption, but suffer from a long conversion. The same with pulse-shrinking TDCs. Vernier TDCs offer a good resolution at the cost of higher silicon and power consumption. Cyclic vernier TDCs removes the drawbacks of conventional vernier TDCs. Pipelined TDCs offer a superior resolution, a high conversion rate, low power consumption. Their range is rather small. Flash TDCs exhibit comparable performance as delay-line TDCs do. Sampling offset flash TDCs provide the best resolution among all flash TDCs. SA-TDCs offer a good resolution and low power consumption. As the choice of TDCs is largely dictated by applications, the familiarisation of the advantages and limitations of sampling TDCs will enable designers to make a better informed decision when choosing TDCs for a specific application.

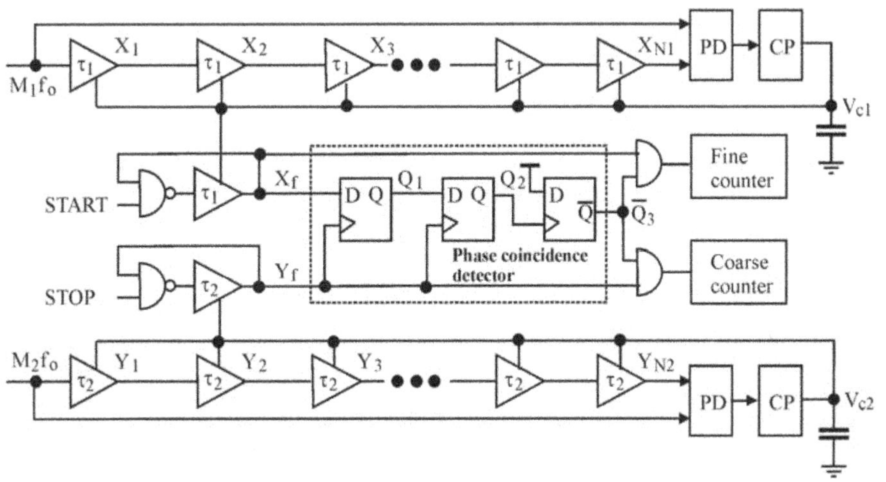

Fig. 13 *Cyclic vernier TDCs [29]*

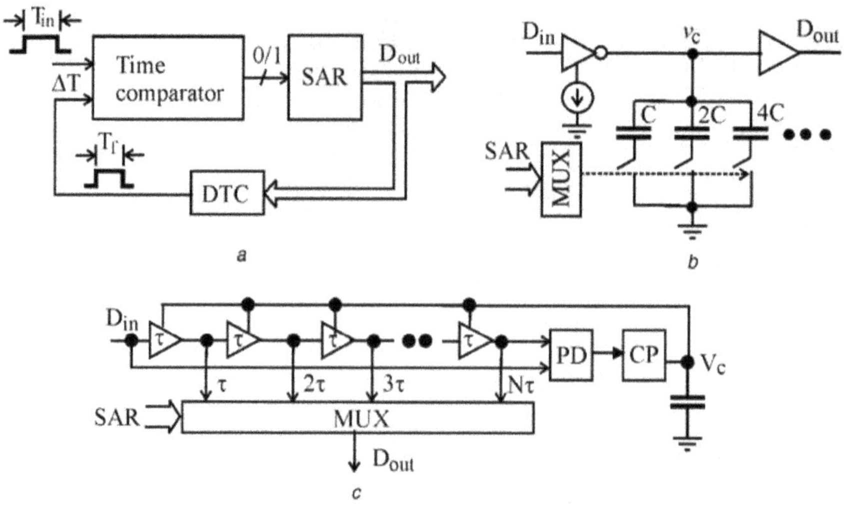

Fig. 14 *Successive approximation TDCs*
a Configuration of SA-TDCs [74]
b Capacitor-array DTCs [74–77]
c Delay-line DTCs [78, 79]

Table 1 Performance comparison of sampling TDCs

References	Type	Tech., μm	Res., ps	Prec., ps	DNL (LSB)	INL (LSB)	Ranges	Rates, MS/s	Power, mW
Ref.	DC3	0.5	312.5	97.5	±0.20	±0.30	80 ns	100	175
	DC4	0.8	31.25	30	–	±0.17	2.5 μs	100	350
Swann *et al.* [10]	DC5	0.8	92	41	±05	±0.6	2 μs	85	100
Raisanen-Ruotsalainen *et al.* [55]	DC5	0.35	30	–	±0.55	−1.5 ~ 1	–	160	50
Mantyniemi *et al.* [47]	DL1	90 nm	21	–	±0.7	±0.7	816 ns	26	6.9
Hwang *et al.* [30]	DL2	1.2	500	–	±0.5	±0.5	10 ns	80	–
Staszewski *et al.* [38]	DL2	1.2	780	3 ns	±0.5	±0.5	10 μs	20	15
Rahkonen and Kostamovaara [28]	DL4	90 nm	4.7	3.3	±0.6	±1.2	–	–	14
Raisanen-Ruotsalainen *et al.* [43]	RR-DL	0.35	12.2	8.1	–	±20 ps	204 μs	5	40
Henzler *et al.* [40]	RR-DL	0.35	8.88	8.6	−0.9 ~ 0.68	−2.8 ~ 0.56	74 μs	20	85
Jansson *et al.* [83]	PST	0.35	50	–	–	±1.1	250 ns	0.1	0.75
Jansson *et al.* [22]	PSH	0.35	68	<54.4	–	–	–	0.1	1.2
Chen *et al.* [4]	PL	0.13	0.63	–	±0.5	±2	1.3 ns	65	10.5
Chen *et al.* [44]	PL	65 nm	1.12	–	±0.6	±1.7	0.573 ns	250	15.4
Seo *et al.* [84]	PL	0.13	1.76	–	±0.6	±1.9	1.8 ns	300	115
Kim *et al.* [82]	*V*1	0.7	30	45	–	–	0.25 ns	125	–
Kim *et al.* [21]	*V*1	0.35	37.5	–	±0.2	±0.35	–	0.1	150
Dudek *et al.* [72]	*V*3	0.18	14.6	–	−1.2	±1.5	50 ns	2	6.4
Chen *et al.* [85]	SA	0.35	1.22	3.2	–	–	327 μs	100	33
Xing *et al.* [29]	CYC	0.13	1.25	–	±0.7	−3 ~ 1	±0.16 ns	100	4.3
Mantyniemi *et al.* [74]	TA	65 nm	3.75	–	±0.9	±2.3	0.48 ns	200	3.6

DL1: tapped delay-line TDC
DL2: pulse-shrinking delay-line TDC
DL3: cyclic pulse-shrinking delay-line TDC
DL4: tapped delay line with resistor interpolation
DL5: array of tapped delay lines
DC1: direct-counter TDC
DC2: direct-counter TDC with delay-line interpolation
DC3: direct-counter TDC with analogue interpolation
DC4: direct-counter TDC with pulse-stretching interpolation
DC5: direct-counter TDC with two-level interpolation
GR: gated ring VCO TDCs
PL: pipelined TDC
PST: pulse-stretching TDC
PSH: pulse-shrinking TDC
RR-DL: reference recycling delay-line TDC
SA: SA-TDC
TA: pulse-train time amplifier TDC
*V*1: vernier TDC
*V*2: two-level vernier TDC
*V*3: cyclic vernier TDC

5 Noise-shaping TDCs

It was shown that sampling TDCs quantise time variables directly with their resolution lower bound by the quantisation noise that is distributed uniformly over the entire spectrum of the TDCs. Increasing sampling frequency lowers the quantisation noise uniformly over the entire spectrum rather than a specific frequency range of the TDCs. As a result, high power consumption is inevitable. Since in many applications such as ADPLLs where TDCs function as a phase detector, we are only interested in the performance of TDCs over the loop bandwidth of the ADPLLs, system-level approaches such as noise-shaping obtained from ΔΣ operations can be utilised to reduce the quantisation noise of TDCs well below that of sampling TDCs over a specific frequency range [86]. These TDCs are termed noise-shaping TDCs [6, 87–90]. Recent advance in TDCs utilises the intrinsic advantages of both resolution-enhancing techniques such as interpolation in sampling TDCs and frequency-dependent noise suppressing techniques such as ΔΣ operation of noise-shaping TDCs simultaneously to improve the resolution of TDCs [91–93]. In this section, we investigate noise-shaping TDCs.

5.1 *Gated ring oscillator TDCs*

TDCs shown in Fig. 15 digitises time variables using a GRO [6]. The operation of GRO is briefly depicted as follows : At the assertion of START, gating signal *EN* is set to logic-1 and the ring oscillator is activated. The number of the cycles of the oscillator within the duration of START = 1 is recorded by the counter. When STOP is asserted, *EN* is reset and oscillation is terminated. The phase of the oscillator remains unchanged during this period of time. Since the frequency of the oscillator is constant during T_{in}, the number of the oscillation cycles of the oscillator during T_{in} is proportional to T_{in} and provides the digital representation of T_{in}. Furthermore, at the end of $(k-1)$th sampling period, the phase of the oscillator is held unchanged during $T_{in}=0$ (neglecting leakage and disturbances coupled to the output nodes of the oscillator), allowing the residual phase of $(k-1)$th sampling period, denoted by $e_f(k-1)$, to be carried over to kth sampling period and become the initial phase of kth sampling period, denoted by $e_i(k)$, that is, $e_i(k)=e_f(k-1)$. The net phase accumulation in kth sampling period is thereby given by $\varphi(k)=K_{vco}v_{in}+[e_f(k)-e_i(k)]$ where K_{vco} is the voltage-to-phase gain of the oscillator. If we let $e(k)$ be the residue phase (quantisation noise) in phase k, that is, $e(k)=e_f(k)$, we have $e_i(k)=e_f(k-1)=e(k-1)$. As a result

$$\phi(k) = K_{vco}v_{in} + [e(k) - e(k-1)] \quad (1)$$

The first-order noise-shaping of quantisation noise is evident from (1). It can be shown that $0 \le e(k) \le 2\pi$. To reduce the quantisation noise 2π to $2\pi/N$ where N is the number of the stages of the oscillator, the transition of the output of each delay stage can be utilised, as shown in Fig. 15 [6]. To avoid the countering error caused by sampling the outputs of the oscillator at transitions, cross-coupled buffers shown in Fig. 15 are typically used at the output of the delay stages.

Since the output nodes of the oscillator during the OFF-state ($T_{in}=0$) are floating, noise-shaping that originates from the continuity of the phase of the oscillator during the OFF-state of the oscillator is sensitive to the leakage of the pn-junctions at the output nodes and charge-sharing during the turn-off of the gating transistors [87]. Although the effect of charge sharing can be mitigated if the gating transistors are small in dimension, this is at the cost of limiting the oscillation frequency [88]. The floating nature of the holding states also makes the phase of the oscillator vulnerable to disturbances such as switching noise and cross-talks [89]. GRO-TDCs also suffer from count-missing caused by the premature reset of the counter by STOP while edge-detection and state-to-phase logic is still in action [89]. This problem can be mitigated by inserting a delay block between STOP and the counter, as shown in Fig. 15. The power consumption of GRO-TDCs can be reduced if asynchronous counters are used, as demonstrated in [89].

5.2 *Switched ring oscillator TDCs*

It was shown in the preceding section that GRO-TDCs are sensitive to charge leakage, charge injection, and switching noise. To eliminate this drawback, Konishi *et al.* showed that freezing the state of the oscillator in the off-state can be replaced by introducing another oscillation state of the oscillator. The oscillator oscillates between two different oscillation states and becomes a switched ring oscillator (SRO), as shown in Fig. 16 [87, 90]. Since the charge of the capacitors at the output nodes of the oscillator at the end of $(k-1)$th phase is carried over in its entirety to the next kth phase, first-order noise-shaping characteristic intrinsic to GRO is preserved in SRO. Unlike GRO, since the frequency of the oscillator in each of the two oscillation states is well defined, there is no floating node. As a result, the issues associated with floating nodes such as leakage, switching noise, and charge injection encountered in GRO-TDCs vanish.

5.3 *Relaxation oscillator TDCs*

Cao *et al.* [95–97] showed that the first-order noise-shaping characteristic of GRO-TDCs is also possessed by RO TDCs, as shown in

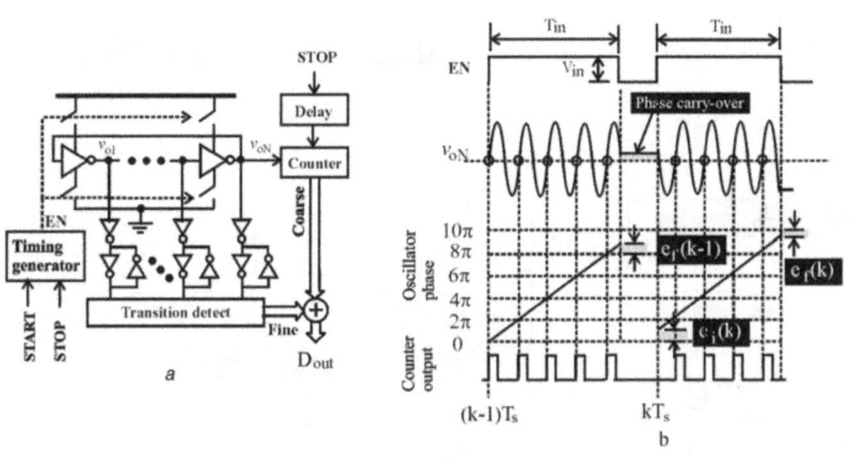

Fig. 15 *Gated ring oscillator TDCs*
a GRO TDCs [6, 94]
b Waveforms

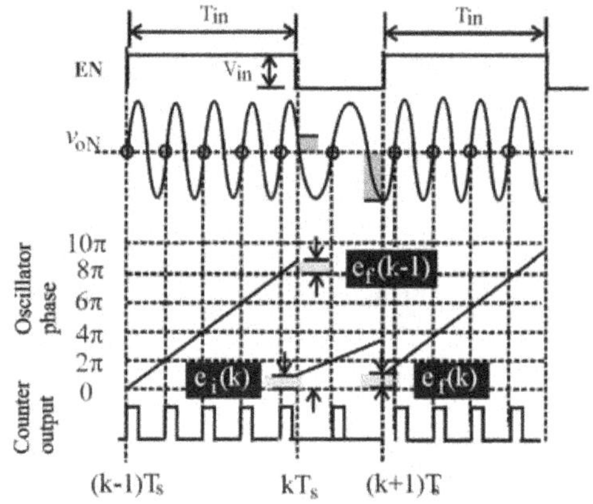

Fig. 16 *Switched ring oscillator TDCs [87, 90]*

Fig. 17. To illustrate this, assume C_1 is being charged while C_2 is discharged at t_1. v_{c1} rises linearly with time while $v_{c2}=0$. At t_2 where $v_{c1}=V_H$, $Q=1$ and $\overline{Q}=0$. C_1 starts to discharge and $v_{c1}=0$ when fully discharged while C_2 starts to be charged. When $v_{c2}=V_H$, $Q=0$ and $\overline{Q}=1$. This process repeats. Since the charge of the capacitors is held unchanged when $T_{in}=0$, the residual phase of the oscillator in $(k-1)$ sampling period becomes the initial phase of kth sampling period, yielding the first-order noise-shaping. Since a distinct characteristic of ROs is their low sensitivity to PVT effect [98], RO TDCs open a door for the realisation of ultra-low-power TDCs for applications such as passive wireless microsystems [99].

5.4 *MASH TDCs*

MASH is an effective means to obtain high-order noise-shaping without sacrificing stability [86]. Fig. 18 shows the 1-1 MASH-TDC proposed by Cao *et al.* [97]. T_{in2} is generated with its START the first rising edge of the counter clock Q_1 and its STOP the same as that of T_{in1}. This quantisation error propagation method was introduced by Konishi *et al.* in [100] and further developed in their subsequent work [88, 101]. Quantisation error propagation is realised using a resettable DFF shown in Fig. 18. Routing v_{o1} to the clock input of the DFF ensures that the DFF is triggered at the first rising edge of v_{o1} while connecting $\overline{T}_{in1}$ ensures that T_{in2} will have the same falling edge as T_{in1} does. To minimise the effect of the metastability of DFFs, two DFFs can be cascaded, as shown in Fig. 18 [101]. Cascading three DFFs was also used to further reduce quantisation noise [88].

5.5 *ΔΣ TDCs*

Although ΔΣ configurations are effective in achieving a superior SNDR, their time-mode realisation is rather difficult because of the lack of time-mode integrators needed to achieve high-order ΔΣ modulators. As a result, ΔΣTDCs are often realised using a partial time-mode partial voltage-mode approach, more specifically, integrators are voltage-mode while quantisers are voltage-controlled oscillator (VCO)-quantisers [102]. Recently, time-difference accumulators that function as time integrators emerged [19]. The core of time-difference accumulators is a time register that holds an input time variable indefinitely and releases the held time variable on the arrival of a triggering signal, as shown in Figs. 19*a* and *b* [103]. The operation of the time register can be briefly depicted as follows: assume C_1 is fully charged initially, when v_{in1} arrives, v_{c1} starts to drop. When v_{in2} arrives, the gated delay cell enters its hold state and v_{c1} remains unchanged indefinitely if no leakage. When T_r is asserted, the gated delay cell is re-activated and v_{c1} starts to drop again. If we assume that the two gated delay cells are identical, then the disrupted discharge process of C_1 is the same as that without disruption. It follows that $T_o=T_{in}$, that is, the time variable T_{in} is stored by the time register and read out on the arrival of the triggering signal T_r. The preceding time register can be utilised to construct a time adder, as shown in Fig. 19*c* [103]. Since $T_{o1}=T_d-T_{in1}$ and $T_{o2}=T_d-T_{in2}$, we have $T_o=T_{o1}-T_{o2}=T_{in1}-T_{in2}$. By reversing the order of the inputs of the second time register, we arrive at $T_o=T_{in1}+T_{in2}$.

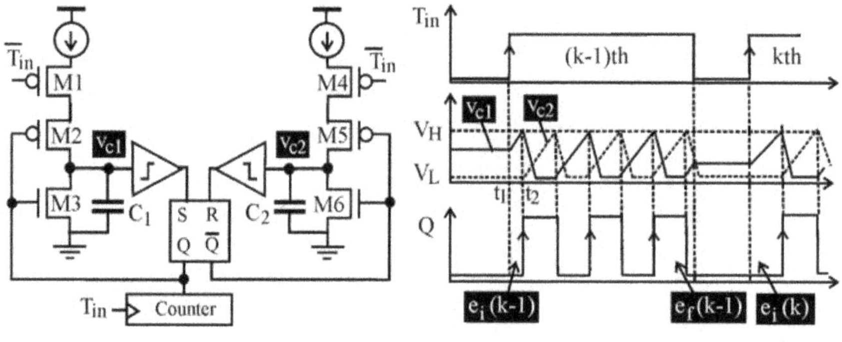

Fig. 17 *MASH TDC with error feedback [95–97]*

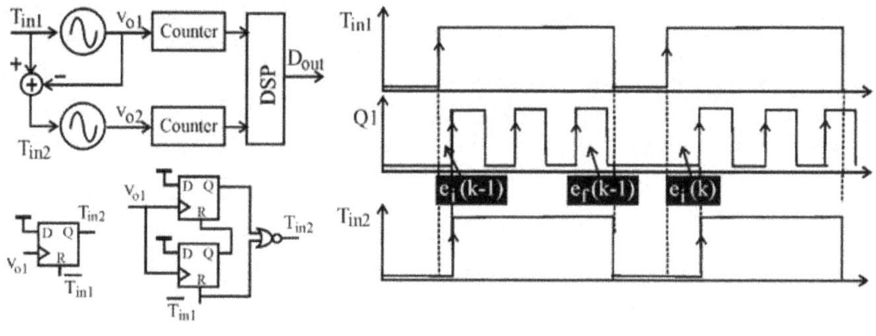

Fig. 18 *1-1 MASH TDC with error feedback [88, 100, 101]. V_L and V_H are lower and upper threshold voltages of the comparators*

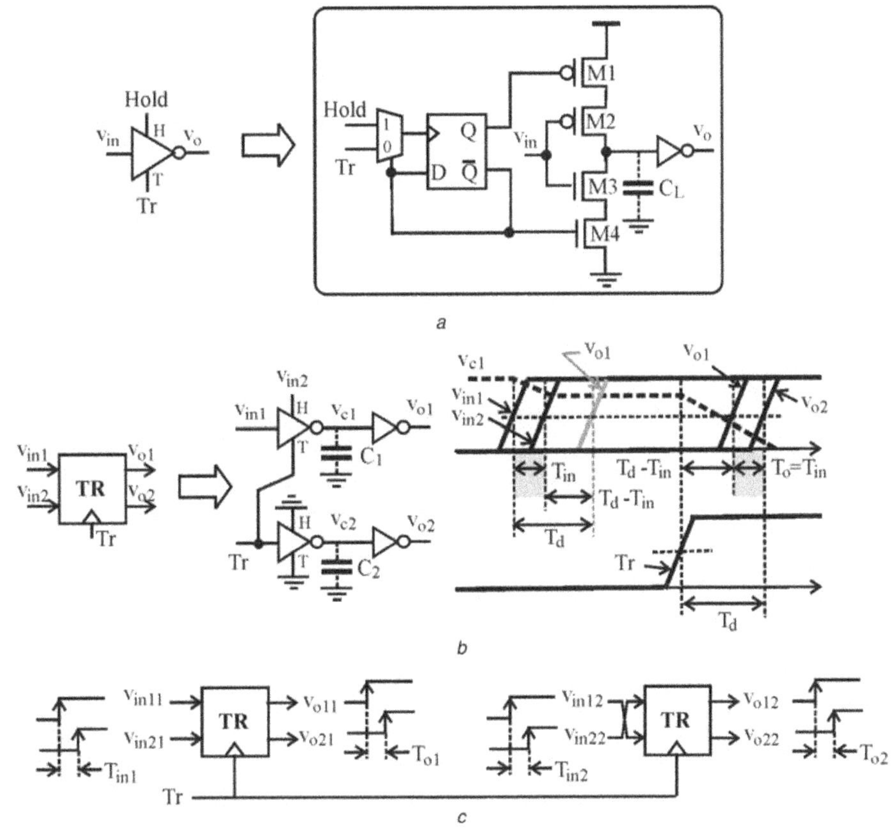

Fig. 19 *Time register proposed in [102]*
a Gated delay cell with hold and trigger (Tr) controls
b Time register
c Time adder

The preceding time adder was utilised in [19] to construct a time integrator to form a first-order time-mode ΔΣ modulator, as shown in Fig. 20. The time accumulator consists of two back-to-back connected time adders to perform time accumulation. $\Delta T = T_{in} - T_f$ is performed using a time adder. ΔT is integrated over the sampling period by the accumulator. The output of the accumulator is digitised by the TDC and the output of the TDC is converted to time variable T_f using a DTC. Since the time accumulator as a first-order integrator shown in Fig. 20, it will provide 20 dB/dec noise-shaping, confirmed by the measurement results in [19].

5.6 *Comparison of noise-shaping TDCs*

Since the resolution of noise-shaping TDCs varies with the order of noise-shaping, the performance matrices used for sampling TDCs such as resolution and precision are generally not used to depict noise-shaping TDCs. The following figure-of-merit (FOM) widely used to quantify the performance of noise-shaping ADCs on the basis of the amount of power per conversion step [104–108] is used to quantify the performance of noise-shaping TDCs

$$\mathrm{FOM} = \frac{P}{2^N(2 \times \mathrm{BW})} \qquad (2)$$

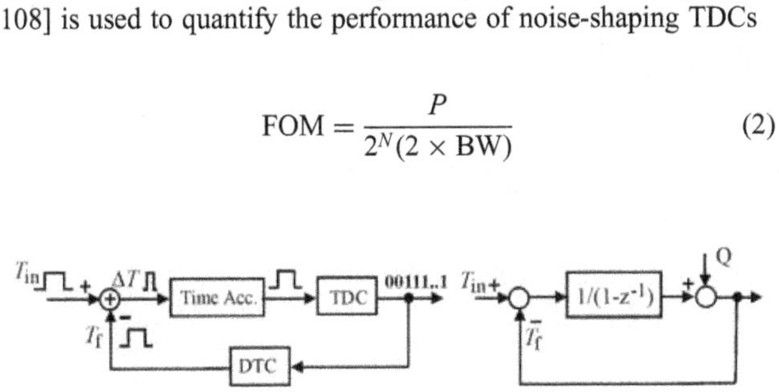

Fig. 20 *Time-mode ΔΣ modulator proposed in [19]*

where BW is the bandwidth of the signal to be digitised, N is the effective number of bits and P is the power consumption of the TDC. N is obtained from $N = (\mathrm{SNDR} - 1.76)/6.02$. Table 2 compares the performance of recently reported noise-shaping TDCs. As compared with Table 1, it is observed that the power consumption of sampling TDCs is generally higher as compared with that of noise-shaping TDCs. This is because the former achieve a high resolution using power greedy complex configurations such as vernier delay lines or multi-level interpolation, whereas the latter lower in-band noise and distortion using feedback with only a moderate increase in power consumption. Noise-shaping TDCs are therefore more attractive for low-power applications. The input frequency of noise-shaping TDCs is generally much lower than that of sampling TDCs because of oversampling constraints. To digitise high-frequency time variables, sampling TDCs are the preferred choice and are more attractive for high-speed applications.

6 Future research of TDCs

TDCs are one of the most studied mixed-mode systems and numerous architectures have been proposed to improve resolution, precision, conversion time and reduce power consumption. A number of stiff challenges, however, are yet to be overcome in order to improve their performance. To improve the resolution of TDCs, advanced design techniques such as cyclic vernier delay lines, multi-level interpolation and time amplification that are effective in reducing quantisation noise can be deployed simultaneously with noise-shaping techniques such as GRO, SRO or ΔΣ. The former lower the quantisation noise by spreading it over the entire frequency spectrum of the TDCs, whereas the latter reduce in-band quantisation noise and distortion through noise-shaping and oversampling. To achieve high-order noise-shaping, time-mode

Table 2 Performance comparison of noise-shaping TDCs

References	Types	Tech., nm	f_{in}, MHz	BW, MHz	f_s, MHz	SNDR, dB	Number of bits	Power, mW	FOM, pJ/step
Straayer and Perrott [7]	GRO	130	26 kHz	1	50	95	15.5	31.5	0.35
Chung *et al.* [94]	GRO	130	7	–	25	–	–	1	–
Hwang and Kim [89]	GRO	180	50 kHz	4	50	–	8	13	6.35
Lu *et al.* [92]	GRO-vernier	90	50	0.8	25	40	6.35	4.5	34.5
Elshazly *et al.* [87]	SRO	90	60 kHz	1	500	80	13	2	0.122
Cao *et al.* [95]	MASH	130	18 kHz	0.1	50	28	4.36	1.7	414
Cao *et al.* [96, 97]	MASH	130	18 kHz	0.1	10	55.2	8.88	0.7	7.43
Konish *et al.* [100]	MASH	40	1	3	100	45	7.28	0.58	0.622
Konish *et al.* [90]	MASH	65	1	3	65	51	8.18	0.271	0.156
Konish *et al.* [88]	MASH	65	61 kHz	0.5	16	61	9.84	0.281	0.307
Okuno *et al.* [101]	MASH	65	1	–	65	62	10	–	–
Ng *et al.* [109]	MASH	65	100	100 kHz	50	31.7	4.97	26	4148
Hong *et al.* [19]	ΔΣ	32	1 kHz	0.1	–	28.95	4.52	0.25	54.5
Gande *et al.* [102]	ΔΣ	130	500 kHz	2.81	90	67	10.84	2.58	0.25

GRO: gated ring oscillators
SRO: switched ring oscillators

ΔΣ modulators are the choice. The recent emerge of time-mode accumulators has fueled the interest in searching for high-order time-mode integrators needed for high-order time-mode ΔΣ modulators without deploying MASH whose performance is degraded because of mismatches. Since VCO-based quantisers are multi-bit quantisers, the feedback of time-mode ΔΣ TDCs is a multi-bit DTC and its non-linearity critical affects the performance of the TDCs manifesting as both in-band harmonic tones and a rising noise floor, DTCs with a superior linearity or time-mode dynamic element matching techniques especially those with noise-shaping characteristics are to be developed. As the deployment of TDCs in mixed-mode signal processing is to combat the performance degradation caused by technology scaling, mixed-mode building blocks such as voltage comparators, voltage-mode integrators, charge pumps etc. whose performance scales poorly with technology should be avoided.

7 Conclusions

A comprehensive treatment of the principles, architectures and design techniques of CMOS TDCs has been provided. Key performance matrices of TDCs such as resolution, precision, non-linearity, voltage and temperature sensitivities, conversion time and conversion range have been examined. Sampling TDCs that digitise time variables have been studied with an emphasis on their advantages and limitations. It has been shown that the resolution of sampling TDCs is lower bound by quantisation noise as its quantisation noise is uniformly distributed over the entire spectrum of the TDCs. The resolution of direct-counter TDCs is the period of the sampling clock. Resolution can be improved to one buffer delay if tapped delay-line interpolation is used to further digitise the quantisation errors of direct-counter TDCs. Although the longer the tapped delay line, the better the interpolation resolution, the non-linearity of tapped delay lines deteriorates with the increase in the length of the delay lines. To overcome the poor non-linearity of long delay lines, two-level delay-line interpolation can be used. Resolution can be further improved to below one buffer delay by using pulse-shrinking delay lines, vernier delay lines, pulse-stretching TDCs or offset-sampling flash TDCs. Cyclic configurations are also favoured in improving linearity and minimising silicon cost.

Noise-shaping TDCs that achieve a better resolution over a specific frequency range have also been investigated. Unlike sampling TDCs, GRO-TDCs offer an intrinsic characteristic of first-order noise-shaping. The noise-shaping of these TDCs, however, is undermined by charge leakage and cross-talk in the phase-holding state. Switched ring oscillator TDCs outperform GRO-TDCs by eliminating the effect of leakage, charge sharing and switching noise in the phase-holding stage. RO TDCs offer the key advantage of low sensitivity to PVT effect. Time-mode ΔΣ TDCs that utilise time accumulators as time integrators remove the need for voltage-mode integrators whose performance scales poorly with technology. High-order time integrators, time quantisers with low quantisation noise and DTCs with a superior linearity critical to noise-shaping TDCs are yet to be developed.

8 Acknowledgment

The author is grateful to reviewers for their invaluable comments and suggestions. This paper could not be in its present form without the comments and suggestions of the reviewers.

9 References

[1] Yoshiaki T., Takeshi A.: 'Simple voltage-to-time converter with high linearity', *IEEE Trans. Instrum. Meas.*, 1971, **20**, (2), pp. 120–122

[2] Porat D.: 'Review of sub-nanosecond time-interval measurements', *IEEE Trans. Nucl. Sci.*, 1973, **NS-20**, pp. 36–51

[3] Park K., Park J.: '20 ps resolution time-to-digital converter for digital storage oscillator'. Proc. IEEE Nuclear Science Symp., 1998, vol. **2**, pp. 876–881

[4] Chen P., Chen C., Shen Y.: 'A low-cost low-power CMOS time-to-digital converter based on pulse stretching', *IEEE Trans. Nucl. Sci.*, 2006, **53**, (4), pp. 2215–2220

[5] Chen C., Chen P., Hwang C., Chang W.: 'A precise cyclic CMOS time-to-digital converter with low thermal sensitivity', *IEEE Trans. Nucl. Sci.*, 2005, **52**, (4), pp. 834–838

[6] Straayer M., Perrott M.: 'A 12-bit, 10-MHz bandwidth, continuous-time ΔΣ ADC with a 5-bit, 950-MS/s VCO-based quantizer', *IEEE J. Solid-State Circuits*, 2008, **43**, (4), pp. 805–814

[7] Straayer M., Perrott M.: 'A multi-path gated ring oscillator TDC with first-order noise shaping', *IEEE J. Solid-State Circuits*, 2009, **44**, (4), pp. 1089–1098

[8] Park M., Perrott M.: 'A single-slope 80 Ms/s ADC using two-step time-to-digital conversion', Proc. IEEE Int. Symp. Circuits Syst., 2009, pp. 1125–1128

[9] Yousefzadeh B., Sharifkhani M.: 'An audio band low voltage CT-ΔΣ modulator with VCO-based quantizer'. Proc. IEEE Int. Conf. Electronics, Circuits Systems, 2011, pp. 232–235

[10] Swann B., Blalock B., Clonts L., *ET AL.*: 'A 100-ps time-resolution CMOS time-to-digital converter for positron emission tomography imaging applications', *IEEE J. Solid-State Circuits*, 2004, **39**, (11), pp. 1839–1852

[11] Raisanen-Ruotsalainen E., Rahkonen T., Kostamovaara J.: 'An integrated time-to-digital converter with 30-ps single-shot precision', *IEEE J. Solid-State Circuits*, 2000, **35**, (10), pp. 1507–1510

[12] Guttman M., Roberts G.: 'K-locked-loop and its application in time mode ADC'. Proc. IEEE Int. Symp. Integrated Circuits, 2009, pp. 101–104

[13] Ravinuthula V.: 'Time-mode circuits for analog computations'. PhD dissertation, University of Florida, 2006

[14] Aouini S., Chuai K., Roberts G.: 'Anti-imaging time-mode filter design using a PLL structure with transfer function DFT', *IEEE Trans. Circuits Syst. I*, 2012, **59**, (1), pp. 66–79

[15] Chen P., Chen C., Tsai C., Lu W.: 'A time-to-digital-converter-based CMOS smart temperature sensor', *IEEE J. Solid-State Circuits*, 2005, **40**, (8), pp. 1642–1648

[16] Tokairin T., Okada M., Kitsunezuka M., Maeda T., Fukaishi M.: 'A 2.1-to-2.8-GHz low-phase-noise all-digital frequency synthesizer with a time-windowed time-to-digital converter', *IEEE J. Solid-State Circuits*, 2010, **45**, (12), pp. 2582–2590

[17] Hsu C., Straayer M., Perrott M.: 'A low-noise wide-BW 3.6-GHz digital ΔΣ fractional-N frequency synthesizer with a noise-shaping time-to-digital converter and quantization noise cancellation', *IEEE J. Solid-State Circuits*, 2008, **43**, (12), pp. 2776–2786

[18] Ghaffari A., Abrishamifar A.: 'A novel wide-range delay cell for DLLs'. Proc. IEEE Int. Electrical and Computer Engineering Conf., 2006, pp. 497–500

[19] Hong J., Kim S., Liu J., ET AL.: 'A 0.004 mm^2 250 μW ΔΣ TDC with time-difference accumulator and a 0.012 mm^2 2.5 mW bang-bang digital PLL using PRNG for low-power SoC applications'. IEEE Int. Conf. Solid-State Circuits Digest of Technical Papers, 2012, pp. 240–242

[20] Rashdan M., Yousif A., Haslett J., Maundy B.: 'A new time-based architecture for serial communication links'. Proc. IEEE Int. Conf. Electronics, Circuits, Systems, 2009, pp. 531–534

[21] Kim J., Yu W., Cho S.: 'A digital-intensive multi-mode multi-band receiver using a $sinc^2$ filter-embedded VCO-based ADC', *IEEE Trans. Microw. Theory Tech.*, 2012, **60**, (10), pp. 3254–3262

[22] Jansson J., Koskinen V., Mantyniemi A., Kostamovaara J.: 'A multichannel high-precision CMOS tie-to-digital converter for laser-scanner-based perception systems', *IEEE Trans. Instrum. Meas.*, 2012, **61**, (9), pp. 2581–2590

[23] Kostamovaara J., Myllyla R.: 'Time-to-digital converter with an analog interpolation circuit', *Rev. Sci. Instrum.*, 1986, **57**, (11), pp. 2880–2885

[24] Kalisz J.: 'Review of methods for time interval measurements with picosecond resolution', *Metrologia*, 2004, **41**, pp. 17–32

[25] Henzler S.: 'Time-to-digital converters' (Springer, New York, 2010)

[26] Rahkonen T., Kostamovaara J., Saynajakangas S.: 'Time interval measurements using integrated tapped CMOS delay lines'. Proc. IEEE Mid-West Symp. Circuits Systems, 1990, pp. 201–205

[27] Aria Y., Matsumura T., Endo K.: 'A CMOS four-channel × 1 K memory LSI with 1-ns/b resolution', *IEEE Trans. Circuits Syst. II*, 1992, **27**, (3), pp. 359–364

[28] Rahkonen T., Kostamovaara J.: 'The use of stabilized CMOS delay lines in the digitization of short time intervals', *IEEE J. Solid-State Circuits*, 1993, **28**, (8), pp. 887–894

[29] Xing N., Woo J., Shin W., Lee H., Kim S.: 'A 14.6 ps resolution, 50 ns input-range cyclic time-to-digital converter using fractional difference conversion method', *IEEE Trans. Circuits Syst. I.*, 2010, **57**, (12), pp. 3064–3072

[30] Hwang C., Chen P., Tsao H.: 'A high-precision tie-to-digital converter using a two-level conversion scheme', *IEEE Trans. Nucl. Sci.*, 2004, **51**, (4), pp. 1349–1352

[31] Levine P., Roberts G.: 'A calibration technique for a high-resolution flash time-to-digital converter'. Proc. IEEE Int. Symp. Circuits Systems, 2004, vol. 1, pp. 253–256

[32] Levine P., Roberts G.: 'High-resolution flash time-to-digital conversion and calibration for system-on-chip testing', *IEE Proc. Comput. Digit. Tech.*, 2005, **152**, (3), pp. 415–426

[33] Gutnik V., Chandrakasan A.: 'On-chip picosecond time measurement'. Symp. VLSI Circuits Digest of Technical Papers, 2000, pp. 52–53

[34] Minas N., Kinniment D., Heron K., Russell G.: 'A high resolution flash time-to-digital converter taking into account process variability'. Proc. IEEE Int. Symp. Asynchronous Circuits and Systems, 2007, pp. 163–174

[35] Yamaguchi T., Komatsu S., Abbas M., Asada K., Maikhanh N., Tandon J.: 'A CMOS flash TDC with 0.84-1.3 ps resolution using standard cells'. Proc. IEEE RFIC, 2012, pp. 527–530

[36] Zanuso M., Levantino S., Puggelli A., Samori C., LacaitA A.: 'Time-to-digital converter with 3-ps resolution and digital linearization algorithm'. Proc. IEEE ESSCIRC, 2010, pp. 262–265

[37] Yao C., Jonsson F., Chen J., Zheng L.: 'A high-resolution time-to-digital converter based on parallel delay elements'. Proc. IEEE Int. Symp. Circuits Systems, 2012, pp. 3158–3161

[38] Staszewski R., Vemulapalli S., Vallur P., Wallberg J., Balsara P.: '13 V 20 ps time-to-digital converter for frequency synthesis in 90 nm CMOS', *IEEE Trans. Circuits Syst. II.*, 2006, **53**, (3), pp. 220–224

[39] Knotts T., Chu D., Sommer J.: 'A 500 MHz time digitizer IC with 15.625 ps resolution'. IEEE Int. Solid-State Circuits Conf. Digest of Technical Papers, 1994, pp. 58–59

[40] Henzler S., Koeppe S., Kamp W., Schmitt-Landsiedel D.: '90 nm 4.7 ps-resolution 0.7-LSB single-shot precision and 19 pJ-per-shot local passive interpolation time-to-digital converter with on-chip characterization'. IEEE Int. Solid-State Circuits Conf. Digest of Technical Papers, 2008, pp. 548–635

[41] Jang T., Kim J., Yoon Y., Cho S.: 'A highly-digital VCO-based analog-to-digital converter using phase interpolator and digital calibration', *IEEE Trans. VLSI Syst.*, 2012, **20**, (8), pp. 1368–1372

[42] Christiansen J.: 'An integrated high resolution CMOS timing generator based on an array of delay locked loop', *IEEE J. Solid-State Circuits*, 1996, **31**, (7), pp. 952–957

[43] Raisanen-Ruotsalainen E., Rahkonen T., Kostamovaara J.: 'A low-power CMOS time-to-digital converter', *IEEE J. Solid-State Circuits*, 1995, **30**, (9), pp. 984–990

[44] Chen P., Liu S., Wu J.: 'A CMOS pulse-shrinking delay element for time interval measurement', *IEEE Trans. Circuits Syst. II.*, 2000, **47**, (9), pp. 954–958

[45] Liu Y., Vollenbruch U., Chen Y., ET AL.: 'Multi-stage pulse shrinking time-to-digital converter for time interval measurements'. Proc. European Conf. Wireless Technology, 2007, pp. 347–350

[46] Liu Y., Vollenbruch U., Chen Y., ET AL.: 'A 6 ps resolution pulse shrinking time-to-digital converter as phase detector in multi-mode transceiver'. Proc. IEEE Radio and Wireless Symp., 2008, pp. 163–166

[47] Mantyniemi A., Rahkonen T., Kostamovaara J.: 'A high-resolution digital CMOS time-to-digital converter based on nested delay locked loops'. Proc. IEEE Int. Symp. Circuits Systems, 1999, vol. **2**, pp. 537–540

[48] Huang H., Wu S., Tsai Y.: 'A new cycle-time-to-digital converter with two level conversion scheme'. Proc. IEEE Int. Symp. Circuits Systems, 2007, pp. 2160–2163

[49] Kalisz J., Pawlowski M., Pelka R.: 'Error analysis and design of the nutt time-interval digitiser with picosecond resolution', *J. Phys. Sci. Instrum.*, 1987, **20**, pp. 1330–1341

[50] Maatta K., Kostamovaara J.: 'A high-precision time-to-digital converter for pulsed time-of-flight laser radar applications', *IEEE Trans. Instrum. Meas.*, 1998, **47**, (2), pp. 521–536

[51] Keranen P., Maatta K., Kostamovaara J.: 'Wide-range time-to-digital converter with 1 ps single-shot precision', *IEEE Trans. Instrum. Meas.*, 2011, **60**, (9), pp. 3162–3172

[52] Nutt R.: 'Digital time intervalometer', *Rev. Sci. Instrum.*, 1968, **39**, pp. 1342–1345

[53] Kurko B.: 'A picosecond resolution time digitizer for laser ranging', *IEEE Trans. Nucl. Sci.*, 1978, **NS-25**, (1), pp. 75–80

[54] Park K., Park J.: 'Time-to-digital converter of very high pulse stretching ratio for digital storage oscilloscope', *Rev. Sci. Instrum.*, 1999, **70**, (2), pp. 1568–1574

[55] Raisanen-Ruotsalainen E., Rahkonen T., Kostamovaara J.: 'An integrated time-to-digital converter with 30 ps single-shot precision', *IEEE J. Solid-State Circuits*, 2000, **35**, (10), pp. 1507–1510

[56] Dehlaghi B., Magierowski S., Belostotski L.: 'Highly-linear time-difference amplifier with low sensitivity to process variations', *IET Electron. Lett.*, 2011, **47**, (13), pp. 743–745

[57] Kwon H., Lee J., Sim J., Park H.: 'A high-gain wide-input-range time amplifier with an open-loop architecture and a gain equal to current bias ratio'. Proc. IEEE Asian Solid-State Circuits Conf., 2011, pp. 325–328

[58] Abas A., Bystrov A., Kinnimnt D., Maevsky O., Russell G., Yakovlev A.: 'Time difference amplifier', *IEE Electron. Lett.*, 2002, **38**, (23), pp. 1437–1438

[59] Abas A., Russell G., Kinniment D.: 'Embedded high-resolution delay measurement system using time amplification', *IET Comput. Digit. Tech.*, 2002, **1**, (2), pp. 77–86

[60] Oulmane M., Roberts G.: 'CMOS time amplifier for femto-second resolution timing measurement'. Proc. IEEE Int. Symp. Circuits Systems, 2004, pp. 509–512

[61] Tong B., Yan W., Zhou X.: 'A constant-gain time-amplifier with digital self-calibration'. Proc. IEEE Int. ASIC Conf., 2009, pp. 1133–1136

[62] Lee M., Abidi A.: 'A 9B, 1.25 ps resolution coarse-fine time-to-digital converter in 90 nm CMOS that amplifies a time residue', *IEEE J. Solid-State Circuits*, 2008, **43**, (4), pp. 769–777
[63] Alahmadi A., Russell C., Yakovlev A.: 'Time difference amplifier design with improved performance parameters', *IET Electron. Lett.*, 2012, **48**, (10), pp. 562–563
[64] Rashidzadeh R., Muscedere R., Ahmadi M., Miller W.: 'A delay generation technique for narrow time interval measurement', *IEEE Trans. Instrum. Meas.*, 2009, **58**, (7), pp. 2245–2252
[65] Lin C., Syrzycki M.: 'Pico-second time interval amplification'. Proc. IEEE Int. SoC Design Conf., 2010, pp. 201–204
[66] Nakura T., Mandai S., Ikeda M., Asada K.: 'Time difference amplifier using closed-loop gain control'. Symp. VLSI Circuits Digest of Technical Papers, 2009, pp. 208–209
[67] Mandai S., Nakura T., Ikeda M., Asada K.: 'Cascaded time difference amplifier using differential logic delay cell'. Proc. Int. SoC Design Conf., 2009, pp. 299–304
[68] Mandai S., Charbon E.: 'A 128-channel, 8.9-ps LSB, column-parallel two-stage TDC based on time difference amplification for time-resolved imaging', *IEEE Trans. Nucl. Sci.*, 2012, **59**, (5), pp. 2463–2470
[69] Lee S., Seo Y., Park H., Sim J.: 'A 1 GHz ADPLL with a 125 ps minimum-resolution sub-exponent TDC in 0.18 μm CMOS', *IEEE J. Solid-State Circuits*, 2010, **45**, (12), pp. 2827–2881
[70] Baron R.: 'The vernier time-measuring technique', *Proc. IRE*, 1957, **45**, (1), pp. 21–30
[71] Ljuslin C., Christiansen J., Marchioro A., Klingsheim O.: 'An integrated 16-channel CMOS time to digital converter', *IEEE Trans. Nucl. Sci.*, 1994, **41**, (4), pp. 1104–1108
[72] Dudek P., Szczepanski S., Hatfield J.: 'A high-resolution CMOS time-to-digital converter utilizing a Vernier delay line', *IEEE J. Solid-State Circuits*, 2000, **35**, (2), pp. 240–247
[73] Li G., Chou H.: 'A high resolution time-to-digital converter using two-level vernier delay line technique'. Proc. IEEE Nuclear Science Symp. Conf. Record, 2007, pp. 276–280
[74] Mantyniemi A., Rahkonen T., Kostamovaara J.: 'A CMOS time-to-digital converter (TDC) based on a cyclic time domain successive approximation interpolation method', *IEEE J. Solid-State Circuits*, 2009, **44**, (11), pp. 3067–3078
[75] Nagaraj G., Miller S., Stengel B., ET AL.: 'A self-calibrating sub-picosecond resolution digital-to-time converter'. Proc. IEEE Int. Microwave Symp., 2007, pp. 2201–2204
[76] Choi Y., Yoo S., Yoo H.: 'A full digital polar transmitter using a digital-to-time converter for high data rate system'. Proc. IEEE Int. Symp. Radio-Frequency Integration Technology, 2009, pp. 56–59
[77] Al-Ahdab S., Mantyniemi A., Kostamovaara J.: 'A 12-bit digital-to-time converter (DTC) for time-to-digital converter (TDC) and other time domain signal processing applications'. Proc. IEEE NORCHIP, 2010, pp. 1–4
[78] Roberts G., Ali-Bakhshian M.: 'A brief introduction to time-to-digital and digital-to-time converters', *IEEE J. Solid-State Circuits*, 2010, **57**, (3), pp. 153–157
[79] Li S., Salthouse C.: 'Digital-to-time converter for fluorescence lifetime imaging'. Proc. IEEE Int. Instrumentation and Measurement Technology Conf., 2012, pp. 894–897
[80] Seo Y., Kim J., Park H., Sim J.: 'A 0.63 ps resolution, 11 b pipeline TDC in 0.13 μm CMOS'. Symp. VLSI Circuits Digest of Technical Papers, 2012, pp. 152–153
[81] Kim J., Seo Y., Suh Y., Park H., Sim J.: 'A 300-MS/s 1.76-ps-resolution 10b asynchronous pipelined time-to-digital converter with on-chip digital background calibration in 0.13 μm CMOS', *IEEE J. Solid-State Circuits*, 2013, **48**, (2), pp. 516–526
[82] Kim K., Yu W., Cho S.: 'A 9b 1.12 ps resolution 2.5b/stage pipelined time-to-digital converter in 65 nm CMOS using time-register'. Symp. VLSI Circuits Digest of Technical Papers, 2013, pp. 136–137
[83] Jansson J., Mantyniemi A., Kostamovaara J.: 'A CMOS time-to-digital converter with better than 10 ps single-shot precision', *IEEE J. Solid-State Circuits*, 2006, **41**, (6), pp. 1286–1296
[84] Seo Y., Kim J., Park H., Sim J.: 'A 0.63 ps resolution 11 b pipeline TDC in 0.13 μm CMOS'. Symp. VLSI Circuits Digest of Technical Papers, 2011, pp. 152–153
[85] Chen P., Chen C., Zheng J., Shen Y.: 'A PVT insensitive vernier-based time-to-digital converter with extended input range and high accuracy', *IEEE Trans. Nucl. Sci.*, 2007, **54**, (2), pp. 294–302
[86] Schreier R., Temes G.: 'Understanding delta–sigma data converters' (John Wiley & Sons, Hoboken, NJ, 2005)
[87] Elshazly A., Rao S., Young B., Hanumolu P.: 'A 13b 315 $f_{s,\mathrm{rms}}$ 2 mW 500 MS/s 1 MHz bandwidth highly digital time-to-digital converter using switched ring oscillators'. Int. Solid-State Circuits Conf. Digest of Technical Papers, 2012, pp. 464–465
[88] Konishi T., Okumo K., Izumi S., Yoshimoto M., Kawaguchi H.: 'A 61 dB SNDR 700 μm second-order all-digital TDC with low-jitter frequency shift oscillator and dynamic flipflops'. Symp. VLSI Circuits Digest of Technical Papers, 2012, pp. 190–191
[89] Hwang K., Kim L.: 'An area efficient asynchronous gated ring oscillator TDC with minimum GRO stages'. Proc. IEEE Int. Symp. Circuits Systems, 2010, pp. 3973–3976
[90] Konishi T., Okumo K., Izumi S., Yoshimoto M., Kawaguchi H.: 'A 51 dB SNDR DCO-based TDC using two-stage second-order noise shaping'. Proc. IEEE Int. Symp. Circuits Systems, 2012, pp. 3170–3173
[91] Lu P., Wu Y., Andreani P.: 'A 90 nm CMOS digital PLL based on vernier-gated-ring-oscillator time-to-digital converter'. Proc. IEEE Int. Symp. Circuits Systems, 2012, pp. 2593–2596
[92] Lu P., Andreani P., Liscidini A.: 'A 3.6 mW 90 nm CMOS gated-vernier tie-to-digital converter with an equivalent resolution of 3.2 ps', *IEEE. Solid-State Circuits*, 2012, **47**, (7), pp. 1626–1635
[93] Lu P., Andreani P., Liscidini A.: 'A 2-D GRO vernier time-to-digital converter with large input range and small latency'. Proc. IEEE RFIC, 2013, pp. 151–154
[94] Chung S., Hwang K., Lee W., Kim L.: 'A high resolution metastability-independent two-step gated ring oscillator TDC with enhanced noise shaping'. Proc. IEEE Int. Symp. Circuits Systems, 2010, pp. 1300–1303
[95] Cao Y., Leroux P., Cock W.D., Steyaert M.: 'A 0.7 mW 11b 1-1-1 MASH ΔΣ time-to-digital converter'. IEEE Int. Solid-State Circuits Conf. Digest of Technical Papers, 2011, pp. 480–481
[96] Cao Y., Leroux P., Cock W.D., Steyaert M.: 'A 0.7 mW 13b temperature-stable MASH ΔΣ TDC with delay-line assisted calibration'. Proc. IEEE Asian Solid-State Circuits Conf., 2011, pp. 361–364
[97] Cao Y., Cock W.D., Steyaert M., Leroux P.: '1-1-1 MASH ΔΣ time-to-digital converters with 6 ps resolution and third-order noise-shaping', *IEEE J. Solid-State Circuits*, 2012, **47**, (9), pp. 2093–2106
[98] Yuan F., Soltani N.: 'A low-voltage low vdd sensitivity relaxation oscillator for passive wireless microsystems', *IET Electron. Lett.*, 2009, **45**, (21), pp. 1057–1058
[99] Yuan F.: 'CMOS circuits for passive wireless microsystems' (Springer, New York, 2010)
[100] Konishi T., Okumo K., Izumi S., Yoshimoto M., Kawaguchi H.: 'A 40-nm 640-μm^2 45-dB opampless all-digital second-order MASH ΔΣ ADC'. Proc. IEEE Int. Symp. Circuits Systems, 2011, pp. 518–521
[101] Okuno K., Konishi T., Izumi S., Yoshimoto M., Kawaguchi H.: 'A 62 dB SNDR second-order gated ring oscillator TDC with two-stage dynamic D-type flipflips a a quantization noise propagator'. Proc. IEEE NEWCAS, 2012, pp. 289–292
[102] Gande M., Maghari N., Oh T., Moon U.: 'A 71 dB dynamic range third-order ΔΣ TDC using charge-pump'. Symp. VLSI Circuits Digest of Technical Papers, 2012, pp. 168–169
[103] Kim S.: 'Time domain algebraic operation circuits for high performance mixed-mode system'. MS thesis, Korean Advanced Institute of Science and Technology, 2010
[104] Wismar U., Wisland D., Andreani P.: 'A 0.2 V 0.44 μW 20 kHz analog to digital ΔΣ modulator with 57 fJ/conversion FoM'. Proc. IEEE European Solid-State Circuits Conf., 2006, pp. 187–190
[105] Li G., Tousi Y., Hassibi A., Afshari E.: 'Delay-line-based analog-to-digital converters', *IEEE Trans. Circuits Syst. II*, 2009, **56**, (6), pp. 464–468
[106] Tousi Y., Afshari E.: 'A miniature 2 mW 4 bit 1.2 GS/s delay-line-based ADC in 65 nm CMOS', *IEEE J. Solid-State Circuits*, 2011, **46**, (10), pp. 2312–2325
[107] Taylor G., Galton I.: 'A mostly-digital variable-rate continuous-time delta-sigma modulator ADC', *IEEE J. Solid-State Circuits*, 2010, **45**, (12), pp. 2634–2646
[108] Rao S., Young B., Elshazly A., Yin W., Sasidhar N., Hanumolu P.: 'A 71 dB SFDR open loop VCO-based ADC using 2-level PWM modulation'. Symp. VLSI Circuits Digest of Technical Papers, 2011, pp. 270–271
[109] Ng A., Zheng S., Luong H.: 'A 4.1 GHz-6.5 GHz all-digital frequency synthesizer with a 2nd-order noise-shaping TDC and a transformer-coupled QVCO'. Proc. IEEE ESSCIRC, 2012, pp. 189–192

12

An enhanced role for an energy storage system in a microgrid with converter-interfaced sources

Konstantinos O. Oureilidis, Charis S. Demoulias

Department of Electrical and Computer Engineering, Aristotle University, Thessaloniki, 54124, Greece
E-mail: chdimoul@auth.gr

Abstract: An enhanced role for the energy storage system (ESS), strategically placed at the point of common coupling (PCC) of the microgrid with the utility grid, is proposed. During island operation, the ESS ensures that the frequency and magnitude of the voltage will remain within the limits specified by the Standard EN 50160. By implementing an adjustable droop control method, the distributed energy resources (DERs) adjust their active and reactive powers in order to fulfil the load demand. When the grid is recovered, the ESS detects its presence and achieves a seamless synchronisation of the microgrid with the main grid, without any kind of communication. In grid-connected mode, the DERs deliver their available active power, whereas their reactive power is determined by a zero-sequence voltage. This voltage is injected by the ESS and aims to the zeroing of the amount of reactive power at the PCC. In this way, a reduction of power losses in the distribution lines of the microgrid is achieved. The effectiveness of the proposed control method in all operation modes, without any physical communication means, is demonstrated through detailed simulation in a representative microgrid with DERs fed by photovoltaics.

1 Introduction

The increase of grid-connected renewable energy resources has contributed to the upcoming decentralised approach of the power grid, where clusters of microsources and loads form entities called microgrids. The microgrids should operate either connected with the upstream utility network or in island mode, in case of grid unavailability [1]. Each distributed energy resource (DER) is interfaced to the microgrid through DC/AC or AC/DC/AC converters. Consequently, the energy management of the microgrid can be carried out through proper control of the converters. In both grid-connected and island mode, the voltage within the microgrid should always comply with the limits imposed by the EN 50160 Standard [2].

When the island operation condition is identified [3, 4], several DERs share the produced power in order to fulfil the load demands. The power sharing among the DERs can be carried out either using signals via physical communication means [1, 5] or using local voltage measurements as communication parameters [6, 7]. The latter method is also referred to as 'wireless' method.

In grid-connected mode, the control strategy of the DERs is adjusted to controllable current sources [8], in order to transfer the available active power to the common AC bus. Regarding the adjustment of the reactive power of the DERs, different methods have been proposed, depending on the existence or not of communication between the DERs. In the 'wireless' method, only compensation of the reactive power of local loads or local voltage regulation is performed [6]. On the other hand, in the communication-based method, the reactive power of the DERs can also be adjusted in terms of optimal power flow [9].

Apart from the DERs, the coordination of the microgrid with an energy storage system (ESS) is considered necessary for several reasons [10]. The most important include the reinforcement of the dynamic behaviour [11], the contribution to the seamless transition from grid-connected to island mode [12] and the improvement of the operation during faults or transient situations [13, 14].

In this paper, a converter-dominated microgrid with an ESS strategically placed at the point of common coupling (PCC) of the microgrid with the utility grid is investigated, as shown in Fig. 1. Additionally, it is assumed that there is no physical communication between the DERs and the ESS. The control of the ESS converter performs the following operations: (i) during the synchronisation process, the ESS adjusts the magnitude, phase and frequency of the common AC bus voltage by regulating the active and reactive powers of the microgrid. Contrary to the current practice [8, 15, 16], no communication between the grid and the DERs is necessary. (ii) During the islanded operation of the microgrid, the ESS absorbs the mismatches between the power consumption and production [17]. Furthermore, a variable virtual impedance is introduced in the control of the DERs, so that they always operate at power factor (PF) larger than 0.8, while at the same time the minimisation of the circulating reactive currents among the DERs is achieved. (iii) In grid-connected mode, the ESS regulates the sharing of the reactive power among the DERs in a novel way. Specifically, the ESS injects a 7th harmonic zero-sequence voltage into the microgrid, with a magnitude depending on the amount of the reactive power at the PCC. The aim is to zero the exchange of reactive power at this point. Thus, the reactive power load demand is covered by its own DERs, leading to the reduction of the circulating reactive current, and as a result a reduction of the power losses. Furthermore, if a DER and a load are connected at the same local bus, the priority is to cover the reactive power of the local load.

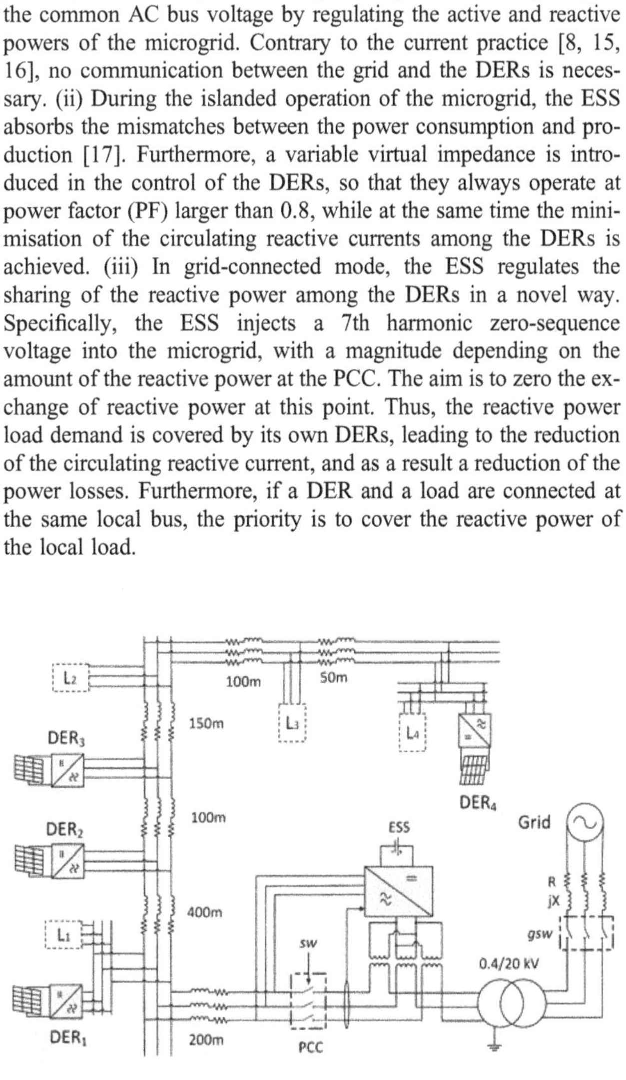

Fig. 1 *Microgrid simulation model*

The injection of a zero-sequence voltage was selected in conjunction with a Dy transformer at the PCC, since the infinite zero-sequence impedance of the delta winding will restrict this voltage only in the microgrid side. To avoid any false detection because of the presence of 50 Hz zero-sequence voltage components under unbalanced conditions, the 7th harmonic was finally selected. This harmonic, especially in zero-sequence form, naturally exists only in very small magnitudes, while it can also be easily detected by the DERs. The magnitude of the injected 7th harmonic voltage should comply with the restrictions imposed by [2].

2 Control strategy in island mode

In island operation mode, the active and reactive powers of each DER are produced by the frequency and the magnitude of the voltage of the common AC bus, emulating the synchronous machine operation philosophy. The droop control equations can be summarised as follows

$$f = f^* - m \cdot P - m_{\mathrm{d}} \frac{\mathrm{d}P}{\mathrm{d}t} \tag{1}$$

$$V_n = V^* - n \cdot Q - n_{\mathrm{d}} \frac{\mathrm{d}Q}{\mathrm{d}t} \tag{2}$$

where f^*, V^* are the frequency and magnitude of the output voltage at no load, m, n are the droop coefficients, P, Q are the average active and reactive powers and m_{d}, n_{d} are the derivative droop coefficients.

If a DER is a renewable one (e.g. photovoltaic (PV) system), the available active power may vary with the available primary energy source (e.g. solar irradiance). In this case, in order to ensure the system stability and the allocation of the active power of the DERs in proportion to the available power of the primary source, the droop coefficients m and n should be accordingly adapted

$$m = \frac{\Delta f_{\max}}{P_{\mathrm{pot}}} \tag{3}$$

$$n = \frac{\Delta V_{\max}}{P_{\mathrm{pot}}} \tag{4}$$

where $\Delta f_{\max}$, $\Delta V_{\max}$ are the maximum deviations of the frequency and the voltage magnitude according to [2], respectively. The power P_{pot} corresponds to the maximum power point (MPP) of the renewable source. When the primary source is a conventional one, P_{pot} is equal to the apparent power rating of the respective converter [17].

In the aforementioned droop control method, an adaptive virtual impedance [7, 18] is added, in order to minimise the effects of different apparent impedances at the terminals of each DER. The proposed virtual impedance ensures that each DER operates at PF larger than 0.8. This statement is based on the engineering judgement that the majority of the loads in a microgrid operates in this range of PF. The virtual impedance is determined by

$$L_{\mathrm{v}} = K_{\mathrm{pv}}(0.8 - \mathrm{PF}) + K_{\mathrm{iv}} \int (0.8 - \mathrm{PF})\, \mathrm{d}t \tag{5}$$

where L_{v} is the virtual impedance, PF is the measured power factor of the DER and K_{pv}, K_{iv} are the proportional and integral parameters of the PI controller. The control strategy is illustrated in Fig. 2.

The ESS role in island operation mode is the compensation of the mismatches in the active and reactive between the loads and the DERs, as described in [17]. The state of charge (SoC) of the ESS is defined as

$$\mathrm{SoC} = \frac{Q_o - \int i_d\, \mathrm{d}t}{Q_{\mathrm{tot}}} \tag{6}$$

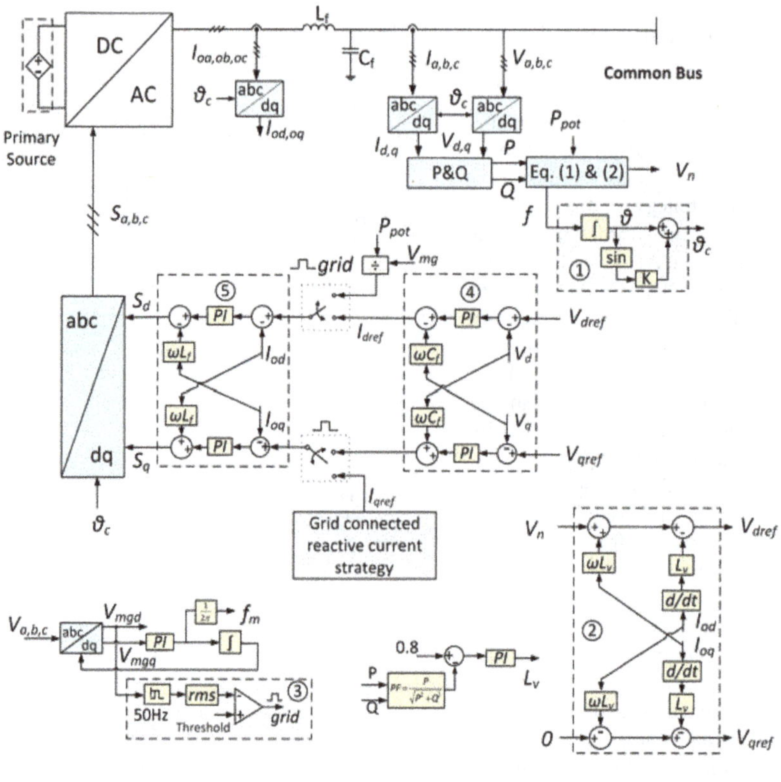

Fig. 2 *DER control strategy scheme: (1) θ_c calculation, (2) virtual impedance control, (3) grid detection algorithm, (4) island voltage control and (5) current control*

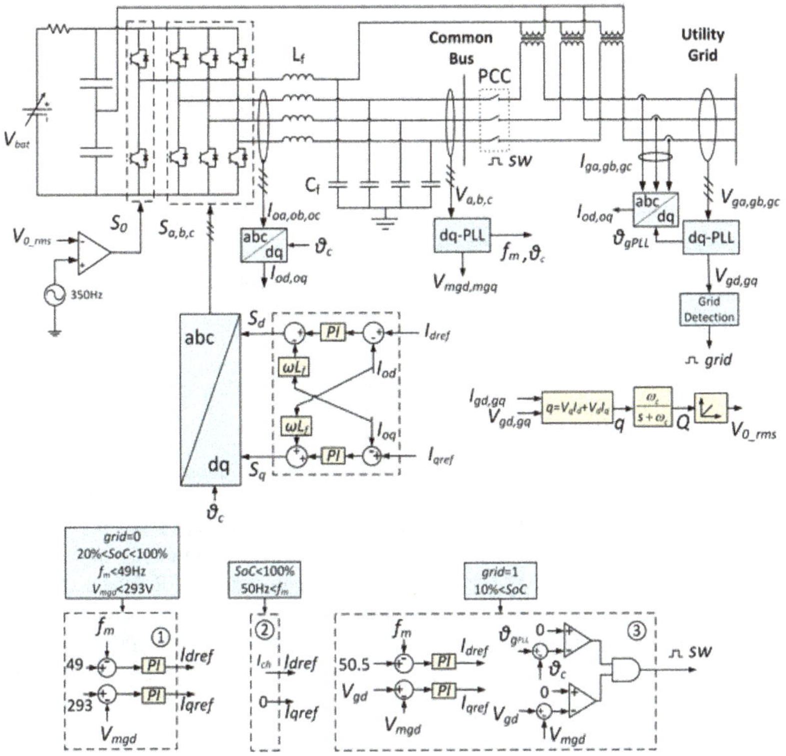

Fig. 3 *ESS control strategy: (1) discharge, (2) charge, (3) synchronisation and (4) current control*

where Q_o is the initial charge of the battery, Q_{tot} is the rated charge of the battery and i_d is the output active current of the battery converter. This paper uses a battery as ESS, taking into consideration the limits of the battery current and its rate of change. The ESS control strategy is shown in more detail in block 1 in Fig. 3.

3 Grid detection and synchronisation

The recovery of the main grid is detected by a phase locked loop (PLL) circuit incorporated into the control system of the ESS. This PLL circuit constantly monitors the voltage at the grid side of the switch *sw*, shown in Figs. 1 and 3. To connect the microgrid with the utility grid with a seamless transient effect, the microgrid magnitude and frequency should reach the respective values of the grid voltage. For this reason, the ESS increases the microgrid frequency by 0.5 Hz higher than the grid frequency. The value of 0.5 Hz results from the difference in the frequency limits among the grid-connected and island operation mode. The active and reactive powers of the battery converter are adjusted by the direct and quadrature axis components of the current according to

$$I_{\mathrm{dref}} = K_{\mathrm{pd}}(f_{\mathrm{g}} + 0.5 - f_{\mathrm{m}}) + K_{\mathrm{id}}\int (f_{\mathrm{g}} + 0.5 - f_{\mathrm{m}})\,\mathrm{d}t \qquad (7)$$

$$I_{\mathrm{qref}} = K_{\mathrm{pq}}(V_{\mathrm{gd}} - V_{\mathrm{mgd}}) + K_{\mathrm{iq}}\int (V_{\mathrm{gd}} - V_{\mathrm{mgd}})\,\mathrm{d}t \qquad (8)$$

where I_{dref}, I_{qref} are the direct and quadrature reference currents, f_{m} is the microgrid frequency, V_{gd} is the grid voltage magnitude, V_{mgd} is the microgrid voltage magnitude, K_{pd}, K_{id}, K_{pq} and K_{iq} are the proportional and integral parameters of the PI controller, as described in block 3 in Fig. 3. The proportional and integral parameters of the PI controller are designed for the worst case, where the grid frequency has its maximum value of 50.5 Hz and the DERs serve the maximum load in island operation mode with 49 Hz frequency. In this extreme case, it is assumed that the time response of the PI controller does not exceed 1 s (50 cycles). This response time affects the stress imposed on the battery discharge without having any other critical influence on the system. Thus, it could be set to any other larger value.

To implement the synchronisation, the following two conditions should be satisfied

1. $V_{\mathrm{gd}} = V_{\mathrm{mgd}}$.
2. $\theta_{\mathrm{gPLL}} = \theta_{\mathrm{c}}$, where θ_{gPLL} is extracted from the grid-side PLL and θ_{c} is the angle of the microgrid voltage.

To calculate the required battery capacity for achieving the synchronisation, the worst-case scenario should be considered. According to this, the voltage waveform at the microgrid side of PCC is assumed to differ almost 2π rad, compared with the grid-side voltage waveform. As the frequency difference between the microgrid and the grid voltage is 0.5 Hz, the required time for synchronising the two voltage waveforms is 100 cycles. Therefore, the aggregated time for synchronisation t_{sync} is set to 150 cycles. The capacity of the battery Q_{sync} can be calculated as

$$Q_{\mathrm{sync}} = I_{\mathrm{dref\_max}} \cdot t_{\mathrm{sync}} \qquad (9)$$

The maximum $I_{\mathrm{dref\_max}}$ is determined by the maximum possible microgrid load. The reasoning is as follows: in the worst case, where the DERs feed the maximum microgrid load, the microgrid frequency is 49 Hz as determined by the respective droop curves. Then, the battery is discharged in order to increase the microgrid frequency to 51 Hz (assuming, in the worst case, the grid frequency to be 50.5 Hz). Following the droop curves, the DERs will supply zero active power at 51 Hz, while the whole load is covered by the battery converter.

After the synchronisation accomplishment, the DERs detect the utility grid, without using any physical communication. The grid detection method is based on indirectly measuring the microgrid impedance. When the grid is connected, the measured microgrid impedance suddenly reduces. The impedance measurement is based on slightly distorting the injected current of each DER by modifying its control angle θ, which is calculated by the frequency

f according to the respective droop. The modified angle θ_c is defined by

$$\theta_c = \theta + k \sin \theta \quad (10)$$

where θ_c is the new control angle and the gain k is considered to be smaller than 0.05.

In this way, the distorted currents cause a respective distortion of the microgrid voltage. The latter is analysed in two components V_{mgd} and V_{mgq} by means of a Park-Transformation-based PLL. The root-mean-square (rms) value at 50 Hz of V_{mgq} is proportional to microgrid impedance, as it is accordingly described for a single-phase PV system in [4]. As a result, when the grid is absent, the rms value of the 50 Hz component of V_{mgq} remains above a predefined threshold (6 V in this paper). On the contrary, in grid presence, this value drops well below this threshold.

It is now obvious that the grid detection and the synchronisation process are accomplished in a 'wireless' way, contrary to the current practice. For this reason, the placement of the ESS at the PCC is crucial.

4 Power management in grid-connected mode

4.1 Active power management

In grid-connected mode, the DERs within the microgrid are controlled in order to inject all the available active power. In this operation mode, the reference value of the active component of the current of each DER I_{dref} is determined by

$$I_{dref} = \frac{P_{pot}}{V_{mgd}} \quad (11)$$

where V_{mgd} is the direct axis voltage component, as shown in Fig. 2.

If the battery is not fully charged, the ESS charges the battery. The optimum battery charging current I_{ch}, which is considered to 1/10 of the nominal battery capacity.

4.2 Reactive power management

A novel approach for the reactive power management in grid-connected mode is proposed in this paper. The aim is the zeroing of the reactive power exchange at the PCC by fulfilling the reactive power demand of the loads by the DERs within the microgrid. This approach will enable the microgrid operators to avoid any costs associated with the exchange of reactive power with the utility grid and will additionally result in the reduction of the power losses in the distribution lines within the microgrid. In the case that the DER is connected at the same local bus with a load (e.g. $DER_1 + L_1$ and $DER_4 + L_4$ in Fig. 1), the priority is to cover the reactive power of the local load. If the DER can deliver any additional reactive power, it will contribute to the reactive power management. The individual DERs contribute with reactive power, in order to fulfil the reactive power of the loads and the reactive power consumed by the distribution lines of the microgrid.

The ESS constantly measures the reactive power exchange at the PCC and injects a zero-sequence voltage drop at the 7th harmonic in series with the utility grid by means of three transformers, forming in this way a zero-sequence component. The magnitude of the injected voltage is determined in relation to the measured reactive power by a proposed curve, shown in Fig. 4. The maximum reactive power, Q_{max_PCC}, corresponds to the maximum aggregated reactive power of the loads within the microgrid, whereas V_{0_rms} is the maximum magnitude of the zero-sequence 7th harmonic voltage. The ESS converter is designed as a four-leg converter (Fig. 3), where the three of the legs are controlled for the injection of active and reactive powers, as mentioned above, while the fourth leg is used for the production of a sinusoidal voltage at the 7th harmonic. To avoid large zero-sequence currents because of this

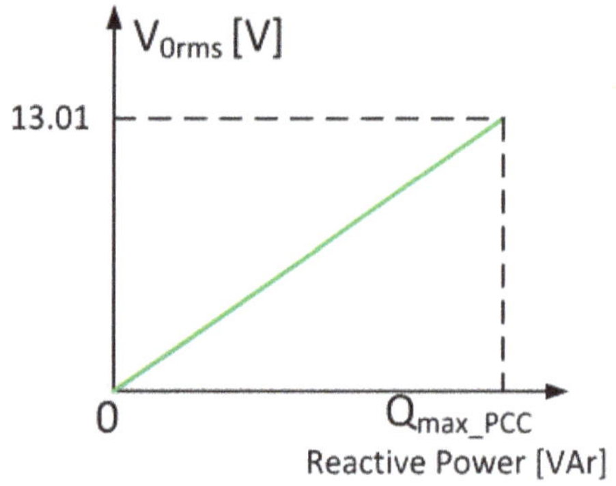

Fig. 4 *V0_rms – Q control curve*

voltage, a power transformer with a delta winding on the grid side should be used at the PCC. The delta winding creates infinite zero-sequence impedance on the grid side, forcing any zero-sequence currents to flow only within the microgrid.

Subsequently, each DER detects only the zero-sequence component at the 7th harmonic of the measured voltage by means of a *dq0* PLL. The amount of reactive power is determined through a PI controller, which tries to zero the 7th harmonic zero-sequence voltage component by injecting larger reactive current, as follows

$$I_{qref} = K_{pqr} \cdot V_{0\_rms}(t) + K_{iqr} \int V_{0\_rms}(t)\, dt \quad (12)$$

where I_{qref} is the reference reactive current, V_{0_rms} is the rms value of the zero-sequence 7th harmonic voltage, K_{pqr}, K_{iqr} are the proportional and integral parameters of the PI controller. The proportional parameter K_{pqr} should be determined in order to ensure that the reactive power of each DER is proportional to its rated apparent power. Therefore it can be calculated as

$$K_{pqr} = c \cdot S_{nom} \quad (13)$$

where c is a constant parameter set equal to 0.025 in this paper, while S_{nom} is in kilovolt amperes (kVA). The integral parameter K_{iqr} is selected in combination with K_{pqr} in order to adjust the injected reactive power to the proper value within ~25 cycles.

The signal for zeroing the reactive power is selected to be zero-sequence of the 7th harmonic, because of numerous reasons. Firstly, it could not belong to a triple harmonic (3rd, 9th etc.), because these harmonics can be normally produced by non-linear loads and behave as zero-sequence. As a result, a possible signal at these frequencies would mix with the respective harmonics of the non-linear loads. Secondly, the solution of interharmonics or even harmonics is also rejected, since [2] permits only very restricted magnitudes of them. Furthermore, the signal cannot be of relatively high harmonic order (higher than the 25th), because of interference with the filters or possible resonance inside the microgrid. Additionally, as the harmonic grows in order, EN50160 Standard reduces its permissible magnitude, which may cause problems in its detection from the DERs. Therefore the available harmonic order can be the 5th, 7th and 11th harmonics, since any resonances at these frequencies seem very unlikely because of the absence of significant capacitances in low-voltage microgrids.

The selection among the available 5th, 7th and 11th harmonics is made under the criteria of their permissible values [2] and the attenuation through the connection lines inside the microgrid. Taking into account (i) the fact that the attenuation (in percent) of the injected voltage is almost the same for each frequency (as is shown in the simulation results), (ii) the permissible limits are 6, 5 and 3.5% of the nominal voltage for the 5th, 7th and 11th harmonics [2], (iii) the necessity to easily detect the injected voltage

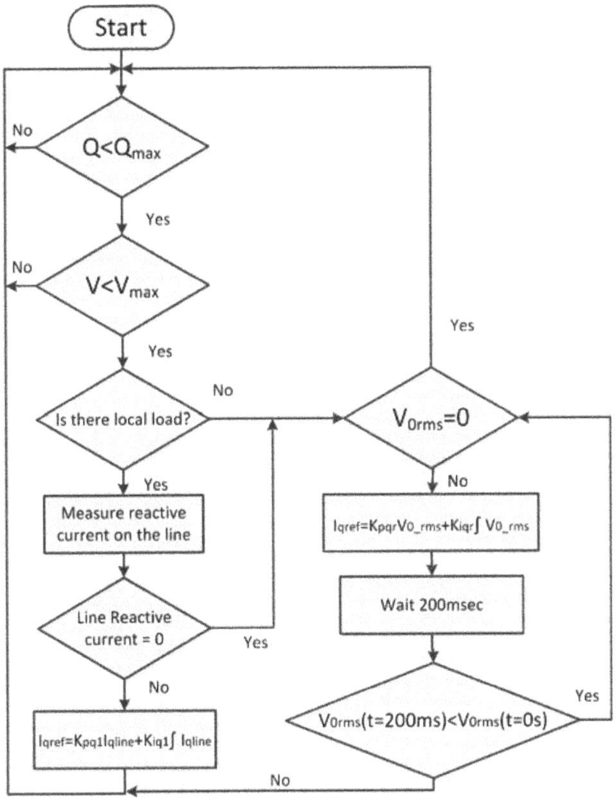

Fig. 5 *Flowchart showing the reactive power control strategy of each DER in grid-connected mode*

and (iv) the likelihood of the natural existence of a respective zero-sequence voltage [19] makes the selection of the 7th harmonic a reasonable trade-off.

Finally, the maximum magnitude of the injected 7th harmonic zero-sequence voltage is selected to be 4% instead of 5% of the nominal voltage, allowing another 1% to any non-linear asymmetrical loads within the microgrid.

Since each DER should provide the microgrid with reactive power, even under maximum active power conditions, P_{pot}, the apparent power of each converter should be oversized. Therefore, the apparent power should be determined as

$$S_{nom} = a \cdot P_{pot} \tag{14}$$

with α being the oversize factor. Since it was assumed that the average PF of the total microgrid load equals to 0.8, the oversize factor a corresponds to 1.25.

Fig. 5 shows in flowchart form the control strategy of the reactive power of each DER in grid-connected mode. The rms value of the zero-sequence voltage is compared with its rms value after 200 ms. If there is no reduction, it means that there is a 7th harmonic zero-sequence component because of non-linear asymmetrical loads. In such a case, the DER stops injecting reactive power.

5 Simulation tests

The simulation model of the microgrid, implemented in the Powersim software (PSIM) platform, consists of four inverter-based DERs, a battery at the PCC and constant-power loads, as seen in Fig. 1. The primary sources of all DERs are PV panels, connected in series and parallel in order to produce the necessary DC power and voltage. Each string was modelled with its *I–V* characteristic by calculating the MPP for given solar irradiance *G* [20]. The MPP model was further modified in this paper so that it can drive the PV system in power levels other than the available maximum one, a feature that is necessary in the island operation mode. Each of DER_1, DER_2 and DER_3 has two PV strings, whereas DER_4 has three PV strings connected. The aggregated nominal DC-power P_{nom} of DER_1, DER_2, DER_3 and DER_4 is 10, 8, 10 and 15 kW_p, respectively, under standard testing conditions (STCs). A boost DC/DC converter is used in order to raise the MPP voltage to 800 V. Table 1 depicts the parameters of the inverters. The distribution lines within the microgrid were assumed to be overhead aluminium conductor steel reinforced (ACSR) 16 mm^2 with $R = 1.268$ Ω/km and $X = 0.422$ Ω/km. A Dy5 20/0.4 kV power transformer of 100 kVA was selected at the PCC. The utility grid was simulated with its Thevenin equivalent, that is, a voltage source of 20 kV (line voltage) in series with a complex impedance of $R = 0.333$ Ω and $X = 1.35$ Ω, corresponding to 250 MVA short-circuit apparent power.

The battery is modelled as a variable DC voltage source in series with a resistance, where the DC voltage is a function of the battery SoC [21]. The battery bank is assumed to consist of 400 series connected 2 V cells. The capacity for performing the synchronisation is calculated by (9) equal to 232.5 A s, considering the maximum current $I_{dref} = 77.5$ A and $t_{sync} = 3$ s. The total battery capacity was selected equal to $Q_f = 1600$ A s, which represents a simulation-scaled value. All the loads in the microgrid were simulated as constant-power loads (Table 2). The maximum available reactive power from the DERs, operating with nominal active power, is 32.25 kVAr. This value corresponds to the maximum reactive power of all microgrid loads, thus $Q_{max\_PCC} = 32.25$ kVAr. The series transformers were assumed ideal, with turns ratio equal to 100/3. The equations for the PV and battery model and the transformer data are described in Appendix.

Table 1 Inverter parameters

Items	Inverter #1, 3	Inverter #2	Inverter #4	Battery
rated power *S*, kVA	12.5	10	18.75	–
nominal DC-power P_{nom}, kW	10	8	15	–
filter inductance L_f, mH	2	2	2	2
filter capacitance C_f, μF	15.5	15.5	15.5	15.5
derivative coefficient m_d, n_d	10^{-5}	10^{-5}	10^{-5}	–
gain of PI K_{pqr}	0.3	0.25	0.45	–
time constant of PI K_{pqr}	0.01	0.01	0.01	–
switch transition frequency f_{tr}, kHz	9.95	9.95	9.95	9.95

5.1 *Islanded operation, grid detection and synchronisation*

Initially, the microgrid operates in island mode with each load having active and reactive powers, as in Table 1. The solar irradiance is assumed to be 900 W/m^2, instead of 1000 W/m^2 under STC, which corresponds to P_{pot} equal to 9 kW for DER_1 and DER_3, 7.2 kW for DER_2 and 13.5 kW for DER_4. In this case, the aggregated available active power from the DERs is 38.7 kW, while the loads demand only 29 kW. The combination of all droop characteristics results in a common microgrid frequency of 49.48 Hz, which in turn determines the desired active power of

Table 2 Load active and reactive powers

Items	Load #1	Load #2	Load #3	Load #4
active power, kW	7	8	6	8
reactive power, kVar	3	5	2	5

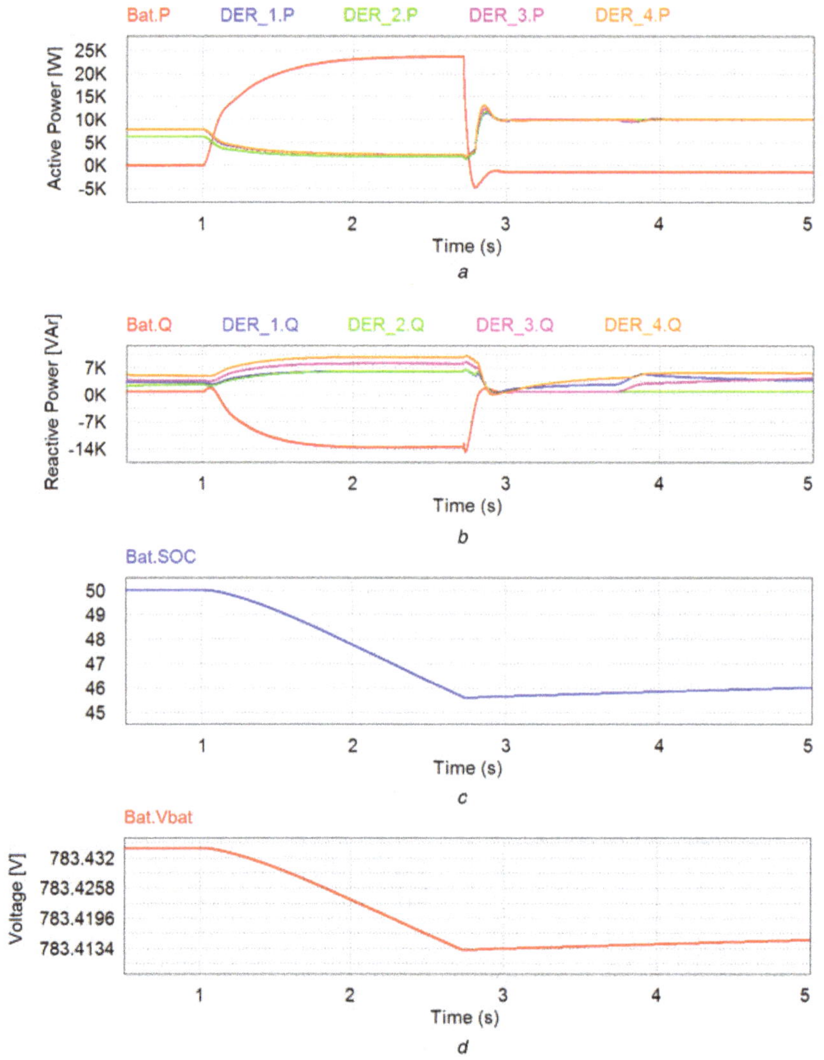

Fig. 6 *Islanded operation, grid detection and synchronisation*
a Active power [W] of the DERs and the battery, denoted as DER_1.P, DER_2.P, DER_3.P and DER_4.P, Bat.P, respectively
b Reactive power [VAr] of the DERs and the battery, denoted as DER_1.Q, DER_2.Q, DER_3.Q and DER_4.Q, Bat.Q
c SoC [%] of the battery
d Battery internal voltage V_{bat} [V]

the respective DER. The signal of this active power is then used by the presented control algorithm of the PV systems, in order to adjust their voltage away from the MPP voltage. Fig. 6*a* shows (from t = 0.5 to 1 s) the active power of each DER of the microgrid, where it is evident that the active power of the total load is covered by the DERs in proportion to their available active power, P_{pot}.

Through the combination of the respective droop characteristics, the reactive power of the loads is shared among the DERs, as shown in Fig. 6*b*, so that the microgrid voltage settles at all nodes within the limits imposed by [2]. With the additional constraint that PF > 0.8 in each DER (see Fig. 2), a large disproportion in the sharing of the reactive power among the DERs is avoided.

At t = 1 s, the grid is initially detected by the grid-side PLL of the ESS, which in simulation terms corresponds to the closing of switch *gsw*. The ESS immediately initialises the synchronisation process, by adjusting the voltage magnitude and frequency of the microgrid. As shown in Figs. 6*c* and *d*, the ESS discharges in order to increase the microgrid frequency to 50.5 Hz. For this reason, the SoC decreases from 50 to 45.6%, resulting in a drop of the internal battery voltage. In a similar way, the reactive power of the ESS changes, so that the microgrid single-phase rms voltage at the PCC changes from 233.26 to 230.7 V. Fig. 7*a* shows the microgrid and grid voltages just before and after the synchronisation. At t = 2.72 s, the synchronisation conditions are satisfied, having as a result the closing of switch *sw*. The DERs realise the grid presence by the decrease of the rms value of the 50 Hz component $V_{mgq\_50\_rms}$, as shown in Fig. 7*b*, and change their control strategy from droop control to grid-connected, when $V_{mgq\_50\_rms}$ drops below the threshold of 6 V. In this case, the synchronisation process took about 1.72 s.

5.2 *Grid-connected operation mode*

When the DERs detect the grid presence, they switch to grid-connected control mode and inject active power equal to P_{pot} (Fig. 6*a*), after ~t = 3 s. After synchronisation, DER_1 and DER_4 provide the reactive power demanded by their local loads, Fig. 6*b* from t = 2.9 s to 3.72 s. For this reason, the reactive power absorbed by the grid is reduced as shown in Fig. 8*a*. At t = 3.72 s, the ESS starts injecting the zero-sequence voltage, whose maximum value is 2.36 V for 5.76 kVAr absorbed from the grid at the PCC, as it appears in Fig. 8*b*. Consequently, all DERs adjust their reactive power according to the detected zero-sequence voltage, which finally results in zeroing the reactive power exchange at t = 3.9 s. In this final condition, the power losses in the distribution lines within the microgrid are calculated to be 680 W. If the reactive

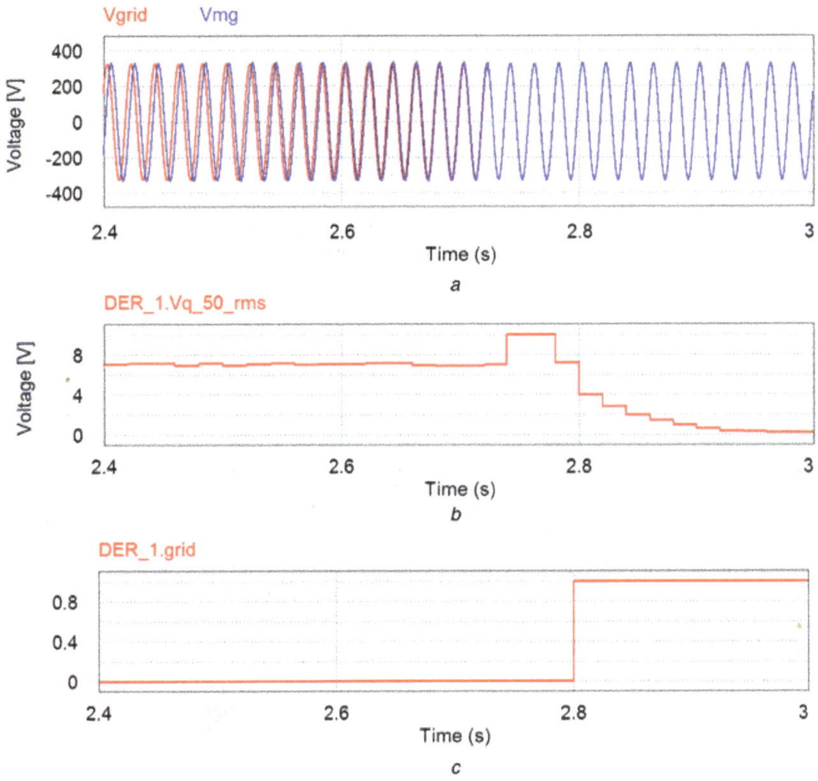

Fig. 7 *Microgrid and grid voltages just before and after the synchronisation*
a Voltage profile at the microgrid side V_{mg} [V] and at the grid side V_{grid} [V]
b RMS value at 50Hz of q-axis voltage of DER_1 denoted as DER_1.Vq_50_rms[V]
c Grid Signal of DER_1

power of the loads were covered by the grid, that is, the DERs delivered no reactive power, the losses in the distribution lines within the microgrid would be 1485 W. Thus, apart from the zeroing of the reactive power at the PCC, a partial optimisation regarding the power losses is achieved with the proposed method. A full optimisation regarding the power losses would require a communication between the DERs, the ESS and the grid and an optimal power flow analysis. The battery starts charging with a constant current after the grid is connected, because its SoC was lower than 100%. As can be seen from Figs. 6*c* and *d*, the internal voltage starts increasing according to the SoC.

To compare the effectiveness of the 7th harmonic selection with respect to other harmonics, the simulation test is conducted for the 5th and the 11th harmonic and the results are presented in Table 3. The 7th harmonic presents less attenuation compared with other harmonics, reinforcing its selection.

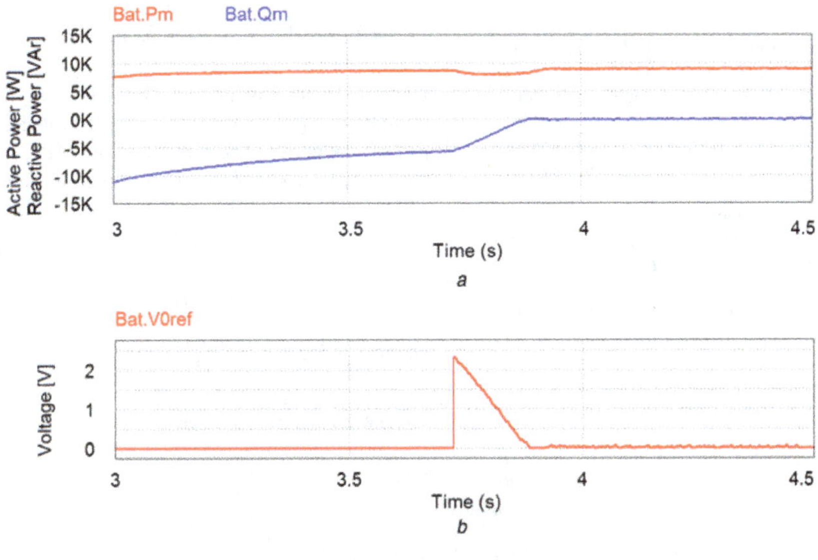

Fig. 8 *Power management strategy results in grid-connected mode*
a Active power [W] and reactive power [VAr] at PCC, denoted as Bat.Pm and Bat.Qm respectively
b Amplitude of zero-sequence voltage component at 350 Hz V_{0ref} [V] sent by the battery

Table 3 Attenuation of the rms value of the zero-sequence voltage

Node	5th, V	%ΔV/V, %	7th, V	%ΔV/V, %	11th, V	%ΔV/V, %
PCC	11.50	–	9.20	–	5.75	–
DER_1	11.19	−2.69	9.16	−0.43	5.68	−1.22
DER_2	11.03	−4.09	9.07	−1.41	5.53	−3.83
DER_3	10.89	−5.30	8.97	−2.50	5.45	−5.22
DER_4	10.76	−6.43	8.78	−4.57	5.30	−7.83

6 Conclusion

A new wireless decentralised energy management method for a microgrid is proposed. It is implemented by strategically placing an ESS at the PCC with the utility grid. The ESS uses an additional PLL to detect the grid presence and guide the microgrid to synchronisation, without requiring any further communication among the grid and the DERs within the microgrid. During the islanding operation, a method for dynamically adjusting the droop characteristic of the DER, when fed by renewable energy sources, is also proposed. Additionally, a constraint imposed on the operating PF of each DER is shown to avoid any misallocation of reactive power among the DERs, regardless of the distribution line impedances and the number and location of the DERs.

A new control method of the reactive power within the microgrid in grid-connected mode is also proposed. The aim is to zero the exchange of reactive power at the PCC and give priority to cover the reactive power of a load locally. A fourth leg in the ESS converter is controlled so as to inject a 7th harmonic zero-sequence voltage at the PCC through series transformers. The DERs adjust their reactive power according to the magnitude of the detected zero-sequence voltage. A Dy 20/0.4 kV power transformer should be placed at the PCC, so that the zero-sequence voltage is confined within the microgrid. It was shown that apart from zeroing the reactive power at the PCC – thus reducing the relative financial charge – may lead to a reduction of the power losses in the distribution lines of the microgrid. This injected zero-sequence voltage could also be used by the microgrid in order to provide ancillary services, such as lagging or leading reactive power to the utility grid. In such a case, the magnitude of the zero-sequence voltage could be determined via an appropriate signal by the utility grid operator. The proposed control methods were verified by detailed simulation.

7 References

[1] Katiraei F., Iravani R., Hatziargyriou N.: 'Microgrids management', *IEEE Power Energy Mag.*, 2008, **6**, (3), pp. 54–65
[2] Voltage characteristics of electricity supplied by public distribution system, standard EN 50160:2004
[3] Yu B., Matsui M., Yu G.: 'A review of current anti-islanding methods for photovoltaic power system', *Sol. Energy*, 2010, **84**, (5), pp. 745–754
[4] Ciobotaru M., Agelidis V.G., Teodorescu R.: 'Accurate and less-disturbing active antiislanding method based on PLL for grid-connected converters', *IEEE Trans. Power Electron.*, 2010, **25**, (26), pp. 1576–1584
[5] Dimeas A.L., Hatziargyriou N.D.: 'Operation of a multi-agent system for microgrid control', *IEEE Trans. Power Syst.*, 2005, **20**, (3), pp. 1447–1455
[6] Katiraei F., Iravani M.R.: 'Power management strategies for a microgrid with multiple distributed generation units', *IEEE Trans. Power Syst.*, 2006, **21**, (4), pp. 1821–1831
[7] Guerrero J.M., Matas J., de vicuña L.G., Castilla M., Miret J.: 'Wireless-control strategy for parallel operation of distributed-generation inverters', *IEEE Trans. Ind. Electron.*, 2006, **53**, (5), pp. 1461–1470
[8] Balaguer I.J., Lei Q., Yang S., Supatti U., Peng F.Z.: 'Control for grid-connected and intentional islanding operations of distributed power generation', *IEEE Trans. Ind. Electron.*, 2011, **58**, (1), pp. 147–157
[9] Levron Y., Guerrero J.M., Beck Y.: 'Optimal power flow in microgrids with energy storage', *IEEE Trans. Power Syst.*, 2013, **28**, (3), pp. 3226–3234
[10] Tan X., Li Q., Wang H.: 'Advances and trends of energy storage technology in microgrid', *Int. J. Electr. Power Energy Syst.*, 2013, **44**, (1), pp. 179–191
[11] Zhou H., Bhattacharya T., Tran D., Siew T.S.T., Khambadkone A.M.: 'Composite energy storage system involving battery and ultracapacitor with dynamic energy management in microgrid applications', *IEEE Trans. Power Electr.*, 2011, **26**, (3), pp. 923–930
[12] Kyebyung L., Son K.M., Gilsoo J.: 'Smart storage system for seamless transition of customers with intermittent renewable energy sources into microgrid'. 31st Int. Telecommunications Energy Conf., 2009, pp. 1–5
[13] Guo W., Liye X., Shaotao D.: 'Enhancing low-voltage ride-through capability and smoothing output power of DFIG with a superconducting fault-current limiter-magnetic energy storage system', *IEEE Trans. Energy Convers.*, 2012, **27**, (2), pp. 277–295
[14] Jayawarna N., Barnes M., Jones C., Jenkins N.: 'Operating microgrid energy storage control during network faults'. IEEE Int. Conf. on System of Systems Engineering, 2007, pp. 1–7
[15] Lee C., Ruei-Pei J., Po-Tai C.: 'A grid synchronization method for droop-controlled distributed energy resource converters', *IEEE Trans. Ind. Appl.*, 2013, **49**, (2), pp. 954–962
[16] Vandoorn T.L., Meersman B., De Kooning J.D.M., *et al.*: 'Transition from islanded to grid-connected mode of microgrids with voltage-based droop control', *IEEE Trans. Power Syst.*, 2013, **28**, (3), pp. 2545–2553
[17] Oureilidis K., Demoulias C.: 'Microgrid wireless energy management with energy storage system'. 47th IEEE Int. Universities Power Engineering Conf., September 2012, pp. 1–6
[18] He J., Yun Wei L.: 'Analysis, design, and implementation of virtual impedance for power electronics interfaced distributed generation', *IEEE Trans. Ind. Appl.*, 2011, **47**, (6), pp. 2525–2538
[19] Bastião F., Lebre A., Jorge L., Blanco A., Leiria A., Alves J.: 'Characterization of voltage harmonic distortion in the Portuguese medium and low voltage grids'. 22nd Int. Conf. and Exhibition on Electricity Distribution (CIRED 2013), June 2013, pp. 1–4
[20] Papaioannou I., Alexiadis M., Demoulias C., Labridis D.P., Dokopoulos P.S.: 'Modeling and field measurements of photovoltaic units connected to LV grid. Study of penetration scenarios', *IEEE Trans. Power Deliv.*, 2011, **26**, (2), pp. 979–987
[21] Bakirtzis E., Demoulias C.: 'Control of a micro grid supplied by renewable energy sources and storage batteries'. XXth Int. Conf. on Electrical Machines, 2012, pp. 2053–2059

8 Appendix

Electrical data of the oil 0.4/20 kV transformer:

$S = 100$ kVA, $P_{Fe} = 320$ W, $P_{Cu} = 1750$ W, $u_k = 4\%$, no-load current = 1.44 A referred to 0.4 kV.

The equations of the PV panel model are

$$I_{sc} = \beta \cdot G,\ V_{oc} = \gamma \cdot G + \delta,\ I_{MPP} = \lambda \cdot G,\ V_{MPP} = \frac{\zeta}{\xi - \sigma \cdot G}$$

where G [W/m^2] is the input irradiance, I_{sc} [A] is the short-circuit current, V_{oc} [V] is the open-circuit voltage, I_{MPP} [A] and V_{MPP} [V] are the current and voltage of the MPP, respectively. The parameters β, γ, δ, λ, ζ, ξ and σ are given in Table 4.

The DC voltage of each 2 V battery is described as

$$V_{oc} = 1.958 + 1.155 \times 10^{-3} \cdot x + 2.946 \times 10^{-5} \cdot x^2 - 2.112 \times 10^{-7} \cdot x^3$$

where x is the SoC of the battery.

Table 4 PV model parameters

kW_p	B, 10^{-3}	γ, 10^{-3}	δ	λ, 10^{-3}	ζ	ξ	σ, 10^{-3}
8	17.2	37	537	15.48	94584.72	190.7	7.74
10	17.6	73.4	650.9	15.84	700703.2	1127.8	7.92
15	26.4	73.4	650.9	23.76	700703.2	1127.8	7.92

MPD model for radar echo signal of hypersonic targets

Xu Xuefei, Liao Guisheng

National Laboratory of Radar Signal Processing, Xidian University, Xi'an 710071, People's Republic of China
E-mail: iexuxuefei@hotmail.com

Abstract: The stop-and-go (SAG) model is typically used for echo signal received by the radar using linear frequency modulation pulse compression. In this study, the authors demonstrate that this model is not applicable to hypersonic targets. Instead of SAG model, they present a more realistic echo signal model (moving-in-pulse duration (MPD)) for hypersonic targets. Following that, they evaluate the performances of pulse compression under the SAG and MPD models by theoretical analysis and simulations. They found that the pulse compression gain has an increase of 3 dB by using the MPD model compared with the SAG model in typical cases.

1 Introduction

It is a challenging task for radar systems to detect the hypersonic targets with velocities more than 20 Mach and located in long range. To detect this kind of dim target, coherent processing of echo signal received by radar in order to obtain the maximum signal-to-noise ratio is the most important problem.

The stop-and-go (SAG) model is the most popular to describe the range history of echo signal of a moving target [1–3]. The SAG model ignores the effect of the target's motion within pulse duration and only considers the time delay and Doppler caused by the motion between different pulses. Hence, the SAG model can simplify the processing of synthetic aperture radar (SAR) imaging [1–3]. However, the SAG model will bring model error when applied to the received echo pulses of hypersonic targets, since the received echo pulses will be affected by the Doppler caused by the motion during the pulse duration and they are different from the transmitted pulses in waveform and spectrum. This issue becomes severe when radar systems have large time-bandwidth product and high resolution. This issue has been analysed in some practical applications. In [4], the effect of the model error on imaging and interferometric processing was analysed and a formula for effective range error was then presented in the form of position, velocity and acceleration parameters of satellites [4]. The error caused by the SAG model in the frequency modulated continuous wave (FM-CW) with a 100% duty-cycle and large pulse duration was analysed in [5]. A method for space-borne SAR geolocation-based on continuously moving geometry instead of the SAG model was proposed in [6].

In the application of detection and tracking of moving targets, the SAG model for echo is still frequently used. There are few literatures about how the SAG model affects the moving target detection. Although some Doppler parameter compensation methods were presented for detecting high-speed moving targets, they are all based on the SAG model and ignore the effects during pulse durations [7, 8].

In this paper, we demonstrate that the SAG model will cause the offset and broadening of mainlobe and increase sidelobe in pulse compression process, based on theoretical analysis and numerical simulations. To solve this problem, a new echo model called 'moving-in-pulse duration (MPD)' is proposed as a more realistic echo model, which can accurately describe the echo signal of hypersonic targets. Pulse compression performances under the SAG and the MPD models have been compared in the scenarios of point targets and extended targets. We find that, under the situations of hypersonic targets and large time-bandwidth product of radar systems, the MPD model has better features than the SAG model in terms of improving the pulse compression gains. Furthermore, we demonstrate the necessity of using the MPD model to describe the echo signal of hypersonic targets. In typical cases, the pulse compression gain has an increase of 3 dB when using the MPD model, compared with the SAG model.

The remaining contents of this paper are organised as follows. Section 2 gives the MPD model with the SAG as the background. Detailed analysis and comparison of the two models in terms of pulse compression are provided in Section 3. Section 4 shows the simulation results of pulse compression of hypersonic targets in the scenarios of point target and high resolution range profile (HRRP). The conclusion is given in Section 5.

2 Model of echo signal

Suppose that a linear FM (LFM) signal is transmitted

$$S(t) = \mathrm{rect}\left(\frac{t}{T_\mathrm{p}}\right)\exp(\mathrm{j}\pi K t^2) \quad (1)$$

where T_p is the pulse duration, K is the FM rate, t is the time and

$$\mathrm{rect}(u) = \begin{cases} 1 & |u| \le 0.5 \\ 0 & \text{else} \end{cases}$$

In the literature, LFM signal was considered to have a relatively high tolerance for Doppler caused by the motion of a moving target. The SAG model for echo of LFM signal is equivalent to making the assumption that a moving target keeps stationary from the beginning of a pulse illuminating the moving target to the end of the pulse duration, namely, the moving target did not move during the pulse duration. Therefore the SAG model can only reflect the time delay and Doppler among different pulses, but does not consider the range variation within single pulse duration. For the SAG model, the range of a hypersonic target changes from one pulse to another. When the mth pulse illuminates a target, the range can be expressed as $R(t_m)$

$$R(t_m) = \sum_{j=0}^{\infty} a_j t_m^j = R_0 - v_\mathrm{r} t_m - \frac{1}{2} a_\mathrm{r} t_m^2 - \cdots \quad (2)$$

where R_0 is the initial range, $m = 0, 1, 2, \ldots$, v_r is the radial velocity, v_r is positive when the target moving forward to the radar, a_r is the radial acceleration, a_r is positive when the direction of a_r is towards the radar and $t_m = mT_\mathrm{r}$ is the slow-time.

During a single pulse, the time delay τ_m is constant depending on $R(t_m)$

$$\tau_m = \frac{2R(t_m)}{c} \tag{3}$$

the mth pulse return can be expressed as [1]

$$S_r(t, t_m) = A_0 \text{rect}\left(\frac{t-\tau_m}{T_p}\right) \exp\left(j\pi K(t-\tau_m)^2\right) \exp(-j2\pi f_0 \tau_m) \tag{4}$$

where A_0 is an amplitude constant, c is the velocity of light and f_0 is the carrier frequency.

When detecting a hypersonic target whose velocity is as high as dozens of Mach, it is necessary to use the LFM signal with large time-bandwidth product to meet the requirement of radar maximal range, resolution and detection performance. In this situation, the motion of the hypersonic target in pulse duration cannot be neglected. Thus, the SAG model is not applicable to the hypersonic target, since it cannot reflect the variation in echo pulses. In this paper, a novel echo model for hypersonic targets, named the MPD echo model, is proposed as follows.

The most significant difference between the MPD mode and the SAG model is the time delay. In the MPD model, the range $R(t)$ in a pulse is a continuous function which varies with t rather than constant in the SAG model. Then, the time delay τ_{mpd} of received echo pulse could be written as

$$\tau_{\text{mpd}} = \frac{2R(t)}{c} = \tau_m + \frac{2(R(t) - R(t_m))}{c} \tag{5}$$

The MPD model of echo signal can be expressed as

$$S_r'(t, t_m) = A_0 \text{rect}\left(\frac{t-\tau_{\text{mpd}}}{T_p}\right) \exp\left(j\pi K\left(t-\tau_{\text{mpd}}\right)^2\right) \times \exp\left(-j2\pi f_0 \tau_{\text{mpd}}\right) \tag{6}$$

The difference between (6) and (4) is the time delay expressed by (5). Both $R(t)$ and τ_{mpd} are functions of t. Therefore, the MPD model can reflect the details in the time delay of the echo signal from a moving target.

By substituting (5) in (6), the received echo pulse under the MPD model is written as

$$\begin{aligned} S_r'(t, t_m) = {} & A_0 \text{rect}\left(\frac{t - \tau_m - ((2(R(t) - R(t_m)))/c)}{T_p}\right) \\ & \times \exp\left(j\pi K\left(t - \tau_m - \frac{2(R(t) - R(t_m))}{c}\right)^2\right) \\ & \times \exp\left(-j2\pi f_0\left(\tau_m + \frac{2(R(t) - R(t_m))}{c}\right)\right) \end{aligned} \tag{7}$$

It is noted that the MPD model will be the same with the SAG model when we do not consider the target's motion within the pulse duration, that is, when $\tau_{\text{mpd}} = \tau_m$ (constant), the MPD model is the same as the SAG model.

For convenience, we provide a special case. Assuming the hypersonic target moving with a constant radial velocity v_r, the range from the target to radar can be expressed as

$$R(t) = R_0 - v_r t = R(t_m) - v_r(t - t_m) \tag{8}$$

Substituting (8) in (7), we obtain the expression of echo pulse as

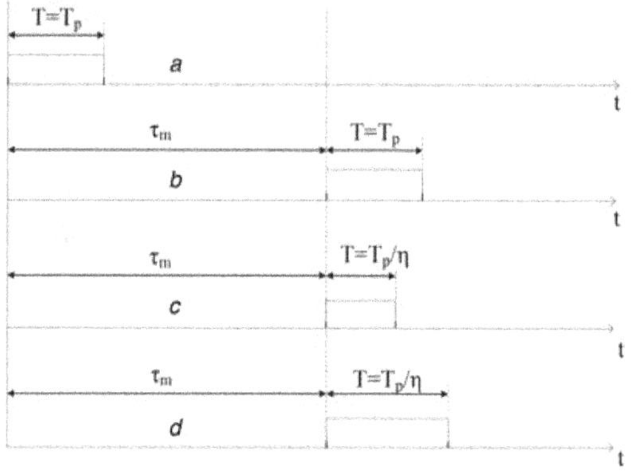

Fig. 1 *Changes in echo pulse duration in the SAG model and the MPD model*
a Transmit pulse duration
b Echo pulse duration of SAG
c Echo pulse duration of MPD ($v_r > 0$)
d Echo pulse duration of MPD ($v_r < 0$)

follows

$$\begin{aligned} S_r'(t, t_m) = {} & A_0 \text{rect}\left(\frac{\eta(t-\tau_m)}{T_p}\right) \exp\left(j\pi K\left(\eta(t-\tau_m)\right)^2\right) \\ & \times \exp\left(-j2\pi f_0\left(\tau_m \eta - \frac{2v_r t}{c}\right)\right) \end{aligned} \tag{9}$$

Define $\eta = (1 + (2v_r/c))$ as the compression/stretching factor. We can obtain the variable range of τ_{mpd} in the time domain: $\tau_{\text{mpd}} \in [\tau_m, \tau_m + T_p/\eta]$.

Fig. 1 shows the comparisons in echo pulse duration between the SAG model and the MPD model. As shown in the second line for the SAG model, echo pulse has the same pulse duration T_p as the transmitted pulse. However, the third and the fourth lines demonstrate that the pulse duration in the MPD model will change because of the Doppler caused by the target's motion. When the radial velocity is positive, the pulse duration of the received echo pulse will be compressed with the factor η, and vice versa.

3 Analysis of pulse compression performance

The following section discusses the pulse compression performance of hypersonic targets in the above-mentioned two models. For convenience, we consider that the velocities of hypersonic targets are constant. We first obtain the frequency domain of echo signals in the two models and then compare the performances in the two models in terms of pulse compression. It is straightforward to extend the results to the targets with non-constant velocities.

3.1 Frequency domain of the two models

Owing to the large time-bandwidth product of transmitted pulse, the principle of stationary phase (POSP) [1–3] can be used to obtain the frequency domain of $S_r(t, t_m)$ and $S_r'(t, t_m)$ (details are in Appendix).

The frequency domain of the SAG model can be written as

$$\begin{aligned} S_r(f, t_m) &= \int_{-\infty}^{+\infty} S_r(t, t_m) \exp(-j2\pi f t)\, dt \\ &= S(f) \exp\left(-j\frac{4\pi}{\lambda} R(t_m)\right) \exp\left(-j\frac{4\pi}{c} f R(t_m)\right) \end{aligned} \tag{10}$$

where $S(f)$ is the emitted pulses in frequency domain

$$S(f) = \text{rect}\left(\frac{f}{KT_p}\right)\exp\left(-j\pi\frac{f^2}{K}\right) \tag{11}$$

Similarly, we can obtain the emitted pulses in frequency domain when considering the MPD model (details are in Appendix)

$$\begin{aligned} S_r'(f, t_m) &= \text{rect}\left(\frac{f-(2v_r/\lambda)}{KT_p\eta}\right)\exp\left(-j\pi\frac{(f-(2v_r/\lambda))^2}{K\eta^2}\right) \\ &\times \exp\left(-j2\pi\frac{f-(2v_r/\lambda)}{\eta}\left(\frac{2R(t_m)}{c}+\frac{2v_r t_m}{c}\right)\right) \\ &\times \exp\left(-j\frac{4\pi}{\lambda}R(t_m)-j\frac{4\pi}{\lambda}v_r t_m\right) \end{aligned} \tag{12}$$

By comparing (11) and (12) of the MPD model with the SAG model, we can find that when describing echo signal with the more realistic MPD model the bandwidth and the centre of the frequency become $KT_p\eta$ instead of KT_p, and $f_0+(2v_r/\lambda)$ instead of f_0. Fig. 2 shows the differences between bandwidths in detail.

Fig. 2 shows the comparisons of bandwidth between the SAG model and the MPD model. The second line shows that the echo pulse has the same bandwidth KT_p as the transmitted pulse in the SAG model. However, the third and the fourth lines indicate that the bandwidth in the MPD model is changed because of the Doppler caused by the high-speed motion of the hypersonic target. The bandwidth of the received echo pulse will be stretched with the factor η and the centre of frequency will offset with $2v_r/\lambda$ at the same time when the radial velocity is positive, and vice versa. In addition, when the time duration is compressed with the factor η, its bandwidth will also be stretched with the factor η.

3.2 Comparisons in terms of pulse compression under the SAG and the MPD model

Owing to the requirements of long range and high resolution for radar system, pulse compression is a commonly used method to compress the emitted pulses into narrow pulses. In general, the original emitted signal is used to obtain the response of the matched filters for the echo signal in pulse compression. If the received echo signals are described by the SAG model, namely, $S_r(f, t_m)$ in (10) is the signal in frequency domain and $S_r(t, t_m)$ in (4) is the

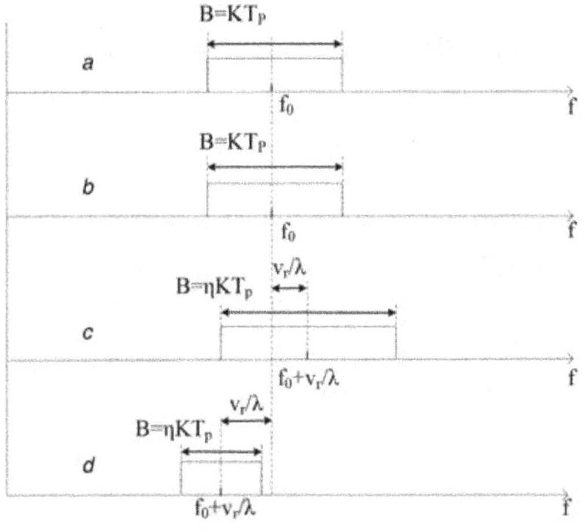

Fig. 2 *Comparison of bandwidth in the SAG model and the MPD model*
a Bandwidth of emitted signal
b Bandwidth in the SAG model
c Bandwidth in the MPD model ($v_r > 0$)
d Bandwidth in the MPD model ($v_r < 0$)

signal in time domain, we can obtain the response of the matched filter as (13)

$$H(f) = S^*(f) = \text{rect}\left(\frac{f}{KT_p}\right)\exp\left(j\pi\frac{f^2}{K}\right) \tag{13}$$

Then the pulse compression results in frequency domain under the SAG model can be expressed as

$$X_{pc}(f, t_m) = S_r(f, t_m)H(f) \tag{14}$$

Substituting (13) into (14), we obtain the pulse compression results of echo pulse as follows

$$X_{pc}(f, t_m) = \text{rect}\left(\frac{f}{KT_p}\right)\exp\left(-j\frac{4\pi}{c}fR(t_m)\right)\exp\left(-j\frac{4\pi}{\lambda}R(t_m)\right) \tag{15}$$

Taking the inverse Fourier transform on (15), we obtain

$$x(t, t_m) = \int_{-\infty}^{+\infty} X_{pc}(f, t_m)\exp(j2\pi ft)\,df \tag{16}$$

We can obtain the expression of pulse compression results after arranging

$$x(t, t_m) = A\,\text{sinc}\left(KT_p\left(t-\frac{2R(t_m)}{c}\right)\right)\exp\left(-j\frac{4\pi}{\lambda}R(t_m)\right) \tag{17}$$

However, the SAG model is no longer applicable to the hypersonic targets; in contrast, the MPD model is more accurate for describing the hypersonic targets. In the MPD model, the matched filter for echo signals of the MPD model should not be the same as in the SAG model, otherwise the pulse compression performance will not be as good as expected. The reasons are as follows.

If we still use the matched filter in the SAG model to match with the echo signal described in the MPD model, we can obtain the result after matching

$$X_{pc}'(f, t_m) = S_r'(f, t_m)H(f) \tag{18}$$

Submitting (12) and (13) into (18), we can obtain

$$\begin{aligned} X_{pc}'(f, t_m) &= \text{rect}\left(\frac{f-((v_r/\lambda)-|v_r/2c|)}{KT_p+(v_r/c)-(2|v_r|/\lambda)}\right) \\ &\times \exp\left(-j\pi\frac{f^2}{K}\left(\frac{1}{\eta^2}-1\right)\right) \\ &\times \exp\left(j2\pi f\frac{-(2R(t_m)/c)-((2v_r t_m)/c)+((2v_r/\lambda)/K\eta)}{\eta}\right) \\ &\times \exp\left(-j\frac{4\pi}{\lambda}R(t_m)-j\frac{4\pi}{\lambda}v_r t_m\right) \\ &\times \exp\left(j2\pi\left(\frac{(2v_r/\lambda)}{\eta}\left(\frac{2R(t_m)}{c}+\frac{2v_r t_m}{c}\right)-\frac{(2v_r^2/\lambda^2)}{K\eta^2}\right)\right) \end{aligned} \tag{19}$$

Comparing (19) with (15), we can see that both the width and the centre of the window were changed, after pulse compression as in the SAG model, the centre of frequency offset with $2\pi(((2v_r/\lambda)/$

$K\eta) - ((2v_r t_m)/c))$. However, the key point is that the matched filter in the SAG model cannot eliminate the quadratic term in the phase of MPD model, and the quadratic term becomes

$$\exp\left(-j\pi\frac{f^2}{K}\left(\frac{1}{\eta^2}-1\right)\right)$$

Its value is dependent on the difference between $1/\eta^2$ and 1, and the residual quadratic term becomes smaller when $1/\eta^2$ is close to 1.

Applying the inverse Fourier transform on (19), we can obtain the pulse compression results in the time domain

$$x'(t,\ t_m) = \int_{-\infty}^{+\infty} X'_{\mathrm{pc}}(f,\ t_m)\exp(\mathrm{j}2\pi ft)\,\mathrm{d}f \tag{20}$$

Owing to the quadratic term, the pulse compression result will not be the sinc function.

Moreover, in the MPD model with the SAG model, the quadratic term is eliminated totally when $(1/\eta^2) \simeq 1$. In this case, the unmatched pulse compression results without quadratic term can be written as

$$\begin{aligned} X'_{\mathrm{pc}}(f,t_m) &\simeq \widehat{X}'_{\mathrm{pc}}(f,t_m) \\ &= \mathrm{rect}\left(\frac{f-((v_r/\lambda)-|v_r/2c|)}{KT_p+(v_r/c)-(2|v_r|/\lambda)}\right) \\ &\times \exp\left(\mathrm{j}2\pi f\left(\frac{-((2R(t_m))/c)-((2v_r t_m)/c) + (((2v_r/\lambda))/K\eta)}{\eta}\right)\right) \\ &\times \exp\left(-\mathrm{j}\frac{4\pi}{\lambda}R(t_m)-\mathrm{j}\frac{4\pi}{\lambda}v_r t_m\right) \\ &\times \exp\left(\mathrm{j}2\pi\left(\frac{(2v_r/\lambda)}{\eta}\left(\frac{2R(t_m)}{c}+\frac{2v_r t_m}{c}\right)-\frac{((2v_r^2)/\lambda^2)}{K\eta^2}\right)\right) \end{aligned} \tag{21}$$

Applying inverse Fourier transform on (21), we can obtain the pulse compression results in time domain

$$\begin{aligned} x'(t,\ t_m) &= \mathrm{sinc}\left(\left(t-\frac{2R(t_m)}{c}-\frac{2v_r t_m}{c}+\frac{(2v_r/\lambda)}{K\eta}\right)\right. \\ &\times \left.\left(KT_p+\frac{v_r}{c}-\frac{2|v_r|}{\lambda}\right)\right) \\ &\times \exp\left(-\mathrm{j}\frac{4\pi}{\lambda}R(t_m)-\mathrm{j}\frac{4\pi}{\lambda}v_r t_m\right) \\ &\times \exp\left(\mathrm{j}2\pi\left(\frac{(2v_r/\lambda)}{\eta}\left(\frac{2R(t_m)}{c}+\frac{2v_r t_m}{c}\right)-\frac{(2v_r^2/\lambda^2)}{K\eta^2}\right)\right) \end{aligned} \tag{22}$$

The process above could not obtain the accurate results for MPD model, instead a matched filter designed for the MPD model should be used in pulse compression process. With the analysis above, we can conclude that if the velocity of a target is known, the matched filter response for the MPD model could be obtained

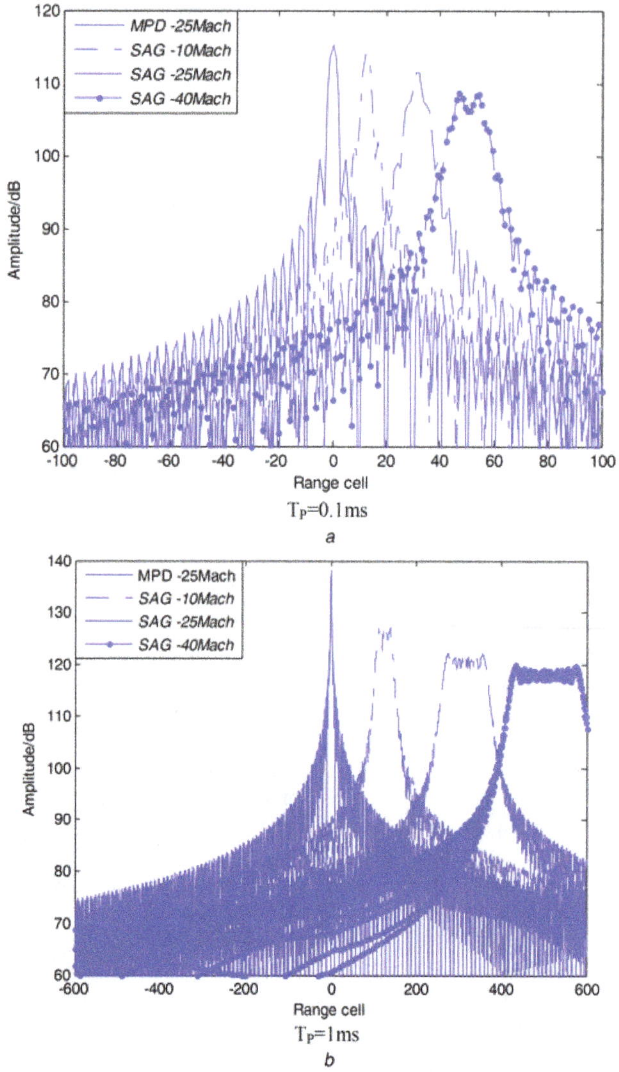

Fig. 3 *Comparisons of pulse compression results with different velocities and pulse durations*
a $T_P = 0.1$ ms
b $T_P = 1$ ms

according to (12), and expressed as follows

$$\begin{aligned} \hat{H}(f) &= \mathrm{rect}\left(\frac{f-(2v_r/\lambda)}{KT_p\eta}\right)\exp\left(\mathrm{j}\pi\frac{(f-(2v_r/\lambda))^2}{K\eta^2}\right) \\ &\times \exp\left(-\mathrm{j}2\pi\frac{(2v_r/\lambda)}{K\eta}\right) \end{aligned} \tag{23}$$

After matching $\hat{H}(f)$ with the MPD model, and applying inverse Fourier transform, we can obtain the pulse compression results in

Table 1 Peak values of pulse compression results

T_P	v_r			
	−25 Mach		−10 Mach	−40 Mach
	MPD	SAG	SAG	SAG
0.1 ms	115.1	111.5	114.1	108.7
1 ms	138.2	121.8	126.8	119.7

time domain

$$\hat{x}''(t,\ t_m) = \int_{-\infty}^{+\infty} S_r'(f,\ t_m)\hat{H}(f)\exp(\mathrm{j}2\pi f t)\,\mathrm{d}f \tag{24}$$

Then, the results could be arranged into a sinc function as follows

$$\begin{aligned}\hat{x}''(t,\ t_m) = {} & \operatorname{sinc}\left(\left(t - \frac{2R(t_m)}{c} - \frac{2v_r t_m}{c}\right)\left(KT_p\eta\right)\right) \\ & \times \exp\left(\mathrm{j}2\pi\left(\frac{(2v_r/\lambda)}{\eta}\left(\frac{2R(t_m)}{c} + \frac{2v_r t_m}{c}\right) - \frac{(2v_r^2/\lambda^2)}{K\eta^2}\right)\right) \\ & \times \exp\left(-\mathrm{j}\frac{4\pi}{\lambda}R(t_m) - \mathrm{j}\frac{4\pi}{\lambda}v_r t_m\right)\end{aligned} \tag{25}$$

From (25) we can conclude that when using the matched filter expressed as (23), the quadratic term in the MPD model could be eliminated totally. Therefore, a sinc function can be obtained in the MPD model after inverse Fourier transform.

From the analysis above, the expected result of sinc function either in the SAG model or the MPD model can be obtained when the matched filter matches the received signal accurately.

4 Simulations

In the following, we give pulse compression results for echo signals of both hypersonic point targets and hypersonic extended targets. For the extended target, we consider the effect of hypersonic velocity on high-resolution range profile in SAR including hypersonic SAR and space-borne SAR.

4.1 Timewidth

4.1.1 Point target case: Assume hypersonic point targets have a constant radical velocity. Numerical experiments with the target velocity of −10, −25 and −40 Mach are done to demonstrate the effect on pulse compression with different target velocities in the SAG model. Then, a comparison of pulse compression performance for echo signals under the two models of a point target with the velocity of −25 Mach is given.

In the experiment of point targets, parameters are set up as follows: the pulse durations (T_p) are 0.1 and 1 ms and the signal bandwidth (B) of radar system is 300 MHz.

Fig. 3 indicates that the pulse compression results become worse with increasing target velocity, and the differences between pulse compression results are more obvious when the pulse duration becomes longer.

Peak values (in dB) of pulse compression are shown in Table 1.

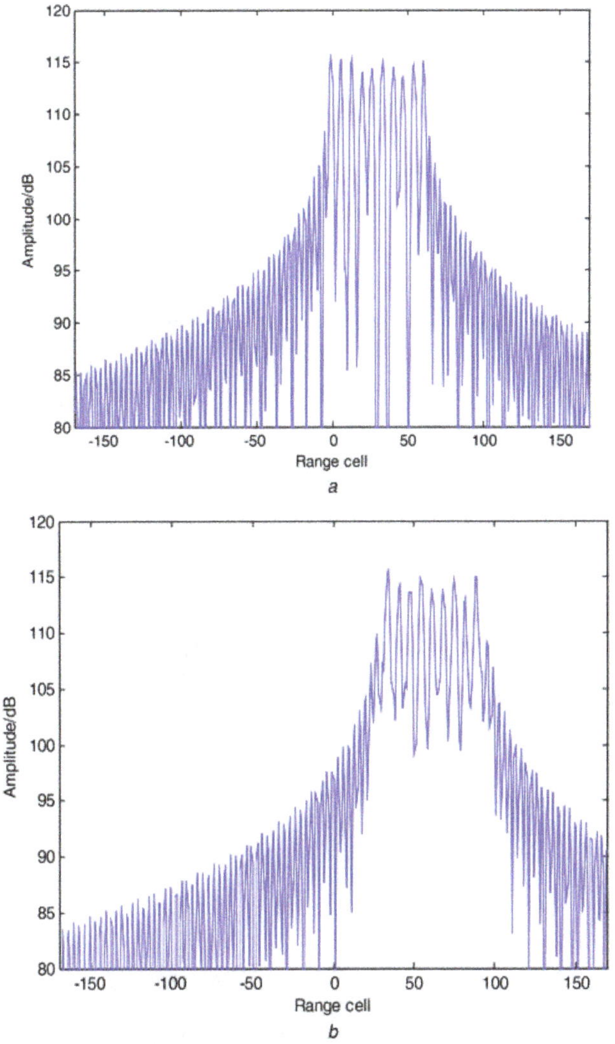

Fig. 4 *Pulse compression results under the two models with the pulse duration of 0.1 ms, velocity of −25 Mach*
a MPD model
b SAG model

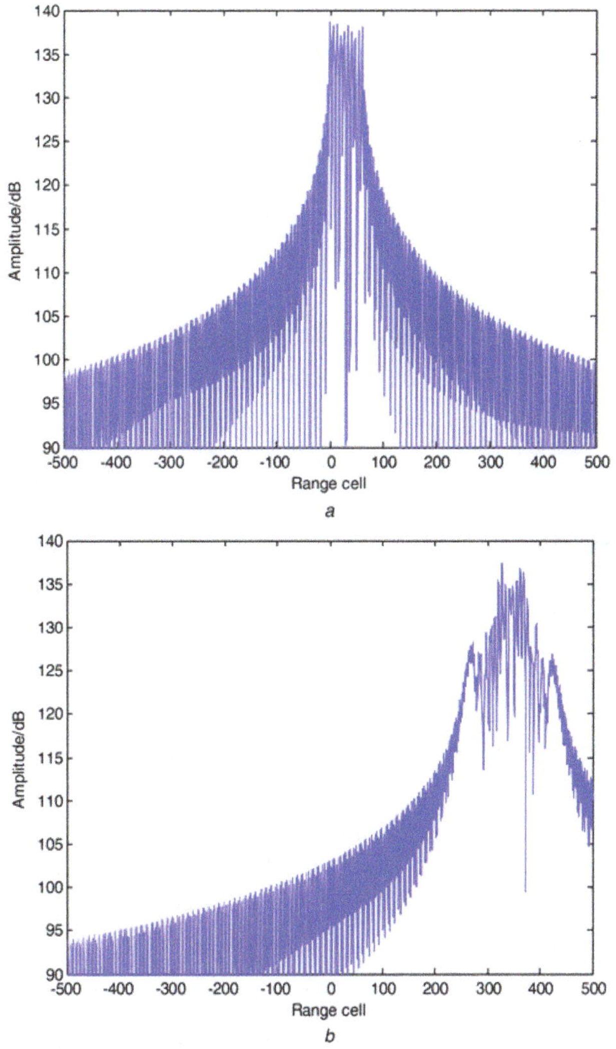

Fig. 5 *Pulse compression results under the two models with the pulse duration of 1 ms, velocity of −25 Mach*
a MPD model
b SAG model

From Table 1, we can see the differences in pulse compression gains between MPD and SAG are 3.6 and 16.4 dB, respectively, when the velocity is −25 Mach. This means there will be a significant loss in pulse compression gain if the SAG model is used in this situation. We can also see that, in the SAG model, the pulse compression gains are lower when the velocity becomes larger. On the other hand, the pulse compression peaks of different velocities have different positions, which will cause errors in measuring range. The better pulse compression performances obtained under the realistic MPD model, rather than that under the imprecise SAG model, are shown in Table 1.

4.1.2 Extended target case: Assume that a hypersonic target has ten extended points located along a straight line with 1 m internal distance. The velocity of this target is −25 Mach, the bandwidth of transmitted signal is 300 MHz and the pulse durations are 0.1 and 1 ms.

Figs. 4 and 5 show the pulse compression results when the time durations are 0.1 and 1 ms, respectively. We can see that the one-dimensional images obtained under the SAG model are worse than that obtained under the MPD model. Therefore, the MPD model is more suitable for dealing with imaging processing in space-borne SAR compared with the SAG model.

Table 2 Peak values of pulse compression results

	v_r			
B, GHz	−25 Mach		−10 Mach SAG	−40 Mach SAG
	MPD	SAG		
1	115.1	107.9	113.0	106.1
2	124.3	113.7	118.1	111.0

4.2 Bandwidth (high resolution)

A comparison of pulse compression under the SAG and the MPD models is simulated for the high-resolution radar. In experiments, parameters are set up as follows: the pulse duration (T_p) is 0.05 ms, the signal bandwidth (*B*) of radar system is 1 and 2 GHz and other parameters can be seen in the following figures.

4.2.1 Point target case: Fig. 6 indicates the simulation results for point target case that is similar to Fig. 3. Peak values (in dB) of pulse compression are shown in Table 2. From Table 2, we can see when the velocity is −25 Mach the differences in pulse

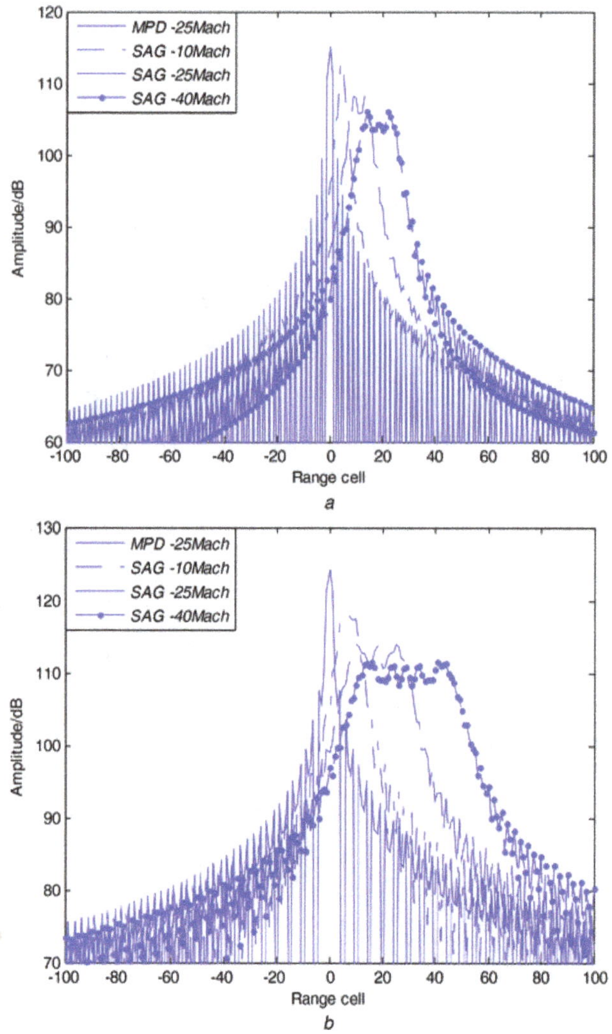

Fig. 6 *Comparisons of pulse compression results with different velocities and bandwidths*
a $B = 1$ GHz
b $B = 2$ GHz

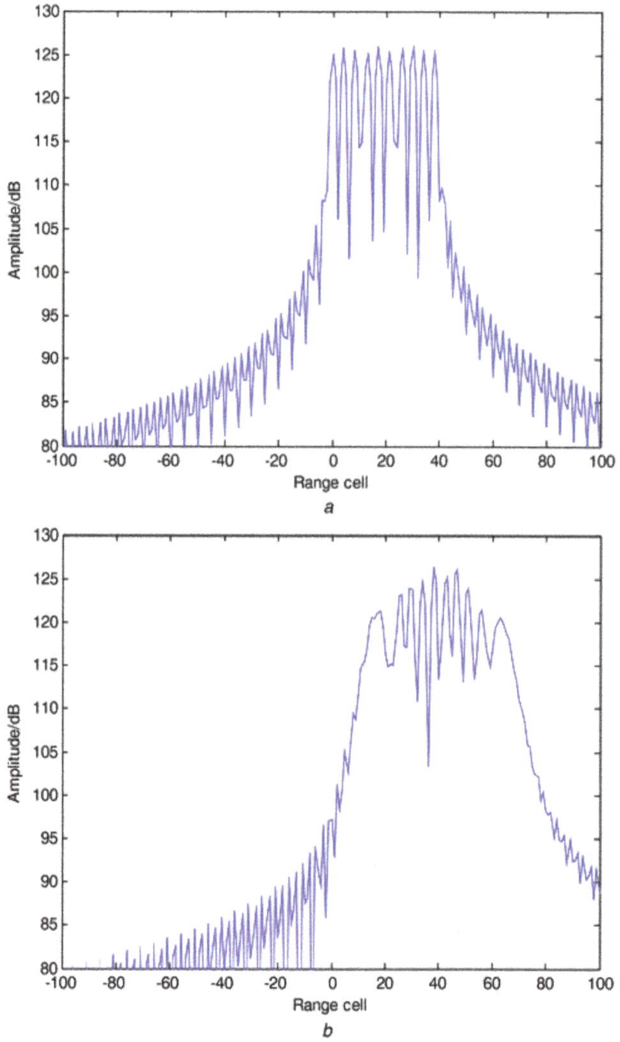

Fig. 7 *Pulse compression results under the two models with the bandwidth of 2 GHz, velocity of −25 Mach*
a MPD model
b SAG model

compression gains between the MPD and the SAG for 1 and 2 GHz bandwidth are 7.2 and 10.6 dB, respectively.

4.2.2 Extended target: Assume that a hypersonic target has ten extended points located along a straight line with 0.13 m apart from one to another. The velocity of this target is −25 Mach, the pulse duration of transmitted signal is 0.05 ms and the bandwidth is 2 GHz.

In Fig. 7, it is shown that pulse compression under the MPD model is better than that of the SAG model.

5 Conclusions

We propose an MPD model for hypersonic targets in this paper. Theoretical analysis and simulation results show that the MPD model has a better performance, in terms of pulse compression compared with the SAG model, when dealing with hypersonic targets or space-borne SAR. The MPD model is suggested to be the echo signal model for accurately describing hypersonic targets or hypersonic SAR.

6 Acknowledgments

The authors thank Dr. Aifei Liu for the careful revision. This work is supported by the National Natural Science foundation of China (no. 61231017).

7 References

[1] John C.C., Robert N.M.: 'Synthetic aperture radar: system and signal processing' (John Wiley & Sons, Hoboken, 1991)
[2] Ian G.C., Frank H.W.: 'Digital signal processing of synthetic aperture radar data: algorithms and implementation' (Artech House Incorporated, Norwood, 2005)
[3] Bao Z., Xing M.D., Wang T.: 'Radar imaging technology' (Publishing House of Electronics Industry, Beijing, 2005)
[4] Huang H.F., Dong Z., Liang D.N.: 'Error analysis of "stop and go" hypothesis of spaceborne bistatic SAR', *Chin. J. Radio Sci.*, 2006, **21**, (6), pp. 863–878
[5] Geng S.M.: 'Research on key technologies of FM_CW synthetic aperture radar signal processing', PhD thesis, National University of Defense Technology, 2008
[6] Qiu X.L., Han C.Z., Liu J.Y.: 'A method for spaceborne SAR geolocation based on continuously moving geometry', *J. Rad.*, 2013, **2**, (1), pp. 54–59
[7] Chen J.J., Wang S.L.: 'Echo model of ultra-high speed moving target and it influence on radar detection', *Mod. Rad.*, 2007, **29**, (8), pp. 60–63
[8] Tao R., Zhang N., Wang N.: 'Analysing and compensating the effects of range and Doppler frequency migrations in linear frequency modulation pulse compression radar', *IET Radar Sonar Navig.*, 2011, **5**, (1), pp. 12–22

8 Appendix

The POSP can be used to obtain the signal in frequency domain under the SAG model and the MPD model. To make clear comparison, the frequency of the MPD model is expressed with the time delay τ_m and the range $R(t_m)$, which is the same as the SAG model.

The signal in frequency domain under the SAG model can be deduced as follows.

According to Fourier transform, the signal in frequency domain under the SAG model is expressed as

$$\begin{aligned} S_r(f,\ t_m) &= \int_{-\infty}^{+\infty} S_r(t,\ t_m)\exp(-j2\pi ft)\,dt \\ &= \int_{-\infty}^{+\infty} \mathrm{rect}\left(\frac{t-((2R(t_m))/c)}{T_p}\right)\exp\left(j\pi K\left(t-\frac{2R(t_m)}{c}\right)^2\right) \\ &\quad \times \exp\left(-j2\pi f_0\frac{2R(t_m)}{c}\right)\exp(-j2\pi ft)\,dt \\ &= \int_{-\infty}^{+\infty} \mathrm{rect}\left(\frac{t-((2R(t_m))/c)}{T_p}\right) \\ &\quad \times \exp\left(j\pi\left(K\left(t-\frac{2R(t_m)}{c}\right)^2-2f_0\frac{2R(t_m)}{c}-2ft\right)\right)dt \end{aligned} \tag{26}$$

Define $\theta(t)$ as

$$\theta(t) = \pi\left(K\left(t-\frac{2R(t_m)}{c}\right)^2 - 2f_0\frac{2R(t_m)}{c} - 2ft\right) \tag{27}$$

According to the POSP, the signal phase resides at the time when the differential is zero. The phase changes slowly in the area of these points, but fast in other areas. Let $d\theta(t)/dt = 0$, we can obtain the time–frequency relationship of these points as

$$f = K\left(t-\frac{2R(t_m)}{c}\right),\quad t = \frac{f}{K}+\frac{2R(t_m)}{c}$$

Substituting these points into $\theta(t)$, the frequency domain of the SAG model could be expressed as

$$\begin{aligned} S_r(f,\ t_m) &= \mathrm{rect}\left(\frac{f}{KT_p}\right)\exp\left(-j\pi\frac{f^2}{K}\right)\exp\left(-j\frac{4\pi}{\lambda}R(t_m)\right) \\ &\quad\times\exp\left(-j\frac{4\pi}{c}f\,R(t_m)\right) \end{aligned} \tag{28}$$

After rearranging (28), $S_r(f,\ t_m)$ could be expressed as

$$S_r(f,\ t_m) = S(f)\exp\left(-j\frac{4\pi}{\lambda}R(t_m)\right)\exp\left(-j\frac{4\pi}{c}fR(t_m)\right) \tag{29}$$

Similarly, the signal in frequency domain under the MPD model is obtained as

$$\begin{aligned} S_r'(f,\ t_m) &= \int_{-\infty}^{+\infty} S_r'(t,\ t_m)\exp(-j2\pi ft)\,dt \\ &= \int_{-\infty}^{+\infty}\mathrm{rect}\left(\frac{t-((2R(t))/c)}{T_p}\right) \\ &\quad\times\exp\left(j\pi K\left(t-\frac{2R(t)}{c}\right)^2 - 2j\pi f_0\frac{2R(t)}{c}\right) \\ &\quad\times\exp(-j2\pi ft)\,dt \\ &= \int_{-\infty}^{+\infty}\mathrm{rect}\left(\frac{t-((2R(t))/c)}{T_p}\right) \\ &\quad\times\exp\left(j\pi K\left(t-\frac{2R(t)}{c}\right)^2 \right.\\ &\quad\left. - 2j\pi f_0\frac{2R(t)}{c} - j2\pi ft\right)dt \end{aligned} \tag{30}$$

Define $\theta'(t) = \pi K(t-((2R(t))/c))^2 - 2\pi f_0((2R(t))/c) - 2\pi ft$, (30) could be rewritten as

$$S_r'(f,\ t_m) = \int_{-\infty}^{+\infty}\mathrm{rect}\left(\frac{t-((2R(t))/c)}{T_p}\right)\exp(j\theta'(t))\,dt \tag{31}$$

Let $d\theta'(t)/dt = 0$, the time–frequency relationship of the MPD model is expressed as

$$f = K\left(t\left(1+\frac{2v_r}{c}\right)^2 + \left(1+\frac{2v_r}{c}\right)\frac{2R(t_m)+v_r t_m}{c}\right) + \frac{2v_r}{\lambda} \quad (32)$$

$$t = \frac{f-(2v_r/\lambda)}{K\left(1+(2v_r/c)\right)^2} + \frac{((2R(t_m))/c)+((2v_r t_m)/c)}{\left(1+(2v_r/c)\right)} \quad (33)$$

The frequency of the MPD model could be expressed as

$$\begin{aligned} &= \mathrm{rect}\left(\frac{f-(2v_r/\lambda)}{KT_p\left(1+(2v_r/c)\right)}\right)\exp\left(-j\pi\frac{\left(f-(2v_r/\lambda)\right)^2}{K\left(1+(2v_r/c)\right)^2}\right) \\ &\times \exp\left(-j2\pi\frac{f-(2v_r/\lambda)}{\left(1+(2v_r/c)\right)}\left(\frac{2R(t_m)}{c}+\frac{2v_r t_m}{c}\right)\right) \\ &\times \exp\left(-j\frac{4\pi}{\lambda}R(t_m) - j\frac{4\pi}{\lambda}v_r t_m\right) \end{aligned} \quad (34)$$

14

Performance evaluation of a transformerless multiphase electric submersible pump system

Ahmed A. Hakeem[1], *Ahmed Abbas Elserougi*[1], *Ayman Samy Abdel-Khalik*[1], *Shehab Ahmed*[2], *Ahmed Mohamed Massoud*[1,3]

[1]*Department of Electrical Engineering, Alexandria University, Alexandria, Egypt*
[2]*ECEN Department, Texas A&M University at Qatar, Doha, Qatar*
[3]*Electrical Department, Qatar University, Doha, Qatar*
E-mail: ahmed.abbas@spiretronic.com

Abstract: Using of low-voltage variable-frequency drive followed by a step-up transformer is the most preferable way to feed an electrical submersible pump motor. The existence of long feeder between the motor and drive systems usually causes over-voltage problems because of the travelling wave phenomenon, which makes the employment of filter networks on the motor or inverter terminals mandatory. The so-called boost-inverter inherently can solve this problem with filter-less operation as it offers a direct sinusoidal output voltage. As boost inverters have voltage boosting capability, it can provide a transformer-less operation as well. This study investigates the performance of a five-phase modular winding induction machine fed from a boost-inverter through a long feeder. A simulation study using a 1000 Hp system and experimental investigation on a 1 Hp prototype machine are used to support the presented approach.

Nomenclature

V_{dc}	input DC voltage
V	amplitude of the inverter phase voltage
f	fundamental frequency
f_s	switching frequency
R_f	resistance of the feeder per kilometres
L_f	inductance of the feeder per kilometres
C_f	capacitance of the feeder per kilometres
L_i	input inductance
R_i	input resistance (series with each inductor)
C_i	boost-inverter output capacitance
I_L	inductor current
$\hat{v}_c$	peak of the capacitor voltage
$\hat{I}_{Ph}$	peak of the load current
$I_{L\ max}$	peak of the inductor current
Δv_C	capacitor voltage ripple
ΔI_L	inductor current ripple
D	duty cycle of DC–DC converter
R_s	stator resistance per-phase
R_r	rotor resistance per-phase
L_s	stator inductance per-phase
L_r	rotor inductance per-phase
L_m	magnetising inductance per-phase

1 Introduction

Electrical submersible pump (ESP) systems are one of the most important systems in offshore oil platform facilities [1]. In ESP systems, variable-frequency drive (VFD) through a medium voltage long feeder is used to supply the electric power to the ESP motor [2, 3]. An ESP system generally consists of multiple centrifugal pump stages, a submersible electrical motor and an armor-protected long feeder [4]. The motors have long stacks and limited diameters and are generally designed with very special form factors [5, 6]; the winding of such machines is one of the costly and time-consuming manufacturing steps. Owing to their odd form factor, ESP motors are manually wound induction machines. A five-phase modular induction machine alternative was proposed in [7] as a potential means to reduce production time by enabling automated winding of the ESP motor when segmented stators are utilised. The machine architecture also overcomes the limitations of modular winding three-phase induction machines [8]. Moreover, multiphase machines offer additional degrees of freedom that have promoted them as serious contenders in various applications [9].

The distributed LC line impedance and the fast rise/fall time of the inverter voltage lead to a reflected wave phenomenon that causes doubling and ringing of the motor terminal voltage at the long feeder's receiving end (RE) [10, 11]. Furthermore, starting problems because of the voltage drop along the feeder are more significant with high-inertia motor loads [2]. More problems arise, such as bearing currents and electromagnetic interference, because of the high d*v*/d*t* of the inverter voltage [12, 13]. To overcome over-voltage problems, the filters are commonly employed. Several topologies have been proposed in the literature to mitigate the effect of long motor leads [10, 13]. Two topologies have commonly emerged; an RC filter at the RE of the long feeder [14], and an RLC filter at the sending end (SE) of the long feeder [15]. In addition to these two topologies, more topologies have been discussed qualitatively in [16–18], where two filter topologies have been developed to reduce the common-mode voltage problem mentioned before [16]. Additionally, an LC resonant filter with a diode bridge may be used, but is limited to low-voltage drives [17]. A common-mode transformer, with a conventional inverter output filter mitigates common-mode currents [18, 19]. An integrated differential-mode and common-mode filter have been presented in [13].

In the field of ESP systems, motors are typically fed from a step-up transformer and a low-voltage VFD based on a classical voltage source inverter (VSI) topology. The so-called boost-inverter [20–22] naturally generates, in a single stage, a boosted sinusoidal output voltage and allows for transformer-less operation with relatively small passive elements instead. Additionally, the above-mentioned problems associated with a VSI are inherently mitigated as the inverter generates a direct sinusoidal voltage output. Hence, feeding a submersible pump motor with a modular induction machine using a five-phase boost-inverter is proposed as a viable new topology with both application and economic benefits. This topology offers potential economic benefits, transformer-less operation and concurrent natural mitigation of long feeder

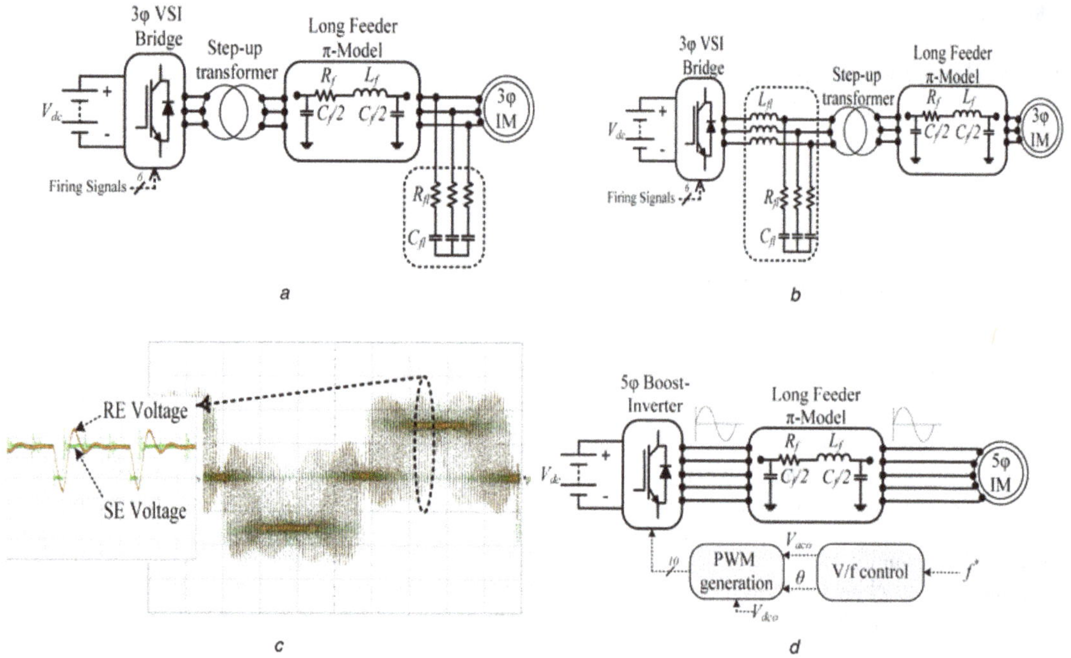

Fig. 1 *Three-phase VFD-ESP system*
a Using RC filter at the long feeder RE
b Using RLC filter at the long feeder SE
c Experimental results of the over-voltage problem because of the long feeder in the VSC-based VFD-ESP system (50 V/div)
d Five-phase boost-inverter-based VFD-ESP system diagram

consequences. A prototype system is built and used for experimental validation using a 1 Hp five-phase modular induction machine, a five-phase boost-inverter and a long feeder emulated using the conventional π-network model.

2 VSI-based VFD-ESP system

A conventional VSI-based VFD-ESP system consists of a low-voltage VFD, a step-up transformer, a downhole cable and an ESP motor. Two filter topologies are generally employed, as shown in Figs. 1*a*, and *b*. The reflected wave phenomenon associated with the presence of the long feeder in the ESP system with the VSI is shown in Fig. 1*c*. This figure shows the RE phase voltage across the prototype motor terminals when fed from a conventional VSI without filters. The machine experiences a notable voltage rise because of the distributed impedance nature of the feeder and the fact that the motor impedance appears as an open circuit relative to the feeder impedance [10].

3 Proposed boost-inverter-based VFD-ESP system

In this section, the proposed ESP system is presented. The general system block diagram is depicted in Fig. 1*d*. In this system, the low-voltage VSI, step-up transformer and the required filter in Figs. 1*a*, and *b* are replaced by a boost-inverter, whereas the three-phase induction machine is replaced by a five-phase modular induction machine. The detailed discussion of each part is given in the subsequent sections.

3.1 Five-phase modular induction machine

In [7], a detailed comparative performance and design study between conventional multiphase induction machines with distributed windings and five-phase modular induction machines is carried out. The general conclusions drawn from such a study prove that the modular multiphase induction machine can be a viable alternative to its three-phase counterpart, but with a power density penalty. Fig. 2 shows the stator winding layout of a

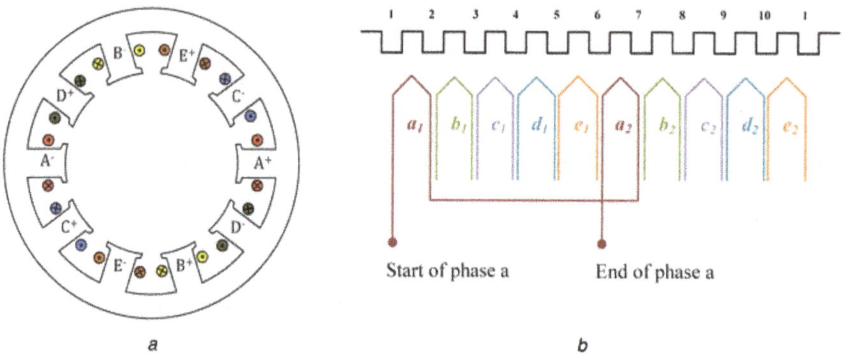

Fig. 2 *Five-phase modular machine stator winding layout*
a Stator layout
b Winding layout

2-pole five-phase modular induction machine. This machine is shown to have the following features [7]:

- The modular construction offers potentially lower winding production cost for long stack, limited diameter applications.
- The additional degrees of freedom because of the higher number of phases offer higher fault tolerant capability.
- For the same power rating and same phase voltage, a five-phase machine draws less line current. Consequently, both cable voltage drop and losses are reduced.
- There are two available operating sequences, namely primary and secondary sequences. The secondary sequence gives a synchronous speed of one third that obtained when the primary sequence is applied, whereas phase current magnitudes corresponding to the primary and secondary sequences are close in value for the same load. In other words, this winding offers two speed operation by only changing the applied voltage sequence.

On the other hand, the main drawback of this machine configuration is that the copper volume used is approximately double that of a conventional three-phase induction machine, thus requiring a slightly higher outer diameter because of the increase in the total slot area to accommodate the increased number of turns.

3.2 Five-phase boost-inverter

A single-stage five-phase boost-inverter provides not only DC-to-AC five-phase conversion but also voltage boosting capability, and inherent output filtering. The system consists of five DC–DC bi-directional boost converters sharing the same common point (O) as shown in Fig. 3*a*. The load is connected to phases *a*, *b*, *c*, *d* and *e*. These converters produce a DC-biased sine wave output via controlling the duty cycle (D_j) of each DC–DC converter. The AC component of each converter output is 72° out of phase. The common point of the capacitors is connected to the negative terminal of the DC-link. The five-phase motor is connected to the inverter terminals which is electrically isolated from the capacitors' common point.

In this topology, the output voltage reference of each DC–DC converter is composed of AC and DC components as in (1), where V_{dco} and V_{aco} represent the DC and AC components of the converter output voltage, respectively.

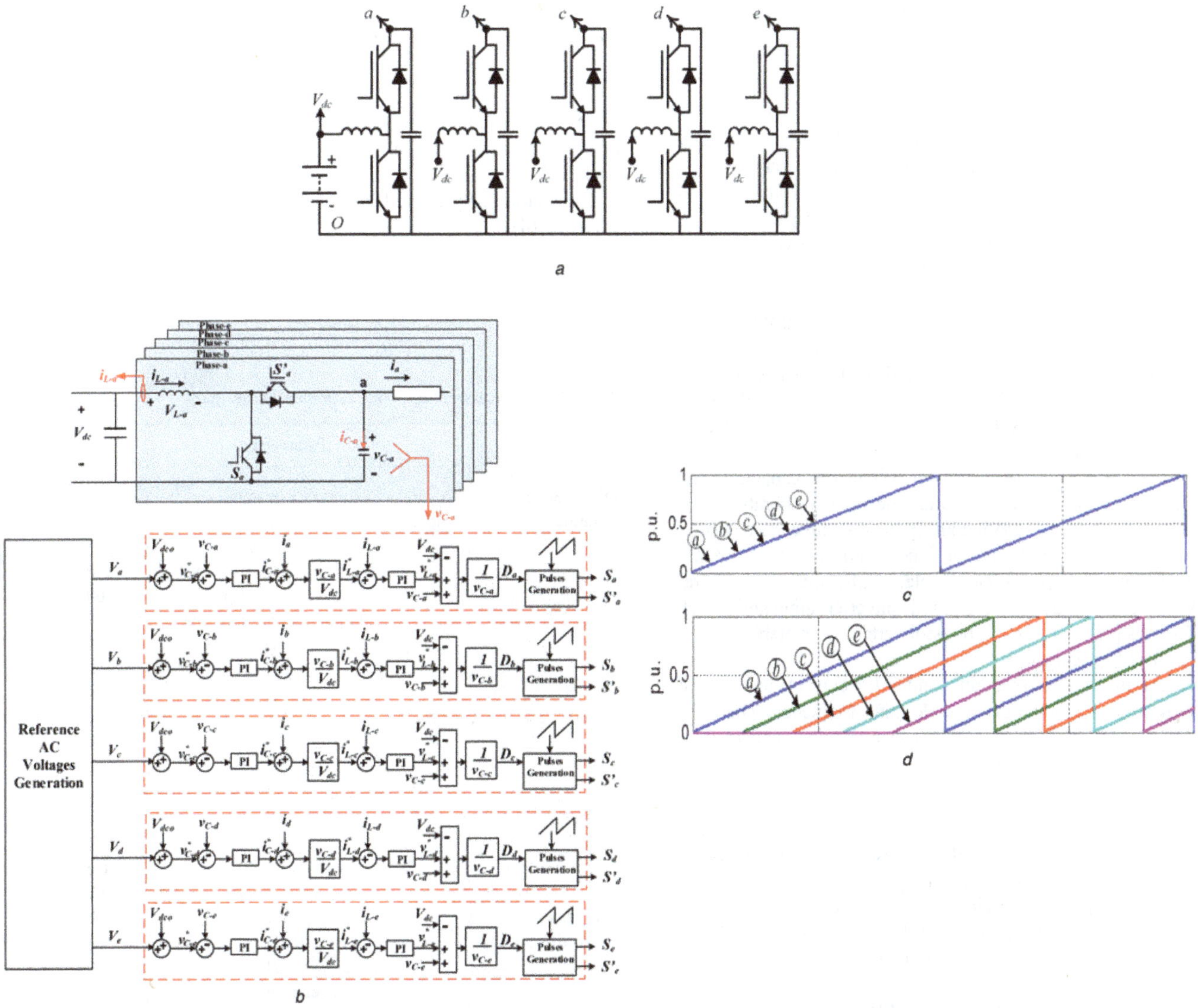

Fig. 3 *Five-phase boost-inverter*
a Topology
b Boost-inverter controller
c Un-shifted carrier of a five-phase boost-inverter
d Shifted carrier of a five-phase boost-inverter

The DC component should be the same for all phases and greater than or equal to the sum of the AC component peak and the DC input voltage (V_{dc}) to avoid operating at zero duty cycle. The AC side terminals are trapped between the capacitors connected to the upper insulated gate bipolar transistors (IGBTs). For equal DC components, the five line voltages are sinusoidal. Equation (2) shows the converter's output line voltage with eradicated DC components.

Each boost DC–DC converter has to be controlled to ensure robust performance during different operating conditions. The closed-loop control schemes for the single-phase boost-inverter are proposed by [20, 21] to ensure robust performance during different operating conditions

$$\begin{aligned} v_{aO}(t) &= V_{dco} + V_{aco}\sin(\omega t) \\ v_{bO}(t) &= V_{dco} + V_{aco}\sin\left(\omega t - \frac{2\pi}{5}\right) \\ v_{cO}(t) &= V_{dco} + V_{aco}\sin\left(\omega t - \frac{4\pi}{5}\right) \\ v_{dO}(t) &= V_{dco} + V_{aco}\sin\left(\omega t - \frac{6\pi}{5}\right) \\ v_{eO}(t) &= V_{dco} + V_{aco}\sin\left(\omega t - \frac{8\pi}{5}\right) \end{aligned} \quad (1)$$

$$v_{ab}(t) = v_{aO}(t) - v_{bO}(t) = 1.1756\, V_{aco}\sin\left(\omega t + \frac{3\pi}{10}\right) \quad (2)$$

The double-loop regulation scheme for a five-phase boost-inverter is shown in Fig. 3*b*. The reference voltages are biased by V_{dco}. The DC bias can be obtained by adding the input voltage V_{dc} to the peak output amplitude. The voltages across capacitors $V_{C\text{-}a}$, $V_{C\text{-}b}$, $V_{C\text{-}c}$, $V_{C\text{-}d}$ and $V_{C\text{-}e}$ are controlled to track the voltage references using proportional–integral (PI) controllers. The currents through L_a, L_b, L_c, L_d and L_e are also controlled by PI controllers to achieve stable operation by estimating suitable instantaneous duty cycles for each DC-to-DC converter (D_a, D_b, D_c, D_d and D_e). The duty cycle signals are compared with carrier waveforms to generate gate pulses for the semiconductor devices (IGBTs). It has to be noted that it is better to use shifted carriers (as in Fig. 3*d*) to reduce the inverter DC current ripples.

Selection of boost-inverter parameters (inductors, capacitors and switching frequency) is a very important design consideration. The inductors are designed for continuous operation with certain acceptable inductor current ripple; on the other hand, the capacitors are selected to limit the output voltage ripple. Finally, the switching frequency is selected based on the required converter rating and switch type. The design steps for the boost-inverter parameters are presented in detail in Appendix.

When the machine is operating under *V*/*f* control, the desired speed is used to determine the suitable input frequency. Then, the machine voltage amplitude is selected to maintain a constant *V*/*f* ratio.

3.3 Long feeder model

The long feeder can be functionally represented by an equivalent π-network model using lumped constants as shown in Fig. 1. This model is capable of giving acceptable results during both steady state and transient periods [10]. More sophisticated models can improve system modelling as presented in [11].

4 Five-phase boost-inverter-based VFD-ESP system design

Since the target applications for the proposed machine are mainly in the high-power arena, a case study has been conducted for the proposed five-phase boost-inverter-based VFD-ESP system on a 2400 V/1000 Hp five-phase induction machine with the rating and parameters listed in Table 1. The feeder utilised in this case is AWG 1/0 whose parameters are listed in Table 1.

4.1 Boost-inverter design

In this system, the boost-inverter output voltage has to be higher than the machine rated voltage to overcome the voltage drop across the long feeder. At machine full-load current, the voltage drop across the feeder is 30 V_{rms}. Therefore, the root-mean-squared (rms) value of the inverter output voltage must be kept at ~2430 V to guarantee delivery of 2400 V at the machine terminals when operating at a frequency of 60 Hz.

If a DC-link voltage of 2 kV is used, the DC component level must be selected to ensure that

$$V_{dco} > 2000 + (2430\sqrt{2})$$

that is, the DC component must be >5437 V, so 5500 V is assigned to V_{dco}.

The full parameters and ratings of the boost-inverter are listed in Table 1. The maximum instantaneous value of the duty cycle is shown in (3)

$$D_{max} = 1 - (V_{dc}/\hat{v}_c) = 1 - \left(\frac{V_{dc}}{V_{dco} + V_{aco}}\right) \cong 0.78 \quad (3)$$

From (8), the inductor current peak is ~500 A. Practically, the inductor current ripple is selected to be 20% of the inductor current peak, that is, $\Delta I_L = 100$ A.

The relation between inductor current ripple and its inductance is given by (4)

$$\Delta I_L = \frac{(V_{dc} - I_{L\,max}\,R_i)D_{max}}{L_i f_s} \cong \frac{1520}{L_i f_s} \quad (4)$$

Table 1 Parameters of high-power simulated case study

	Parameters	Values
five-phase modular machine	number of poles	4
	f, Hz	60
	connection	Y
	rated phase voltage, V rms	2400
	rated power, Hp	1000
	rated phase current, A rms	70
	R_s, Ω	0.8755
	X_s, Ω	4
	R_r', Ω	0.4116
	X_r', Ω	3.246
	X_m, Ω	195.6
	total inertia	40 kg m^2
	friction coefficient	0.5 N m/rad/s
feeder	L_f, mH/km	0.296
	R_f, Ω/km	0.426
	C_f, μF/km	0.3
	feeder length, km	1.0
five-phase boost-inverter	L_i, mH	3
	R_i, Ω	0.1
	C_i, μF	50
	inductor rating, A	600
	capacitor rating, kV	10
	f_s, kHz	5
	V_{dco}, V	5500
	V_{aco}, V	$2430\sqrt{2}$
	semiconductor switches rating	600 A, 10 kV

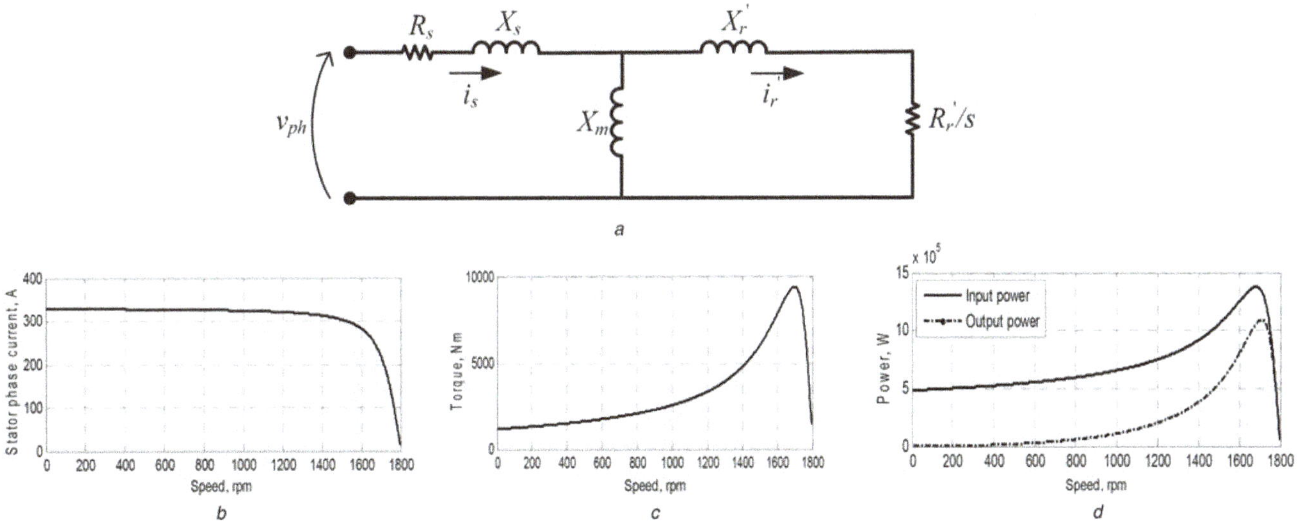

Fig. 4 *Five-phase modular machine characteristics*
a Equivalent per-phase circuit
b Current-speed characteristic
c Torque-speed characteristic
d Power-speed characteristic

If a 5 kHz switching frequency is used, a 3 mH inductor is a proper selection.

On the other hand, if the desired output voltage ripple is 3.5%, the voltage ripple is 312 V, the relation between capacitor voltage ripple and converter capacitance is given by (5)

$$\Delta v_C = \frac{D_{max}\hat{I}_{ph}}{C_i f_s} = \frac{0.78\,(70\sqrt{2})}{C_i f_s} \quad (5)$$

For a 5 kHz switching frequency, a 50 μF capacitor is needed.

4.2 Steady-state analysis

Since 60 Hz sinusoidal voltages are fed from the five-phase boost-inverter to the modular machine terminals, a steady-state analysis has been carried out using the machine's per-phase equivalent circuit shown in Fig. 4*a*. The speed is changed from zero to no-load speed (1800 rpm) to calculate the corresponding current, torque and power based on the machine per-phase equivalent circuit. The machine characteristics are shown in Figs. *b–d*. Based on Fig. 4*d*, the efficiency of the machine is 92% at full-load (the rotational losses are neglected).

4.3 Simulation

A MATLAB/SIMULINK model has been built to represent the performance of the proposed system during transients as well as steady-state periods. First, the machine is started at no-load, then the full-load torque is applied at $t = 0.5$ s, the simulation results are shown in Fig. 5. Fig.5*a* shows the voltages and currents of the presented system. It is obvious that the motor phase voltage fed from the boost-inverter is sinusoidal (i.e. the need for filters is eliminated). As a result, the full-load phase currents are also

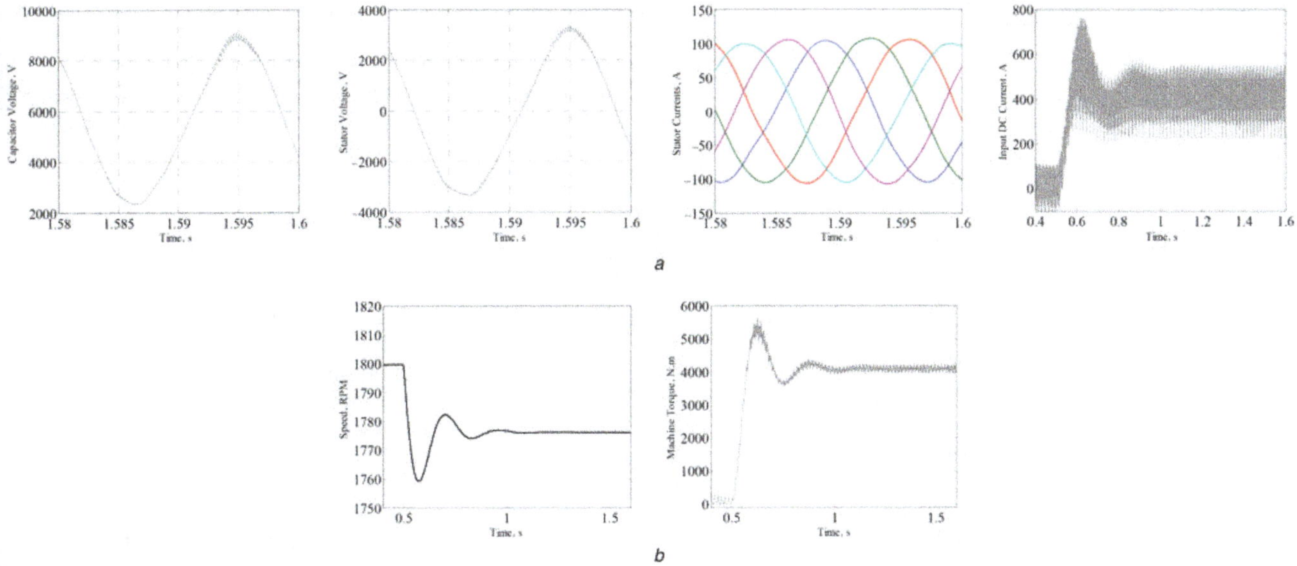

Fig. 5 *Simulation results*
a Electrical variables
b Mechanical variables

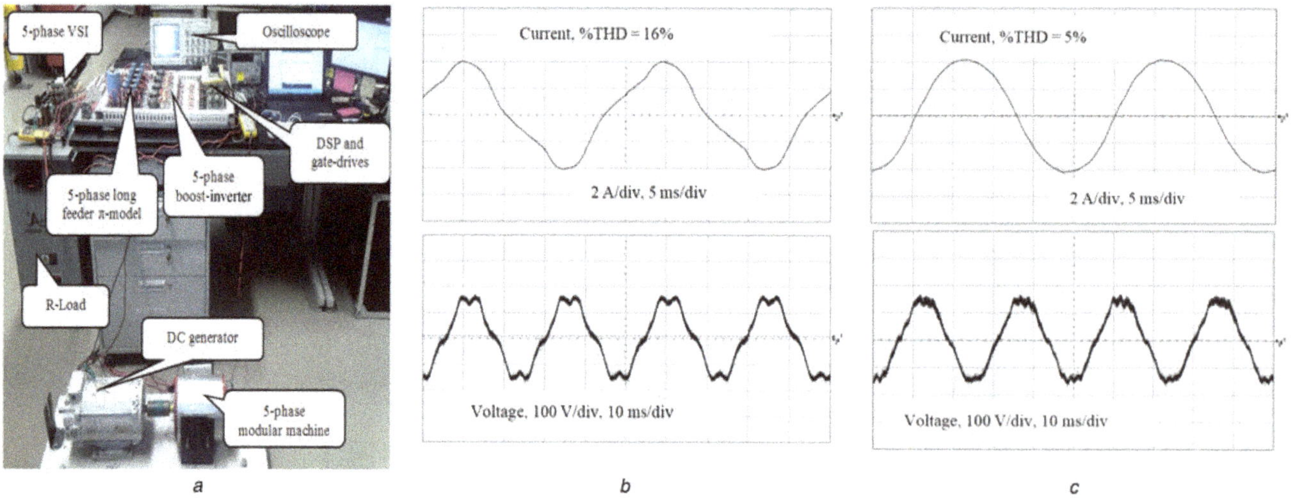

Fig. 6 *Experimental work*
a Experimental setup
b No load stator phase current/voltage
c Full-load Stator phase current/voltage

sinusoidal. In the presented case, the DC input current has a mean of 405 A at full-load. In this simulation, the circuit is fed from a constant DC voltage. Practically, the input current ripples can be significantly reduced with proper DC-link capacitor design. Fig. 5*b* shows the motor speed and torque during no-load as well as full-load conditions.

5 Experimental setup

A small-scale experimental prototype of the proposed system has been built for experimental validation. This is as shown in Fig. 6*a*. A 1 Hp five-phase modular induction machine and boost-inverter are built with the parameters given in Table 2. The machine phase voltage is 110 V rms at 40 Hz. The machine is fed from a five-phase boost-inverter operating at a 5 kHz switching frequency with a 150 V DC supply. The machine is coupled to a PM DC-generator of the same power rating, which acts as a mechanical load. The output power of the generator is measured and added to its losses to estimate the induction machine output mechanical power. A conventional π-network model for each phase with parameters given by Table 2 is implemented to emulate the feeder. The conventional *V*/*f* control is used to control the machine speed. A Texas Instruments TMS320F28335 digital signal processor is used to carry out the control algorithm.

Table 2 Parameters of experimental prototype

	Parameters	Values
five-phase modular machine	rated phase voltage, V, rms	110
	rated phase current, A, rms	3.1
	rated power, Hp	1
	rated frequency, Hz	40
	number of poles	4
	R_s, Ω	8.7
	R_r, Ω	3
	L_s, mH	15.7
	L_r, mH	15.7
	L_m, mH	172
feeder	L_f, mH/km	1.0
	R_f, Ω/km	0.3
	C_f, μF/km	1.0
	feeder length, km	1.0
five-phase boost-inverter	L_i, mH	1.0
	R_i, Ω	0.1
	C_i, μF	30
	V_{dco}, V	310
	V_{aco}, V	160
	f_s, kHz	5
	f, Hz	40

6 Experimental results

This section presents a summary of the main experimental results conducted on the prototype. First, the waveforms of different variables are illustrated under different operating conditions. Next, the machine characteristic curves are presented.

Figs. 6*b* and *c* show the stator phase current/voltage waveforms at no-load as well as full-load conditions, respectively, with inverter rms reference voltage set to (110 V, 40 Hz) using the fundamental sequence. It is noted that the no-load current suffers a high third harmonic component which decreases significantly as the machine is loaded. This third harmonic mainly appears because of core saturation of the prototype machine. However, this problem becomes negligible when the machine is loaded because the internally induced electromotive force (EMF) decreases because of the relatively high stator drop caused by the high stator impedance of the prototype machine. This is common in low-power rating induction machines. As the internally induced EMF decreases with loading, the machine flux decreases, and the magnitude of the third harmonic component significantly decreases. The saturation problem during no-load causes a distortion in the motor terminal phase voltage, as shown in Fig. 6*c*.

Fig. 7 shows the five-phase inverter output currents, the SE line-voltage, the RE line-voltage, the boost-inverter capacitor voltage and the DC input current at full-load condition.

The effect of a sudden change in reference speed at no-load is shown in Fig. 8*a*. It is assumed that the reference frequency is changed from 10 to 40 Hz and the voltage is changed accordingly to maintain a constant *V*/*f* ratio. Another case is shown in Fig. 8*b*, where the machine is suddenly loaded.

It has to be noted that the maximum efficiency of the prototype is relatively low mainly because of the high-power loss in the stator resistance, which is an expected characteristic in low-rating machines. For high-power machines the total machine loss is much smaller, as given by the case study in Section 4.

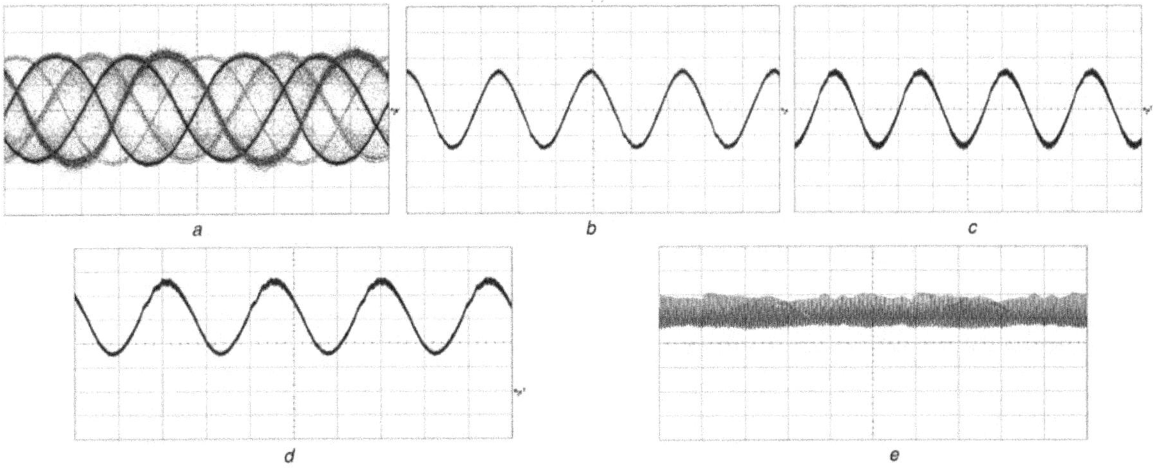

Fig. 7 *Different results of the prototype at full-load conditions*
a Inverter output five-phase currents (2 A/div, 5 ms/div)
b SE line voltage (200 V/div, 10 ms/div)
c RE line voltage (200 V/div, 10 ms/div)
d Boost-inverter capacitor voltage (100 V/div, 10 ms/div)
e Input DC current (10 A/div, 5 ms/div)

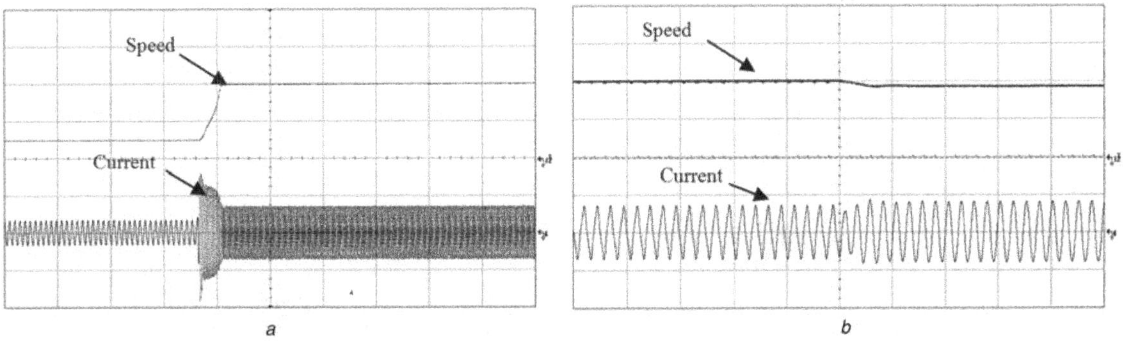

Fig. 8 *Transient current and speed response for a sudden change in the speed*
a From 10 to 40 Hz at no-load (600 rpm/div, 5 A/div, 1 s/div)
b Sudden loading (600 rpm/div, 5 A/div, 0.1 s/div)

7 Conclusion

This paper proposes a new VFD-ESP system which provides potential, economical and technical advantages over the current state-of-the-art. The proposed system can potentially reduce manufacturing cost/time through automated winding, provides transform-less operation and suppresses both differential-mode and common-mode voltages. The system consists of a five-phase modular induction machine at the end of a long feeder fed by a five-phase boost-inverter. The reflected wave phenomenon associated with long feeders is inherently mitigated because of the direct sinusoidal output of the proposed boost-inverter. Hence, the required filter in conventional VSI-based VFD-ESP systems is dispensed with. The additional degrees of freedom offered by the five-phase system add to system features and increase its fault tolerant capability. Additionally, the multiphase machines are generally advantageous with their lower torque ripples and space harmonics, hence, improved drive performance. The performance of the proposed system has been evaluated using an experimental prototype system.

The efficiency of the prototype machine used in this paper is low because of the high stator copper loss; this is not the case with the high-power rated modular machines as shown in the high-power motor design case study. The availability of the boosting function in the boost-inverter makes it possible to use it directly in high-power ESP systems without using a transformer to step up the voltage or install filters.

8 Acknowledgments

This publication was made possible by NPRP grant (NPRP 4-250-2-080) from the Qatar National Research Fund (a member of Qatar Foundation). The statements made herein are solely the responsibility of the authors.

9 References

[1] Al-Haiki Z., Shaikh-Nasser A.: 'Power transmission to distant offshore facilities', *IEEE Trans. Ind. Appl.*, 2011, **47**, pp. 1180–1183
[2] Liang X., Ilochonwu O.: 'Induction motor starting in practical industrial applications', *IEEE Trans. Ind. Appl.*, 2011, **47**, pp. 271–280
[3] Liang X., Faried S., Ilochonwu O.: 'Subsea cable applications in electrical submersible pump systems', *IEEE Trans. Ind. Appl.*, 2010, **46**, pp. 575–583
[4] Liang X., Ilochonwu O., Adedun R.: 'Dynamic response of variable frequency drives in electrical submersible pump systems'. IEEE Industrial and Commercial Power Systems Technical Conf. (I&CPS), 2012, pp. 1–11

[5] Verma V., Singh B., Chandra A., Al-Haddad K.: 'Power conditioner for variable-frequency drives in offshore oil fields', *IEEE Trans. Ind. Appl.*, 2010, **46**, pp. 731–739
[6] Jankowski T.A., Prenger F.C., Hill D.D., ET AL.: 'Development and validation of a thermal model for electric induction motors', *IEEE Trans. Ind. Electron.*, 2010, **57**, pp. 4043–4054
[7] Abdel-Khalik A.S., Ahmed S.: 'Performance evaluation of a five-phase modular winding induction machine', *IEEE Trans. Ind. Electron.*, 2012, **59**, pp. 2654–2669
[8] EL-Rafaie A.M., Shah M.R.: 'Comparison of induction machine performance with distributed and fractional-slot concentrated windings'. Conf. Record IEEE IAS Annual Meeting, 2008, pp. 1–8
[9] Levi E.: 'Multiphase electric machines for variable-speed applications', *IEEE Trans. Ind. Electron.*, 2008, **55**, pp. 1893–1909
[10] Abdelsalam A.K., Masoud M.I., Finney S.J., Williams B.W.: 'Vector control PWM-VSI induction motor drive with a long motor feeder: performance analysis of line filter networks', *IET Electr. Power Appl.*, 2011, **5**, pp. 443–456
[11] Wang L., Ngai-Man Ho C., Canales F., Jatskevichm J.: 'High-frequency modeling of the long-cable-fed induction motor drive system using TLM approach for predicting overvoltage transient', *IEEE Trans. Power Electron.*, 2010, **25**, pp. 2653–2664
[12] Tallam R., Skibinski G., Shudarek T., Lukaszewski R.: 'Integrated differential-model and common-model filter to mitigate the effects of long motor leads on AC drives', *IEEE Trans. Ind. Appl.*, 2011, **47**, pp. 2075–2083
[13] Hava A., Ün E.: 'A high-performance PWM algorithm for common-mode voltage reduction in three-phase voltage source inverters', *IEEE Trans. Power Electron.*, 2011, **26**, pp. 1998–2008
[14] von Jouanne A., Rendusara D.A., Enjeti P.N., Gray J.W.: 'Filtering techniques to minimize the effect of long motor leads on PWM inverter-fed AC motor drive systems', *IEEE Trans. Ind. Appl.*, 1996, **32**, pp. 919–926
[15] von Jouanne A., Enjeti P.N.: 'Design considerations for an inverter output filter to mitigate the effects of long motor leads in ASD applications', *IEEE Trans. Ind. Appl.*, 1997, **33**, pp. 1138–1145
[16] Rendusara D.A., Enjeti P.N.: 'An improved inverter output filter configuration reduces common and differential modes DV/DT at the motor terminals in PWM drive systems', *IEEE Trans. Power Electron.*, 1998, **13**, pp. 1135–1143
[17] Habetler T.G., Naik R., Nondahl T.A.: 'Design and implementation of an inverter output LC filter used for DV/DT reduction', *IEEE Trans. Power Electron.*, 2002, **17**, pp. 327–331
[18] Chen X., Xu D., Liu F., Zhang J.: 'A novel inverter-output passive filter for reducing both differential- and common-mode d*v*/d*t* at the motor terminals in PWM drive systems', *IEEE Trans. Ind. Electron.*, 2007, **54**, pp. 419–426
[19] Moreira A.F., Santos P.M., Lipo T.A., Venkataramanan G.: 'Filter networks for long cable drives and their influence on motor voltage distribution and common-mode currents', *IEEE Trans. Ind. Electron.*, 2005, **52**, pp. 515–522
[20] C'aceres R.O., Barbi I.: 'A boost DC–AC converter: analysis, design, and experimentation', *IEEE Trans. Power Electron.*, 1999, **14**, pp. 134–141
[21] Sanchis P., Ursæa A., Gubía E., Marroyo L.: 'Boost DC–AC inverter: a new control strategy', *IEEE Trans. Power Electron.*, 2005, **20**, pp. 343–353
[22] Cecati C., Dell'Aquila A., Liserre M.: 'A novel three-phase single-stage distributed power inverter', *IEEE Trans. Power Electron.*, 2004, **19**, pp. 1226–1233

10 Appendix

Boost converter design

(1) *Inductor selection:* During the instant of inductor current peak (i.e. maximum duty cycle), the power balance equation for the boost converter circuit is given by (6)

$$P_i = P_{L\,Loss} + P_o \tag{6}$$

where P_i is the input DC power at this instant $= I_{L\,max} V_{dc}$, $P_{L\,Loss}$ is the power loss in the inductor $= R_i I^2_{L\,max}$, P_o is the output power at this instant $= \hat{v}_c \hat{I}_{Ph}$. That is

$$(-V_{dc} I_{L\,max} + R_i I^2_{L\,max} + \hat{v}_c \hat{I}_{Ph} = 0) \tag{7}$$

By solving this second-order algebraic equation, the inductor current peak is given by (8)

$$I_{L\,max} = \frac{V_{dc} - \sqrt{V^2_{dc} - 4R_i \hat{v}_c \hat{I}_{Ph}}}{2R_i} \tag{8}$$

From DC–DC boost converter basics, the expression for inductor current ripple is given by (9)

$$\Delta I_L = \frac{(V_{dc} - I_{L\,max}\, R_i) D_{max}}{L_i f_s} \tag{9}$$

where

$$D_{max} = 1 - \frac{V_{dc}}{\hat{v}_C} \tag{10}$$

For a desired switching frequency and desired level of inductor current ripples, a suitable inductor can be selected. Practically, the inductor current ripple is selected to be 20% of the inductor current peak [20].

(2) *Capacitor selection:* From DC–DC boost converter basics, the output voltage ripple is given by (11)

$$\Delta v_C = \frac{D_{max} \hat{I}_{ph}}{C_i f_s} \tag{11}$$

For a desired switching frequency and desired level of output voltage ripple, a suitable capacitor can be selected. Practically the output voltage ripple is selected to be lower than 5% of $\hat{v}_c$ [20].

15

Square pulse emission with ultra-low repetition rate utilising non-linear polarisation rotation technique

Sin Jin Tan[1,2], Zian Cheak Tiu[1,2], Sulaiman Wadi Harun[1,2], Harith Ahmad[2]

[1]Department of Electrical Engineering, Faculty of Engineering, University of Malaya, 50603 Kuala Lumpur, Malaysia
[2]Photonics Research Center, University of Malaya, 50603 Kuala Lumpur, Malaysia
E-mail: swharun@um.edu.my

Abstract: The generation of nanosecond square pulse and microsecond harmonic pulse in a passively mode-locked fibre ring laser is demonstrated by inserting a 20 km long single mode fibre in the cavity. The laser operates in anomalous region based on the non-linear polarisation rotation process. The square pulse generation is because of the dissipative soliton resonance effect, which clamps the peak intensity of the laser and broadens the pulse width. The pulse width can be tuned from 28.2 to 167.7 ns. It was found that the square pulse can deliver higher pulse energy compared with the harmonic pulse. The highest recorded pulse energy is 249.8 nJ under the maximum available pump power of 125 mW without pulse breaking.

1 Introduction

A resonator that is kilometres long offers lower fundamental repetition rate in the kilohertz and hence allows the deliverance of higher pulse energy. However, there is a challenge that needs to be overcome if the oscillator is long. The combined action of both Kerr non-linearity and dispersion generally leads to pulse break up (multi-pulse) after the accumulated non-linear phase has exceeded a certain level. Pulse breaking leads to higher repetition rate and lower pulse energy compared with single pulse operation. Apart from the dissipative soliton (DS) with steep spectral edges, a new approach, namely the DS resonance (DSR) has been suggested to increase the pulse energy from a fibre laser. The formation of DSR is based on certain parameters selection within the frame of complex Ginzburg–Landau equation, where its pulse energy can be increased infinitely. DSR is recognised as a square pulse with flat top and steep edges, and thus its pulse duration is rather broad. It normally operates in the nanosecond region. Since the first demonstration of the square pulse emission by Matsas *et al.* [1], research effort on this topic has been lacking.

Recently, DSR pulse with higher pulse energy has been demonstrated by many researchers [2–6]. The formation of square pulse is theoretically independent of the sign of cavity dispersion and has been proven where square pulse can be formed in the positive [4] and negative [2] dispersion regions. Based on the previously published reports, the lowest attainable repetition rate was 173.05 kHz, which was obtained by inserting 1.16 km of highly non-linear fibre (HNLF) [6]. DSR pulse is also observed to be able to maintain single pulse operation, compared with the conventional DS pulse where DS pulse broke into five pulses at maximum available pump power [7]. The majority of the experiments conducted adopted a long piece of fibre such as single mode fibre (SMF) and HNLF. Therefore it is predicted that the formation of DSR requires large dispersion and high non-linearity.

Recently, some reports showed that mode-locked fibre lasers can deliver pulses with different duration simultaneously. For instance, Mao *et al.* [8] demonstrated a mode-locked fibre laser delivering both conventional and DSs pulses. More recently, Han *et al.* [9] demonstrated simultaneous generation of picosecond and femtosecond solitons using a carbon nanotubes (CNTs) saturable absorber. In this paper, nanosecond DSR square pulse generation with an ultra-low repetition rate of 10.2 kHz is demonstrated by inserting a 20 km long SMF in a simple ring resonator. The proposed laser can deliver nanosecond square pulse and microsecond harmonic pulse by adjusting a polarisation controller (PC). It is worth noting that although the cavity length is significantly long, the fibre laser still operates at its fundamental repetition rate without pulse breaking. By manipulating the polarisation state in the cavity, the proposed laser can also be adjusted to operate in harmonic mode. The performance of this harmonic laser with $sech^2$ shape is also investigated for comparison purposes. It is found that the pulse energy produced by DSR square pulse is much higher compared with the harmonic pulse.

2 Experimental setup

The experimental setup of the proposed mode-locked fibre laser is schematically shown in Fig. 1. It uses a 4.5 m long erbium-doped fibre (EDF) with an erbium concentration of 2000 ppm, cut-off wavelength of 910 nm, a pump absorption coefficient of 24 dB/m at 980 nm and a dispersion coefficient of −21.64 ps/nm km at $\lambda = 1550$ nm, as the gain medium. The EDF is pumped with a 1480 nm laser diode through a 1480/1550 nm wavelength division multiplexer. A polarisation dependent isolator (PDI) is used to ensure unidirectional propagation of light in the cavity and at the same time to generate linear light polarisation. A PC is employed to adjust the polarisation of light. A 20 km spool of SMF constitutes the long cavity and also serves to increase the non-linearity and dispersion. The dispersion parameter of the SMF is 17 ps/nm km. About 50% of the circulating light is taken out of the cavity via a 3 dB coupler and then fed into another 3 dB coupler. The second coupler splits the light for simultaneous monitoring, one part into an optical spectrum analyser (OSA) and the other into an oscilloscope and radio-frequency spectrum analyser together with a high-speed photodetector. The cavity is operating in a large negative dispersion region because of the long SMF.

3 Results and discussion

The mode-locked laser is generated based on non-linear polarisation rotation (NPR) effect in the ring cavity. The polarising isolator placed beside the PC acts as the mode-locking element in the proposed laser. It plays the double role of an isolator and a polariser, such that light leaving the isolator is linearly polarised. Consider a linearly polarised pulse just after the isolator. The polarisation state evolves non-linearly during the propagation of the pulse inside the EDF and SMF because of self-phase modulation and cross-phase modulation effects in the ring cavity. The state of

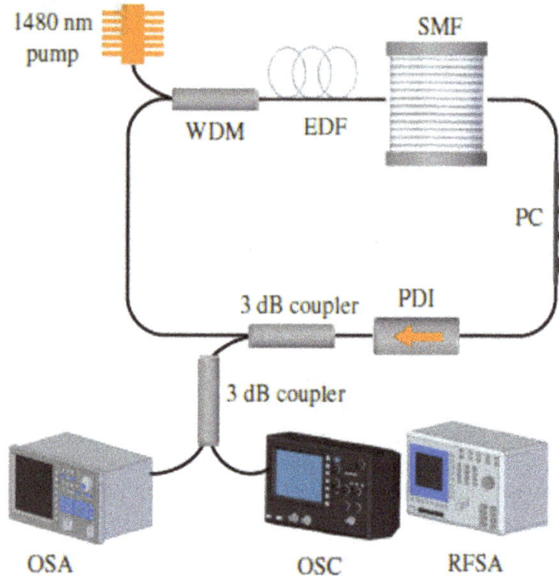

Fig. 1 *Experimental setup of the proposed DSR laser*

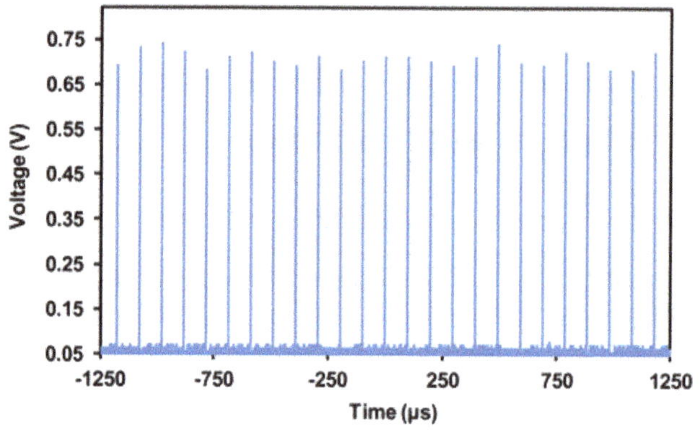

Fig. 3 *Fundamental repetition rate at 10.2 kHz*

polarisation is non-uniform across the pulse because of the intensity dependence of the non-linear phase shift. The PC is adjusted so that it forces the polarisation to be linear in the central part of the pulse. The polarising isolator lets the central intense part of the pulse pass but blocks (absorbs) the low-intensity pulse wings. The net result is that the pulse is slightly shortened after one round trip inside the ring cavity, an effect identical to that produced by a fast saturable absorber (SA). In another words, the PDI, working together with the birefringence fibres, generates an intensity dependent loss mechanism in the cavity that contributes to mode-locked square pulse generation in the cavity.

Apart from NPR technique, SA such as CNTs and graphene can also be used to initiate mode-locking. For instance, Liu *et al.* [10] reported a highly stable mode-locked fibre laser with a multi-wavelength output based on CNTs. In another work, mode-locked fibre laser emitting both dissipative and conventional solitons is also demonstrated by using a more complex nanomaterial based saturable absorber, which was obtained by mixing graphene and CNTs [11]. However, the square pulse phenomenon is not observed in both experiments when CNTs or graphene is used. This could be because of the cavity loss, which is drastically increased when these SAs are inserted into the laser cavity.

In the proposed experiment, by careful adjustment of the PC, stable square pulse starts to form at pump power of 108 mW. Fig. 2 shows the optical spectrum of the typical square pulse emission from the laser at three different pump powers of 108, 112 and 125 mW. At the maximum pump power of 125 mW, the laser operates at 1568.7 nm with the peak power of −17.2 dBm and 3 dB bandwidth of about 1 nm. Fig. 3 shows the oscilloscope trace of a square pulse train. The pulse train has an ultra-low repetition rate at 10.2 kHz, as determined by the cavity length. Fig. 4 focuses on a single pulse at two different pump powers. As shown in the figure, the square pulse has the distinct characteristic of steep leading and trailing edges and its pulse width can be tuned by changing the pump power. At 120 mW, the measured pulse width is 120.0 ns, whereas at 125 mW pump power, the pulse width increases to 167.7 ns. At the maximum pump power, the pulse still has a square shape while keeping the peak power almost constant. With the orientations of the wave-plates fixed, it is observed that the peak power of the square pulse is maintained while the pulse width increases with pump power. The ripple structures on the top of the pulse are probably because of insufficient gain to compensate for the loss in the ultra-long cavity. A cleaner square pulse structure is expected at higher pump power.

As shown in Fig. 2, the shape and 3 dB bandwidth of the mode-locked spectra are almost invariable with pump power. It is believed that the square pulse formed here has the characteristic of a square shape which undergoes pulse broadening with constant peak power and also invariable 3 dB optical bandwidth spectra which resembles the DSR theory that is predicted by Chang *et al.* [12, 13]. The theory of DSR indicates that the pulse energy could be boosted up to an infinitely large value, whereas the square pulse duration will broaden with increasing pump power while pulse amplitude converges to a given plateau value when the cavity parameters are chosen near to the resonance curve.

The evolution of pulse width with respect to pump power is presented in Fig. 5. The pulse width can be tuned from approximately 28.2 ns to 167.7 ns without pulse breaking by increasing pump power from 108 to 125 mW. The generation of square pulse in

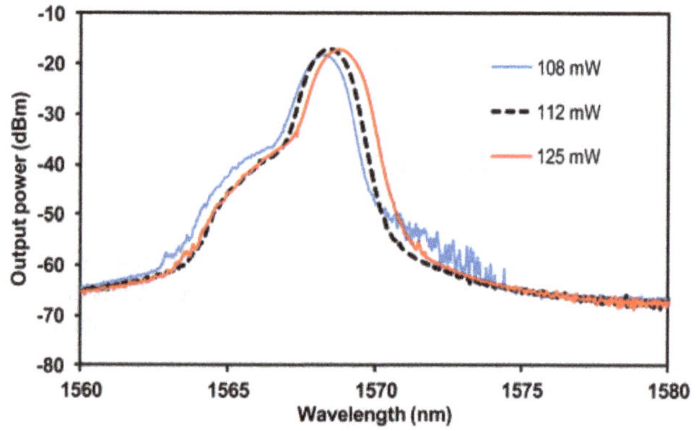

Fig. 2 *Optical output spectra of pulse laser at three different pump powers*

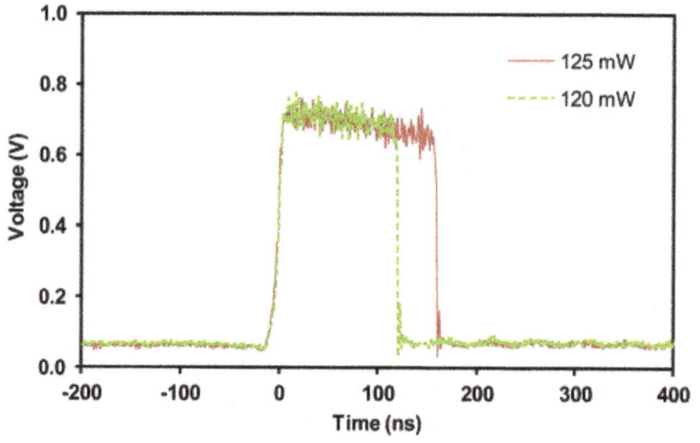

Fig. 4 *Oscilloscope of single square pulse at different pump powers*

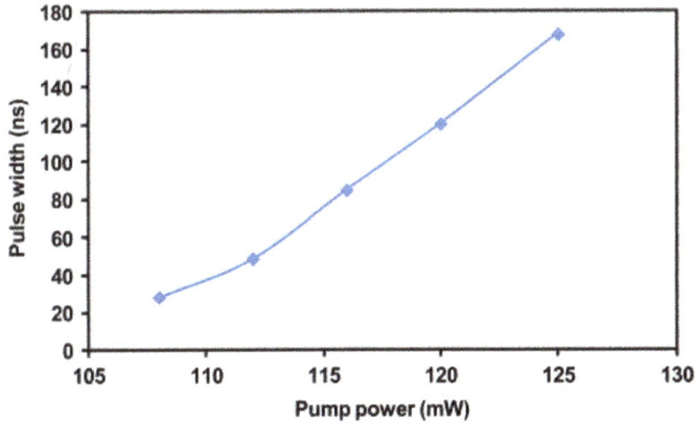

Fig. 5 *Pulse width of the square pulse against pump power*

the long cavity is most probably because of the DSR phenomenon in the long cavity laser. After the generation of square pulse, the peak amplitude is kept almost constant and does not increase with pump power anymore. The excess power circulating in the cavity now accounts for the increase in the pulse width rather than the peak intensity. Owing to the increment of pulse width, the pulse energy could be increased greatly as opposed to other soliton operation regions.

Fig. 6 shows the RF spectrum of the mode-locked fibre laser (at pump power of 125 mW) for both square and harmonic pulses, which reveals the repetition rates of 10.2 and 20.4 kHz, respectively. The signal-to-noise ratio (SNR) is obtained from the intensity ratio of the fundamental peak to the pedestal extinction, estimated to be ~32 and 40 dB for square and harmonic pulses, respectively, which indicates the stability of the laser. However, the SNR value is lower compared with other mode-locked fibre laser, which usually has an SNR of about 50 dB [14]. This is attributed to the cavity length used, which is significantly longer.

The square pulse can be switched to harmonic pulse operating in microsecond region by careful adjustment of the PC while maintaining all other cavities' parameters. Self-starting harmonic mode-locking can be realised by an appropriate adjustment of polarisation of light and at an adequate pump power. When pump power is raised to ~100 mW, mode-locked pulse is formed with repetition rate of 10.2 kHz which corresponds to the fundamental frequency. As the pump power is increased to 108 mW, pulse breaking is observed where its repetition rate doubles to 20.4 kHz, representing the second harmonic order pulse. Harmonic mode-locking is regarded as a phenomenon when a single circulating pulse breaks into multiple pulses with constant temporal spacing. This technique is often adopted for high repetition rates in multi-gigahertz fibre lasers [15]. The typical pulse train of the mode-locked fibre laser is shown in Fig. 7 for two different pump powers of 100 and 108 mW. The attainable pulse widths are 14.2 and 8.1 μs at 100 and 108 mW, respectively. The optical spectrum of the harmonic pulse is illustrated in Fig. 8, when the pump power is set at 125 mW. As shown in the figure, the harmonic laser operates at 1569.36 nm with peak output power of −18.4 dBm and 3 dB bandwidth of 1.7 nm.

Fig. 9 depicts the relationship between the output power and pump power for both DSR and harmonic pulses obtained from

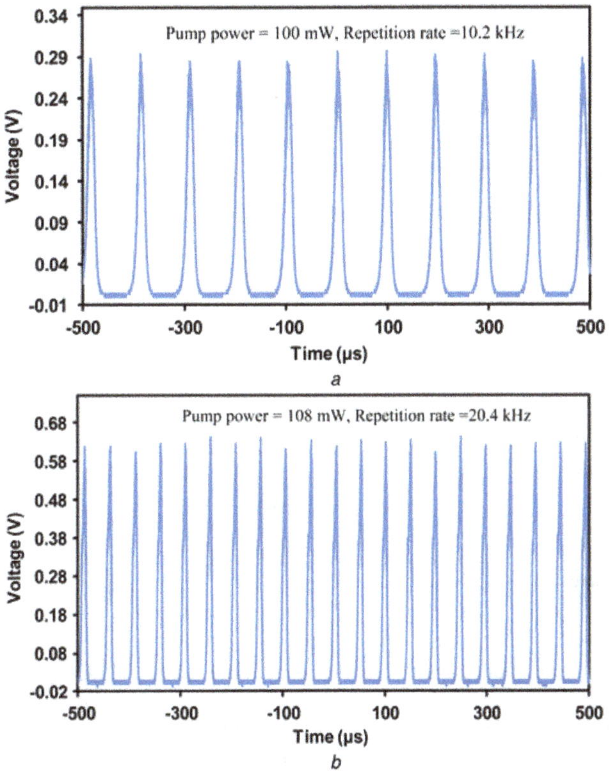

Fig. 7 *Typical pulse train of the mode-locking pulse at two different pump powers*
a 100 mW
b 108 mW

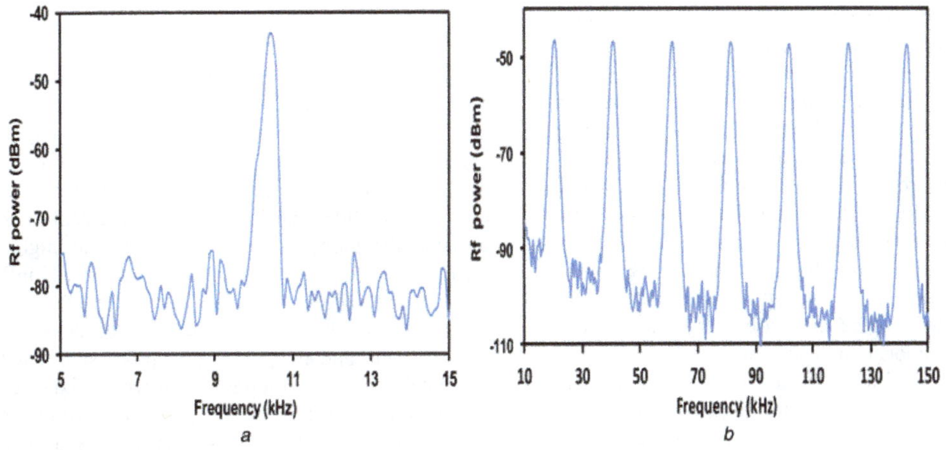

Fig. 6 *RF spectrum of the generated DSR pulses*
a Square
b Harmonic

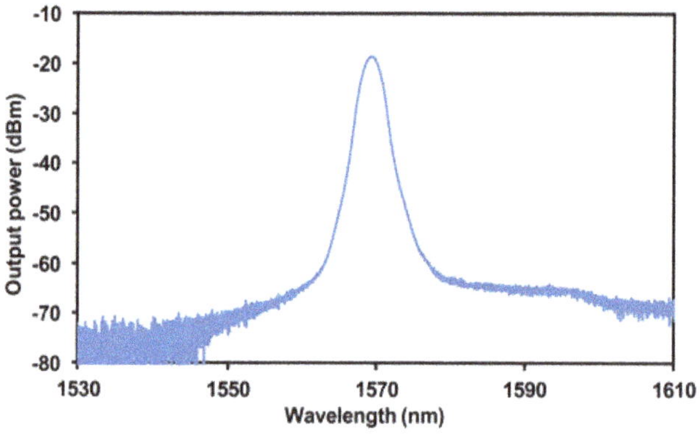

Fig. 8 *Output spectrum of the harmonic mode-locked EDFL*

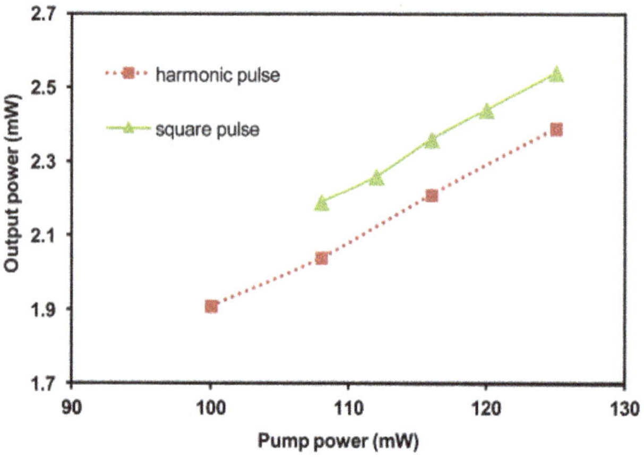

Fig. 9 *Measured output power for square and harmonic pulses at various pump powers*

the proposed mode-locked erbium-doped fibre laser (EDFL) of Fig. 1. The output power is measured at the 50% port of second 3 dB coupler, which channels the output light into OSA. It is observed that the output power increases linearly with pump power for both lasers. As expected, the square pulse recorded higher output power compared with harmonic pulse. The output power for square pulse varies from 2.19 to 2.54 mW as the pump power increases from 108 to 125 mW. On the other hand, for harmonic pulse, the highest measured output power is 2.39 mW at a pump power of 125 mW. Fig. 10 shows how the pulse energy changes with the increment of pump power for both lasers. Both square and harmonic pulses exhibit relatively high pulse energy in the nano-Joule range because of the long cavity length used in the laser setup. Pulse energy of both lasers is found to be increasing with pump power. By increasing the pump power from 108 to 125 mW, the pulse energy of the square pulse increases from 215.3 to 249.8 nJ, whereas the pulse energy for the harmonic pulse improves from 0.215 to 0.249 pJ. It is observed that the pulse energy reduces drastically from 188.7 nJ (100 mW) to 100 nJ (108 mW) as the fundamental pulse breaks into the second harmonic order pulse. When the fundamental pulse breaks, the repetition rate doubles from 10.2 to 20.4 kHz. This will reduce the pulse energy as the pulse energy is inversely proportional to repetition rate. Consequently, the pulse energy of square pulse can be further increased by optimising the laser parameters and employing higher pump power.

The maximum attainable pulse energy for the square pulse is higher than second harmonic pulse by 53.1%. Lower pulse energy is observed for harmonic pulse because of the occurrence of pulse breaking phenomena, where a single pulse breaks into

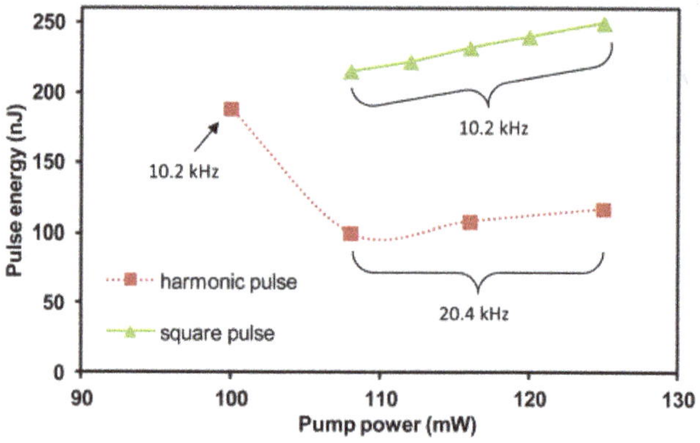

Fig. 10 *Pulse energy of produced pulse for square and harmonic pulses*

many pulses; two pulses in this experiment. After a single pulse is formed and traverses in the cavity, it encounters high non-linear effects and dispersion, introduced by the long SMF. A single pulse will break into many pulses where overtaking of different parts of a pulse will lead to optical wave breaking. It can be concluded that a laser should remain in single pulse operation in order to realise high pulse energy. Square pulse, which has steeper leading and trailing edges along with flat top in the temporal domain, can better withstand pulse breaking compared with the Gaussian or $sech^2$ shape pulse. It can be presumed that the square pulse has very low-frequency chirps across the central region of pulse and has non-linear pulse chirping at the pulse edges [16]. The non-linear chirp at the pulse edges can resist the dispersion and non-linearity effects in the cavity and thus can maintain its wave breaking free pulse.

Neither square pulse nor harmonic pulse is observed when the 20 km long SMF is removed. Without the long SMF, only the conventional pulse with Kelly side-bands is obtained. The insertion of 20 km long SMF brings the cavity's parameter near to the resonance curve, thereby producing wave breaking free square pulse. From the experimental results, the laser operates in two different operating regimes, harmonic and square pulse, by changing the pump power and also the polarisation of light. The change of light polarisation leads to different saturable absorption strength and intrinsic spectral filtering, which affects the intra-cavity non-linear gain and transmittivity. As a result, various kinds of pulse shapes can be formed. Both square and harmonic pulses generated by the mode-locked EDFL are stable. If there is no perturbation introduced to the laser, both square and harmonic pulses EDFL can last several hours under normal laboratory conditions.

4 Conclusion

A nanosecond square pulse laser source operating at 1568.7 nm with ultra-low repetition rate is demonstrated by inserting a 20 km long SMF in the ring cavity. A stable square pulse is obtained at the fundamental repetition rate of 10.2 kHz based on the NPR process. The pulse width can be easily tuned from 28.2 to 167.7 ns while maintaining almost constant peak power by increasing the pump power. Pulse energy as high as 249.8 nJ is recorded at the maximum pump power of 125 mW. This square pulse in nanosecond time scale is beneficial for numerous applications, especially in all optical square wave clocks for photonics integrated circuits and all optical signal processing. The DSR pulse can be switched to the harmonic pulse with a fixed repetition rate of 10.2 kHz by adjusting the PC. The pulse width of the harmonic pulse can be tuned from 14.2 to 8.08 μs as the pump power increases from 108 to 125 mW. Compared with the harmonic laser, the DSR can acquire higher non-linear effects without pulse breaking and therefore higher pulse energy could be delivered.

5 Acknowledgments

This project was funded by the Ministry of Education and University of Malaya under various grant schemes (grant no. D000009-16001 and PG068-2013B).

6 References

[1] Matsas V.J., Newson T.P., Zervas M.N.: 'Self starting passively mode-locked fiber ring laser exploiting nonlinear polarization switching', *Opt. Commun.*, 1992, **92**, (1–3), pp. 61–66

[2] Duan L., Liu X., Mao D., Wang L., Wang G.: 'Experimental observation of dissipative soliton resonance in an anomalous dispersion fiber laser', *Opt. Express*, 2012, **20**, (1), pp. 265–270

[3] Wang S.-K., Ning Q.-Y., Luo A.-I., Lin Z.-B., Luo Z.-C., Xu W.-C.: 'Dissipative soliton resonance in a passively mode-locked figure eight fiber laser', *Opt. Express*, 2013, **21**, (2), pp. 2402–2407

[4] Wu X., Tang D.Y., Zhang H., Zhao L.M.: 'Dissipative soliton resonance in an all normal dispersion erbium doped fiber laser', *Opt. Express*, 2009, **17**, (7), pp. 5580–5584

[5] Yang J., Guo C., Ruan S., Ouyang D., Lin H., Wu Y., Wen R.: 'Observation of dissipative soliton resonance in a net normal dispersion figure of eight fiber laser', *IEEE Photonics J.*, 2013, **5**, (3), p. 1500806

[6] Zhang X., Gu C., Chen G., ET AL.: 'Square wave pulse with ultra wide tuning range in a passively mode-locked fiber laser', *Opt. Lett.*, 2012, **37**, (8), pp. 1334–1336

[7] Wang L., Liu X., Gong Y., Mao D., Duan L.: 'Observation of four types of pulses in a fiber laser with large net normal dispersion', *Opt. Express*, 2011, **19**, (8), pp. 7616–7624

[8] Mao D., Liu X., Han D., Lu H.: 'Compact all fiber laser delivering conventional and dissipative solitons', *Opt. Lett.*, 2013, **38**, (16), pp. 3190–3193

[9] Han D.D., Liu X.M., Cui Y.D., Wang G.X., Zeng C., Yun L.: 'Simultaneous picoseconds and femtosecond solitons delivered from a nanotube mode-locked all fiber laser', *Opt. Lett.*, 2014, **39**, (6), pp. 1565–1568

[10] Liu X., Han D., Sun Z., ET AL.: 'Versatile multi-wavelength ultrafast fiber laser mode-locked by carbon nanotubes', *Sci. Rep.*, 2013, **3**, p. 2718

[11] Cui Y., Liu X.: 'Graphene and nanotube mode-locked fiber laser emitting dissipative and conventional solitons', *Opt. Express*, 2013, **21**, (16), pp. 18969–18974

[12] Chang W., Ankiewwicz A., Soto-Crespo J.M., Akhmediev N.: 'Dissipative soliton resonances', *Phys. Rev. A*, 2008, **78**, (2), p. 023830

[13] Chang W., Soto-Crespo J.M., Ankiewicz A., Akhmediev N.: 'Dissipative soliton resonances in the anomalous dispersion regime', *Phys. Rev. A*, 2009, **79**, (3), p. 033840

[14] Sotor J., Sobon G., Krzempek K., Abramski K.M.: 'Fundamental and harmonic mode locking in erbium doped fiber laser based on graphene saturable absorber', *Opt. Commun.*, 2012, **285**, (13), pp. 3174–3178

[15] Zhang Z.X., Zhan L., Yang X.X., Luo S.Y., Xia Y.X.: 'Passive harmonically mode-locked erbium doped fiber laser with scalable repetition rate up to 1.2 GHz', *Laser Phys. Lett.*, 2007, **4**, (8), pp. 592–596

[16] Liu X.: 'Mechanism of high energy pulse generation without wave breaking in mode-locked fiber lasers', *Phys. Rev. A*, 2010, **82**, (5), p. 053808

16

Complex-valued interferometric inverse synthetic aperture radar image compression base on compressed sensing

Liechen Li[1,2], Daojing Li[1], Bo Liu[3], Qingjuan Zhang[4]

[1]Science and Technology on Microwave Imaging Laboratory, Institute of Electronics, Chinese Academy of Sciences, Beijing 100190, People's Republic of China
[2]University of Chinese Academy of Sciences, Beijing 100049, People's Republic of China
[3]China Academy of Space Technology, Beijing 100094, People's Republic of China
[4]Patent Examination Cooperation Center Jiangsu Center of the Patent Office, SIPO, Suzhou 215011, People's Republic of China
E-mail: chrislee365@hotmail.com

Abstract: Complex-valued interferometric inverse synthetic aperture radar (InISAR) image compression is discussed in this study. The target scene has its continuity and is compressible. However, because of the random phase of each resolution cell, the frequency spectrum of an ISAR image is wide and the complex-valued image is hard to compress. A complex-valued ISAR image compression approach is proposed. Using two or more antennas and interferometry processing, the random phase of image pixel can be cancelled and the frequency spectrum becomes sparse. Therefore the theory of compressed sensing can be introduced to the process of the complex-valued image compression. Hence, the complex-valued InISAR image compression and reconstruction can be completed. Results on real data are presented to validate the method. In comparison with results of the conventional compression techniques, the proposed method shows the better ability to preserve both the imaging magnitude and interferometric phase.

1 Introduction

Inverse synthetic aperture radar (ISAR) imaging is an important tool in many military and civilian applications. However, while the volume of data collected is increasing rapidly, the ability to transmit it, or to store it, is not increasing as fast. An ISAR system faces the challenge of storage and transmission of mass data. An ISAR system may collect data at a high rate that easily exceeds the capacity of the downlink channel or the volume of the mass storage medium. Moreover, the volume of data doubles or more in the ISAR interferometry case. The situation has become even more severe in the past few years with the increased requirements of modern ISAR systems, including high resolution, multi-polarisation, three-dimensional (3D) imaging, multi-frequency and multi-operation mode. As a result, effective data compression becomes necessary [1, 2].

There are mainly two approaches to compress ISAR data: compression of ISAR raw data and compression of the ISAR image acquired in real time. For ISAR raw data compression, algorithms developed can generally be divided into three categories: scalar compression algorithms, vector compression algorithms and transform domain compression algorithms [3–5]. For image compression, it is much more difficult in the ISAR case because an ISAR image differs from an optical image in several ways. The spectrum of an ISAR image tends to have less spectral rolloff than an optical image. The dynamic range of an ISAR image is typically much higher than an optical image. The ISAR image has much high-frequency energy. These make an ISAR image hard to compress. The existing methods [6–8] may compress and restore a single ISAR image well, but are hard to apply on ISAR interferometry for its high phase loss.

To make the frequency spectrum of an ISAR image narrow, interferometry technique is used. The random phases of each resolution cell are considered as the same if the two antennas are close enough. Therefore the random phase of each resolution cell can be cancelled and the spectrum of an ISAR image becomes narrow and sparse. This property suggests the using of compressed sensing (CS) [9, 10] methods for image compression and restoration. CS, as a favourable sparse reconstruction technique, is a new and attractive method for image compression and restoration. It aims at minimising the number of measurements to be taken from signals while still retaining the information necessary to approximate them well. Compared with conventional compression techniques, CS exploits the sparsity of the signal.

The main contribution of this paper is a presentation of a new complex-valued ISAR image compression method based on CS with ISAR interferometry techniques. We proposed the idea that using the information of two antennas to shrink the range of the image frequency spectrum by introducing the interferometry techniques to cancel the random phase of each resolution cell. To compress more and restore with lower loss, we apply CS to compression processing by constructing a specially designed dictionary. Both the amplitude and phase of complex-valued ISAR images are reconstructed with overwhelming probability. Therefore, the proposed method can preserve the image phase while obtaining the magnitude at the same time. By compressing the data volume, the proposed method can release the pressure on record devices and shorten the data transmission time.

The structure of this paper is organised as follows. The next section discusses about the characters of complex-valued ISAR images and interferometry. Section 3 gives a brief introduction of CS theory. Section 4 presents the implementation the ISAR compression and restoration based on CS. In Section 5, real data experiments are made to verify the feasibility of the proposed method. Theoretical analysis and experiment results are given to assess the performance between the CS-based method and conventional low pass filter (LPF) method. Finally, the conclusions of this paper and discussion of the proposed method are provided in Section 6.

2 ISAR image characters

ISAR images can provide information on the surface properties of the detected targets which are always continuous in the target scene. However, the magnitude of the radar image resolution is much larger than that of wavelength, resulting in the random phase of

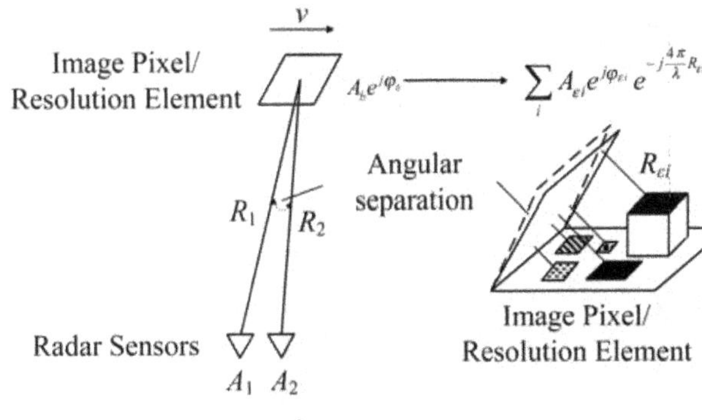

Fig. 1 *Random phase mechanism*

pixels which makes the scene discontinuous and spectrum wide. Fig. 1 shows the mechanism of the random phase.

As Fig. 1 depicts, pixels in radar image are a complex phasor representation of the coherent backscatter from the resolution element on the target and the propagation phase delay [11]. The propagation phase delay is determined by R_i ($i = 1, 2$) which denotes the distance between radar sensors and targets. Backscatter phase delay is coherent sum of contributions from all elemental scatterers in the resolution element with backscatter and their differential path delays $R_{\varepsilon i}$, that is

$$\varphi_b = \arg\left\{\sum_i A_{\varepsilon i} e^{j\varphi_{\varepsilon i}} e^{-j(4\pi/\lambda)R_{\varepsilon i}}\right\} \tag{1}$$

where λ is the wavelength. The total phase delay can be written as

$$\varphi = 2\frac{2\pi}{\lambda}R + \varphi_b \tag{2}$$

The distance R_i is continuous in most cases and varies slowly, so it contributes to the low-frequency energy. Backscatter phase delay ϕ_b is random for a single ISAR image and contributes to the high-frequency energy, which makes the spectrum wide and compression hard to implement.

When the system has two antennas as is shown in Fig. 1, two images will be attained. The phase delay of each antenna can be written as

$$\begin{cases} \varphi_1 = 2\dfrac{2\pi}{\lambda}R_1 + \varphi_{b1} \\ \varphi_2 = 2\dfrac{2\pi}{\lambda}R_2 + \varphi_{b2} \end{cases} \tag{3}$$

If the view angle is much less than 1°, the coherent sum is nearly unchanged. We can suppose the random phases of two images are the same, that is, $\phi_{b1} = \phi_{b2}$. Pixels in two radar images observed from nearby vantage points have nearly the same complex phasor representation of the coherent backscatter from a resolution element on the target, but a different propagation phase delay. Although the backscatter phase delay is random for a single image, it can be cancelled using conjugate multiplication. After co-registration of two images, construct a new image similar to obtaining an interferogram as

$$s = A_2 e^{j\varphi_2} e^{-j\varphi_1} = A_2 e^{-j(4\pi/\lambda)\Delta R} \tag{4}$$

where $\Delta R = R_1 - R_2$. The new image removes the random phase and keeps the differential phase which can be further used. The phase delay is proportional to the range difference which is mostly continuous in the target scene. Hence, the spectrum of the complex-valued image becomes sparse and CS theory can be introduced.

3 CS basis

CS is a new theory that focuses on sparse signal compression and reconstruction. For the sparse signal, CS measures M ($K < M \ll N$) projections of $\boldsymbol{x}$ and reconstructs the sparse signal from this small set of non-adaptive linear measurements. Each measurement can be viewed as an inner product with the signal $\boldsymbol{x}$ and some vector $\boldsymbol{\psi}_i$. If we collect M measurements in this way, we may then consider the $M \times N$ measurement matrix whose row are the vectors ψ_i^{T}. The sparse recovery problem can be considered as the recovery of the K-sparse signal $\boldsymbol{x}$ from its measurement vector $\boldsymbol{y} = \boldsymbol{\Psi x}$. A direct formulation of this problem is to solve the ℓ_0-minimisation problem

$$\min_{\boldsymbol{x} \in \boldsymbol{R}^N} \|\boldsymbol{x}\|_0, \quad \text{s.t. } \boldsymbol{y} = \boldsymbol{\Psi x} \tag{5}$$

However, ℓ_0-minimisation is computationally difficult to solve, as it involves NP-hard enumerative search. Fortunately, recent work in CS has shown that the convex relaxation approach relies on the fact that, besides the ℓ_0 norm, the ℓ_1 norm also promotes sparsity in a solution. The relaxed version of the problem can be written as

$$\min_{\boldsymbol{x} \in \boldsymbol{R}^N} \|\boldsymbol{x}\|_1, \quad \text{s.t. } \boldsymbol{y} = \boldsymbol{\Psi x} \tag{6}$$

The ℓ_1-minimisation approach provides uniform guarantees and stability and relies on methods in linear programming. Equation (6) requires a condition on the measurement matrix $\boldsymbol{\Psi}$ stronger than the simple injectivity on sparse vectors, but many kinds of matrices have been shown to satisfy this condition number of measurements $M \geq K \log N$. Candès and Tao showed that under a slightly stronger condition, which is known as restricted isometry property (RIP), basis pursuit (BP) can recover every K-sparse signal by solving (6) [12]. The RIP requires

$$(1 - \delta)\|\boldsymbol{x}\|_2 \leq \|\boldsymbol{\Psi x}\| \leq (1 + \delta)\|\boldsymbol{x}\|_2 \tag{7}$$

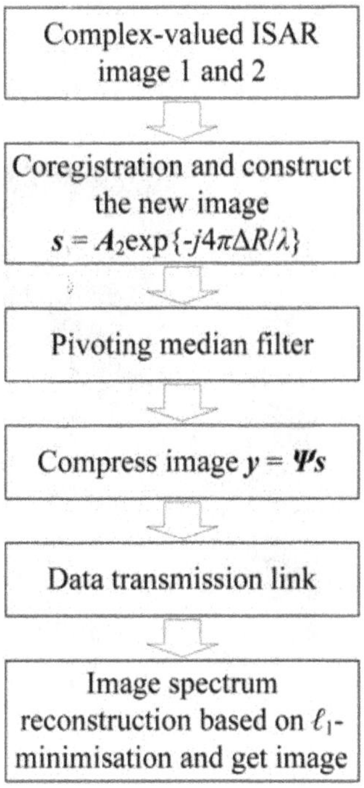

Fig. 2 *Flowchart of the proposed method*

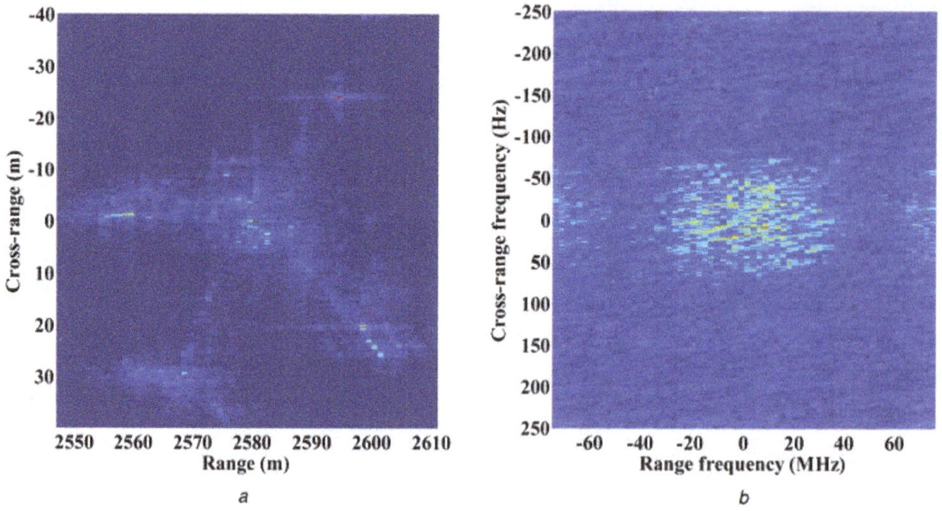

Fig. 3 *Original image and frequency spectrum*
a The original ISAR image
b The original ISAR frequency spectrum

The RIP is closely related to an incoherency property. It has been shown that with exponentially high probability, random Gaussian, Bernoulli and partial Fourier matrices satisfy the RIP with number of measurements nearly linear in the sparsity level.

When the signal $\boldsymbol{y}$ is noisy, the signal representation problem becomes a signal approximation problem. The modified convex problem can be described as

$$\min_{x \in \boldsymbol{R}^N} \|\boldsymbol{x}\|_1, \quad \text{s.t. } \|\boldsymbol{y} - \boldsymbol{\Psi x}\|_2 \leq \varepsilon \tag{8}$$

where ε bounds the amount of noise in measured data. Many approximate algorithms for the CS reconstruction given by (5) have also been developed, such as BP and the family of matching pursuit algorithms [13]. Note that it is known that Fourier measurements represent good projections for CS of sparse point like signals when representing random undersampling of the spatial frequency data. This suggests a natural application to the image compression problem.

4 Interferometric ISAR image compression

According to the CS theory, if the image is sparse in frequency domain, sampling in time domain does not need to satisfy the Nyquist sampling theorem. Only a few measurements are needed to reconstruct the image. As the new complex-valued ISAR image in (4) is sparse in frequency domain, it can be compressed by using the CS-based method.

The process of the complex-valued image compression can be expressed as

$$\boldsymbol{y} = \boldsymbol{\Phi s} + \boldsymbol{n} = \boldsymbol{\Phi F \sigma} + \boldsymbol{n} = \boldsymbol{\Psi \sigma} + \boldsymbol{n} \tag{9}$$

where $\boldsymbol{y}$ is the compressed image, $\boldsymbol{s}$ is the new image to compress which takes the form as (4), $\boldsymbol{\sigma}$ is the image frequency spectrum to be reconstructed and $\boldsymbol{n}$ denotes the noise. To decrease the effect of speckle noise, pivoting median filter [14] is used on the phase of image before the compression. It is worth noting that since CS theory deals with 1D problem, 2D variables $\boldsymbol{s}$ and $\boldsymbol{y}$ must reshape into vector form.

In (9), $\boldsymbol{\Phi}$ is the projection matrix and $\boldsymbol{F}$ is a transform matrix. $\boldsymbol{\Phi}$ is designed incoherent to $\boldsymbol{F}$ and determined by the sampling sequence. For example, if the sampling sequence is [100101], where 1 stands for samplings and 0 stands for discarded samplings, $\boldsymbol{\Phi}$ can be

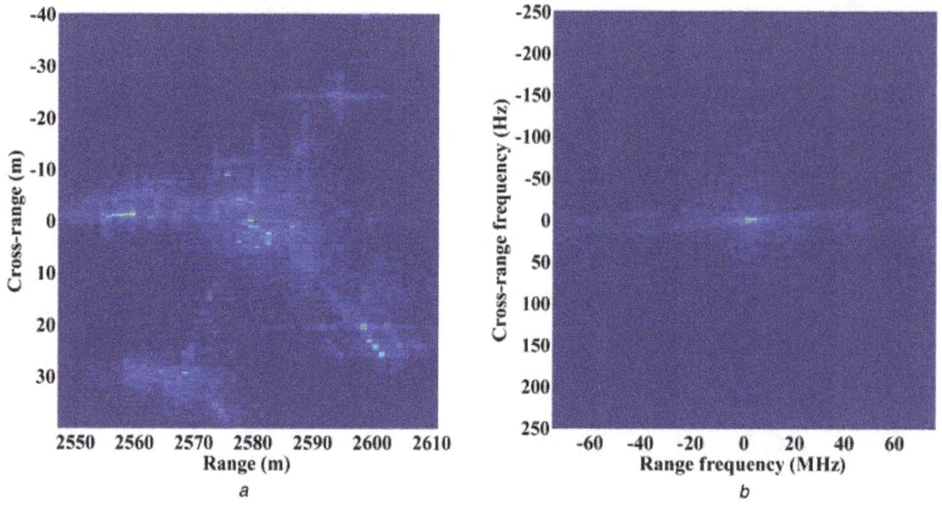

Fig. 4 *Image without random phase and frequency spectrum*
a The ISAR image without random phase
b The frequency spectrum of the InISAR image

Table 1 Reconstruction comparison

	MSE1	MSE2	MPE, °
CS image 1	0.0633	0.0218	35.0035
LPF image	0.0496	0.0628	25.0685
CS image 2	0.0076	0.0197	13.6379

expressed as

$$\boldsymbol{\Phi} = \begin{pmatrix} 1 & 0 & 0 & 0 & 0 & 0 \\ 0 & 0 & 0 & 1 & 0 & 0 \\ 0 & 0 & 0 & 0 & 0 & 1 \end{pmatrix} \quad (10)$$

With the projection matrix $\boldsymbol{\Phi}$, an $N \times 1$ reshaped image $\boldsymbol{s}$ is projected on $\boldsymbol{R}^M$ space, where $M \ll N$, so the compression is achieved. With $\boldsymbol{F}$, the image can be sparse represented. To reconstruct the signal with high probability, the matrix product $\boldsymbol{\Psi} = \boldsymbol{\Phi F}$ has to satisfy RIP in (7). Here Fourier transform is chosen to represent the image. Fourier basis is a simple and popular dictionary. As aforementioned, the new image $\boldsymbol{s}$ can be sparsely represented under Fourier basis. Fourier basis is also highly incoherent with the matrix $\boldsymbol{\Phi}$, which provides a good condition for correct reconstruction.

To restore the image, the CS-based method does not reconstruct the image directly, but reconstruct the frequency spectrum of the image first and then the image [15]. The spectrum can be reconstructed by solving the optimisation problem as follows

$$\min_{\sigma \in R^N} \|\boldsymbol{\sigma}\|_1, \quad \text{s.t. } \|\boldsymbol{y} - \boldsymbol{\Psi\sigma}\|_2 \le \varepsilon \quad (11)$$

where ε denotes the noise level in measured data and is determined by the noise energy. After frequency spectrum of the image is reconstructed, the image can be restored simply by 2D inverse Fourier transform. After restoring the image, multiply with the phase of image 1 if needed. The main procedures are shown in Fig. 2.

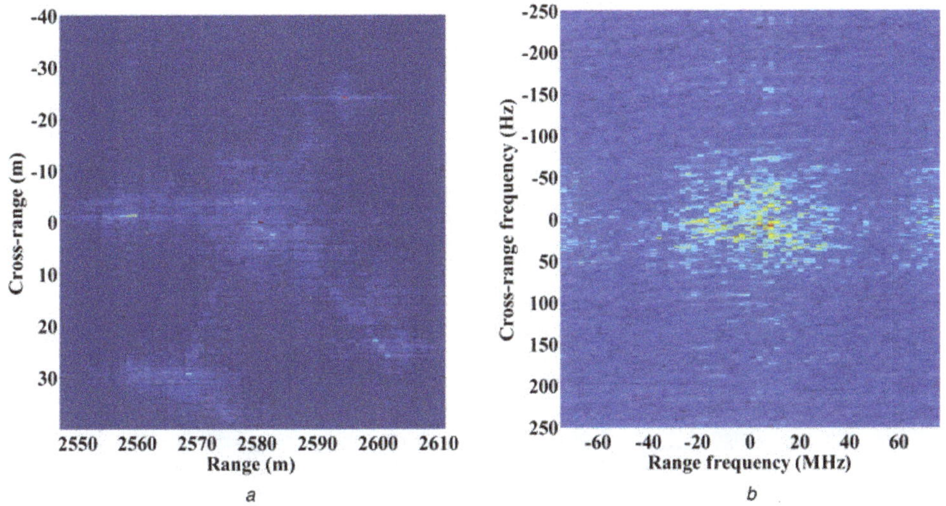

Fig. 5 *Reconstruction of the original image by CS*
a The reconstructed ISAR image
b The reconstructed frequency spectrum

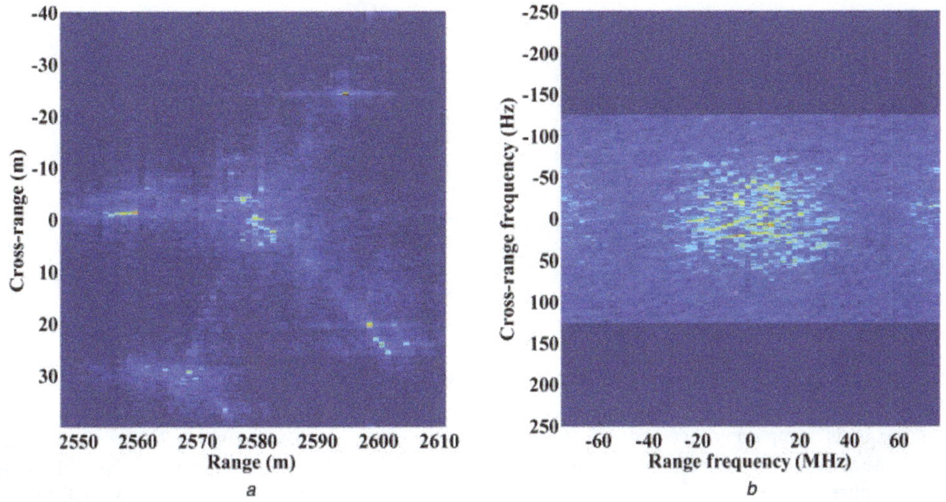

Fig. 6 *Reconstruction of the original image by LPF*
a The reconstructed ISAR image
b The frequency spectrum after LPF

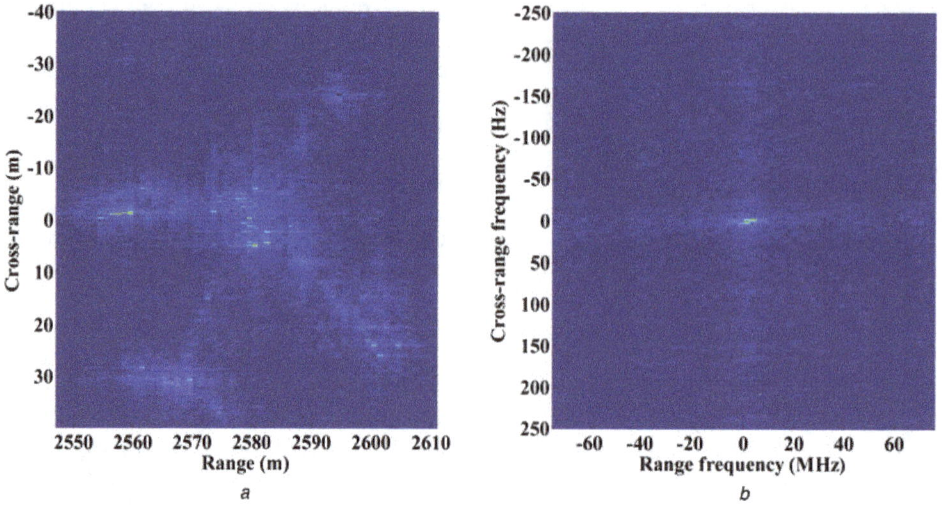

Fig. 7 *Reconstruction of the image without the random phase by CS*
a The reconstructed InISAR image
b The reconstructed InISAR image frequency spectrum

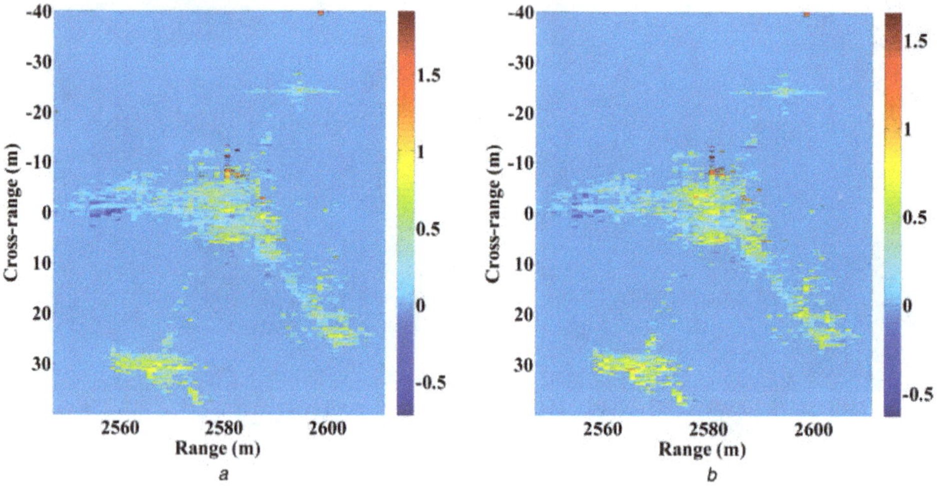

Fig. 8 *Original interferometric phase and reconstructed interferometric phase by CS*
a The original interferometric phase
b The reconstructed interferometric phase

5 Real data experiments

To illustrate the feasibility of the method introduced in this paper, this section presents some results based on data obtained by a millimetre wave prototype ISAR with three antennas which is developed and operated by Institute of Electronics, Chinese Academy of Sciences. The radar works on Ka-band and the baseline between two antennas which we use is 0.4 m.

The aeroplane ISAR image shown in Fig. 3 is used as reference to compare the imaged quality of reconstructed image, where (a) is the ISAR image magnitude and (b) is the frequency spectrum of the image. The image is formed using ωK algorithm after parameter estimation. The image constructed as (4) without random phases is shown in Fig. 4 whose frequency spectrum becomes sparse. A 5×3 window is used in pivoting median filter. About 50% of the image is sampled to reconstruct the full frequency spectrum of the image. Gaussian random sampling and LPF are used to compress the image. To reconstruct the spectrum, BP, one of the most commonly studied ℓ_1-minimisation approach, is used. Mean square error (MSE) and mean phase error (MPE) shown in (12) and (13) are used to judge the reconstruction quality

$$\mathrm{MSE} = \sqrt{\frac{1}{N_a N_r} \sum_{i=1}^{N_a} \sum_{j=1}^{N_r} \left(A_{ij} - A'_{ij} \right)^2} \tag{12}$$

$$\mathrm{MPE} = \frac{1}{N_a N_r} \sum_{i=1}^{N_a} \sum_{j=1}^{N_r} \left| P_{ij} - P'_{ij} \right| \tag{13}$$

where A stands for the amplitude of the image or spectrum and P denotes the phase of image. The subscripts i and j denote the ith pixel in cross-range direction and the jth pixel in range direction and the superscripts denote the reconstructed image or spectrum. Results of comparison between image to compress and reconstructed image have been listed in Table 1. MSE1 stands for the MSE of frequency spectrum and MSE2 stands for the MSE of image. CS images 1 and 2 stand for reconstruction of the original complex-valued image by CS and reconstruction of the image without the random phase by CS, respectively.

Fig. 5 shows the reconstruction of an original ISAR image by CS. Fig. 6 shows the reconstruction of the image by LPF. Both results perform poorly because of the random phases. The frequency spectrum shown in Fig. 7 is sparse which satisfies the limitation of CS. The reconstruction image quality of Fig. 7 is better than those of Figs. 5 and 6. The MSE of reconstructed image is much less than those of the other two. Image without random phases can be compressed and reconstructed with lower loss. Fig. 8*a* shows the original interferometric phase and Fig. 8*b* shows the reconstructed interferometric phase. The compression phase loss is about 10° and can be further used in ISAR interferometry field such as moving targets angle measurement and positioning.

6 Conclusions

In this paper, a CS-based approach is presented for complex-valued ISAR image compression and restoration. First the spectrum of the complex-valued ISAR image is made sparse by using the interferometric techniques. Then CS method can be introduced into compression. When the compressed image needs to be restored, it solves an optimisation problem to reconstruct the spectrum and then transform to image instead of reconstructing the image directly. Real data are used to verify the feasibility of the method proposed in this paper. The results demonstrate that the presented CS-based method can achieve better restored image quality than conventional LPF method and sparseless ISAR original images. The proposed method keeps both the amplitude and phase in the compression process. The reconstructed image can be further used in interferometry. The reconstructed complex-valued image can be further used in ISAR interferometry field. Although only the complex-valued ISAR image is investigated in this paper, the proposed concept can be also used in SAR image compression.

7 Acknowledgments

This work was supported by the Researching Program of Chinese Academy of Sciences and the National Natural Science Foundation of China (61271422).

8 References

[1] Zeng Z., Cumming I.G.: 'SAR image data compression using a tree-structured wavelet transform', *IEEE Trans. Geosci. Remote Sens.*, 2001, **39**, (3), pp. 546–552

[2] Hua B., Qi H.M., Zhang P., Li X.: 'Vector quantization for saturated SAR raw data compression', *Adv. Space Res.*, 2010, **45**, pp. 1330–1337

[3] Gergič B., Planinšič P., Banjanin B., *ET AL.*: 'A comparison between SAR data compression in Cartesian and polar coordinates', *Int. J. Remote Sens.*, 2004, **25**, (10), pp. 1987–1994

[4] Benz U., Strodl K., Moreira A.: 'A comparison of several algorithms for SAR raw data compression', *IEEE Trans. Geosci. Remote Sens.*, 1995, **33**, (5), pp. 1266–1276

[5] Qiu X.L., Lei B., Ge Y.P., *ET AL.*: 'Performance evaluation of two compression methods for SAR raw data', *J. Electron. Inf. Technol.*, 2010, **32**, (9), pp. 2268–2272

[6] Baxter R.A.: 'SAR image compression with the Gabor transform', *IEEE Trans. Geosci. Remote Sens.*, 1999, **37**, (1), pp. 574–588

[7] El Assad S., Morin X., Barba D., Slavova V.: 'Compression of polarimetric synthetic aperture radar data', *Prog. Electromagn. Res.*, 2003, **39**, pp. 125–145

[8] McGinley B., O'Halloran M., Conceicao R.C., Higgins G., Jones E., Glavin M.: 'The effects of compression on ultra wideband radar signals', *Prog. Electromagn. Res.*, 2011, **117**, pp. 51–65

[9] Donoho D.: 'Compressed sensing', *IEEE Trans. Inf. Theory*, 2006, **52**, (4), pp. 5406–5425

[10] Candès E., Romberg J., Tao T.: 'Robust uncertainty principles: exact signal reconstruction from highly incomplete frequency information', *IEEE Trans. Inf. Theory*, 2006, **52**, (2), pp. 489–509

[11] Rosen P.A., Hensley S., Joughin I.R., *ET AL.*: 'Synthetic aperture radar interferometry', *Proc. IEEE*, 2000, **88**, (3), pp. 333–382

[12] Candès E., Tao T.: 'Decoding by linear programming', *IEEE Trans. Inf. Theory*, 2005, **51**, (12), pp. 4203–4215

[13] Needell D.: 'Topics in compressed sensing', PhD thesis, University of California, CA, USA, Davis, 2009

[14] Meng D., Sethu V., Ambikairajah E., Ge L.: 'A novel technique for noise reduction in insar images', *IEEE Geosci. Remote Sens. Lett.*, 2007, **4**, (2), pp. 226–230

[15] Li L., Li D., Liu B., Zhang Q., Wei L.: 'Three-aperture inverse synthetic aperture radar moving targets imaging processing based on compressive sensing'. Proc. ISICT 2012, London, UK, 2012, pp. 210–214

Heat-pump performance: voltage dip/sag, under-voltage and over-voltage

William J.B. Heffernan[1], Neville R. Watson[2], Jeremy D. Watson[2]

[1]Electric Power Engineering Centre, University of Canterbury, Christchurch 8140, New Zealand
[2]Electrical & Computer Engineering Department, University of Canterbury, Private Bag 4800, Christchurch 8140, New Zealand
E-mail: neville.watson@canterbury.ac.nz

Abstract: Reverse cycle air-source heat-pumps are an increasingly significant load in New Zealand and in many other countries. This has raised concern over the impact wide-spread use of heat-pumps may have on the grid. The characteristics of the loads connected to the power system are changing because of heat-pumps. Their performance during under-voltage events such as voltage dips has the potential to compound the event and possibly cause voltage collapse. In this study, results from testing six heat-pumps are presented to assess their performance at various voltages and hence their impact on voltage stability.

1 Introduction

Rapid introduction of heat-pumps has occurred in New Zealand, as well as many other countries, because of the desire to use energy more efficiently [1, 2]. Subsidies for heat-pumps have also aided their uptake. Most of the heat-pumps entering the system are inverter-based. This will alter the nature of the loading on the power system because of their characteristics [3–5], in a similar way that compact fluorescent lamps have changed the nature of lighting loads [6]. An earlier contribution concentrated on the harmonic performance of six different types of heat-pump at nominal voltage [7] in conjunction with relevant standards [8–10]. This paper looks at their performance when subjected to voltage variations (both steady-state (SS) voltage and voltage dips).

The characteristics of the grid transmitting power into the Auckland region from southward generation, coupled with the concentration of loading in this region, makes voltage collapse in the northern part of the North Island a real threat. To assess the impact of this change of load type on the voltage stability, the same six heat-pumps tested in [7] were subjected to a range of further tests. In [7], each of the six heat-pump circuits was presented and classified, with laboratory verification, and the same designation is used in this paper. Five of the heat-pumps (designated A25, A50, B, D and E) employed inverter-driven compressor motors, with one direct-on-line unit (designated C). Unit C, along with units A50, B, D and E, all from different manufacturers, had rated heating and cooling power of about 5–6 kW. Unit A25 had a rated heating and cooling power of about 3 kW and was from the same manufacturer as A50.

2 SS voltage test

The applied voltage was slowly varied from 1.1 per unit (pu) down to the cut-out voltage while the real, reactive and apparent powers, current total harmonic distortion (THD*i*), power factor (PF) and displacement PF (DPF) were recorded. The results of these tests are tabulated in Tables 2–7 for heating mode and Tables 8–13 for cooling mode (Appendix 1). Figs. 1–4 display this information graphically. Positive reactive power implies a lagging/inductive current. Care must be taken interpreting these results as the electrical power drawn by a heat-pump is determined by many factors. Factors include the indoor and outdoor temperatures, the refrigerant temperatures and phases, as well as the heat/cool setting and temperature set-point. For this reason, it is very difficult to exactly reproduce any given operating condition and there is some variability in these parameters when tests have been repeated. For instance, a measured drop in input power as the applied voltage is reduced may be because of the reduced voltage, but may be compounded because of changes in refrigerant temperature, phase etc. that are approximately coincident. The cut-out voltages are also influenced to some extent by the power drawn by the unit at that particular instant, but are generally within a few volts of the figures recorded. A marker showing the nominal input power ratings of each unit is displayed in Fig. 1. This shows that these devices may operate at levels significantly different from nominal, even at nominal voltage. Moreover, a marker showing the nominal current at nominal voltage, assuming unity PF is displayed in Fig. 3, demonstrating that nominal parameters are a very poor guide to predicting actual heat-pump behaviour. Despite the variability, overall patterns emerge for each unit. Note that all units continued to operate down to a voltage of 0.7 pu or below.

All units except C are inverter drive types, although their rectifier circuits differ significantly, as reported in [7]. Unit C is a direct-on-line induction motor type. It was noted in [7] that the units classified as types 1 and 2 (A25, B, E and A50), have two distinct modes of operation: at reasonably high input power levels they draw leading current with reasonably low distortion and acceptable PF. At lower input power levels, they draw current with high distortion and poor PF. The change-over point varies between about 400 and 800 W, with some hysteresis, depending on unit type. This is because of a switchable active PF correction (PFC) circuit, which may be trying to maximise efficiency by staying out of circuit at lower power levels. In practice, this makes such units particularly hard to test in a coherent manner.

Running below its nominal rated input power, the A25 unit operated without its PFC circuit being activated during the tests, the power level being 400–500 W in both cycles, accounting for the poor PF results at all voltage levels. The unit's current never exceeded 4 A during these tests, although its current can approach 6 A as shown in [7]. This unit cuts out below about 162 V (0.7 pu) in heating mode and 155 V (0.67 pu) in cooling mode.

Unit A50 ran with approximately constant input power over the heating mode test, with reasonable leading PF and DPF at this power level, down to about 120 V (0.52 pu), at which point the input current was 9.6 A. This is 160% of the measured current drawn at nominal voltage (and 138% of rated current at nominal voltage). This is the highest current recorded for this unit in any of the SS tests. In cooling mode, A50 again continued to operate down to about 120 V (0.52 pu). The PF was reasonable for this unit, although falling off at higher voltages. Unit B behaved in a

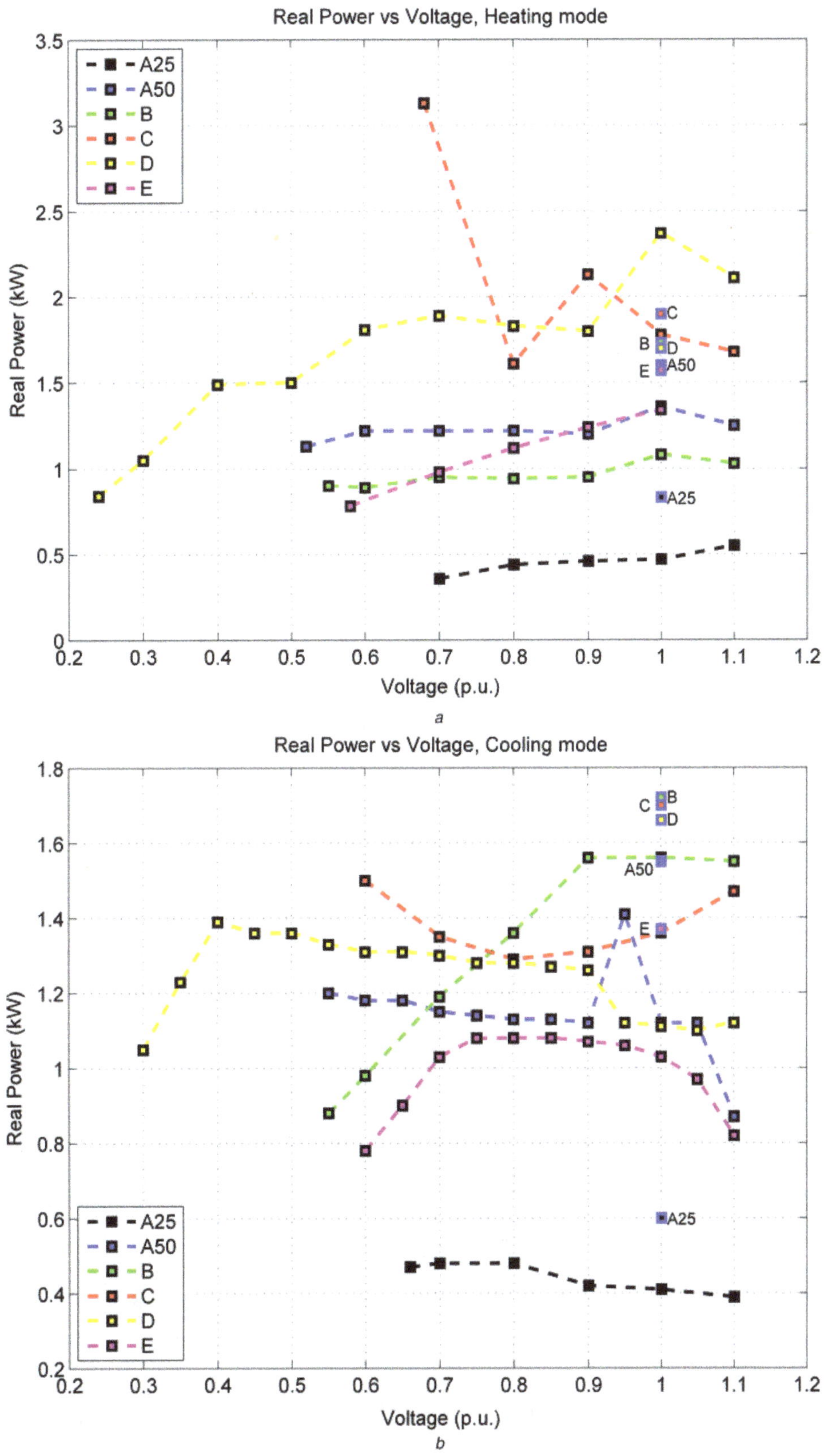

Fig. 1 *Comparison of real power consumption against per unit voltage*
a Heating mode
b Cooling mode

fairly similar fashion to A50 down to about 127 V (0.55 pu) in heating mode, at which point the input current of 7.4 A was 154% of the current at nominal voltage (and 98% of nominal current at nominal voltage). In cooling mode, unit B also ran down to 127 V (0.55 pu), and on this cycle appeared to be very well behaved, apparently being controlled to a current limit (CL) of about 7.5–8 A (causing power to fall off with falling supply voltage below 0.9 pu, well aligned with nominal current of 7.48 A at nominal voltage) with good (leading) PF under all recorded conditions.

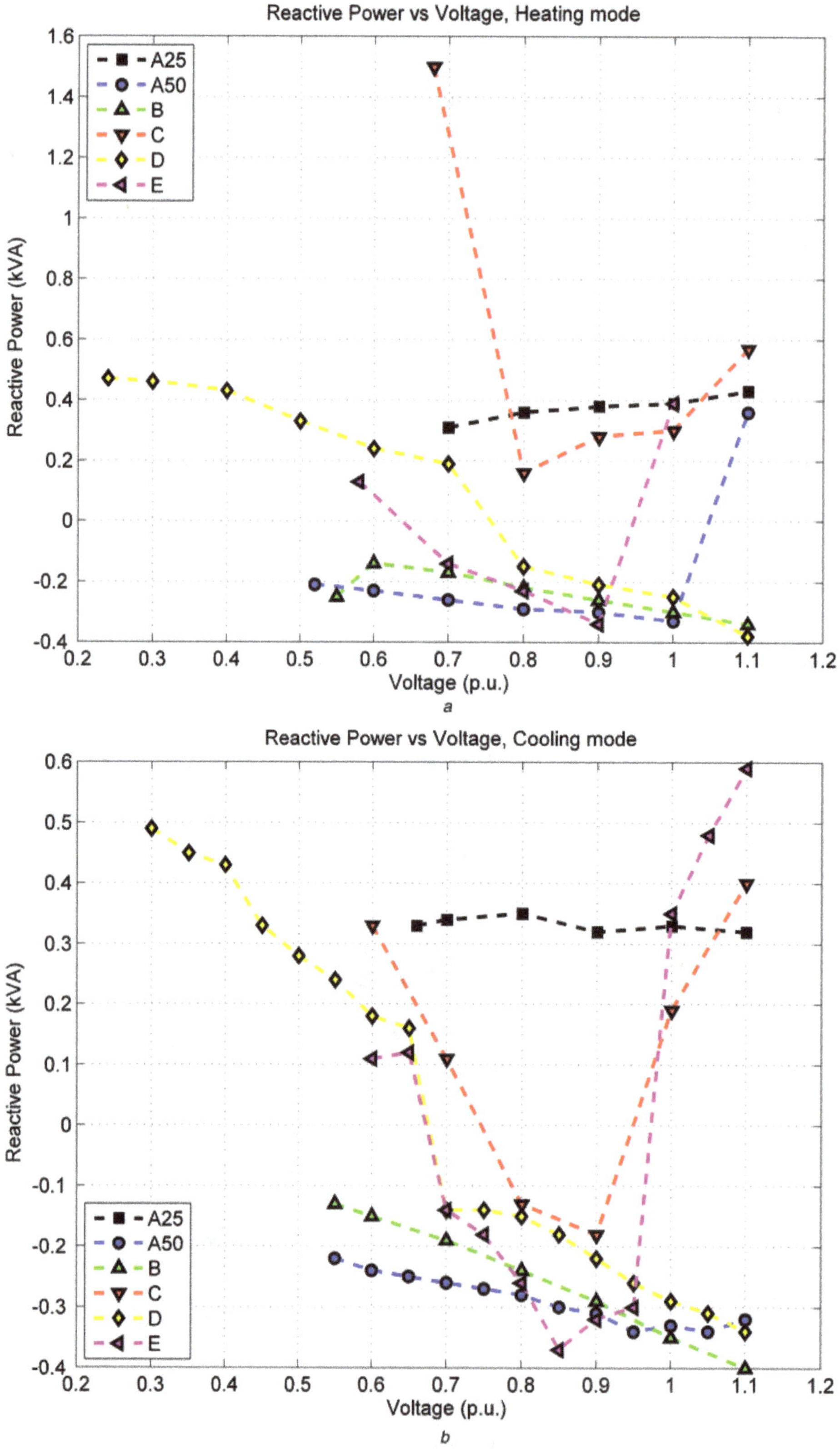

Fig. 2 *Comparison of reactive power against per unit voltage (+ve for lagging current)*
a Heating mode
b Cooling mode

In heating mode, unit C operated down to about 156 V (0.68 pu), at which point it was drawing 285% of the current at nominal voltage (and 269% of nominal current at nominal voltage), being about 22 A, and presenting an increasingly inductive load. Between 0.8 and 1.1 pu, it was reasonably well behaved, although a worsening of PF and DPF is noted at high supply voltage. In cooling mode, unit C operated down to about 138 V (0.6 pu), at which point it was drawing 185% of the current at nominal voltage, being about 11 A. Between 0.7 and 1.1 pu, it was reasonably well behaved, although a slight worsening of PF and DPF is

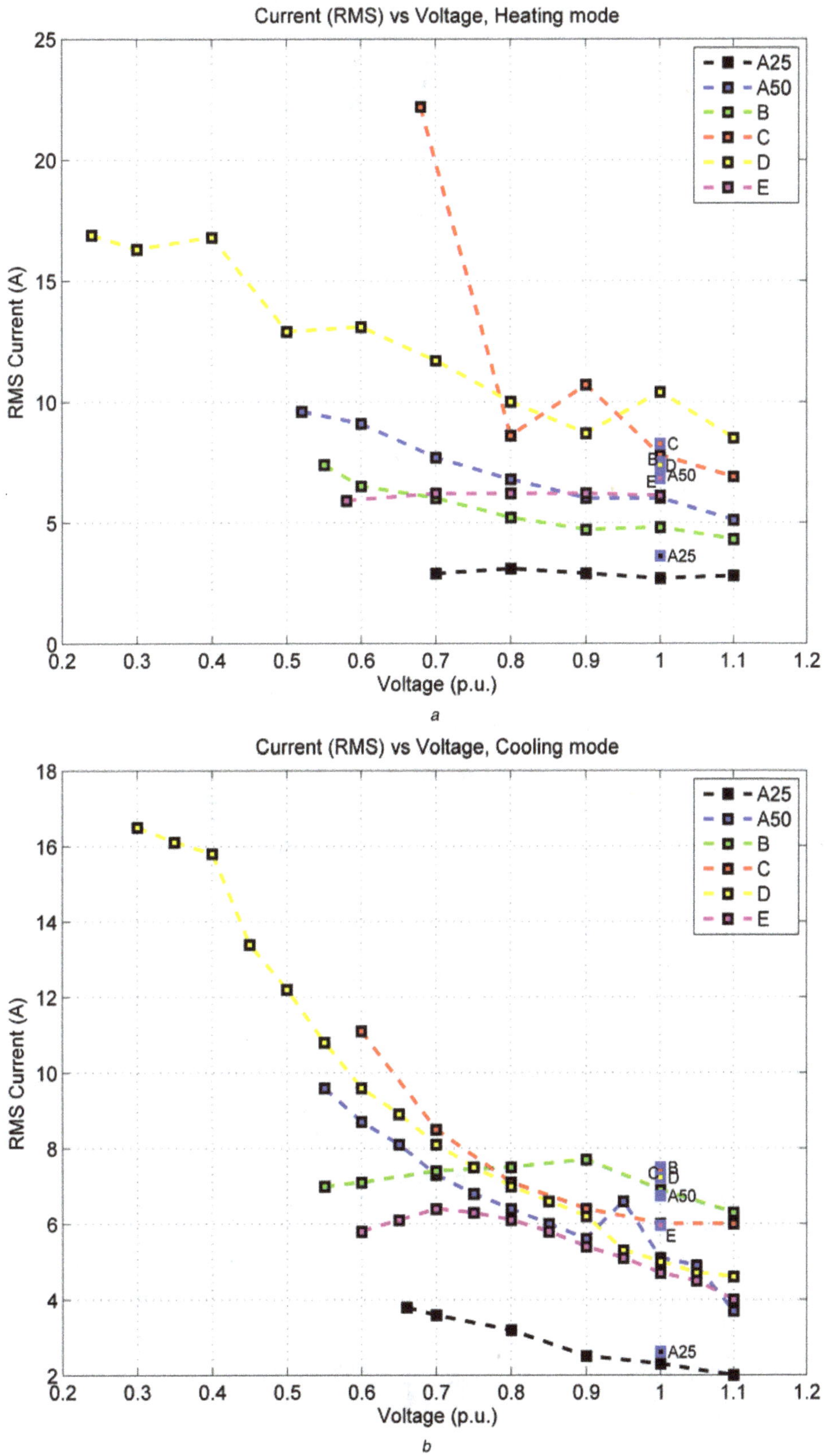

Fig. 3 *Comparison of RMS current against per unit voltage*
a Heating mode
b Cooling mode

noted at high supply voltages. Under different operating conditions (e.g. higher power), it may again show the same highly inductive low supply voltage behaviour as in the heating case.

In heating mode, unit D continued to operate down to an extremely low voltage of 56 V (0.24 pu), at which point the input current of 16.9 A was 163% of the measured current at nominal voltage (and 229% of nominal current at nominal voltage). The PF and DPF are reasonable down to about 0.5 pu, but get worst below this and change from leading to lagging below about 0.75 pu. In cooling mode, unit D again operated down to 56 V (0.24 pu).

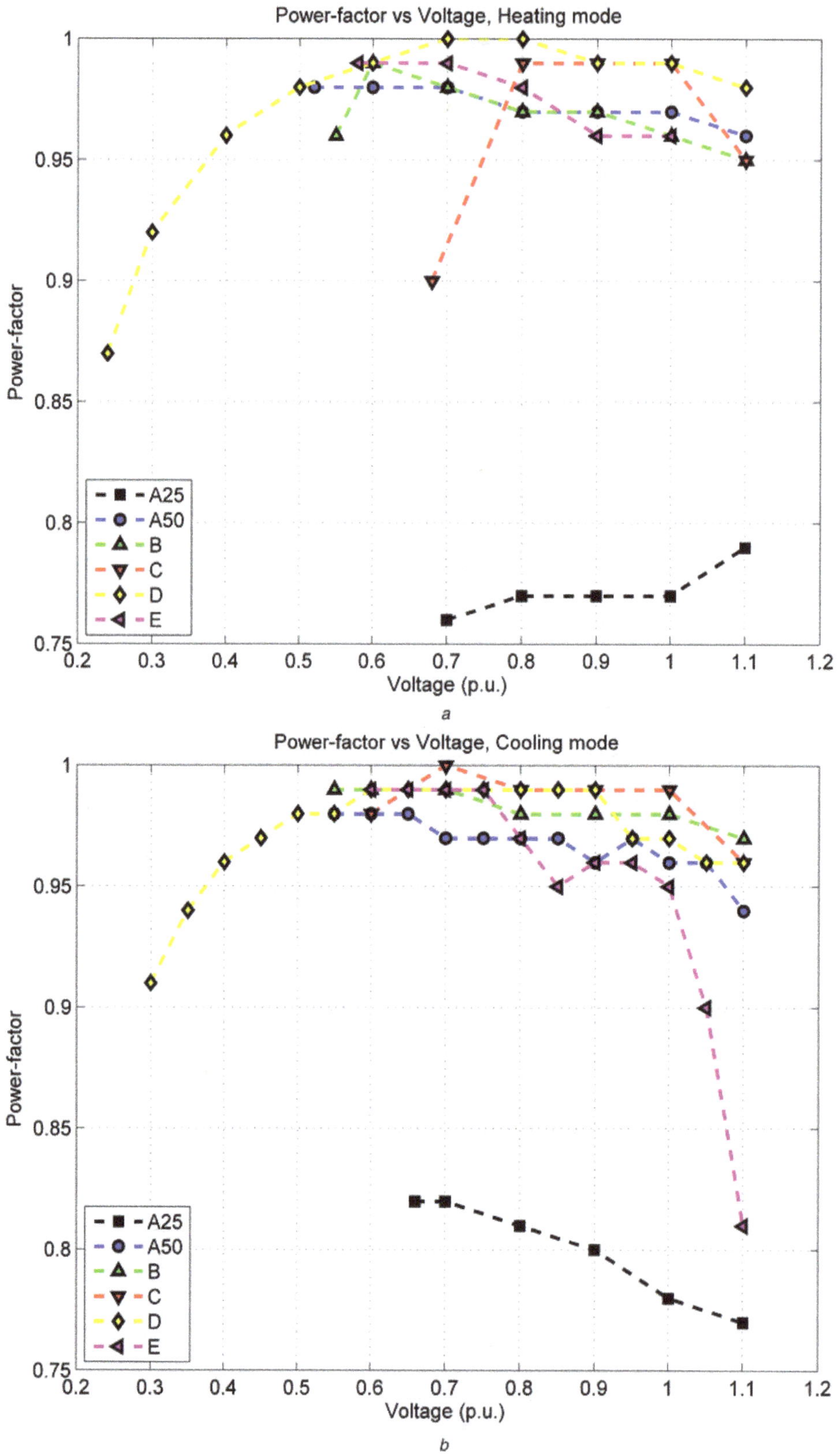

Fig. 4 *Comparison of PF against per unit voltage*
a Heating mode
b Cooling mode

At 0.3 pu, the input current, at 16.5 A, was 330% of the current at nominal voltage (again 229% of nominal current at nominal voltage). The PF and DPF are reasonable between 0.45 and 1 pu, but they get worst outside these limits. (Repeated tests in Figs. 5 and 7 show that this unit is probably controlled to a CL of about 16.5 A. This, in conjunction with its ability to operate below 0.3 pu, indicates that its control circuitry may be set up for 115 V supply.)

In both heating and cooling modes, unit E appears to behave well, apparently operating with a CL of about 6.5 A, well aligned with its nominal current of 6–7 A at nominal voltage. (Again this is very difficult to state categorically without very extensive testing in controlled ambient conditions, to obtain repeatability.) Nevertheless its PF and DPF deteriorate at high supply voltage.

It was found that the cut-in voltage, as the supply voltage was increased again, was approximately the same as the cut-out voltage for all models, but that there was a time delay in starting.

The variability in operating point is demonstrated in Figs. 5–8 where the heat-pumps are tested on different days with whatever the ambient outside temperature was on that day (uncontrolled). Hence, the operating state will be different because of the differing temperatures. The most dramatic difference is the peaks at 0.95 and 0.6 pu on two different runs with the A50 and the absence of such peaks on the third (see Fig. 6). Figs. 6–8 clearly demonstrate a sample of the many different operating regions the heat-pumps can be working in (when the outside and inside temperatures are not controlled).

3 Transient performance

3.1 Switch-on inrush current

The typical inrush current was measured for each of the six heat-pumps when switched on at nominal supply voltage. Note that a slightly different result will be obtained each time as the transients are never absolutely identical. The root mean square (RMS) current magnitude, in amperes, is plotted against historical time, in minutes, in Figs. 9–14. Although the details are slightly different, the inrush currents for heating and cooling modes are similar. Note that the electrical supply was already connected to each heat-pump, the switch on transient being activated by the remote control unit.

All of the units draw a spike of current at the instant of switch on, with the exception of unit D. These spikes are of no greater amplitude than the normal running current, except for unit C, the direct-on-line induction motor model, which draws a major spike of nearly 40 A. Some of the inverter drive units draw a second current spike a little later, in some instances shortly before the main motor current starts. Analysis of the heat-pump electrical circuits [7] shows that, for the inverter drive models, some of the spikes are almost certainly caused by charging of the inverter's direct current (DC) bus capacitors before the inverter starts switching. In the case of units displaying two spikes, the earlier one is likely to be because of charging of DC bus capacitors on power rails supplying some of the ancillary circuitry. However, some of the units may keep one or more ancillary rail powered up during standby operation, in which case inrush current to such a rail would only be observed on supply voltage connection. Unit D seems to avoid an initial spike altogether, probably by means of a relay and its boost converter rectifier, which is capable of charging the inverter bus capacitors with a controlled current. In general, any type 2 or 3 heat-pump [7] should be capable of avoiding inrush current into the inverter DC bus. In all inverter drive cases, the magnitude of inrush into the DC bus capacitors is in any case limited by the PFC inductance [7]. The magnitude of the inrush current spikes will also vary with the instantaneous voltage at turn-on, unless a zero-crossing detection scheme is used. There is perhaps room for further investigation in this area.

In the case of unit C, the much larger current spike observed is a combination of stalled rotor current and magnetising inrush current for the direct-on-line induction motor [11].

After a variable delay time, typically about 2–3 min, depending on model, indoor/outdoor temperatures, refrigerant state etc., the inverter-based heat-pumps ramp up to operating power in a more or less controlled fashion. Unit C's outdoor unit (in which the compressor motor resides) begins operation immediately. As can be seen, units D and E have the most benign start-up behaviour.

3.2 Voltage dip/sag performance, low-impedance system fault (heating)

To assess the transient performance of the heat-pumps, they were subjected to the voltage dip/sag shown in Fig. 15. This is a representative voltage profile for a low-impedance fault on the New

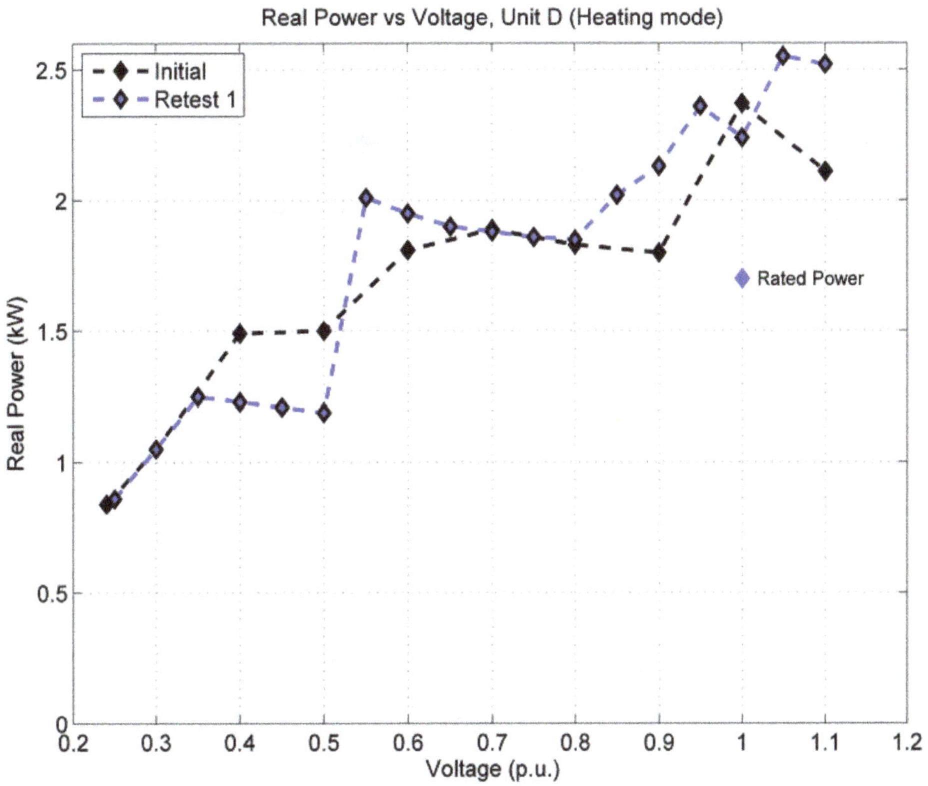

Fig. 5 *Real power against per unit voltage repeated for unit D – heating mode*

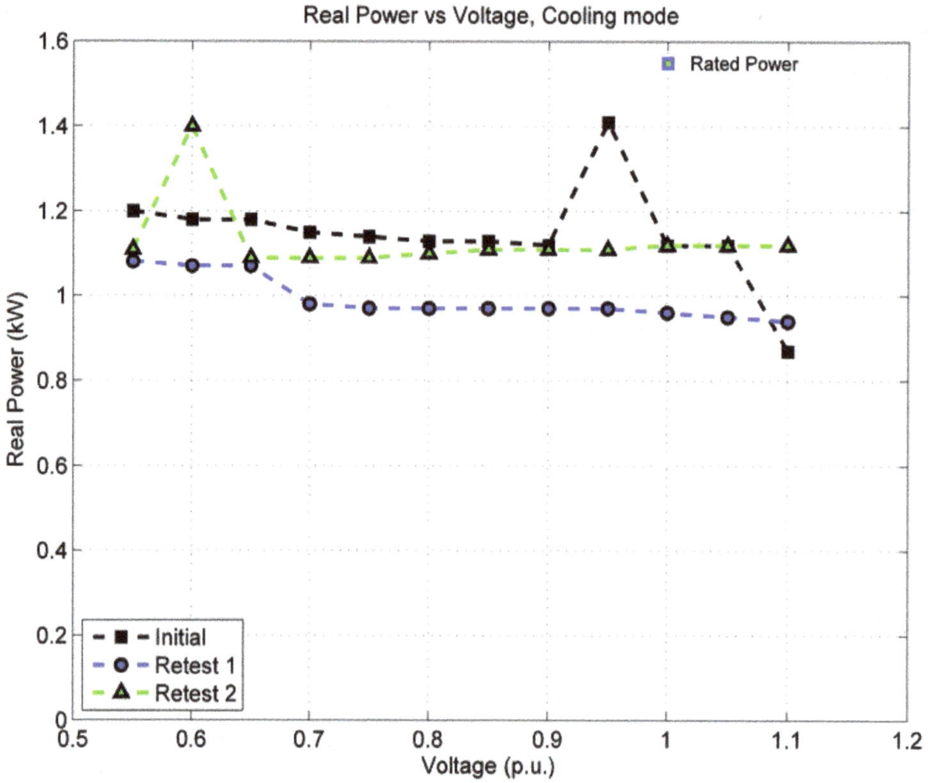

Fig. 6 *Real power against per unit voltage for A50 – cooling mode*

Zealand system and was supplied by the national grid company, Transpower NZ Ltd. This voltage profile was emulated by programming a Chroma AC voltage source to produce the piecewise linear approximation to this waveform shown in Fig. 16 (with expanded view in Fig. 17), in which actual voltage (230 V nominal system voltage) is shown against historical time in seconds. All six heat-pumps, while running on the heating cycle, were subjected to this voltage transient and their current waveforms captured (the voltage waveform is the same in all cases, although the trigger point varies in time). Note that only the expanded current traces

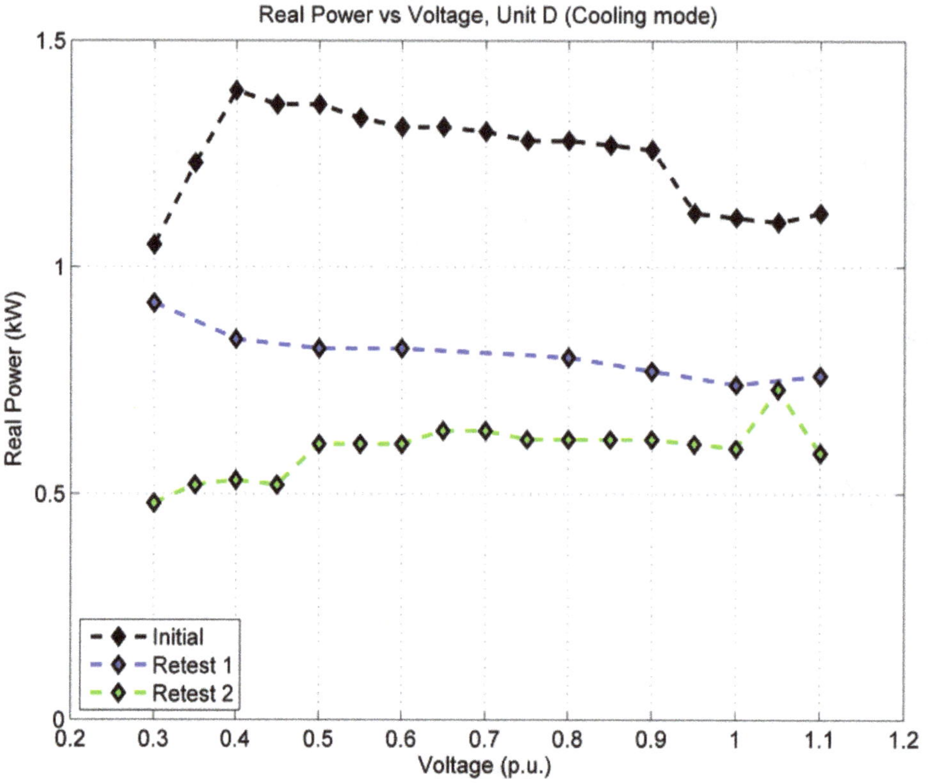

Fig. 7 *Real power against per unit voltage repeated for unit D – cool mode (rated power 1.66 kW)*

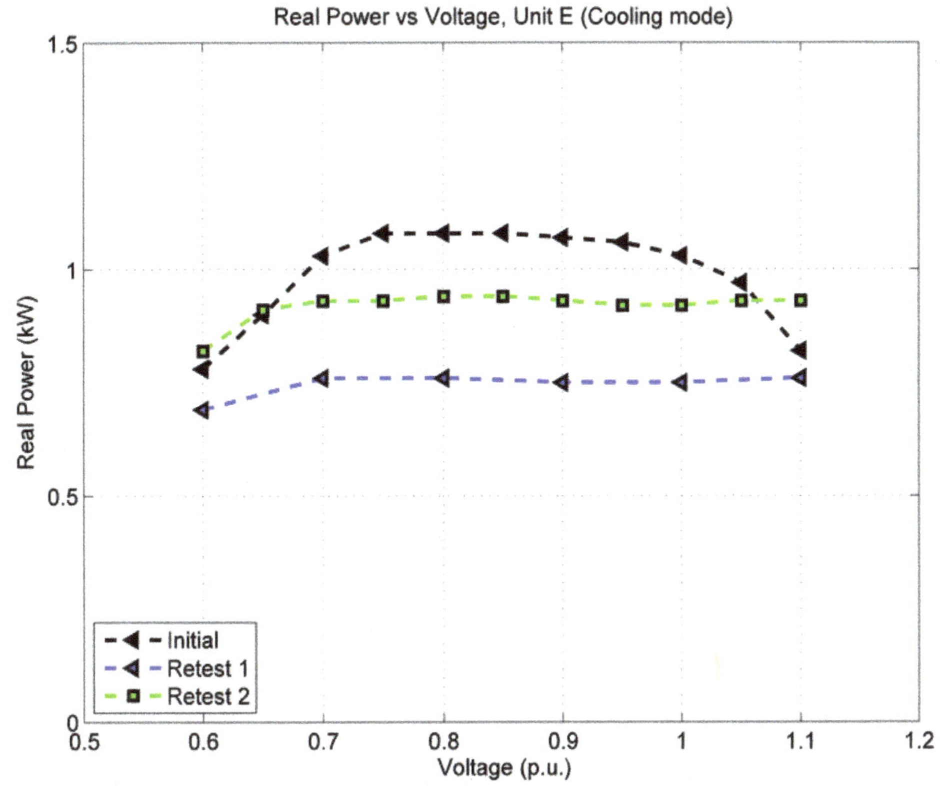

Fig. 8 *Real power against per unit voltage repeated for E – cool mode (rated power 1.37 kW)*

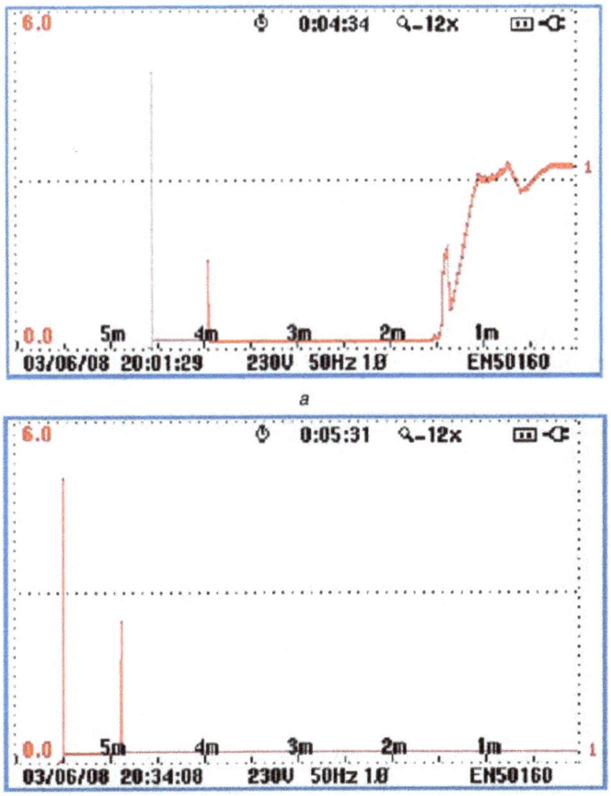

Fig. 9 *Inrush current for A25*
a Heating mode (initial peak: 5 A; run: 3.2 A)
b Cooling mode (initial peak: 5 A; run: 0.2 A (zero cooling power))

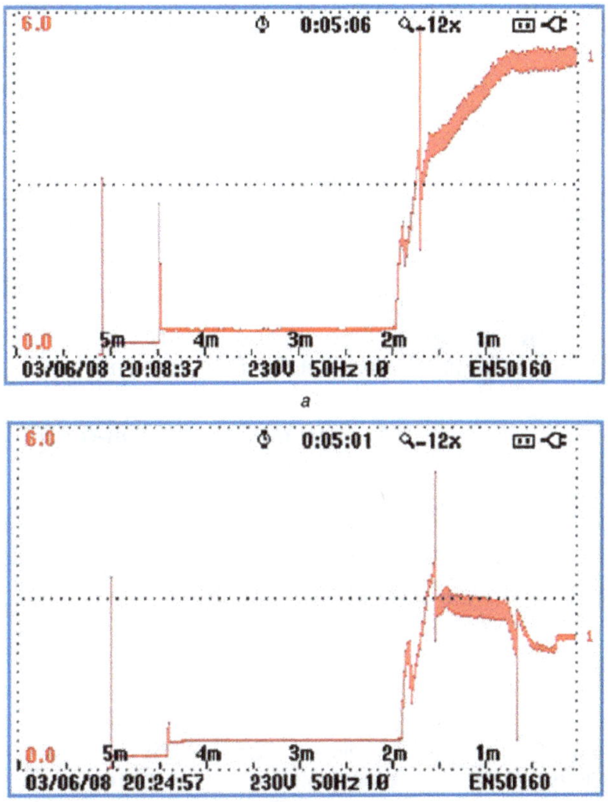

Fig. 10 *Inrush current for A50*
a Heating mode (initial peak: 3.2 A; run: 5.2 A)
b Cooling mode (initial peak: 3.3 A; run: 3 A, then 2.3 A)

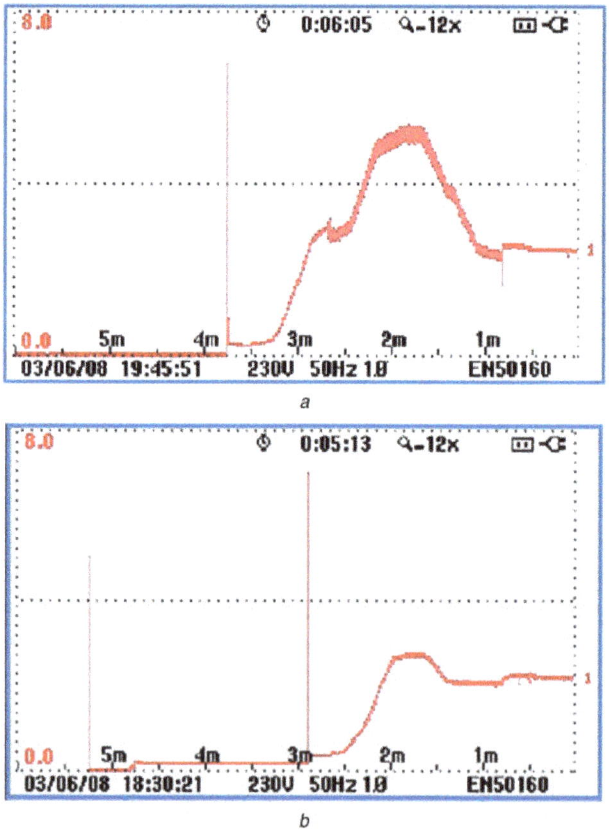

Fig. 11 *Unit B inrush current*
a Heating mode (initial peak: 7 A; run: 5.5 A, then 2.4 A)
b Cooling mode (initial peak: 5.5 A, then 7 A; run: 2.2 A)

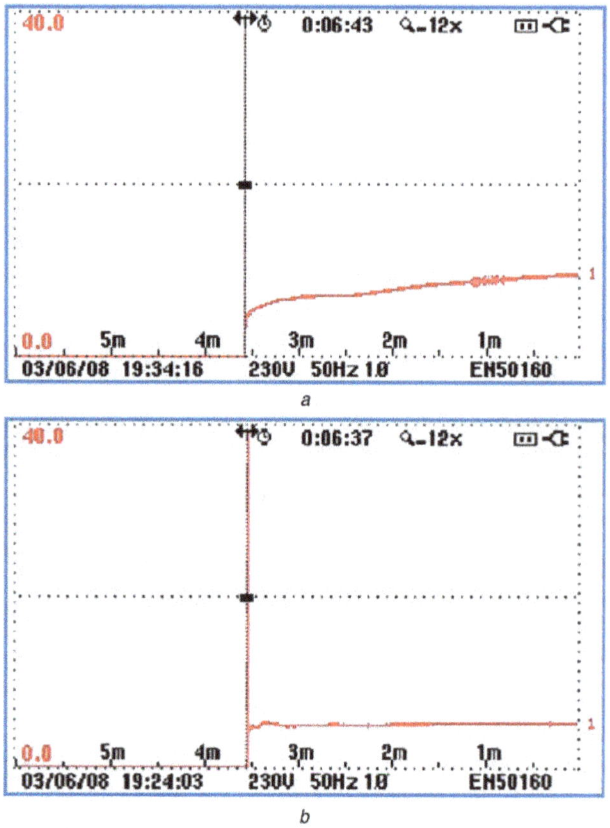

Fig. 12 *Unit C inrush current*
a Heating mode (initial peak: 38.6 A, duration 100 ms; run: 9.4 A)
b Cooling mode (initial peak: 38.9 A, duration 102 ms; run: 5.2 A)

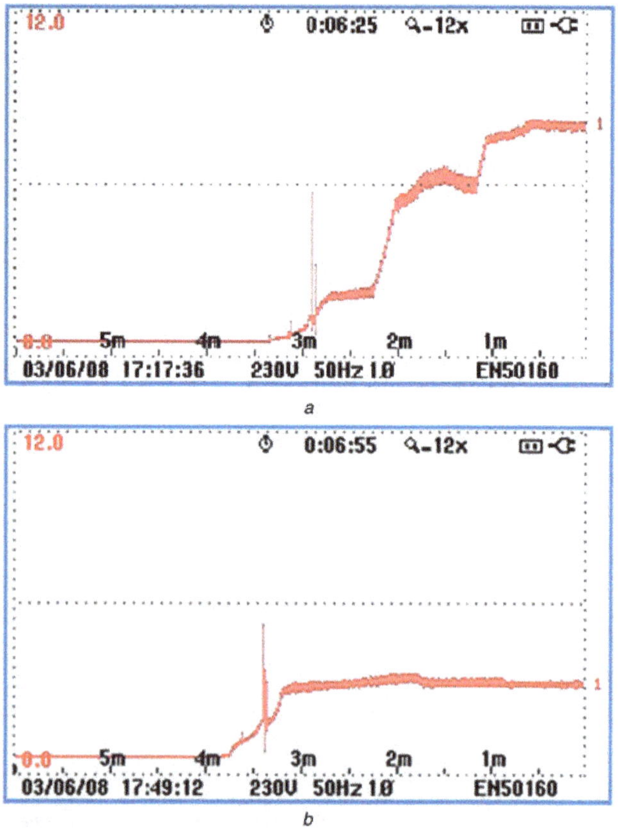

Fig. 13 *Inrush current for unit D*
a Heating mode (initial peak: N/A; run: 8.2 A)
b Cooling mode (initial peak: N/A, then 5.5 A; run: 3.2 A)

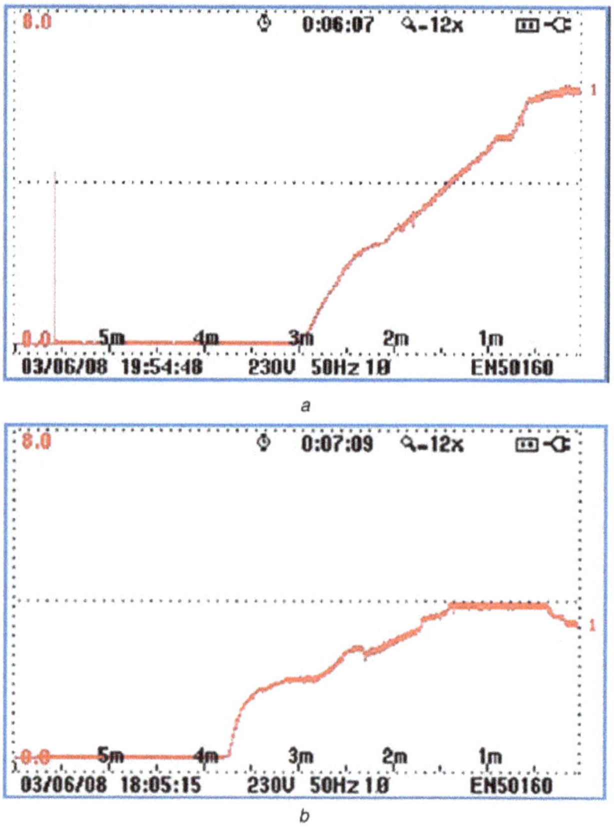

Fig. 14 *Inrush current for unit E*
a Heating mode (initial peak: 4.2 A; run: 6.2 A)
b Cooling mode (initial peak: 6.2 A (not shown); run: 3.9 A)

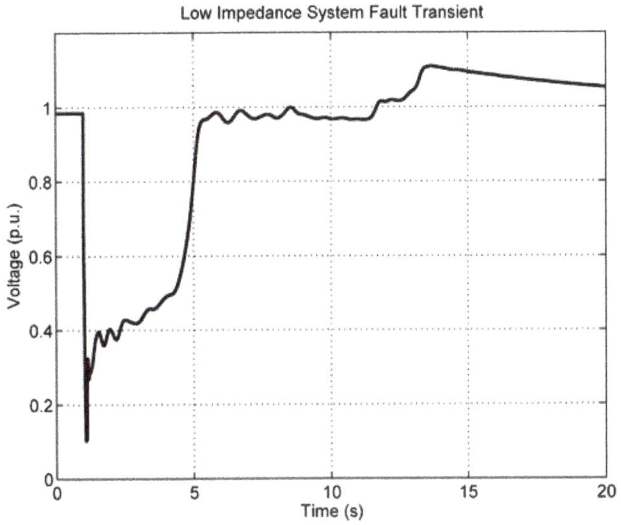

Fig. 15 *Representative voltage profile for a low-impedance fault on the New Zealand system*

are shown, as there is nothing worthy of note in the trailing part of the transient current waveforms.

The recorded waveforms are displayed in Figs. 18–22, in which the heat-pump's RMS current in amperes is shown against historical time in seconds. Also shown below each figure are the nominal rated current, I, at nominal rated voltage, V (230 V) and the approximate SS CL (I_lim) observed. Units A25 and E behave in the most benign fashion, as they do not draw a spike of current as the voltage falls, despite the fact that they are both operating below nominal current at nominal voltage and certainly below any SS CL observed earlier. All the other models tested make some attempt to maintain their input power during the voltage dip, resulting in a significant current spike higher than their nominal current at nominal voltage, although only unit B exceeds its observed SS CL. Note that all the inverter-based units return to a waiting state when the supply voltage comes back up and will not draw significant current again for typically 2–3 min. The non-inverter model (Unit C) could not be tested as the Chroma source could not provide the inrush current necessary to start it (nor could two Chroma sources in parallel), indicating the unit was trying to draw an extremely large current.

3.3 Voltage dip/sag performance, low-impedance system fault (cooling)

The voltage dip/sag tests were repeated with the heat-pumps in cooling mode. The recorded waveforms are displayed in Figs. 23–27, with the same format as in the heating mode.

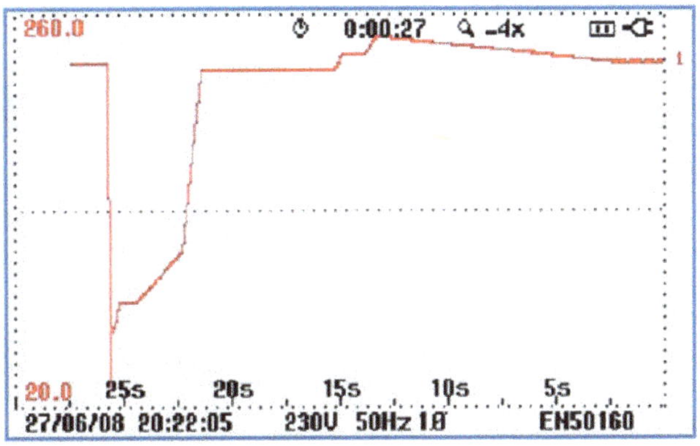

Fig. 16 *Chroma approximation to system fault transient waveform*

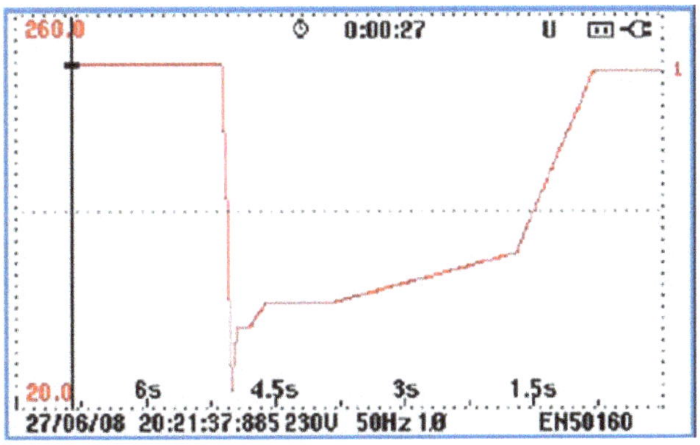

Fig. 17 *Chroma approximation to system fault transient waveform – expanded*

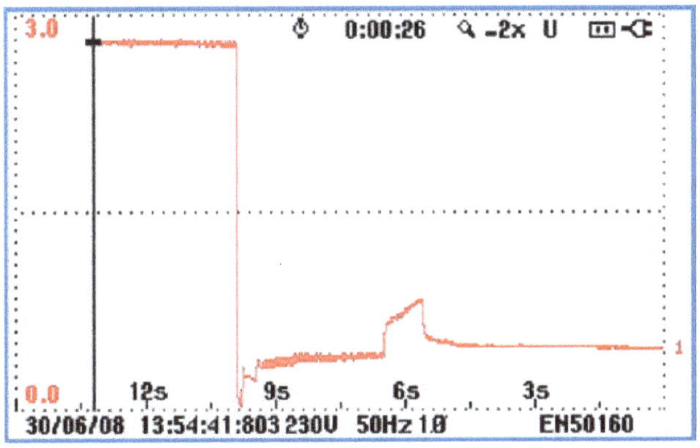

Fig. 18 *Unit A25, heat, current, A – expanded*
Initial current, 2.8 A (nominal *I* at nominal *V*: 3.61 A, SS I_lim ≥ 6 A)

Recorded behaviour is very similar to the heating case, although this time inrush spikes for both units A50 and B exceed their observed SS CLs. Unit C's large starting current again prevented this test from being carried out with the Chroma sources.

3.4 Voltage dip/sag performance, high-impedance system fault (heating)

One concern was whether the heat-pump D, with its low cut-out voltage, would ride through a high-impedance fault event and try to maintain constant power (CP). The representative high-impedance

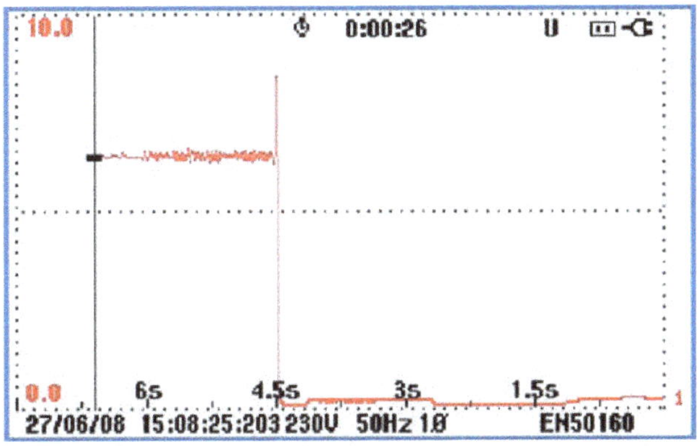

Fig. 19 *Unit A50, heat, current, A – expanded*
Initial current, 6.4 A; peak 8.3 A (nominal *I* at nominal *V*: 6.96 A, SS I_lim ≥ 9.6 A)

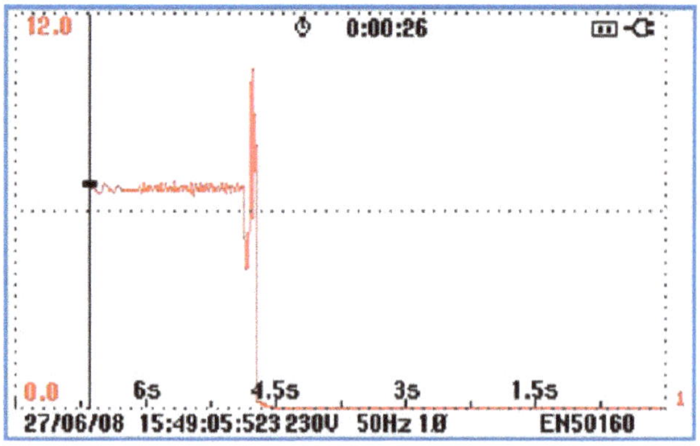

Fig. 20 *Unit B, heat, current, A – expanded*
Initial current, 6.8 A; peak 10.2 A (nominal I at nominal V: 7.52 A, SS I_lim ≃ 8 A)

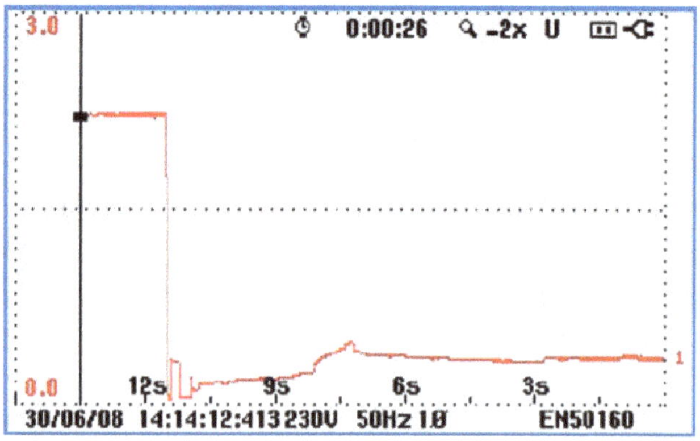

Fig. 23 *Unit A25, cool, current, A – expanded*
Initial current, 2.2 A (nominal I at nominal V: 2.61 A, SS I_lim ≤ 6 A)

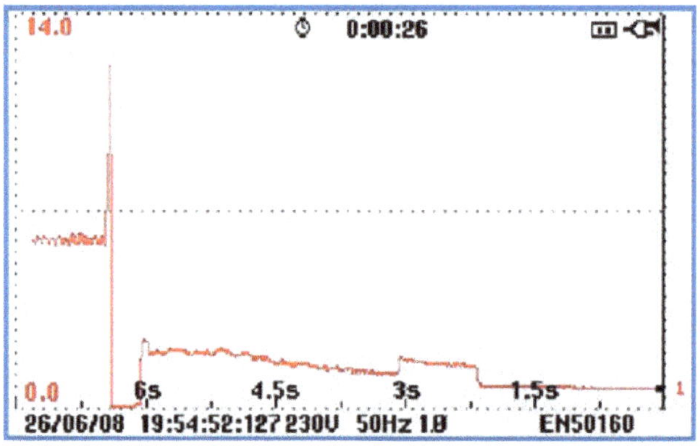

Fig. 21 *Unit D, heat, current, A – expanded*
Initial current, 6.0 A; peak 12.4 A (nominal I at nominal V: 7.39 A, SS I_lim ≃ 16.5 A)

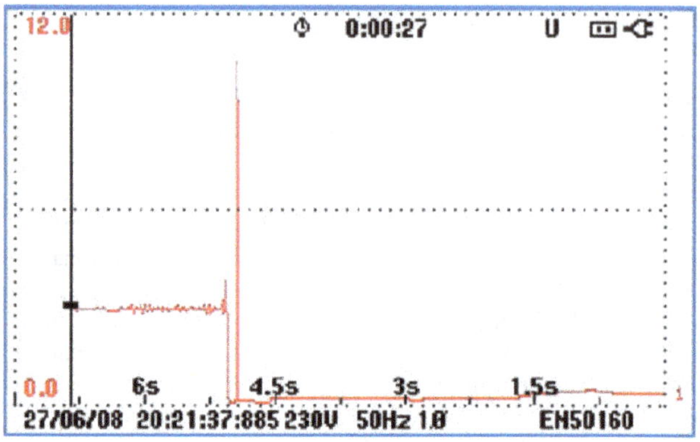

Fig. 24 *Unit A50, cool, current, A – expanded*
Initial current, 3.1 A, peak 10.5 A (nominal I at nominal V: 6.74 A, SS I_lim ≤ 9.6 A)

fault event supplied by Transpower NZ Ltd., shown in Fig. 28, was used for this test. This voltage transient was imposed on the heat-pump while operating at different power levels and the response observed to determine the factors influencing its performance. When the supply current exceeded ~25 A, the heat-pump would trip out on its own internal over-current protection. If this threshold was not reached, the unit would attempt to draw the necessary current to keep its power input approximately constant (in some cases, this would trip the Chroma voltage sources because of the sustained high current). Depending on the minimum voltage during the transient and the power demand from the heat-pump, the unit may either draw up to 20 A or so for about 200 ms or trip out within a few milliseconds. Figs. 29–35 give a selection of responses for these tests. In these figures, the voltage transient, displaying actual RMS voltage (230 V nominal) against historical time in seconds, is juxtaposed with the coincident RMS current transient.

Despite the SS current being below nominal value (7.39 A) in each test, actual current exceeded nominal during every transient, by a factor of up to about 300%.

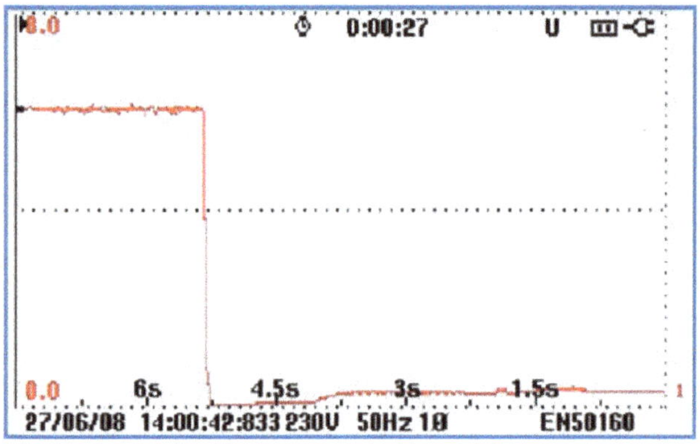

Fig. 22 *Unit E, heat, current, A – expanded*
Initial current, 6.0 A (nominal I at nominal V: 6.83 A, SS I_lim ≃ 7 A)

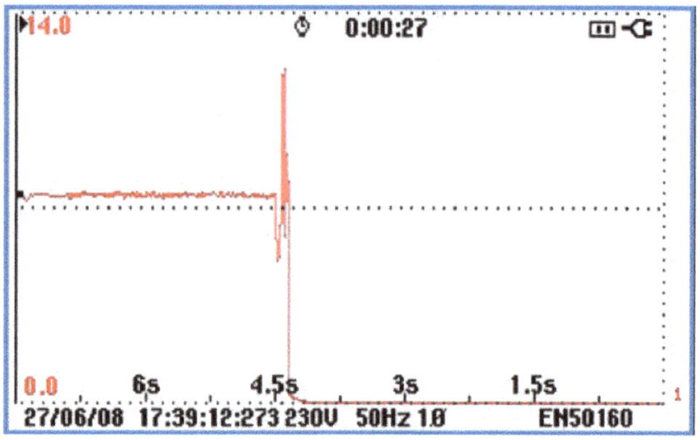

Fig. 25 *Unit B, cool, current, A – expanded*
Initial current, 7.4 A, peak 11.8 A (nominal I at nominal V: 7.48 A, SS I_lim ≃ 8 A)

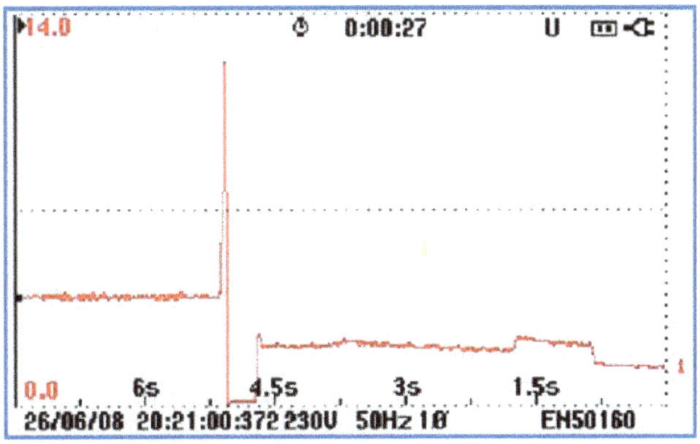

Fig. 26 *Unit D, cool, current, A – expanded*
Initial current, 3.9 A, peak 12.1 A (nominal *I* at nominal *V*: 7.22 A, SS I_lim ≃16.5 A)

4 Discussion

It is not possible to control input power directly, as this is determined by the indoor and outdoor temperatures (and humidity/thermal capacity of the air) and the state of the refrigerant. Hence, it requires many repeated tests to build up an overall picture of each heat-pump's full range of operating behaviour.

4.1 SS system voltage tests

In the SS tests, all the units generally met the harmonic distortion requirements of AS/NZS 61000-3-2 Class A, at nominal voltage [7].

As the system voltage deviates from nominal, both above and below 230 V, the harmonic content of the current drawn varies significantly, resulting in the changes of PF shown in Fig. 4. Harmonic content also varies with input power and is affected by the existing system voltage distortion [7].

Analysis of results obtained indicates that all units have a reasonably well-defined under-voltage cut-out (and cut-in) level. However, for many of the models this is set at a surprisingly low voltage. This can be of concern, because, depending on actual power drawn in the present operating conditions, the units typically operate in CP mode; hence, lower system voltage leads to higher operating current. To illustrate this, Fig. 36 shows the approximate aggregate effect on a system in which one of each of the six units is running in heating mode. In practice, the balance of different units would lead to different results, but in all cases the current–voltage relationship is far from linear and generally leads to increased load current as system voltage falls to 0.7 pu. Fortunately, maximum current in SS operation appears to be limited for all inverter-driven units. However, for at least one model this is set

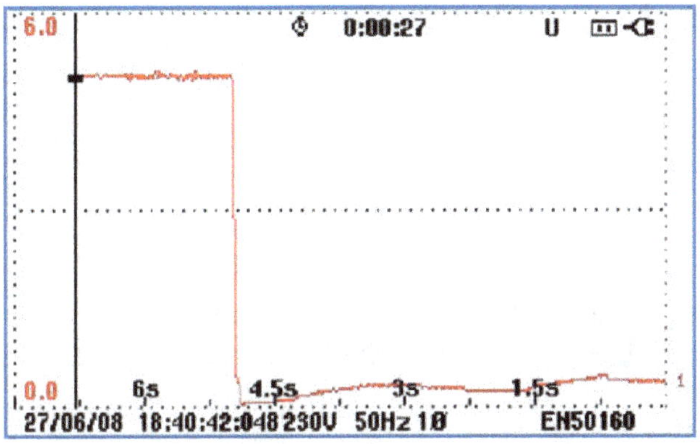

Fig. 27 *Unit E, cool, current, A – expanded*
Initial current, 5.0 A (nominal I at nominal *V*: 5.96 A, SS I_lim ≃ 7 A)

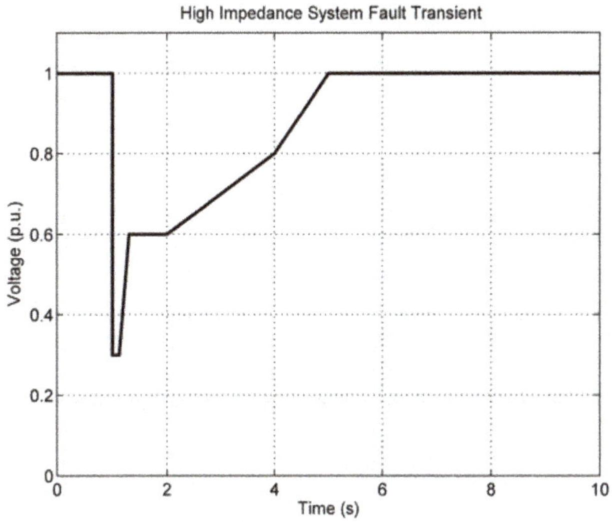

Fig. 28 *High-impedance system fault transient waveform*

at a surprisingly high level. An investigation into the effects on the system of the aggregated current spikes drawn during a system transient has not been carried out, but may be worthy of further study.

4.2 Switch-on inrush current

As expected, the non-inverter model draws a large inrush current of nearly five times its nominal value on start-up in either cycle. All the inverter-driven units draw inrush currents of the same order of magnitude as their nominal currents in either cycle.

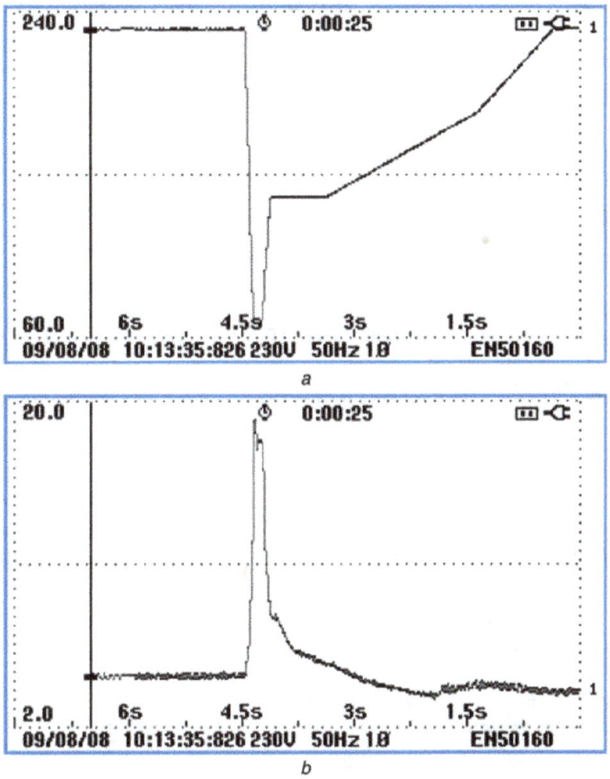

Fig. 29 *Dip to 0.3 pu, 4.8 A current before transient, peak current about 18 A, rides through*
a Voltage
b Current

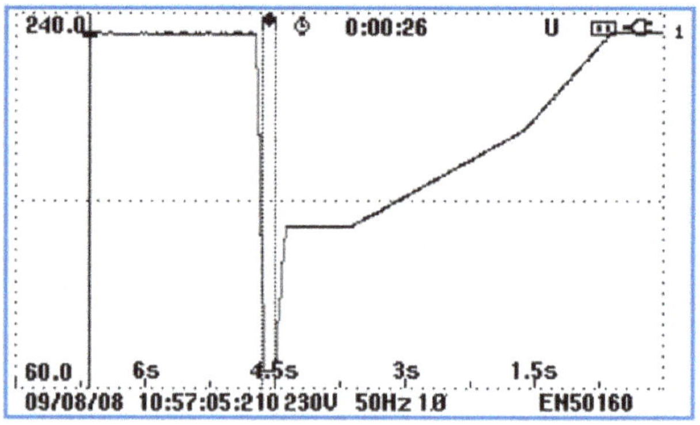

a

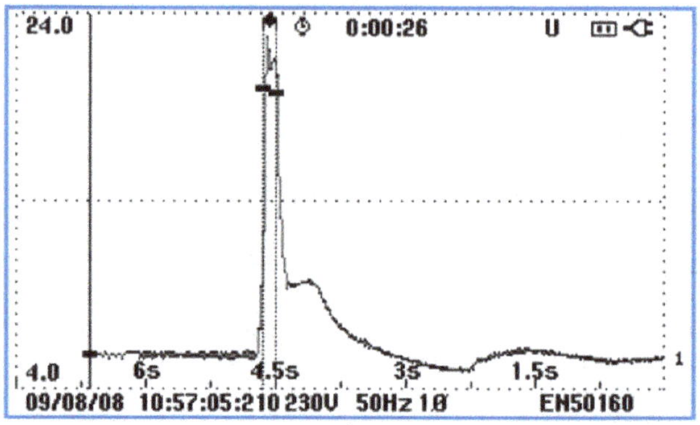

b

Fig. 30 *Dip to 0.3 pu, 5.9 A current before transient, exceeds 20 A for 151 ms, rides through*
a Voltage
b Current

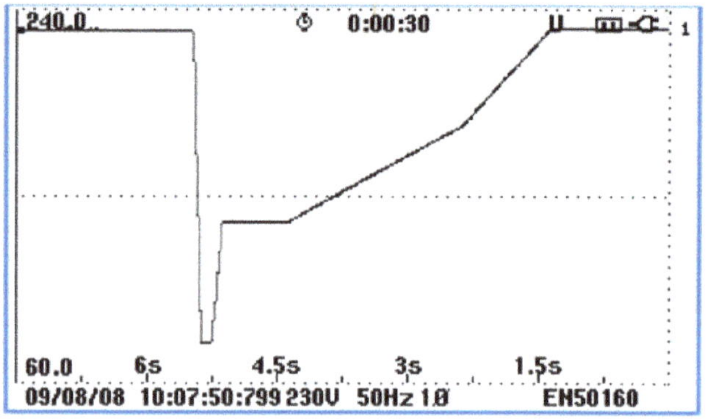

a

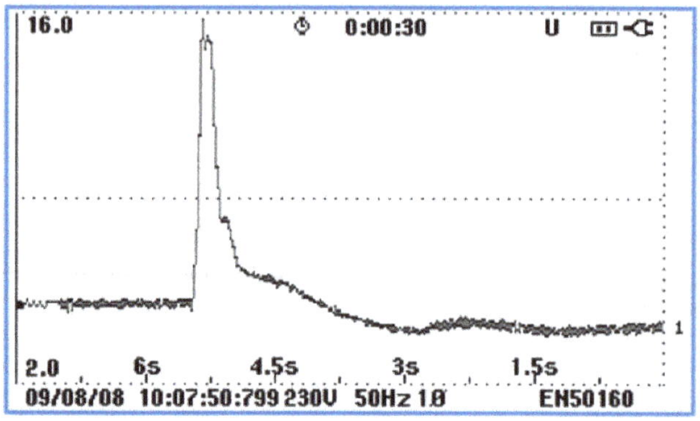

b

Fig. 32 *Dip to 0.35 pu, 5.0 A current before transient, peak current about 15 A, rides through*
a Voltage
b Current

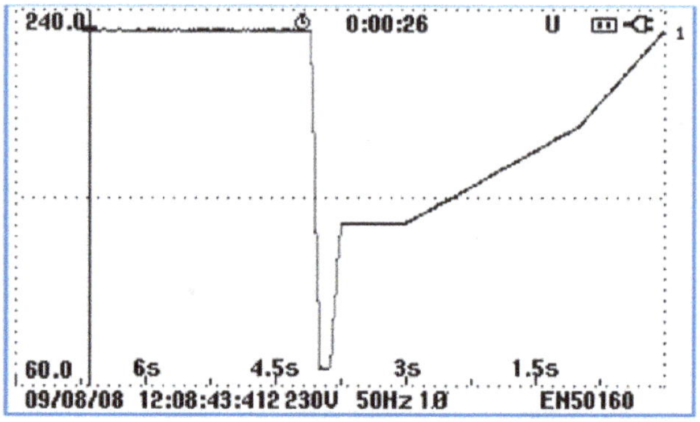

a

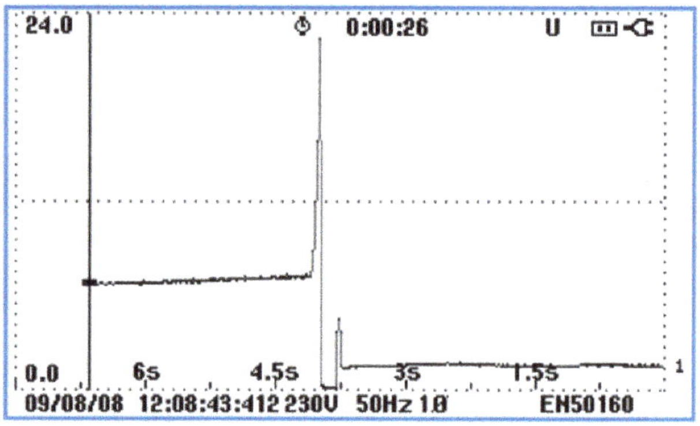

b

Fig. 31 *Dip to 0.3 pu, 6.8 A current before transient, exceeds 25 A instantaneously, trips out*
a Voltage
b Current

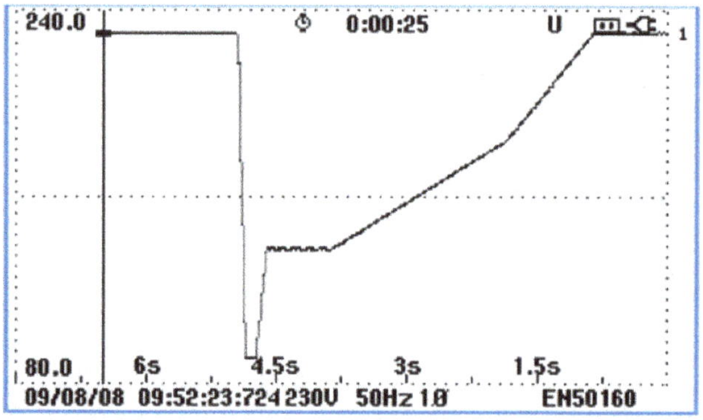

a

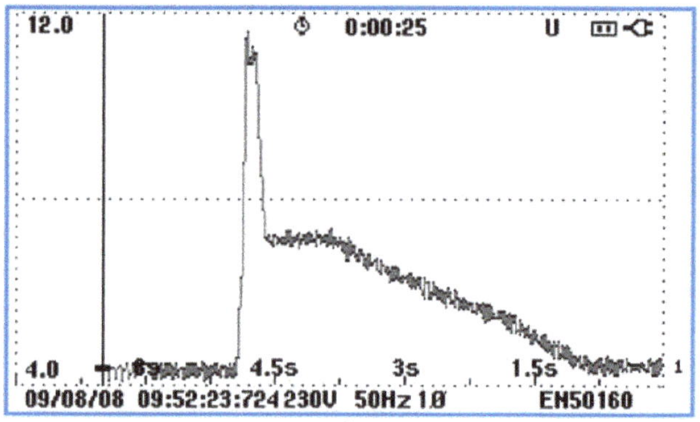

b

Fig. 33 *Dip to 0.4 pu, 4.4 A current before transient, peak current about 11 A, rides through*
a Voltage
b Current

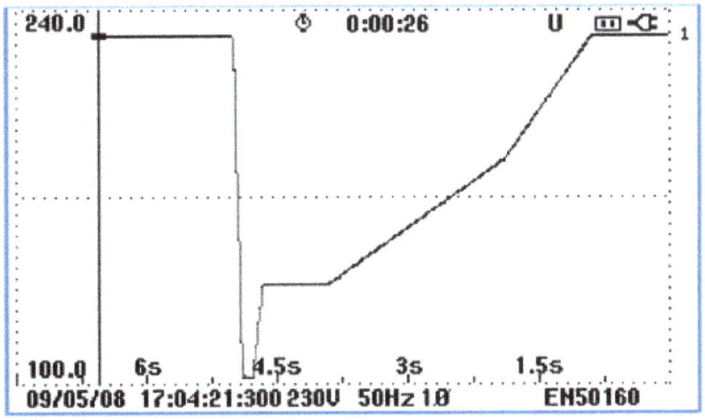

a

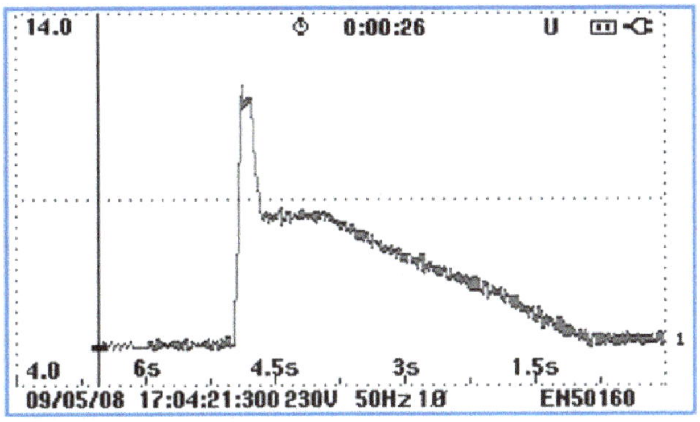

b

Fig. 34 *Dip to 0.45 pu, 5.0 A current before transient, peak current about 12 A, rides through*
a Voltage
b Current

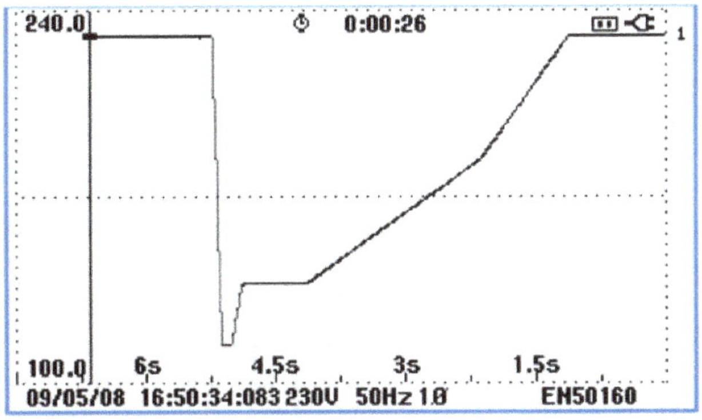

a

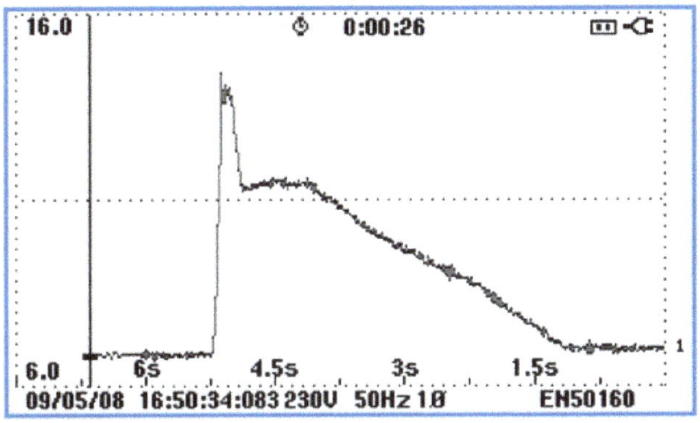

b

Fig. 35 *Dip to 0.5 pu, 6.8 A current before transient, peak current about 14 A, rides through*
a Voltage
b Current

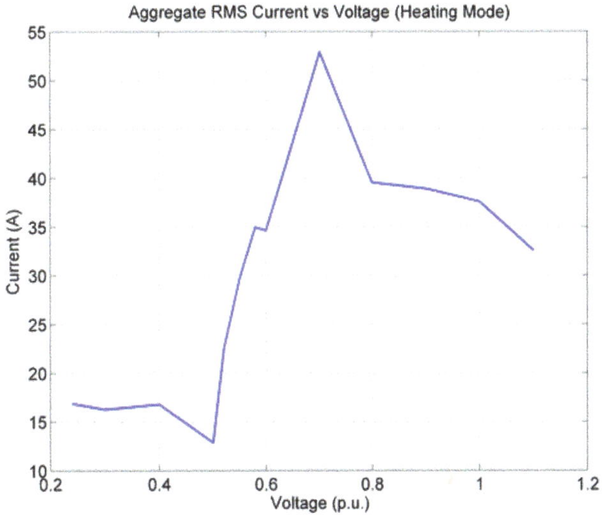

Fig. 36 *Aggregate RMS current drawn by the heat-pumps tested, against voltage, while in heating mode*

4.3 Voltage dip/sag transient tests

The non-inverter model could not be subjected to the system transient tests because of its large inrush current, which could not be supplied by the electronic AC voltage source, even with a second voltage source connected in parallel.

During the low-impedance fault transient, three of the five inverter-driven models drew substantial spikes of current during the collapse of the system voltage, as they tried to keep up their input power, in both heat and cool cycles. These current spikes could be of greater magnitude than the observed SS CL, albeit of relatively short duration (typically <3 cycles). The A25 and the E unit were notable in that they did not draw a current spike, but retired gracefully in both heating and cooling cycles. All five units fell to low power operation after the transient and resumed their previous operating conditions after their normal start-up delay (typically 1–3 min).

A voltage dip/sag transient for a high-impedance fault was applied to heat-pump D, as SS tests had indicated that its cut-out voltage was only slightly above the minimum voltage of the low-impedance fault transient. This confirmed the unit would stay operational provided the peak current did not exceed the heat-pump's instantaneous over-CL, which appears to be about 25 A (distinct from the SS CL of 16.5 A).

5 Conclusions

Six heat-pumps available on the NZ market have been tested to determine their electrical behaviour under both prolonged (SS) voltage sag and swell and transient conditions in both the heating and cooling cycles. Five of these units have inverter-driven three-phase compressor motors. The inverters run from a DC bus, supplied by a single-phase rectifier, as described in [7]. The sixth unit has a direct-on-line single-phase induction motor (with a capacitor-run auxiliary winding).

Overall, each of the heat-pumps has certain shortcomings with respect to its electrical behaviour, with each exhibiting some good and some poor features, as summarised in Table 1.

6 Acknowledgments

The financial support for this research from Transpower New Zealand Ltd, the New Zealand Electricity Engineers' Association and the New Zealand Foundation for Research in Science and Technology (now MBIE) is gratefully acknowledged. The authors would also like to thank Ken Smart (University of Canterbury)

Table 1 Comparison of heat-pump electrical behaviour

Model	A25	A50	B	C	D	E
current harmonics at 1 pu supply voltage near nominal power[a]	poor 25% ≤ I_{THDf} ≤ 61%	fair 17% ≤ I_{THDf} ≤ 18%	fair 17% ≤ I_{THDf} ≤ 18%	good/fair 9% ≤ I_{THDf} ≤ 13%	good 3% ≤ I_{THDf} ≤ 7%	fair/poor 22% ≤ I_{THDf} ≤ 26%
current harmonics over the voltage range 0.7–1.1 pu[a]	fair/poor (depends on power)	fair/poor/very poor (depends on power)	fair/poor (depends on power)	good/fair (depends on power)	good/fair (depends on power)	fair/poor
inrush current	moderate	moderate	moderate	very high	good	moderate
apparent[b] action with reducing voltage	unclear CL ≃ 6 A	CP CL ≃ 10 A	CP CL ≃ 8 A	unclear/motor stall	CP CL ≃ 17 A	CP CL ≃ 7 A
current at low voltage under-voltage cut-off, pu	low good (0.7)	moderate low (0.52)	low low (0.55)	very high moderate (0.6/0.68)	very high very low (0.24)	low low (0.58/0.54)
system transient response	good	fair	fair	unknown	can draw very high current	good

[a]Relative to 'good' performance achievable with best-practice PFC rectifiers (good: $I_{THDf} < 10\%$; fair: $10\% < I_{THDf} \leq 25\%$; poor: $25\% < I_{THDf} \leq 100\%$; and very poor: $I_{THDf} > 100\%$).
[b]'Apparent' action based on tests reported. Key: CP: constant power; CL: observed or inferred SS current limit, if applicable.

and Stewart Hardie & Dudley Smart (Electric Power Engineering Centre) for their help.

7 References

[1] French L.: 'Active cooling and heat-pump use in New Zealand – survey results'. BRANZ Study Report No. 186, 2008

[2] Goetzler W., Zogg R., Lisle H., Burgos J.: 'Ground-source heat-pumps: overview of market status, barriers to adoption, and options for overcoming barriers'. Report for US Department of Energy, Navigant Consulting, February 2009

[3] Mohan N.: 'Electric drives: an integrative approach' (MNPERE, Minneapolis, 2000), ISBN 0-9663530-1-3

[4] Jungreis A.M., Kelley A.W.: 'Adjustable speed drive for residential applications', *IEEE Trans. Ind. Appl.*, 1995, **31**, (6), pp. 1315–1322

[5] Domijan A., Hancock O., Maytrott C.: 'A study and evaluation of power electronic based adjustable speed motor drives for air conditioners and heat-pumps with an example utility case study of the Florida power and light company', *IEEE Trans. Energy Convers.*, 1992, **7**, (3), pp. 396–404

[6] Watson N.R., Scott T., Hirsch S.: 'Implications for distribution networks of high penetration of compact fluorescent lamps', *IEEE Trans. Power Deliv.*, 2009, **24**, (3), pp. 1521–1528

[7] Heffernan W.J.B., Watson N.R., Buehler R., Watson J.D.: 'Harmonic performance of heat-pumps', *IET J. Eng.*, 2013, doi: 10.1049/joe2013.0012

[8] IEC 61000-3-2: 'Title: Electromagnetic compatibility (EMC) – Part 3-2: Limits – Limits for harmonic current emissions (equipment input current ≤ 16 A per phase)'

[9] AS/NZS 61000.3.2:2007: 'Electromagnetic compatibility, Part 3.2: limits-limits for harmonic current emissions' (Standards Australia/Standards NZ)

[10] AS/NZS 3823.1.1:1998: 'Performance of electrical appliances – air conditioners and heat-pumps' (Standards Australia/Standards NZ)

[11] Natarajan R., Misra V.K.: 'Starting transient current of induction motors without and with terminal capacitors', *IEEE Trans. Energy Convers.*, 1991, **6**, (1), pp. 134–139

8 Appendix 1

See Tables 2–15.

Table 2 Unit A25, heat; cut-out voltage ~162 V (0.7 pu)

Voltage pu	0.7	0.8	0.9	1	1.1
current, A RMS	2.9	3.1	2.9	2.7	2.8
real power, kW	0.36	0.44	0.46	0.47	0.55
apparent power, kVA	0.47	0.56	0.6	0.61	0.7
reactive power, kVAR	0.31 lag	0.36 lag	0.38 lag	0.39 lag	0.43 lag
voltage THD, %F	1.5	1.5	1.3	1.2	1.4
current THD, %F	63.8	63.4	67.1	70.8	68.3
PF	0.76	0.77	0.77	0.77	0.79
DPF	0.93	0.94	0.95	0.96	0.96

Table 3 Unit A50, heat; cut-out voltage ~120 V (0.52 pu)

Voltage pu	0.52	0.6	0.7	0.8	0.9	1	1.1
current, A RMS	9.6	9.1	7.7	6.8	6	6	5.1
real power, kW	1.13	1.22	1.22	1.22	1.2	1.36	1.25
apparent power, kVA	1.15	1.25	1.25	1.25	1.3	1.4	1.3
reactive power, kVAR	0.21 lead	0.23 lead	0.26 lead	0.29 lead	0.3 lead	0.33 lead	0.36 lead
voltage THD, %F		2.8	2.5	2.2	2	1.8	1.9
current THD, %F		16.5	18.3	20	22.1	21.1	24.6
PF	0.98	0.98	0.98	0.97	0.97	0.97	0.96
DPF	1	1	0.99	0.99	0.99	1	0.99

Table 4 Unit B, heat; cut-out voltage ~127 V (0.55 pu)

Voltage pu	0.55	0.6	0.7	0.8	0.9	1	1.1
current, A RMS	7.4	6.5	6	5.2	4.7	4.8	4.3
real power, kW	0.9	0.89	0.95	0.94	0.95	1.08	1.03
apparent power, kVA	0.94	0.9	0.96	0.96	0.98	1.12	1.09
reactive power, kVAR	0.25 lead	0.14 lead	0.17 lead	0.22 lead	0.26 lead	0.30 lead	0.34 lead
voltage THD, %F		2.1	2	1.9	1.7	1.9	2.2
current THD, %F		13.6	15.8	18.5	20.8	22	23.8
PF	0.96	0.99	0.98	0.97	0.97	0.96	0.95
DPF	0.99	1	1	0.99	0.99	0.99	0.98

Table 7 Unit E, heat; cut-out voltage ~134 V (0.58 pu)

Voltage, pu	0.58	0.7	0.8	0.9	1.0	1.1
Current, A RMS	5.9	6.2	6.2	6.2	6.1	5.8
real power, kW	0.78	0.98	1.12	1.24	1.34	1.29
apparent power, kVA	0.8	0.99	1.15	1.28	1.4	1.47
reactive power, kVAR	0.13 lag	0.14 lead	0.23 lead	0.34 lead	0.39 lag	0.70 lag
voltage THD, %F		1.6	1.8	2	1.8	2.4
current THD, %F		14.2	16.7	20.8	26	38.7
PF	0.99	0.99	0.98	0.96	0.96	0.88
DPF	1	1	0.99	0.99	0.99	0.95

Table 5 Unit C, heat; cut-out voltage ~156 V (0.68 pu)

Voltage pu	0.68	0.8	0.9	1	1.1
current, A RMS	22.2	8.6	10.7	7.8	6.9
real power, kW	3.13	1.61	2.13	1.78	1.68
apparent power, kVA	3.47	1.61	2.15	1.81	1.77
reactive power, kVAR	1.5 lag	0.16 lag	0.28 lag	0.3 lag	0.57 lag
voltage THD, %F		0.9	1	0.9	1.1
current THD, %F		9.6	12.2	16.4	19.5
PF	0.9	0.99	0.99	0.99	0.95
DPF	0.9	1	1	1	0.97

Table 8 Unit A25, cool; cut-out voltage ~151 V (0.66 pu)

Voltage, pu	0.66	0.7	0.8	0.9	1	1.1
current, A RMS	3.8	3.6	3.2	2.5	2.3	2
P, kW	0.47	0.48	0.48	0.42	0.41	0.39
S, kVA	0.57	0.59	0.59	0.52	0.53	0.5
Q, kVAr	0.33 lag	0.34 lag	0.35 lag	0.32 lag	0.33 lag	0.32 lag
voltage THD, %F		1.6	1.4	1.2	1.2	1.3
current THD, %F		50.7	57.1	67.1	71.9	77.6
PF	0.82	0.82	0.81	0.8	0.78	0.77
DPF	0.91	0.92	0.94	0.96	0.97	0.98

Table 6 Unit D, heat; cut-out voltage ~56 V (0.24 pu)

Voltage pu	0.24	0.3	0.4	0.5	0.6	0.7	0.8	0.9	1	1.1
current, A RMS	16.9	16.3	16.8	12.9	13.1	11.7	10	8.7	10.4	8.5
real power, kW	0.84	1.05	1.49	1.5	1.81	1.89	1.83	1.8	2.37	2.11
apparent power, kVA	0.96	1.14	1.55	1.54	1.82	1.9	1.84	1.82	2.39	2.15
reactive power, kVAR	0.47 lag	0.46 lag	0.43 lag	0.33 lag	0.24 lag	0.19 lag	0.15 lead	0.21 lead	0.25 lead	0.38 lead
voltage THD, %F		12.2	7.7	6.1	4	3.1	2.4	1.8	1.8	1.6
current THD, %F		18.3	10.8	11.3	7.8	7.2	6.9	6	5.2	5.5
PF	0.87	0.92	0.96	0.98	0.99	1	1	0.99	0.99	0.98
DPF	0.92	0.95	0.97	0.99	1	1	1	1	1	0.99

Table 9 Unit A50, cool (31/07/08); cut-out voltage ~119 V (0.52 pu)

Voltage, pu	0.55	0.6	0.65	0.7	0.75	0.8	0.85	0.9	0.95	1	1.05	1.1
current, A RMS	9.6	8.7	8.1	7.3	6.8	6.4	6	5.6	6.6	5.1	4.9	3.7
real power, kW	1.2	1.18	1.18	1.15	1.14	1.13	1.13	1.12	1.41	1.12	1.12	0.87
apparent power, kVA	1.22	1.21	1.21	1.18	1.18	1.16	1.17	1.16	1.45	1.17	1.17	0.93
reactive power, kVAR	0.22 lead	0.24 lead	0.25 lead	0.26 lead	0.27 lead	0.28 lead	0.3 lead	0.31 lead	0.34 lead	0.33 lead	0.34 lead	0.32 lead
voltage THD, %F	<5	<5	<5	<5	<5	<5	<5	<5	<5	<5	<5	<5
current THD, %F	15.9	17	17.8	19	20.3	21.1	22	23.1	19.9	24.4	25.1	31.2
PF	0.98	0.98	0.98	0.97	0.97	0.97	0.97	0.96	0.97	0.96	0.96	0.94
DPF	1	1	0.99	0.99	0.99	0.99	0.99	0.99	0.99	0.99	0.99	0.98

NB. Local mains supply used as voltage source for this test (not sine wave generator)

Table 9a Retest 1 (3/08/08), reducing volts, A50, cool

Voltage, pu	0.55	0.6	0.65	0.7	0.75	0.8	0.85	0.9	0.95	1	1.05	1.1
current, A RMS	8.7	7.9	7.4	6.3	5.9	5.6	5.2	4.9	4.6	4.4	4.2	4
real power, kW	1.08	1.07	1.07	0.98	0.97	0.97	0.97	0.97	0.97	0.96	0.95	0.94
apparent power, kVA	1.11	1.1	1.1	1.01	1.01	1.01	1.01	1.01	1.01	1.01	1.01	1
reactive power, kVAR	0.23 lead	0.25 lead	0.25 lead	0.26 lead	0.27 lead	0.28 lead	0.29 lead	0.3 lead	0.31 lead	0.32 lead	0.33 lead	0.34 lead
voltage THD, %F	4.7	4.6	4.6	4.5	3.7	4.1	4.0	3.9	3.7	4.2	4.2	4.2
current THD, %F	18.6	20.1	20.8	23.4	23.4	23.5	25.3	26.1	27.1	28.2	28.9	30
PF	0.98	0.97	0.97	0.97	0.96	0.96	0.96	0.96	0.95	0.95	0.94	0.94
DPF	0.99	0.99	0.99	0.99	0.99	0.99	0.99	0.99	0.99	0.99	0.98	0.98

NB. Local mains supply used as voltage source for this test (not sine wave generator)

Table 9b Retest 2 (3/08/08), increasing volts, A50, cool

Voltage, pu	0.55	0.6	0.65	0.7	0.75	0.8	0.85	0.9	0.95	1	1.05	1.1
current, A RMS	9	10.3	7.6	7	6.6	6.3	6	5.6	5.3	5.1	4.9	4.6
P, kW	1.11	1.4	1.09	1.09	1.09	1.1	1.11	1.11	1.11	1.12	1.12	1.12
S, kVA	1.13	1.42	1.12	1.13	1.14	1.14	1.15	1.15	1.16	1.17	1.17	1.18
Q, kVAR	0.23 lead	0.26 lead	0.26 lead	0.27 lead	0.28 lead	0.29 lead	0.31 lead	0.32 lead	0.33 lead	0.34 lead	0.34 lead	0.36 lead
voltage THD, %F	4.4	4.2	4.3	4.5	4.4	4.3	4.3	4.1	3.8	3.7	3.8	4
current THD, %F	17.9	15.7	20.1	21.5	22.1	22.7	23.3	24.1	24.1	24.7	25.2	26.2
PF	0.98	0.98	0.97	0.97	0.97	0.97	0.96	0.96	0.96	0.96	0.96	0.95
DPF	0.99	0.99	0.99	0.99	0.99	0.99	0.99	0.99	0.99	0.99	0.99	0.98

NB. Local mains supply used as voltage source for this test (not sine wave generator)

Table 10 Unit B, cool; cut-out voltage ~127 V (0.55 pu)

Voltage, pu	0.55	0.6	0.7	0.8	0.9	1	1.1
current, A RMS	7	7.1	7.4	7.5	7.7	6.9	6.3
real power, kW	0.88	0.98	1.19	1.36	1.56	1.56	1.55
apparent power, kVA	0.89	0.99	1.2	1.39	1.59	1.6	1.6
reactive power, kVAR	0.13 lead	0.15 lead	0.19 lead	0.24 lead	0.29 lead	0.35 lead	0.4 lead
voltage THD, %F		2.2	2.1		2.1	2.1	2.6
current THD, %F		13.2	14.2	15.2	15.9	18	19.8
PF	0.99	0.99	0.99	0.98	0.98	0.98	0.97
DPF	1	1	1	1	1	0.99	0.99

Table 11 Unit C, cool; cut-out voltage ~ 138 V (0.6 pu)

Voltage, pu	0.6	0.7	0.8	0.9	1	1.1
current, A RMS	11.1	8.5	7.1	6.4	6	6
real power, kW	1.5	1.35	1.29	1.31	1.36	1.47
apparent power, kVA	1.53	1.36	1.3	1.33	1.38	1.52
reactive power, kVAR	0.33 lag	0.11 lag	0.13 lead	0.18 lead	0.19 lag	0.4 lag
voltage THD, %F		1.2	0.8	0.8	0.7	0.8
current THD, %F		7.2	8.9	11	14.1	17.7
PF	0.98	1	0.99	0.99	0.99	0.96
DPF	0.98	1	1	1	1	0.98

Table 12 Unit D, cool (31/07/08); cut-out voltage ~56 V (0.24 pu)

V, pu	0.3	0.35	0.4	0.45	0.5	0.55	0.6	0.65	0.7	0.75	0.8	0.85	0.9	0.95	1.0	1.05	1.1
I, A RMS	16.5	16.1	15.8	13.4	12.2	10.8	9.6	8.9	8.1	7.5	7	6.6	6.2	5.3	5.0	4.7	4.6
P, kW	1.05	1.23	1.39	1.36	1.36	1.33	1.31	1.31	1.3	1.28	1.28	1.27	1.26	1.12	1.11	1.10	1.12
S, kVA	1.15	1.31	1.46	1.4	1.38	1.36	1.33	1.32	1.3	1.29	1.29	1.28	1.28	1.15	1.15	1.14	1.17
Q, kVAR	0.49 lag	0.45 lag	0.43 lag	0.33 lag	0.28 lag	0.24 lag	0.18 lag	0.16 lag	0.14 lead	0.14 lead	0.15 lead	0.18 lead	0.22 lead	0.26 lead	0.29 lead	0.31 lead	0.34 lead
V THD, %F	<5	<5	<5	<5	<5	<5	<5	<5	<5	<5	<5	<5	<5	<5	<5	<5	<5
I THD, %F	19.8	14.3	12	11	10.8	10.4	9.9	9.8	9.5	9.1	8.5	7.8	7.2	8.9	9.7	10.7	11.6
PF	0.91	0.94	0.96	0.97	0.98	0.98	0.99	0.99	0.99	0.99	0.99	0.99	0.99	0.97	0.97	0.96	0.96
DPF	0.94	0.96	0.97	0.98	0.99	0.99	1	1	1	1	1	0.99	0.99	0.98	0.97	0.97	0.96

NB. Local mains supply used as voltage source for this test (not sine wave generator)

Table 13 Unit E, cool (31/07/08); cut-out voltage ~ 125 V (0.54 pu)

Voltage, pu	0.6	0.65	0.7	0.75	0.8	0.85	0.9	0.95	1	1.05	1.1
current, A RMS	5.8	6.1	6.4	6.3	6.1	5.8	5.4	5.1	4.7	4.5	4
real power, kW	0.78	0.9	1.03	1.08	1.08	1.08	1.07	1.06	1.03	0.97	0.82
apparent power, kVA	0.79	0.91	1.04	1.09	1.11	1.14	1.12	1.1	1.09	1.09	1.01
reactive power, kVAR	0.11 lag	0.12 lead	0.14 lead	0.18 lead	0.26 lead	0.37 lead	0.32 lead	0.3 lead	0.35 lag	0.48 lag	0.59 lag
voltage THD, %F	<5	<5	<5	<5	<5	<5	<5	<5	<5	<5	<5
current THD, %F	13.1	13.4	13.7	14.9	17.9	22.1	27.4	28.5	31.6	41.5	54.8
PF	0.99	0.99	0.99	0.99	0.97	0.95	0.96	0.96	0.95	0.90	0.81
DPF	1	1	1	1	0.99	0.97	0.99	1	0.99	0.97	0.94

NB. Local mains supply used as voltage source for this test (not sine wave generator)

Table 14 Comparison of power quality, near rated power, at nominal voltage, heating

Type	P_{nom}, kW	P, kW	S, kVA	Q, kVAr	I_{THD}, %fund.	CF	PF	DPF
A25	0.83	0.72	0.92	0.36 lag	56.5	1.97	0.78	0.89
		0.87	0.93	0.25 lead	25.7	1.6	0.93	0.96
A50	1.6	1.55	1.58	0.17 lead	17.5	1.6	0.98	0.99
B	1.73	1.7	1.72	0.12 lead	17.2	1.51	0.98	1
C	1.9	1.79	1.81	0.15 lag	10.3	1.45	0.99	1
		2.02	2.05	0.15 lag	9.2	1.44	0.99	1
D	1.7	1.43	1.45	0.19 lead	6.7	1.52	0.99	0.99
		2	2.1	0.2 lead	3.6	1.5	0.99	0.99
E	1.57	1.6	1.67	0.32 lag	22.3	1.57	0.96	0.98

Table 15 Comparison of power quality, near rated power, at nominal voltage, cooling

Type	P_{nom}, kW	P, kW	S, kVA	Q, kVAR	I_{THD}, %fund.	CF	PF	DPF
A25	0.6	0.59	0.76	0.28 lag	60.9	2.03	0.77	0.9
A50	1.55	1.6	1.63	0.18 lead	17.2	1.6	0.98	0.99
B	1.72	1.65	1.68	0.12 lead	17.5	1.54	0.98	1
C	1.7	1.6	1.62	0.12 lag	12.2	1.44	0.99	1
D	1.66	1.65	1.66	0.22 lead	5.2	1.52	0.99	0.99
E[a]	1.37	1.04	1.11	0.21 lag	30.7	1.67	0.94	0.98

[a]Unit E could not be made to draw rated power on the cooling cycle (possibly because of an internal fault) although on heating mode rated power could be achieved.

18

Hybrid model for throughput evaluation of orthogonal frequency division multiple access networks

Shyam Babu Mahato[1], Tien Van Do[2], Ben Allen[1], Enjie Liu[1], Jie Zhang[3]

[1]Centre for Wireless Research (CWR), University of Bedfordshire, Luton LU1 3JU, UK
[2]Department of Telecommunication, Budapest University of Technology and Economics, Budapest, Hungary
[3]Centre for Wireless Network Design (CWiND), Department of Electronic and Electrical Engineering, University of Sheffield, Sheffield, UK
E-mail: Ben.Allen@beds.ac.uk

Abstract: Data throughput is an important metric used in the performance evaluation of the next generation cellular networks such as long-term evolution (LTE) and LTE-advanced. To evaluate the performance of these networks, Monte Carlo simulation schemes are usually used. Such simulations do not provide the throughput of intermediate call state; instead it gives the overall performance of this network. The authors propose a hybrid model consisting of both analysis and simulation. The benefit of this model is that the throughput of any possible call state in the system can be evaluated. Here, the probability of possible call distribution is first obtained by analysis, which is used as input to the event-driven-based simulator to calculate the throughput of a call state. Comparison is made between throughput obtained from the author's hybrid model with that obtained from event-driven-based simulation. Numerical results are presented and show good agreement between both the proposed hybrid model and the simulation. The maximum difference of relative throughput between their hybrid model and the simulation is found in the interval of (0.04 and 1.06%) over a range of call arrival rates, mean holding times and number of resource blocks in the system.

1 Introduction

With increasing demand for mobile data services, orthogonal frequency division multiplexing (OFDM) has become one of the most promising radio interface technologies for future-generation wireless networks. It has already been adopted by systems such as long-term evolution (LTE) and LTE-advanced (LTE-A) [1]. The goal of LTE is to improve the spectral efficiency and hence increase the network capacity, improve services and lower costs. In an orthogonal frequency division multiple access (OFDMA) network, transmission is achieved by transmitting data via multiple orthogonal channels. The system allocates power and transmission rate adaptively and optimally among the subcarriers to achieve high data throughput. Owing to the use of multiple orthogonal channels, OFDM also performs equalisation and is consequently robust to inter-symbol interference and frequency-selective fading [2, 3].

One of the most important aspects of any commercial mobile network deployment is its information carrying capacity, and data throughput is one fundamental parameter in the capacity planning for cellular system deployment. LTE-A is being standardised, so performance evaluation is essential in order to provide insights into competing contributions prior to deployment. System-level performance studies of emerging broadband wireless networks such as LTE-A is typically simulation-based. Such simulations are oriented towards assessing the base station (BS) performance, and do not consider user performance. To estimate user performance, an appropriate performance metric that reflects the throughput is needed.

Ismail and Matalgah [4] have evaluated the performance of code division multiple access (CDMA) network by simulation approach, whereas Kelif and Alman [5] and Kostas and Lee [6] have evaluated the performance of CDMA and time division multiple access (TDMA) network by analytical approach, respectively. Ahn and Wang [7] have evaluated the performance of the OFDMA network based on analytic and simulation approach separately. However, none of the papers mentioned above have considered the combined approach of analysis and simulation. On the other hand, Wu and Sakurai [8] have considered a hybrid simulation/analysis approach, where a detailed rate distribution obtained via simulation is used as input to a generalised processor sharing queue model, but does not consider the user-level. To the best of the authors' knowledge, none of the papers have considered a hybrid model to evaluate user performance. There may be the case where we need to know the throughput of any particular call state from the user point of view. Simulation approach do not provide the throughput of any intermediate call state because it is not possible to obtain this data, whereas the analytical approach provides the average performance. The benefit of the hybrid model is that the throughput of any possible call state can be evaluated.

In this paper, we present a novel hybrid model of analysis/simulation for determining the average data throughput of a system from the user point of view, where a detailed probability of call distribution is obtained from analytical expressions and used as input in the simulator to evaluate the throughput performance. Such a model has the benefit of evaluating the performance of any specific call state from the user perspective.

The rest of the paper is organised as follows. Section 2 describes the system considered for evaluation. Section 3 describes the analytical model for calculating the probability of call distribution in the system, simulation model for evaluation and our proposed hybrid model. The simulation configuration is described in Section 4 and then Section 5 presents and compares results obtained from both conventional simulation and our hybrid approach. Finally, Section 7 provides conclusions.

2 System description

For OFDMA-based networks, user data are divided and modulated onto a large number of narrow-band subcarriers in the frequency domain, and each of them is modulated by low rate data [9]. The subcarriers are orthogonal to each other, meaning that cross-talk between the subcarriers is eliminated and inter-carrier guard bands are not required. The orthogonality among the subcarriers prevents inter-subcarrier interference because the subcarrier's spectrum has nulls located at the centre frequencies of adjacent subcarriers [9]. A group of consecutive subcarriers is known as a subchannel. Moreover, the time domain is split into consecutive frames that are in turn divided into time slots called OFDM symbols. As a multiple access technique, OFDMA offers the possibility of enhancing the spectral efficiency of networks by assigning distinct OFDM symbols or subchannels to distinct users, thus taking advantage of their diverse time and frequency channel conditions as compared with TDMA and FDMA techniques.

For LTE, downlink transmission is based on OFDMA. The radio resources can be considered as a frequency–time resource grid as illustrated in Fig. 1. In the frequency domain, the radio spectrum is divided into a number of narrow subcarriers of 15 kHz (in addition to 15 kHz subcarrier spacing, a reduced subcarrier spacing of 7.5 kHz with twice OFDM symbol time is also defined for LTE which targets multicast-broadcast single-frequency network-based multicast/broadcast transmissions [10].) In the time domain, a frame of 10 ms duration is divided into ten subframes of 1 ms each. Each subframe is further divided into two time slots of 0.5 ms each. Each time slot then consists of six or seven OFDM symbols depending on the length of cyclic prefix (normal or extended cyclic prefix) [9]. A grid of one subcarrier (15 kHz) in the frequency domain and one OFDM symbol (0.5 ms) in the time domain is known as one resource element, whereas a grid of 12 adjacent subcarriers (12 × 15 = 180 kHz) and one OFDM symbol (0.5 ms) is known as one resource block (RB). Hence, an RB is a rectangular block of resource elements, which spans 12 adjacent subcarriers in the frequency domain and 7 OFDM symbols in the time domain (180 kHz × 0.5 ms). In LTE, an RB is also known as a 'subchannel', and from now on we refer to an RB as a subchannel. Depending on the transmission bandwidth, a downlink carrier comprises a variable number of subchannels in the frequency domain. The minimum bandwidth of 1.4 MHz corresponds to six RBs, whereas the maximum one of 20 MHz corresponds to 110 RB. The assignment of subchannels to users is carried out by the medium access control scheduler, and it is performed on a subframe-by-subframe basis, that is, each 1 ms. The scheduler decides which users are allowed to transmit on which subchannel. It should be noted that the minimum resource scheduling unit [From now on, when we refer to an RB, we refer to this minimum scheduling unit of two consecutive RBs, spanning 1 ms.] that the scheduler can assign to a user is comprised of two consecutive RBs and thus spans an entire subframe.

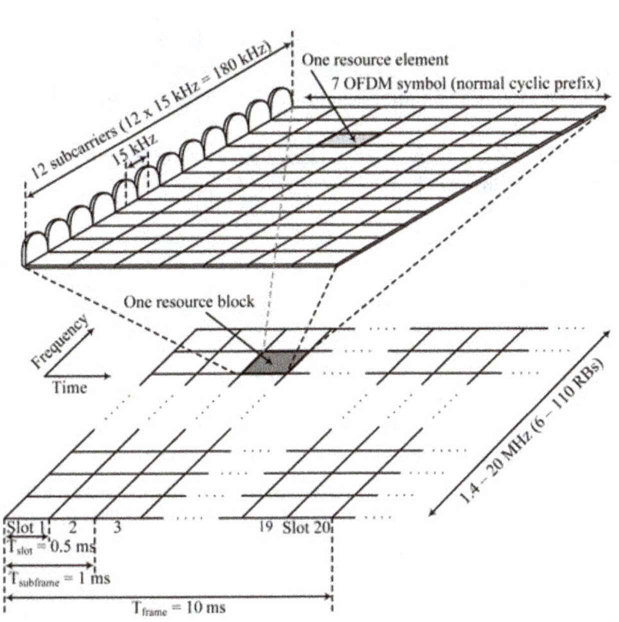

Fig. 1 *Physical layer structure of LTE (3GPP Release 8)*

3 Model description

Consider cellular layout as shown in Fig. 2, where each hexagon is divided into three cells. Each cell is equipped with one transmit antenna and each user equipment (UE) has one receive antenna. Users arrive in a cell according to a Poisson process and are uniformly distributed within the cell.

3.1 Notations

For the sake of clarity, let us introduce some general notations. Let

- J be the total number of cells in the system.
- U be the number of user/subscriber class types, that is, voice, video, streaming etc. For simplicity, we will use call service for all types of service.
- M_j be the maximum number of users in cell j.
- R_j be the number of subchannels in cell $j(1 \leq j \leq J)$.
- $\lambda_{u,j}$ be the mean arrival rate of type-u $(1 \leq u \leq U)$ class user in cell j.
- $\lambda_{\mathrm{nu},j}$ be the mean arrival rate of type-u class new user in cell j.
- $\lambda_{\mathrm{hu},j}$ be the mean arrival rate of type-u class handoff user in cell j.
- $h_{u,j}$ be the mean holding time of type-u class user in cell j.
- $r_{u,j}$ be the cell residence time of type-u class user in cell j and is exponentially distributed with mean $1/r_{u,j}$.
- $\mu_{u,j}$ be the mean service rate of type-u class user in cell j.
- $\mathrm{p}_{ju,ku}$ be the probability that user of type-u class moves from cell j to a neighbouring cell k, given that it moves to a neighbouring cell before the call is completed such that $\sum_{k=1}^{J} \mathrm{p}_{ju,ku} = 1$.
- $X_{u,j}(t)$ denotes a random variable related to the number of type-u class users in progress in cell j at time t.

3.2 Steady-state distribution of calls

We assume that new users are generated in cell j according to a Poisson random process with mean arrival rate $\lambda_{\mathrm{nu},j}$ and the requested call connection time is exponentially distributed with mean $\mu_{\mathrm{nu},j} = 1/t_{u,j}$ [11]. Both the inter-arrival time $(1/\lambda_{u,j})$ and the mean holding time $(h_{u,j})$ are exponentially independent and identically distributed random variables [11–13]. In a cellular network, the handoff traffic is considered more important than a new arriving traffic, because the forced termination of an ongoing call is considered less desirable than the blocking of a new call [14–16].

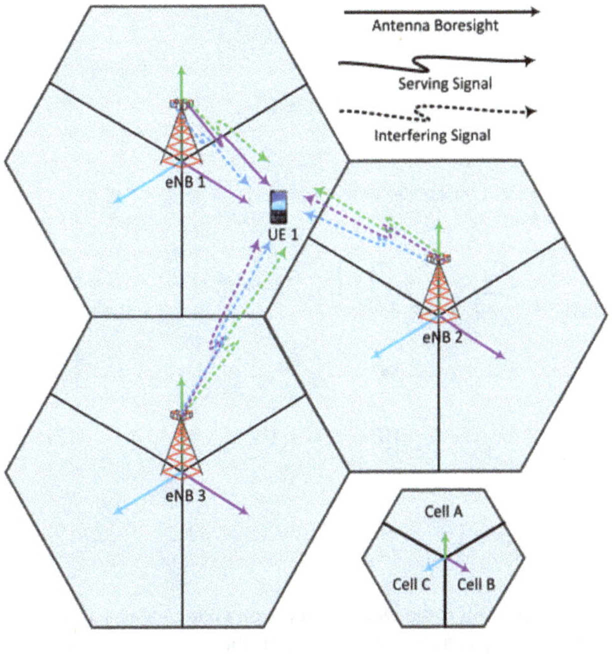

Fig. 2 *System model depicting tri-sector cell*

The priority schemes of how to handle the handoff traffic depend on network designers. One of the popular scheme is 'guard channel scheme' [11, 14, 16, 17], where a fixed number of subchannels in a given cell is reserved for handoff traffic. For example, in a cell with R subchannels, r_g number of subchannels is reserved for handoff traffic.

We assume that the network topology does not change before the steady state is reached. The state of a cell j at time t can be written as

$$X_j(t) = (X_{1,j}(t), X_{2,j}(t), \ldots, X_{U,j}(t)) \tag{1}$$

The state of the network at time t is

$$X(t) = (X_1(t), X_2(t), \ldots, X_J(t))$$

with state space

$$\mathcal{S} = \{(\boldsymbol{n}_1, \boldsymbol{n}_2, \ldots, \boldsymbol{n}_J):\boldsymbol{n}_j = (n_{1,j}, n_{2,j}, \ldots, n_{U,j})\} \tag{2}$$

where $0 \leq \sum_{u=1}^{U} n_{u,j} \leq R_j$, $\boldsymbol{n} = (\boldsymbol{n}_1, \ldots, \boldsymbol{n}_j, \ldots, \boldsymbol{n}_J)$ and $\boldsymbol{n}_j$ is a vector representing the number of calls in cell j. The statistically stationary distribution of calls in the system is given by [11, 18]

$$\begin{aligned} \pi(\boldsymbol{n}) &= \underset{t\to\infty}{Pr}(X_1(t) = \boldsymbol{n}_1, \ldots, X_J(t) = \boldsymbol{n}_J) \\ &= \prod_{j=1}^{J} \pi_j(\boldsymbol{n}_j),\ \boldsymbol{n}_j \in \mathcal{S} \end{aligned} \tag{3}$$

and

$$\pi_j(\boldsymbol{n}_j) = \frac{1}{G_j}\prod_{u=1}^{U}\prod_{l=1}^{n_{u,j}}\frac{\lambda_{u,j}}{l\mu_{u,j}},\quad j = 1, 2, \ldots, J \tag{4}$$

where G_j is the normalisation constant chosen such that the sum of the probabilities of all possible call states in any cell j is 1, and the sum of the product of probabilities of all possible call configurations in the system is 1. That is

$$\sum_{n_j=0}^{M_j} \pi_j(\boldsymbol{n}_j) = 1 \tag{5}$$

and

$$\sum_{\boldsymbol{n}_j\in\mathcal{S}}\prod_{j=1}^{J} \pi_j(\boldsymbol{n}_j) = 1 \tag{6}$$

The normalisation constant is written as [11]

$$G_j = \sum_{0 \leq \sum_{u=1}^{U} n_{u,j} \leq R_j} \prod_{u=1}^{U}\prod_{l=1}^{n_{u,j}}\frac{\lambda_{u,j}}{l\mu_{u,j}},\quad j = 1, 2, \ldots, J \tag{7}$$

where

$$\lambda_{u,j} = \lambda_{\text{nu},j}1[l < R_j - r_g] + \lambda_{\text{hu},j} \tag{8}$$

$$\mu_{u,j} = \left(\frac{1}{t_{u,j}} + \frac{1}{r_{u,j}}\right)l,\quad l = 1, \ldots, R_j \tag{9}$$

in which $1[l < R_j - r_g]$ is the indicator function taking value 1 if the statement $l < R_j - r_g$ is true, else zero; $\lambda_{\text{nu},j}$ is the arrival rate of new user of type-u class and $\lambda_{\text{hu},j}$ is the arrival rate of handoff user of type-u class and is given by Chao and Li [11]

$$\lambda_{\text{hu},j} = \sum_{k=1}^{J}(\lambda_{\text{nu},k} + \lambda_{\text{hu},k})\mathfrak{p}_{ku,ju} \tag{10}$$

where $\mathfrak{p}_{ku,ju}$ is the probability that user u moves from a cell k to cell j.

Based on the above equations, we can compute the probability of cell j, $\pi_j(\boldsymbol{n}_j)$, being in a particular state of call and the probability of the system, $\pi(\boldsymbol{n})$, of a specific state $\boldsymbol{n} = (\boldsymbol{n}_1, \ldots, \boldsymbol{n}_j, \ldots, \boldsymbol{n}_J) \in \mathcal{S}$ of the system, as illustrated in an example in Appendix.

Since the constant term G_j is a function of R_j, we shall write it as $G_j(R_j)$, that is

$$G_j(R_j) = \sum_{0\leq\sum_{u=1}^{U} n_{u,j}\leq R_j}\ \prod_{u=1}^{U} g_{u,j}(n_{u,j}) \tag{11}$$

where for convenience $g_{u,j}$ is defined as

$$g_{u,j}(0) = 1,\quad \text{and}\quad g_{u,j}(n_{u,j}) = \prod_{l=1}^{n_{u,j}}\frac{\lambda_{u,j}}{l\mu_{u,j}} \tag{12}$$

The marginal probability that there are $n_{u,j}$ class u calls in the cell j is [11]

$$\begin{aligned} \pi_{u,j}(n_{u,j}) &= \sum_{0\leq\sum_{v\neq u}^{U} n_{v,j}\leq R_j - n_{u,j}} \pi_j(\boldsymbol{n}_j) \\ &= \frac{G_j^{(u)}(R_j - n_{u,j})}{G_j(R_j)} g_{u,j}(n_{u,j}) \end{aligned} \tag{13}$$

for $n_{u,j} = 1, \ldots, R_j$, where $G_j^{(u)}(R_j - n_{u,j})$ is the normalisation constant of cell j with class u calls removed, that is

$$G_j^{(u)}(R_j - n_{u,j}) = \sum_{0\leq\sum_{v\neq u}^{U} n_{v,j}\leq R_j - n_{u,j}}\ \prod_{v\neq u} g_{v,j}(n_{v,j}) \tag{14}$$

3.3 *Blocking probability*

According to the guard channel scheme, a new call in a cell j gains a subchannel if it finds that there are less than $R_j - r_g$ calls in the cell and that there is at least one subchannel available, otherwise, the new call is blocked in cell j and will be cleared from the system. On the other hand, a handoff call into a cell j gains a subchannel if it finds at least one subchannel available, otherwise, it is blocked.

(1) *New call blocking probability:* a class u new call is accepted to cell j with probability [11]

$$\begin{aligned} &\sum_{l=0}^{R_j - r_g - 1}\ \sum_{\sum_{v\neq u} n_{v,j}\leq R_j - l - 1} \pi_j(\boldsymbol{n}_j) \\ &= \frac{1}{G_j(R_j)}\sum_{l=0}^{R_j - r_g - 1} g_{u,j}(l)G_j^{(u)}(R_j - l - 1) \end{aligned} \tag{15}$$

Then, the blocking probability for a class u call in cell j is given as [11]

$$P_{u,j}(B_N) = 1 - \frac{1}{G_j(R_j)}\sum_{l=0}^{R_j - r_g - 1} g_{u,j}(l)G_j^{(u)}(R_j - l - 1) \tag{16}$$

where $P_{u,j}(B_N)$ represents the blocking probability of class u new call in cell j.

Case 1: no handover and no reserved channel: suppose users move within the serving cell (i.e. no inter-cell mobility) and there is no policy of reserved guard band channels in order to maximise the spectral efficiency. Then, the probability that the user of type-u class moves from cell j to a neighbouring cell k is $p_{ju,ku} = 0$ and the guard band channel, $r_g = 0$. In this case, (8) reduces to $\lambda_{u,j} = \lambda_{\text{nu},j}$, which is the arrival rate of new calls. In this case, the blocking probability of class u new calls can be evaluated as

$$P_{u,j}(B_N) = 1 - \frac{1}{G_j(R_j)} \sum_{l=0}^{R_j - 1} g_{u,j}(l) G_j^{(u)}(R_j - l - 1) \quad (17)$$

For user of voice type service, the blocking probability of new call in a cell can be evaluated as

$$\begin{aligned} P_{u,j}(B_N) &= 1 - \frac{1}{G_j(R_j)} \sum_{l=0}^{R_j - 1} g_{u,j}(l) G_j(l) \\ &= \pi_j(R_j) \end{aligned} \quad (18)$$

(2) *Handoff call blocking probability:* a class u handoff call is accepted to cell j with probability [11]

$$\sum_{\sum_v n_{v,j} \le R_j - 1} \pi_j(\boldsymbol{n}_j) = \frac{G_j(R_j - 1)}{G_j(R_j)} \quad (19)$$

Then, the blocking probability for a class u call to cell j is given as [11]

$$P_{u,j}(B_H) = 1 - \frac{G_j(R_j - 1)}{G_j(R_j)} \quad (20)$$

where $P_{u,j}(B_H)$ represents the blocking probability of class u handoff call to cell j.

3.4 *Channel modelling*

The medium between the transmitting and the receiving antennas is known as the 'channel'. The characteristics of radio signal changes as it travels from the transmitter antenna to the receiver antenna. The characteristics depend on the parameters such as distance between these two antennas, propagation scenario (e.g. outdoor-to-outdoor, outdoor-to-indoor, indoor-to-indoor etc.) and the surrounding environment (e.g. buildings, trees etc.). The received signal can be estimated if we have a suitable model of the medium. This model of the medium is called the 'channel model'. The radio channel propagation is typically modelled as the combination of three main effects: the mean path loss, the shadowing generally characterised as log-normal [19, 20] and the fading typically modelled as Rayleigh [21]. In OFDMA system, the data are multiplexed over a large number of narrow-band subcarriers that are spaced apart at separate frequencies; the subchannel consists of parallel, flat and non-frequency-selective fading. The received signal is then only impacted by slow fading.

(1) *Path loss model:* path loss is the distance dependent mean attenuation of signal as it propagates through space. A suitable model of path loss depends on the parameters such as type of the environment (e.g. macrocell, microcell, indoor etc.), the propagation medium (e.g. outdoor-to-outdoor, outdoor-to-indoor, indoor-to-indoor etc.), the carrier frequency and the distance. The path loss model recommended by 3rd generation partnership project (3GPP) [1] for outdoor macrocells at a carrier frequency of 2 GHz is modelled as

$$(L_{u,j})_{\text{dB}} = 128.1 + 37.6\, log_{10}(d_{u,j}) \quad (21)$$

where $L_{u,j}$ is the path loss in dB from cell j for user u and $d_{u,j}$ is the distance in km from the cell j to the user u.

(2) *Auto-correlation shadow fading model:* in reality, clutters from objects such as buildings, trees, terrain conditions etc. along the path of signal propagation differs for every path, and consequently signal attenuation varies from path to path. Shadow fading is used to model variations in the path loss because of such obstacles between the mobile and the BS. Shadow fading is also known as 'slow fading' [22]. The effect of shadowing is commonly approximated by a log-normal distribution [22, 23]. Accordingly, the shadow fading in the path between a BS and a UE can be priori modelled using a log-normal random variable, $L_{\text{sh}} = \mathcal{N}(\mu, \sigma)$, where μ and σ are the mean and the standard deviation in dB, respectively. However, the modelling of shadow fading when considering the change of user's position is more intricate because of spatial auto-correlation between paths. The shadow fading process is auto-correlated in space, meaning that a moving UE may see similar shadow fading attenuations from the same BS at different but nearby locations.

A widely adopted auto-correlation model for shadow fading is the Gudmundson model [22], which defines the auto-correlation coefficient as follows

$$\rho_a(\Delta x) = e^{(-(|\Delta x|/d_{\text{cor}}) ln\, 2)} \quad (22)$$

where d_{cor} is the decorrelation distance (which is defined as the distance at which the correlation coefficient ρ_a falls to 0.5 [24]), and Δx is the distance between two positions.

The auto-correlation of shadow fading can be implemented as follows. If L_{sh}^1 is the log-normal component, $\mathcal{N}(\mu, \sigma)$, in dB at position P_1 and L_{sh}^2 is the log-normal component in dB at position P_2, which is Δx away from P_1, then L_{sh}^2 can be modelled as a normally distributed random variable, L_{sh}, in dB with mean μ' and standard deviation σ' as [24]

$$L_{\text{sh}} = L_{\text{sh}}^2 = \mathcal{N}(\mu', \sigma') \quad (23a)$$

$$\mu' = \rho_a(\Delta x) L_{\text{sh}}^1 \quad (23b)$$

$$\sigma' = \sqrt{(1 - \rho_a^2(\Delta x))\, \sigma^2} \quad (23c)$$

Table 1 Modulation and coding scheme [25, 26]

MCS	Modulations	Code rates	SINR threshold, dB	RAB efficiency, bits/symbol
MCS_1	QPSK	1/12	−6.5	0.15
MCS_2	QPSK	1/9	−4.0	0.23
MCS_3	QPSK	1/6	−2.6	0.38
MCS_4	QPSK	1/3	−1.0	0.60
MCS_5	QPSK	1/2	−1.0	0.88
MCS_6	QPSK	3/5	−3.0	1.18
MCS_7	16QAM	1/3	6.6	1.48
MCS_8	16QAM	1/2	10.0	1.91
MCS_9	16QAM	3/5	11.4	2.41
MCS_{10}	64QAM	1/2	11.8	2.73
MCS_{11}	64QAM	1/2	13.0	3.32
MCS_{12}	64QAM	3/5	13.8	3.90
MCS_{13}	64QAM	3/4	15.6	4.52
MCS_{14}	64QAM	5/6	16.8	5.12
MCS_{15}	64QAM	11/12	17.6	5.55

Table 2 Simulation parameters [1]

Statistical parameters	Values
call distribution	uniform
call generation process	Poisson
call mean arrival rate	0.1, 0.2, 0.3, 0.4, 0.5, 0.6, 0.7, 0.8, 0.9, 1.0 calls/min/cell
call mean holding time	0.5, 1, 1.5, 2.0, 2.5, 3.0, 3.5, 4.0, 4.5, 5.0 min
call inter-arrival/holding time distance	exponential
LTE system parameters	
site layout	one hexagonal site with three cells
carrier frequency	2 GHz
subcarrier spacing	15 kHz
RB spacing	180 kHz
number of subchannels	2, 3, 4, 5
data symbol per time slot	11 OFDM data symbols
frame duration	1 ms
thermal noise density	−174 dBm/Hz
thermal noise power	−121.4 dBm
resource allocation	one RB/call
eNB and UE parameters	
eNB Tx power	46 dBm
eNB/UE antenna height	32/1.5 m
eNB/UE antenna gain	18/0 dBi
UE noise figure	9 dB
UE receiver sensitivity	−95 dBm
eNB antenna boresight	0/120/240°
eNB antenna pattern	3GPP case1: three-dimensional antenna pattern
propagation parameters	
shadowing model	Gudmunson model
shadowing standard deviation	8 dB
correlation distance of shadowing	50 m

3.5 *Signal-to-interference plus noise ratio*

For simplicity, uniform and equal transmission power is distributed on each subcarrier. Assuming that all subcarriers within a subchannel experience the same channel condition, the downlink signal-to-interference plus noise ratio (SINR) at the UE, u, on subchannel r connected to the cell j is given by

$$\gamma_{u,r,j} = \frac{P_{u,r,j}\, G_{u,r,j}/(L_{u,j}\, L_{\rm sh})}{\sum_{i\in\Psi, i\neq j} P_{u,r,j}\, G_{u,r,j}/(L_{u,j}\, L_{\rm sh})\alpha_{r,i} + N} \tag{24}$$

where $P_{u,r,j}$ and $P_{u,r,i}$ are the transmit power from the cells j and i, respectively; $G_{u,r,j}$ and $G_{u,r,i}$ are the antenna gains from the cells j and i, respectively; $L_{u,j}$ and $L_{u,i}$ are the path loss from the cells j and i, respectively; $L_{\rm sh}$ is the attenuation because of shadowing; N is the thermal noise power; and $\alpha_{r,i}$ is the subchannel allocation indicator, which is given by

$$\alpha_{r,i} = \begin{cases} 1, & \text{if the RB } r \text{ is used in the cell } i \\ 0, & \text{otherwise} \end{cases} \tag{25}$$

3.6 *Radio access bearer (RAB) efficiency*

The RAB is the entity responsible for transporting radio frames of an application over the radio access of the network. From the estimated SINR, a suitable modulation and coding scheme (MCS) is selected for each user provided that the SINR satisfies the threshold for the selected MCS. The higher the SINR, the higher-order MCS is used satisfying the SINR threshold value. The RAB efficiency (defined in 3GPP 36.213 Table 7.2.3-1 [25]) is shown in Table 1. In general, the RAB efficiency of a user on subchannel r is estimated as [27]

$$\xi_r = \mathrm{CR}_k\, log_2(M_k) \tag{26}$$

where ξ_r is the RAB efficiency (bits/symbol) on the subchannel r for the selected MCS, CR_k is the coding rate of the MCS and M_k is the number of constellation points of the MCS_k, where $k \in \{1, 2, \ldots, 15\}$ represents a particular MCS as shown in Table 1.

3.7 *User throughput*

Once an MCS is selected, the bit-rate of the user u over the subchannel r can be estimated as [28]

$$\mathrm{BR}_{u,r} = \frac{\xi_r\, n_r\, \mathrm{symbol}_r}{\tau} \tag{27}$$

where $\mathrm{BR}_{u,r}$ is the bit-rate (bits/s), n_r is the number of subcarriers in the subchannel r, symbol_r is the number of data symbols in the subchannel r in τ duration of a subframe.

Once the bit-rate is known, the throughput of the user u connected to the cell j over the subchannel r can be estimated as [28]

$$T_{u,r,j} = \mathrm{BR}_{u,r}\left(1 - \varepsilon\left(\gamma_{u,r,j}, \mathrm{MCS}_k\right)\right) \tag{28}$$

where $\varepsilon(\gamma_{u,r,j}, \mathrm{MCS}_k)$ represents the block-error rate suffered by the user u over the subchannel r connected to the sector j, which is a function of its both SINR, $\gamma_{u,r,j}$, and RAB MCS_k.

Table 3 System throughput performance with mean holding time, $1/\mu$ = 3 min

Arrival rate, calls/min	Average system throughput, Mbps								Percentage throughput error, %			
	$R=2$		$R=3$		$R=4$		$R=5$		$R=2$	$R=3$	$R=4$	$R=5$
	Sim	Hyb	Sim	Hyb	Sim	Hyb	Sim	Hyb				
0.1	0.548	0.557	0.586	0.613	0.597	0.618	0.595	0.618	1.532	4.626	3.654	3.933
0.2	1.075	0.977	1.239	1.186	1.288	1.205	1.306	1.233	−9.121	−4.294	−6.436	−5.574
0.3	1.310	1.270	1.541	1.557	1.652	1.708	1.694	1.785	−2.986	0.993	3.347	5.360
0.4	1.504	1.543	1.964	1.898	2.173	2.172	2.304	2.211	2.539	−3.335	−0.037	−4.032
0.5	1.621	1.670	2.123	2.161	2.404	2.541	2.574	2.732	3.048	1.804	5.691	6.138
0.6	1.826	1.741	2.473	2.411	2.914	2.833	3.191	3.123	−4.650	−2.483	−2.783	−2.122
0.7	1.949	1.901	2.689	2.603	3.261	3.120	3.616	3.493	−2.453	−3.199	−4.320	−3.396
0.8	1.968	1.936	2.720	2.708	3.330	3.350	3.746	3.788	−1.631	−0.441	0.619	1.132
0.9	1.993	2.002	2.850	2.842	3.512	3.506	4.020	4.045	0.426	−0.284	−0.154	0.624
1	2.079	2.006	2.975	2.912	3.717	3.675	4.353	4.307	−3.540	−2.108	−1.133	−1.057

3.8 Hybrid model

This model combines the analytic and simulation approach, which, unlike traditional simulation approaches, enables us to estimate throughput of any intermediate call state. The probability of call distribution being in a particular state in the system is obtained from analytic expressions and is used as input to the simulation to calculate the throughput of a cell. According to hybrid model, the average throughput for cell j is expressed as follows

$$\sum_{\boldsymbol{n}\in\mathcal{S}}\sum_{i=1}^{s_j}\left[\pi_j(\boldsymbol{n}_j)\underbrace{T_{u,r,j}}_{\text{sim.thr.}}\right] \tag{29}$$

where $\boldsymbol{n}\in\mathcal{S}$ is the possible call configurations in the system, $\sum_{i=1}^{s_j}$ represents the number of users in the cell j, $\pi_j(\boldsymbol{n}_j)$ represents the probability that there are $\boldsymbol{n}_j$ calls in cell j and $T_{u,r,j}$ is the simulation throughput (sim. thr.) of a user u on subchannel r in cell j obtained for a particular state of call in the system.

The average throughput for the system is expressed as follows

$$\sum_{\boldsymbol{n}\in\mathcal{S}}\left[\pi(\boldsymbol{n})\underbrace{T}_{\text{sim.thr.}}\right] \tag{30}$$

where $\pi(\boldsymbol{n})$ represents the probability of state space $\boldsymbol{n}$ in the system, and T is the average system throughput obtained from simulation for a particular call configuration in the system.

4 Simulation configuration

To evaluate the performance of the system and validate our proposed hybrid model, an event-driven dynamic system-level simulation was used. The traffic is modelled by a homogeneous Poisson random process in each cell. To account for the dynamic behaviour of the incoming traffic pattern or service time and hence thereby obtain various traffic loads in the network, we have implemented two approaches. In the first approach, the mean holding time of users is fixed and the inter-arrival time is varied in order to model the dynamic behaviour of incoming traffic, whereas in the second approach, the inter-arrival time is fixed and the mean holding time is varied in order to model the dynamic service time. During the simulation, an event occurs when:

(a) a user arrives and accesses a subchannel to connect to the network
(b) the user moves position randomly
(c) the user leaves the network and the subchannel is freed
(d) the system triggers to log the simulator status indicator

The logged data of the users such as SINR and throughput are obtained on a regular basis. The wireless channel for a user from a BS is selected randomly from available subchannels and remains the same as long as the user stays in the network.

A different level of adaptive MCS is selected from Table 1 when mapping the user's SINR to its achievable data throughput. The parameters in the simulation are consistent with the LTE downlink, and are listed in Table 2.

The process for evaluating the performance of the system in the hybrid model is described as follows:

1. Calculate the probability of the system being in a particular call configuration from the statistically stationary distribution of calls using (3)–(7).
2. Run the event-driven simulation for this particular configuration only and change the user position during the simulation (thus accounting for the effect of different locations when estimating the average throughput).
3. Multiply the simulated throughput for this particular configuration obtained from the simulation with the probability calculated analytically for the system in a specific configuration.
4. Repeat the simulation for all possible configurations of calls in the system.
5. Sum the throughput for all possible configurations for the overall system throughput.

5 Results

To validate the simulation, we have tested the results of probability of call distribution and probability of call blocking obtained from the simulation with that obtained by analysis for different traffic intensities. Fig. 3 shows that both the probability of call distribution

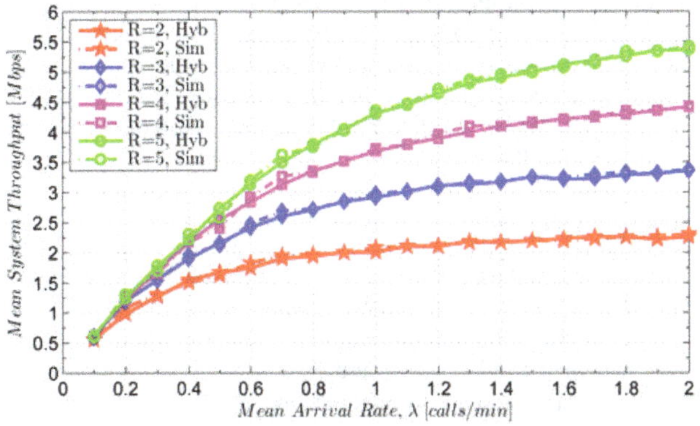

Fig. 4 *System throughput performance with mean holding time, 1/μ = 3 min*

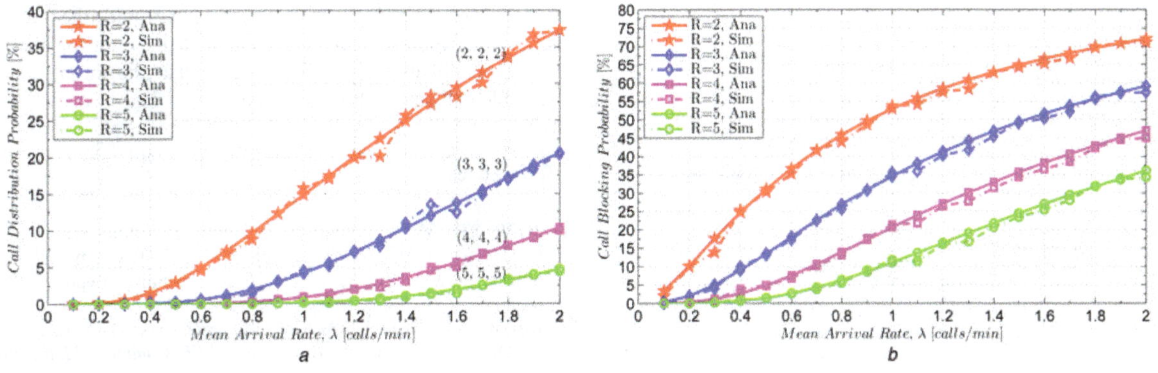

Fig. 3 *Probability comparison between simulation and analysis at mean holding time, 1/μ = 3 min, in terms of*
a Probability of the highest possible call, that is, (R, R, R) (call distribution probability)
b Probability of the call blocking (call blocking probability)

Table 4 System throughput performance with mean arrival rate, $\lambda = 2$ calls/min

Mean holding time, min	Average system throughput, Mbps								Percentage throughput error, %			
	$R=2$		$R=3$		$R=4$		$R=5$		$R=2$	$R=3$	$R=4$	$R=5$
	Sim	Hyb	Sim	Hyb	Sim	Hyb	Sim	Hyb				
0.5	1.402	1.430	1.766	1.684	1.905	1.885	1.998	1.924	2.033	−4.649	−1.034	−3.748
1	1.859	1.826	2.532	2.542	3.036	3.040	3.353	3.376	−1.754	0.383	0.122	0.689
1.5	2.073	2.075	2.960	2.945	3.761	3.703	4.384	4.318	0.068	−0.517	−1.540	−1.503
2	2.167	2.094	3.153	3.144	4.080	4.057	4.887	4.848	−3.391	−0.289	−0.554	−0.796
2.5	2.215	2.226	3.264	3.242	4.271	4.204	5.183	5.184	0.515	−0.680	−1.587	0.023
3	2.297	2.254	3.368	3.345	4.417	4.407	5.407	5.360	−1.876	−0.695	−0.229	−0.858
3.5	2.270	2.320	3.408	3.375	4.469	4.528	5.517	5.515	2.207	−0.963	1.325	−0.047
4	2.320	2.313	3.478	3.443	4.589	4.530	5.678	5.633	−0.267	−0.992	−1.275	−0.791
4.5	2.334	2.385	3.484	3.417	4.627	4.595	5.743	5.737	2.189	−1.923	−0.685	−0.101
5	2.338	2.355	3.473	3.463	4.641	4.585	5.762	5.790	0.710	−0.271	−1.207	0.500

for the highest possible call state and the probability of call blocking obtained by simulation are in-line with that obtained by analysis for different traffic intensities and different number of users. The performance of the system throughput obtained by our hybrid model was evaluated for different scenarios of traffic behaviour by changing the mean arrival rate, the mean holding time and the number of subchannels.

5.1 *Varying arrival rate*

In this case, the performance of the system was evaluated for different call arrival rates for different number of subchannels to account for the variation of inter-arrival traffic, whereas the mean holding time remained fixed. Table 3 and Fig. 4 show the performance of the average system throughput for different arrival rates. It is noted that the system throughput performance by the hybrid model is similar to the simulation for all arrival traffic patterns. The mean of the throughput error between the two methods is found to be −0.88% for $R=2$, −7.60% for $R=3$, −0.36% for $R=4$ and −0.2% for $R=5$.

5.2 *Varying holding time*

In this case, the performance of the system was evaluated for different mean holding times for different number of subchannels to account for the variation of service time, whereas the mean arrival rate remained fixed. Table 4 and Fig. 5 show the performance of the average system throughput for a range of mean holding times. It is noted that the system throughput performance by the hybrid model is similar to the simulation for all arrival traffic patterns. The mean of the throughput error between the two methods is found to be 0.94% for $R=2$, −1.06% for $R=3$, −0.67% for $R=4$ and −0.66% for $R=5$.

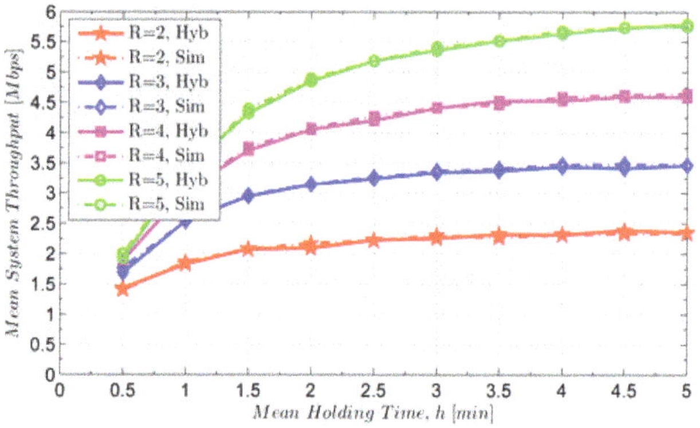

Fig. 5 *System throughput performance with mean arrival rate, $\lambda = 2$ calls/min*

6 Conclusions

We have proposed a hybrid model consisting of analysis and simulation for the evaluation of average system data throughput compared with event-driven-based simulations. Our approach allows throughput of intermediate call state to be evaluated as well as the overall network throughput. To evaluate the performance by the hybrid model, a detailed probability of call distribution in the system is first obtained from analytical expressions of a statistically stationary distribution, which are used as an input to the simulator to calculate the system throughput. We compared the results of the hybrid model with those obtained from simulation. We tested the model for different parameters of user arrival rate, their mean holding time and different numbers of radio subchannels in the network. It has been found that the results of the hybrid model are in-line with the simulation-based results. The maximum difference of mean throughput error performance between the hybrid model and the simulation is found to be in the interval of (0.04 and 1.06%) for different call arrival rates, mean holding times and number of subchannels in the system. It was noted that for a large number of cells and users, the number of possible call configurations in the system is very large. In such a large possible number of call configurations, it is difficult to evaluate the system throughput by the hybrid model from the user point of view, because we need to evaluate the throughput for all possible call configurations. However, there may be the case where we need to know the throughput of any particular call configuration in the system from the user point of view. The simulation does not provide the throughput of any intermediate call state because it does not log the call state. The benefit of the hybrid model is that the throughput of any possible call state can be evaluated.

7 References

[1] 3GPP TR 36.814: 'Evolved universal terrestrial radio access (E-UTRA); further advancements for E-UTRA physical layer aspects (Release 9)', 3GPP TSG RAN, Technical Report, v.9.0.0, March 2010

[2] Koffman I., Roman V.: 'Broadband wireless access solutions based on OFDM access in IEEE 802.16', *IEEE Commun. Mag.*, 2002, **40**, (4), pp. 96–103

[3] Coleri S., Ergen M., Puri A., Bahai A.: 'Channel estimation techniques based on pilot arrangement in OFDM systems', *IEEE Trans. Broadcast.*, 2002, **48**, (3), pp. 223–229

[4] Ismail M.H., Matalgah M.M.: 'Simulation results for the impact of users locations and distribution characteristics on the performance of downlink WCDMA/TDD in fading channels'. IEEE Radio and Wireless Conf., September 2004, pp. 495–498

[5] Kelif J.-E., Alman E.: 'Downlink fluid model of CDMA networks'. IEEE Vehicular Technology Conf. (VTC2005-Spring, IEEE 61st), 30 May–1 June 2005, vol. **4**, pp. 2264–2268

[6] Kostas T.A., Lee C.C.: 'A performance model for data throughput with adaptive modulation', *IEEE Trans. Wirel. Commun.*, 2007, **6**, (1), pp. 79–89

[7] Ahn S., Wang H.: 'Throughput-delay tradeoff of proportional fair scheduling in OFDMA systems', *IEEE Trans. Veh. Technol.*, 2011, **60**, (9), pp. 4620–4626

[8] Wu W., Sakurai T.: 'Flow-level capacity of fractionally loaded OFDMA networks with proportional fair scheduling'. IEEE Vehicular Technology Conf. (VTC2012-Fall, IEEE 76th), September 2010, pp. 1–5

[9] 3GPP TS 36.211: 'Evolved universal terrestrial radio access (E-UTRA); physical channels and modulation (Release 10)', 3GPP TSG RAN Std., v.10.5.0, June 2012

[10] Dahlman E., Parkvall S., Skold J., Beming P.: '3G evolution: HSPA and LTE for mobile broadband' (Elsevier, 2008, 2nd edn.), ASIN B002ZJSVW8

[11] Chao X., Li W.: 'Performance analysis of a cellular network with multiple classes of calls', *IEEE Trans. Commun.*, 2005, **53**, (9), pp. 1542–1550

[12] Sarangan V., Ghosh D., Gautam N., Acharya R.: 'Steady state distribution for stochastic knapsack with bursty arrivals', *IEEE Commun. Lett.*, 2005, **9**, (2), pp. 187–189

[13] Haring G., Marie R., Puigjaner R., Trivedi K.: 'Loss formulas and their application to optimization for cellular networks', *IEEE Trans. Veh. Technol.*, 2001, **50**, (3), pp. 664–672

[14] Xhafa A.E., Tonguz O.K.: 'Handover performance of priority schemes in cellular networks', *IEEE Trans. Veh. Technol.*, 2008, **57**, (1), pp. 565–577

[15] Zhang Y., Soong B.-H.: 'Handoff dwell time distribution effect on mobile network performance', *IEEE Trans. Veh. Technol.*, 2005, **54**, (4), pp. 1500–1508

[16] Dharmaraja S., Trivedi K.S., Logothetis D.: 'Performance modeling of wireless networks with generally distributed handoff interarrival times', *J. Comput. Commun.*, 2003, **26**, (15), pp. 1747–1755

[17] Li W., Fang Y., Henry R.R.: 'Actual call connection time characterization for wireless mobile networks under a general channel allocation scheme', *IEEE Trans. Wirel. Commun.*, 2002, **1**, (4), pp. 682–691

[18] Kelly F.P.: 'Reversibility and stochastic networks' (Cambridge University Press, 2011), ISBN 978-1107401150

[19] French R.C.: 'The effect of fading and shadowing on channel reuse in mobile radio', *IEEE Trans. Veh. Technol.*, 1979, **28**, (3), pp. 171–181

[20] Yeh Y.S., Schwartz S.C.: 'Outage probability in mobile telephony due to multiple log-normal interferers', *IEEE Trans. Commun.*, 1984, **32**, (4), pp. 380–388

[21] Sowerby K.W., Williamson A.G.: 'Outage probability calculations for a mobile radio system having multiple Rayleigh interferers', *Electron. Lett.*, 1987, **23**, (11), pp. 600–601

[22] Gudmundson M.: 'Correlation model for shadow fading in mobile radio systems', *Electron. Lett.*, 1991, **27**, (23), pp. 2145–2146

[23] Monserrat J., Fraile R., Cardona N., Gozalvez J.: 'Effect of shadowing correlation modeling on the system level performance of adaptive radio resource management techniques'. IEEE Int. Symp. Wireless Communication Systems, 2 September 2005, pp. 460–464

[24] Fu I.K., Li C.F., Song T.C., Sheen W.H.: 'Correlation models for shadow fading simulation'. IEEE 802.16, Orland, USA, TGm Evaluation Methodology Development IEEE S802.16 m-07/060, March 2007

[25] 3GPP TS 36.213: 'Evolved universal terrestrial radio access (E-UTRA); physical layer procedures (Release 10)', 3GPP TSG RAN Std. v.10.4.0, December 2011

[26] Lopez-Perez D., Chu X.: 'Inter-cell interference coordination for expanded region picocells in heterogeneous networks'. Int. Conf. Computer Communications and Networks (ICCCN, 20th) Proc., Hawaii, USA, 31 July–4 August 2011, pp. 1–6

[27] Andrews J.G., Ghosh A., Muhamed R.: 'Fundamentals of WiMAX', in Rappaport T.S. (Ed.): (Prentice-Hall, USA, October 2007, 3rd edn.)

[28] Lopez-Perez D., Ladanyi A., Juttner A., Zhang J.: 'OFDMA femtocells: a self-organizing approach for frequency assignment'. IEEE Int. Symp. Personal, Indoor and Mobile Radio Communications (PIMRC, IEEE 20th), September 2009, pp. 2202–2207

8 Appendix

Example: calculation of distribution probability

Consider there are three cells in the system, that is, $J=3$. Assume two subchannels in each cell, that is, $R=2$, then the maximum number of calls in each cell is two. Without loss of generality, for simplicity, suppose each cell has only one type of user service, that is, voice. Then the state space, $\mathcal{S}$, of the system consists of all possible vectors, $\boldsymbol{n}$ ($\boldsymbol{n}=(\boldsymbol{n}_1, \boldsymbol{n}_2, \boldsymbol{n}_3)$, $\boldsymbol{n}_j=(n_{1,j})$).

The possible configurations of calls in the system are as follows:

- $(\boldsymbol{n}=((0), (0), (0))) \Rightarrow$ no calls in the system.
- $(\boldsymbol{n}=((1), (0), (0))) \Rightarrow$ 1 call in cell 1.
- $(\boldsymbol{n}=((2), (0), (0))) \Rightarrow$ 2 calls in cell 1.
- $(\boldsymbol{n}=((0), (1), (0))) \Rightarrow$ 1 call in cell 2.
- $(\boldsymbol{n}=((0), (0), (2))) \Rightarrow$ 2 calls in cell 3.
- $(\boldsymbol{n}=((1), (1), (0))) \Rightarrow$ 1 call in each cell 1 and 2.
- $(\boldsymbol{n}=((2), (1), (0))) \Rightarrow$ 2 calls in cell 1 and 1 call in cell 2.
- ... ...
- $(\boldsymbol{n}=((2), (2), (2))) \Rightarrow$ 2 calls in each cell, which is the highest possible call state in the system.

The number of possible configurations is $3\times3\times3=27$. If $R=3$, the number of possible configurations is $4\times4\times4=64$ etc. In general, the number of possible configurations can be written as $(R+1)^J$. It should be noted that for the large number of cells and subchannels, the number of possible configurations becomes large. For example, in two-tier networks having 19 cell-sites with three cells per site, the total number of cells in the system is $J=57$. Suppose, there are 50 RBs in each cell and each call is allowed only one RB. Then, the maximum number of calls in each cell would be 50. Hence, the total number of possible configurations would be $(R+1)^J=(50+1)^{57}=51^{57}$, which is a very large number. To evaluate the system throughput by hybrid approach for such a large scenario, we need to evaluate the throughput for each possible configuration which is time consuming. Hence, it is difficult to evaluate the system throughput by hybrid approach for large number configurations from the user point of view. Therefore hybrid approach is useful in the place where we consider to evaluate the throughput of any possible call state.

Since for this analysis there is only one type of user service, the subscript of user type can be dropped. The normalisation constant of (7) can then be simplified as

$$G=\sum_{n_j\le R_j}\prod_{l=1}^{n_j}\frac{\lambda_j}{l\mu_j}=\sum_{n_j\le R_j}\frac{\left(\lambda_j/\mu_j\right)^{n_j}}{n_j!},\quad j=1,\ \dots,\ J \tag{31}$$

Arbitrarily, assume the state of the system at a particular time is $(\boldsymbol{n}=((1), (2), (1)))$, $\lambda_j=1.5$ calls/min and $\mu_j=1$ calls/min. Since $R=2$, the possible number of calls in a cell would be 0, 1, ..., R. The normalisation constant can be calculated as $G=(\lambda/\mu)^0/0!+(\lambda/\mu)^1/1!+(\lambda/\mu)^2/2!=3.625$. To check the summation of probability in a cell (say, cell 1), we can write (5) as

$$\sum_{n_1=0}^{2} \pi_1(n_1) = \pi_1(0) + \pi_1(1) + \pi_1(2)$$
$$= \frac{1}{G}\left[\left(\frac{\lambda}{\mu}\right)^0/0! + \left(\frac{\lambda}{\mu}\right)^1/1! + \left(\frac{\lambda}{\mu}\right)^2/2!\right] \quad (32)$$
$$= \frac{1}{3.625}[1 + 1.5 + 1.125] = 1$$

Using (4), we can calculate the probabilities of corresponding calls in each cell as

- $\pi_1(1) = (\lambda/\mu)^1/1!/G = 0.41379 = 41.379\%$.
- $\pi_2(2) = (\lambda/\mu)^2/2!/G = 0.31034 = 31.034\%$.
- $\pi_3(1) = (\lambda/\mu)^1/1!/G = 0.41379 = 41.379\%$.

Hence, we can easily calculate the probability of the system in the state space ($\boldsymbol{n} = ((1), (2), (1))$) using (3) as

$$\pi(n) = \prod_{j=1}^{3} \pi_j(n_j) = \pi_1(1)\,\pi_2(2)\,\pi_3(1)$$
$$= 0.053137 = 5.314\%$$

The probability 5.314% is just for one possible call configuration in the system. If we sum the probabilities of all possible call configurations, we will achieve 100%, according to (6).

Bipolar latch with compensated keep-alive current

Dr. Hans Gustat

Dept. Circuit Design, IHP - Institute for High-Performance Microelectronics, Im Technologiepark 25, D-15236 Frankfurt (Oder)
E-mail: hans@gustat.de

Abstract: A permanent current in addition to the main clocked current is sometimes used to increase the maximum clock rate of a bipolar latch. Although it speeds up the activation of a clocked differential stage, it deteriorates the latch function by the additional current in the inactive phase of each differential stage. Thus, a keep-alive current must be kept small with respect to the main clocked current. In this Letter, a compensation technique is shown avoiding the erroneous output of a keep-alive current. It still speeds up the activation of the main transistor pair, but results in a constant symmetric offset without affecting the differential value of the output voltage. In simulations of flip-flops and clocked comparators, this compensated keep-alive current has a much larger effect on the maximum clock rate than the uncompensated keep-alive current used so far.

1 Introduction

Various methods are known to increase the maximum clock rate of bipolar latches, flip-flops and comparator circuits built from latch circuits. Sometimes, when the usual techniques (such as inductive peaking, optimised load resistors, pairs of emitter followers after the load network) are not enough to increase the maximum clock rate, a permanent bias current ('keep-alive current') is applied to further enhance the clock rate. It can be beneficial even though additional power consumption has to be traded in for higher speed. It may be a useful option in cases where only a small fraction of an integrated circuit should operate at maximum possible clock rate, so that the additional power consumption of that small fraction can be tolerated. Such minor high-frequency parts of a system may involve the first stage of a frequency divider chain, fast pseudo-random binary sequence (PRBS) generators and also clocked comparators, for example, in analogue-to-digital converters (ADCs) with a very small number of comparators, such as certain successive-approximation ADCs employing only one comparator. In such cases, pushing the clock rate limit of a few bipolar latches might improve important system parameters, such as upper frequency limit, conversion rate etc.

2 Keep-alive current

In addition to the main (clocked) operational current, a second current is permanently fed through the bipolar differential stage (BDS). This way, the base–emitter p–n-junctions of the BDS are always kept in forward biasing, even in the inactive clock phase of the BDS. A keep-alive current can be applied in parallel to other speed-enhancement techniques, because these are mostly aimed at enhancing the gain or bandwidth in the active clock phase of the BDS, whereas permanent forward biasing shortens the transition time from inactive to active phase. However, a keep-alive current usually has only a minor impact ([1]: <10%), because the additional permanent current I_{alive} also provides an erroneous contribution of the BDS during its inactive clock phase to the output of the latch. With increasing I_{alive}, the latch function becomes worst, because of activity of the parts which ought to be inactive. To maintain a correct function, I_{alive} shall be limited to a small percentage of the main clocked current.

3 Keep-alive current compensation

To keep the benefits of a permanent forward current while avoiding its drawback, an identical current is added inversely to the outputs of the BDS, as shown in Fig. 1. R3 provides a permanent current I_{alive} to the BDS formed by Q1 and Q2. An equal current is fed by another resistor R3a to a secondary BDS of Q1a and Q2a, which is forming an antiparallel amplifier to the original BDS Q1/Q2. If I_{alive} is split in $\alpha\, I_{alive}$ for Q1 and $(1-\alpha)\, I_{alive}$ for Q2, then it is also split equally for Q1a and Q2a, respectively, because of the parallel inputs. The collector current sum of Q1 and Q2a is always I_{alive}, and also for Q2 and Q1a. As a result, Q1 and Q2 are kept in forward biasing, with only a constant output current of I_{alive} at both load resistors, not altering the differential output voltage. The collector resistors Rc1a and Rc2a in Fig. 1 decouple the parasitic capacitance of the collectors of Q1a and Q2a from their respective load resistors. Compared with the uncompensated keep-alive current, the compensation allows a much larger range for I_{alive}, because it has no detrimental effect for the latch function anymore, and should also increase the effect for the same value of I_{alive}.

A complete latch circuit employing the keep-alive current with the compensation structure of Fig. 1 is shown in Fig. 2. Apart from the keep-alive current and the compensation circuit, the circuit structure is identical to the bipolar latch used in [2]. In the upper part, Q1/Q2 transfers the latch input in the active clock phase to the load resistors R1/R2, with Q1a/Q2a as an antiparallel compensation BDS as in Fig. 1. Two pairs of emitter followers (Q4/Q5, Q6/Q7) are decoupling

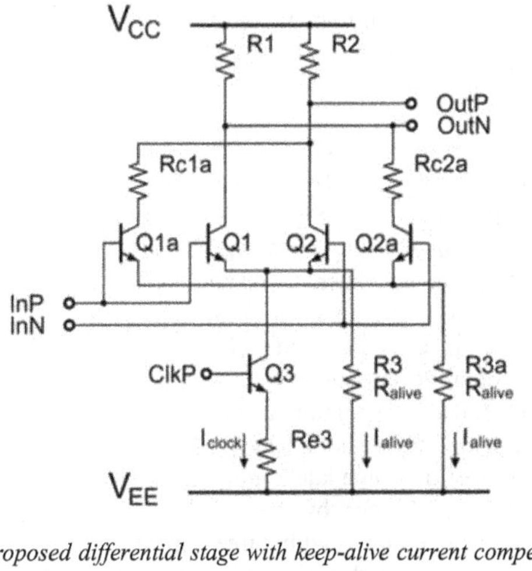

Fig. 1 *Proposed differential stage with keep-alive current compensation*

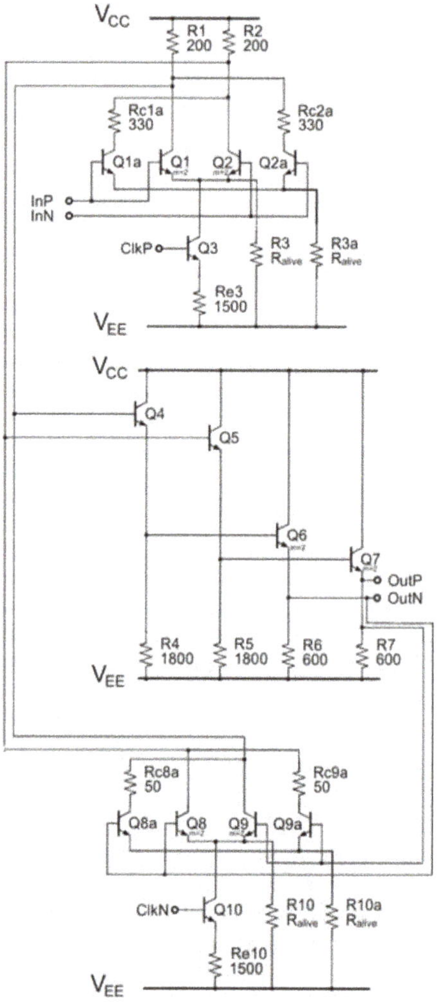

Fig. 2 *Complete bipolar latch circuit with keep-alive current compensation*

output load impedance from R1/R2 and also feed the hold stage (Q8/Q9) with (Q8a/Q9a) as compensation. Applying the compensation circuit of Fig. 1 to both the transfer and the hold BDS of a latch, I_{alive} results only in a symmetric and constant offset at the output load.

4 Verification

The compensation principle of Fig. 1 for a keep-alive current can be applied to any - even non-silicon-based - bipolar technology. As a test vehicle, a mature 250 nm 200 GHz-silicon germanium process [3] was chosen for which the latch circuit in Fig. 2 has been dimensioned. A main reason for this choice is that both design details and measured data of a PRBS generator test chip in this technology were available [4]. These data have been used as a reference point for realistic simulation of the maximum clock rate at $I_{alive} = 0$. By varying R_{alive}, the latch of Fig. 2 in the PRBS application has been simulated with various levels of I_{alive} up to 200% of the main clocked current I_{clock}. Fig. 3*a* shows the results. In contrast to the PRBS generator with uncompensated keep-alive current, about twice the range of I_{alive} can be applied, and the maximum clock rate is increased by a factor of about 1.6, relative to no keep-alive current circuitry. Other applications of the same latch circuit have also been investigated: a frequency divider by two with a single flip-flop (Fig. 3*b*), a Johnson counter with three flip-flops acting as a divider by six (Fig. 3*c*) and a clocked comparator with 2 mV input swing applied (Fig. 3*d*). The maximum clock rates have

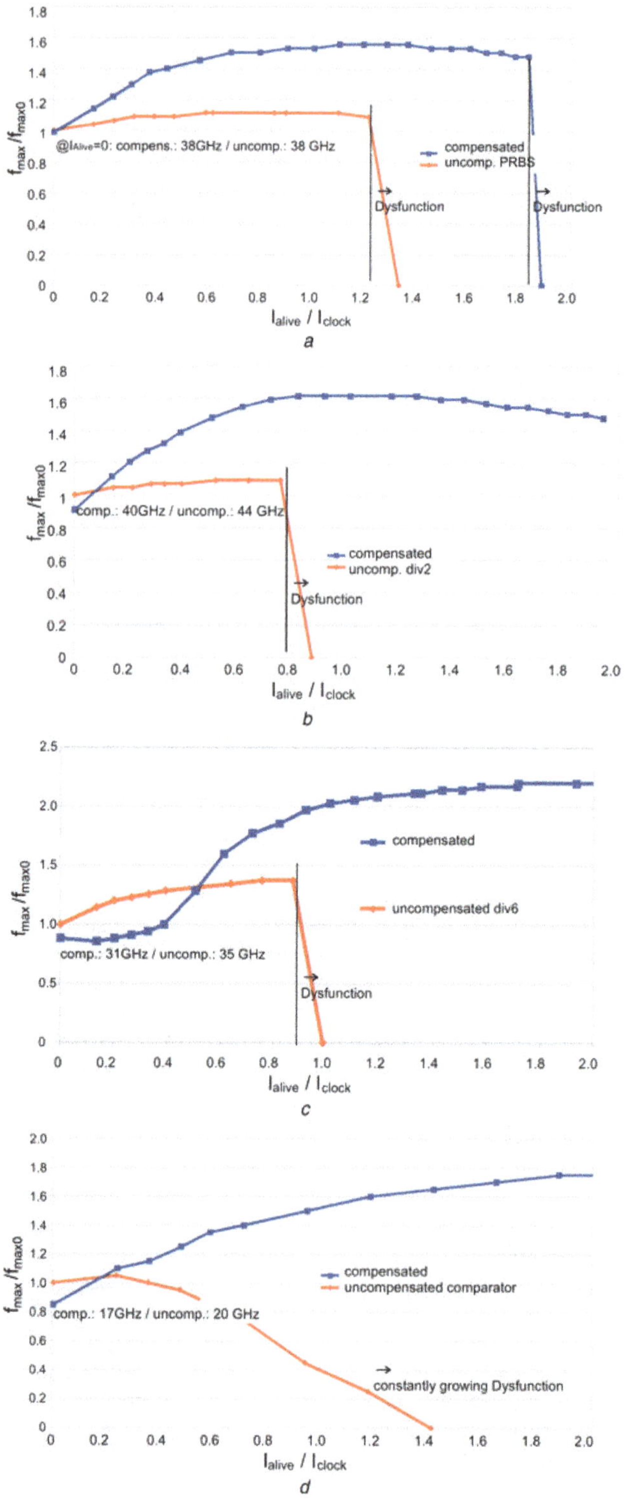

Fig. 3 *Simulated normalised maximum operational frequency against compensated and uncompensated keep-alive current for different latch applications*

a PRBS generator, 2^7–1 states (for $I_{alive} = 0$ as in [4])
b Simple frequency divider by two (one flip-flop)
c Johnson counter as frequency divider by six (three flip-flops)
d Clocked comparator, 2 mV input swing

been normalised to the case of $I_{alive} = 0$ and no compensation circuit. The absolute values for both the compensated and uncompensated latch are also given, for $I_{alive} = 0$. For small values of I_{alive}, the uncompensated case is superior, because of the additional parasitics of the compensation circuit. Although the latch function in the digital circuits ceases rather abruptly, it degrades continuously in the analogue comparator.

5 Conclusion

All investigated latch applications exhibited a much higher clock rate improvement of about 1.6–2.2 with increasing keep-alive current, compared with the uncompensated case, where the increase stays below a limit of 1.1-1.2. With the proposed compensation, a dysfunctional region was detectable at much higher values of I_{alive} or not visible at all. This way, a rather high a keep-alive current may serve as a helpful option to further raise the maximum clock rate, if power-neutral techniques (e.g. inductive peaking) have reached their limits.

6 References

[1] Collins T.E., Manan V., Long S.I.: 'Design analysis and circuit enhancements for high-speed bipolar flip-flops', *IEEE J. Solid-State Circuits*, 2005, **40**, (5), pp. 1166–1174

[2] Rylyakov A.: 'A 51 GHz master–slave latch and static frequency divider in 0.18 μm SiGe BiCMOS'. Proc. of the Bipolar/BiCMOS Circuits and Technology Meeting, 2003, pp. 75–77, doi: 10.1109/BIPOL.2003.1274939

[3] Heinemann B., Rücker H., Barth R., *et al.*: 'Novel collector design for high-speed SiGe:C HBTs'. Int. Electron Devices Meeting, IEDM'02, 2002, pp. 775–778, doi: 10.1109/IEDM.2002.1175953

[4] Gustat H., Borngräber J.: 'NOR/OR register based ECL circuits for maximum data rate'. Proc. of the Bipolar/BiCMOS Circuits and Technology Meeting, 2005, pp. 90–93, doi: 10.1109/BIPOL.2005.1555207

20

Q-switched thulium-doped fibre laser operating at 1900 nm using multi-walled carbon nanotubes saturable absorber

Norazlina Saidin[1,2,3], Dess I.M. Zen[1,4], Fauzan Ahmad[1,2], Hazlihan Haris[1], Muhammad T. Ahmad[1,5], Anas A. Latiff[5], Harith Ahmad[1], Kaharudin Dimyati[4], Sulaiman W. Harun[1,2]

[1]Photonics Research Centre, University of Malaya, 50603 Kuala Lumpur, Malaysia
[2]Department of Electrical Engineering, Faculty of Engineering, University of Malaya, 50603 Kuala Lumpur, Malaysia
[3]Department of Electrical and Computer Engineering, International Islamic University Malaysia, Jalan Gombak, 53100 Kuala Lumpur, Malaysia
[4]Department of Electrical and Electronic Engineering, National Defense University of Malaysia, Kem Sungai Besi, 57000 Kuala Lumpur, Malaysia
[5]Faculty of Electronic and Computer Engineering, Universiti Teknikal Malaysia Melaka, 76100 Durian Tunggal, Melaka, Malaysia
E-mail: swharun@um.edu.my

Abstract: Simple, low-cost and stable passive Q-switched thulium-doped fibre lasers (TDFLs) operating at 1892.4 and 1910.8 nm are demonstrated using 802 and 1552 nm pumping schemes, respectively, in conjunction with a multi-walled carbon nanotubes (MWCNTs) saturable absorber (SA). The MWCNTs composite is prepared by mixing the MWCNTs homogeneous solution into a dilute polyvinyl alcohol (PVA) polymer solution before it is left to dry at room temperature to produce thin film. Then the film is sandwiched between two FC/PC fibre connectors and integrated into the laser cavity for Q-switching pulse generation. The pulse repetition rate of the TDFL configured with 802 nm pump can be tuned from 3.8 to 4.6 kHz, whereas the corresponding pulse width reduces from 22.1 to 18.3 μs as the pump power is increased from 187.3 to 194.2 mW. On the other hand, with 1552 nm pumping, the TDFL generates optical pulse train with a repetition rate ranging from 13.1 to 21.7 kHz with a pulse width of 11.5–7.9 μs when the pump power is tuned from 302.2 to 382.1 mW. A higher performance Q-switched TDFL is expected to be achieved with the optimisation of the MWCNT-SA saturable absorber and laser cavity.

1 Introduction

Q-switched fibre-based laser systems operating in the 'eye-safe' wavelength of 1900 nm region are promising for applications such as light detection and ranging (lidar), differential absorption lidar and as pumps for mid-IR generation. They can be realised using thulium-doped or holmium-doped fibre lasers based on either active or passive methods [1–3]. Compared with the active ones, passively Q-switched fibre lasers feature flexibility of configuration and do not require additional switching electronics. These lasers have been successfully demonstrated using different kinds of saturable absorbers (SAs), such as semiconductor SA mirrors (SESAMs) [3, 4] and single-wall carbon nanotubes (SWCNTs) [5, 6]. However, SESAMs are still expensive and complex to be fabricated. Therefore, more focus has been given on the utilisation of SWCNTs as an SA in recent years especially for all-fibre Q-switched and mode-locked fiber lasers [7–9]. This is because of their inherent advantages, including good compatibility with optical fibers, low saturation intensity, fast recovery time and wide operating bandwidth.

Recently, a new member of carbon nanotubes family, multi-walled carbon nanotubes (MWCNTs) [10, 11] have also attracted many attentions for non-linear optics applications because of their production cost, which is about 50–20% of that of SWCNT material [12]. The growth of the MWCNT material does not need complicated techniques or special growing conditions so that its production yield is high for each growth. Compared with SWCNTs, the MWCNTs have higher mechanical strength, better thermal stability as well as can absorb more photons per nanotube because of its higher mass density of the multi-walls. These favourable features are because of the structure of MWCNTs which take the form of a stack of concentrically rolled graphene sheets. The outer walls can protect the inner walls from damage or oxidation so that the thermal or laser damage threshold of MWCNT is higher than that of the SWCNTs [13, 14]. To date, there are only a few reported works on application of MWCNTs material as an SA. For instance, Zhang *et al.* [12] employs multi-walled MWCNTs based SA for mode locking of an Nd:YVO_4 laser. In another work, Q-switched Nd-YAG laser is demonstrated using the MWCNTs based SA as a Q-switcher.

In this paper, an all-fibre Q-switched thulium-doped fibre laser (TDFL) is demonstrated using a simple and low-cost newly developed MWCNTs based SA as the Q-switcher. The SA employs MWCNTs-polyvinyl alcohol (PVA) film, which is fabricated by mixing a dispersed MWCNTs suspension into a PVA solution. The SA is integrated in the TDFL by sandwiching the film between two fibre connectors that results in a stable pulse train with 4.6 kHz repetition rate, 18.3 μs pulse width and 126.1 nJ pulse energy at 194.2 mW 802 nm pump power. The performance of the TDFL is also investigated with 1552 nm pumping, whereby a stable pulse train with 21.7 kHz repetition rate, 7.9 μs pulse width and 103.4 nJ pulse energy at 382.1 mW pump power.

2 Fabrication and Raman characterisation of MWCNT-PVA film

The MWCNTs material used for the fabrication of the absorber in this experiment is functionalised so that it can be dissolved in water. The diameter of the MWCNTs used is about 10–20 nm and the length distribution is from 1 to 2 μm. The functionaliser solution was prepared by dissolving 4 g of sodium dodecyl sulphate in 400 ml deionised water. About 250 mg MWCNT was added to the solution and the homogenous dispersion of MWCNTs was achieved after the mixed solution was sonicated for 60 min at 50 W. The solution was then centrifuged at 1000 rpm to remove large particles of undispersed MWCNTs to obtain dispersed suspension that is stable

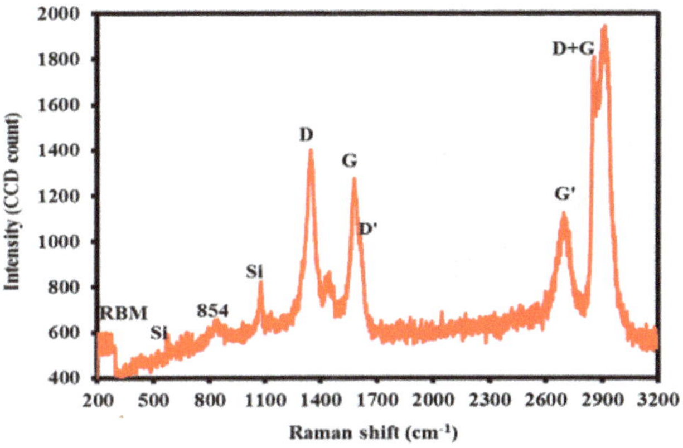

Fig. 1 *Raman spectrum obtained from the MWCNTs-PVA film*

for weeks. MWCNTs-PVA composite was prepared by adding the dispersed MWCNTs suspension into a PVA solution by a three to two ratio. Solution was prepared by dissolving 1 g of PVA (M_w = 89 × 10^3 g/mol) in 120 ml of deionised water. The homogeneous MWCNTs-PVA composite was obtained by a sonification process for more than one hour. The composite was casted onto a glass petri dish and left to dry at room temperature for about one week to produce thin film with a thickness of around 50 μm.

Raman spectroscopy was performed on the prepared MWCNT-PVA film using laser excitation at 532 nm to confirm the presence of the carbon nanotubes. Fig. 1 shows the Raman spectrum, obviously indicating the distinct feature of the MWCNT. It is shown that the Raman spectrum bears a lot of similarity to graphene, which is not too surprising as it is simply a rolled up sheet of graphene. MWCNT has many layers of graphene wrapped around the core tube. We can see well defined G (1580 cm^{-1}) and G' (2705 cm^{-1}) bands in Fig. 1 as there were in graphene and graphite. The G-band originates from in-plane tangential stretching of the carbon–carbon bonds in graphene sheet. We also see a prominent band around 1350 cm^{-1}, which is known as the D-band. The D-band originates from a hybridised vibrational mode associated with graphene edges and it indicates the presence of some disorder to the graphene structure. This band is often referred to as the disorder band or the defect band and its intensity relative to that of the G-band is often used as a measure of the quality with nanotubes. There is another series of bands appearing at the low-frequency end of the spectrum known as radial breathing mode bands. These bands correspond to the expansion and contraction of the tubes and are not clearly present in the MWCNT because of the outer tubes, which restrict the breathing mode. As expected, the prominent D-band is observed in Fig. 1, indicating the nanotubes as a multi-walled type, which has multi-layer configuration

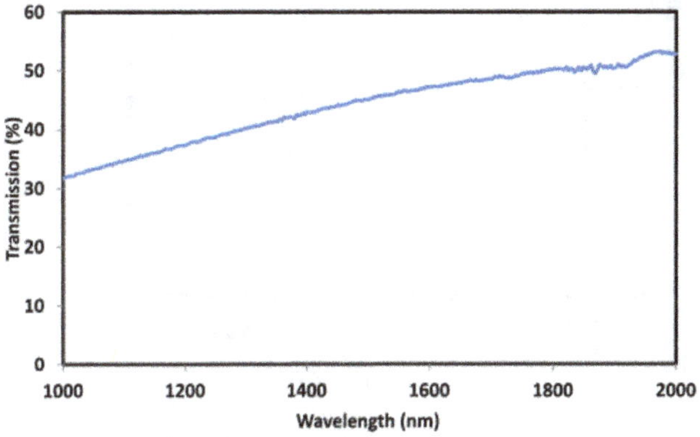

Fig. 2 *Transmission spectrum of the MWCNTs-PVA film*

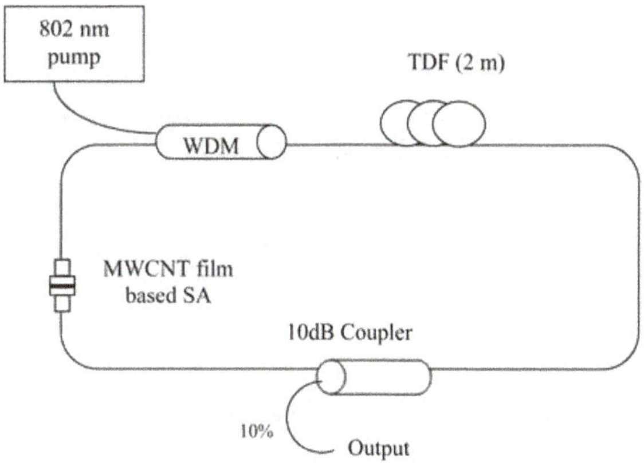

Fig. 3 *Schematic configuration of the Q-switched TDFL with 802 nm pumping*

and disorder structure. The D'-band which is a weak shoulder of the G-band is also observed at 1613 cm^{-1} because of double resonance feature induced by disorder and defect. In addition, other distinguishable features like D + G band (2920 cm^{-1}), a small peak at 854 cm^{-1} and Si were also observed as depicted in Fig. 1. Fig. 2 shows the transmission spectrum of the fabricated MWCNTs-PVA film. As shown in the figure, the transmission is featureless in the near infra-red region, showing the broadband property of our SA. The transmittance is observed to be about 50.8% at 1900 nm region.

3 Configuration of the laser

The schematic of the proposed Q-switched TDFL is shown in Fig. 3. It was constructed using a simple ring cavity, in which a 2 m long TDF with absorption of 27 dB/m at 785 nm was used for the active medium and the fabricated MWCNT-based SA was used as a Q-switcher. The SA is fabricated by cutting a small part of the prepared MWCNTs-PVA film and sandwiching it between two ferrule connector/physical contact (FC/PC) fibre connectors, after depositing an index-matching gel onto the fibre ends. Figs. 4*a* and *b* show the image of the film attached onto a fibre ferule and the constructed SA, respectively. The insertion loss of the SA is measured to be around 3.3 dB at 1900 nm. The thalium-doped fibre (TDF) was pumped by an 802 nm laser diode via an 800/2000 nm wavelength division multiplexer (WDM). The temporal characteristics of the laser output were monitored using a combination of a photo-detector and a real-time oscilloscope. The optical spectrum was measured using an optical spectrum analyser. The cavity length is measured to be ~7.6 m. The performance of the Q-switched TDFL is also investigated for 1552 nm pumping. In the experiment, 802 nm laser diode and 800/2000 nm WDM are replaced with 1552 nm pump and 1550/1900 nm WDM, whereas the TDF length is increased to 5 m for optimum laser performance.

4 Q-switching performance with 802 nm pumping

Fig. 5 shows the output power of both Q-switched and continuous wave (CW) lasers against the input pump power, which are obtained with and without the SA, respectively. Without the SA, a CW laser operates with an efficiency of 3.77% and threshold pump power of 133.1 mW. The efficiency is relatively low since the components used have a considerably high insertion loss at 1900 nm region. As the SA is inserted into the ring cavity, a stable and self-starting Q-switching operation is obtained just by adjusting the pump power over a threshold of 187.3 mW. However, the efficiency of the laser is slightly reduced to 2.68% because of the increased cavity loss. Fig. 6 shows the output spectrum of the TDFL with and without SA at the pump power

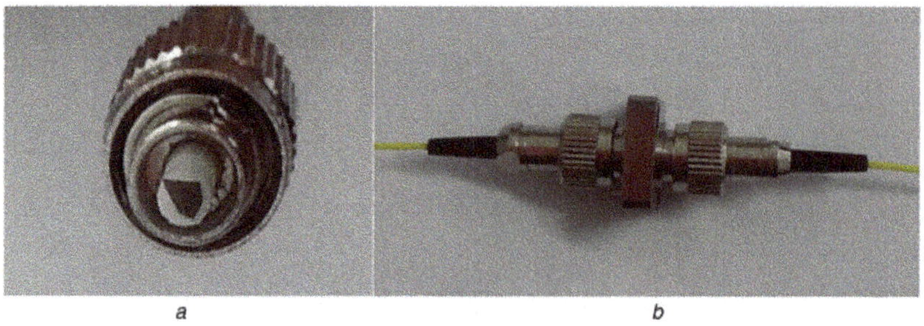

Fig. 4 *MWCNTs-PVA film-based SA*
a Attachment of the film on the fibre ferrule
b Integration of MWCNT composite film in laser cavity

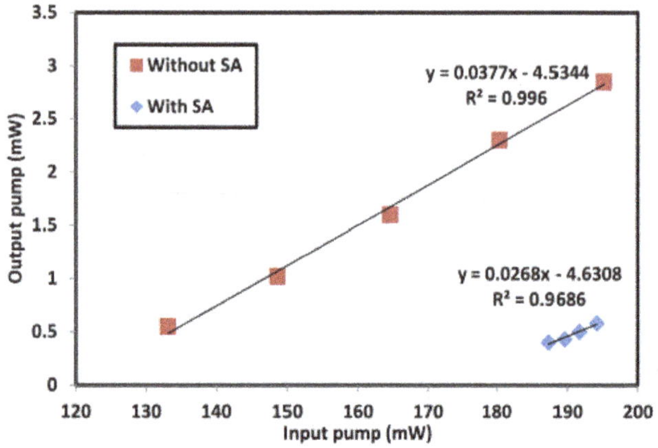

Fig. 5 *Output power characteristic against the pump power with and without the SA*

threshold of 187.3 mW. As can be seen from the figure, the Q-switched laser operates at a wavelength of 1892.4 nm with an optical signal-to-noise ratio (SNR) of more than 30 dB. Compared with the CW laser (without SA), the operating wavelength of the Q-switched laser has shifted to a shorter wavelength. This is attributed to the cavity loss which increases with the incorporation of SA. Thus, the oscillating light in the cavity shifts to a shorter wavelength, which is closer to the peak absorption of the TDF at around 1800 nm to compensate for the loss. The spectrum bandwidth is also broadened in the Q-switched laser because of the self-phase modulation effect in the ring cavity.

Fig. 7 shows the oscilloscope trace of the Q-switched pulse train at the pump power of 191.7 mW. As shown in the figure, there is no

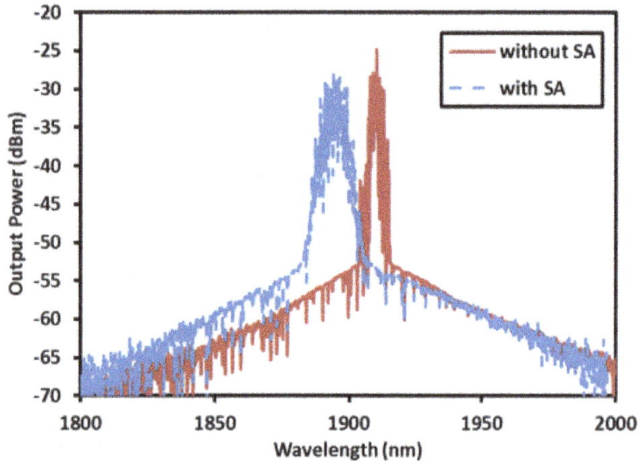

Fig. 6 *Output spectrum of the ring TDFL with and without the SA*

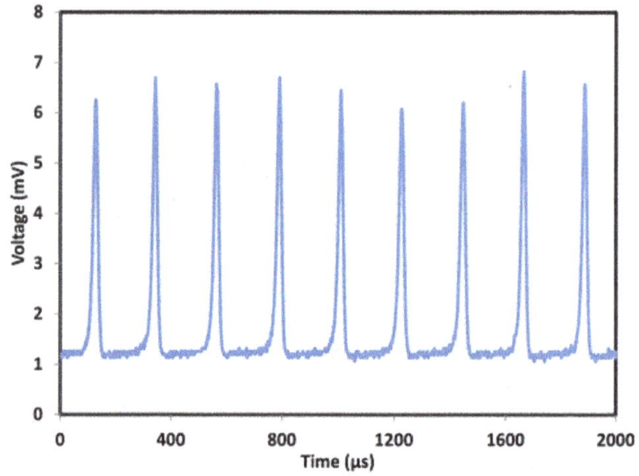

Fig. 7 *Q-switching pulse train at the pump power of 191.7 mW*

distinct amplitude modulation in each Q-switched envelope spectrum, which means that the self-mode locking effect on the Q-switching is weak. At this pump power, the proposed TDFL generates a stable Q-switching pulse with an average output power of 0.5 mW and repetition rate of 4.5 kHz. The pulse energy is calculated to be around 111.1 nJ at this pump power. The pulse energy

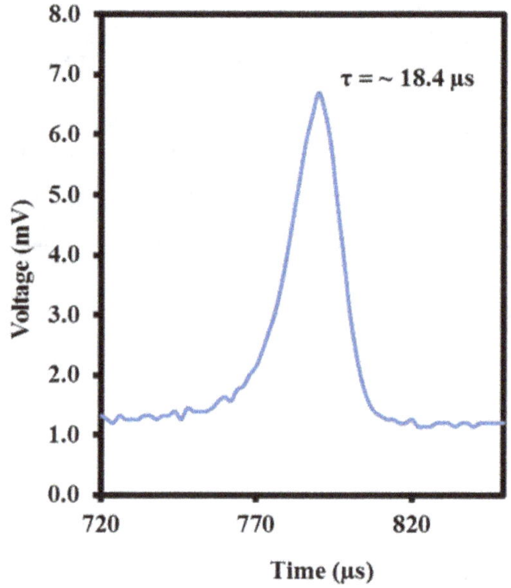

Fig. 8 *Pulse envelope of the Q-switched laser at the pump power of 191.7 mW*

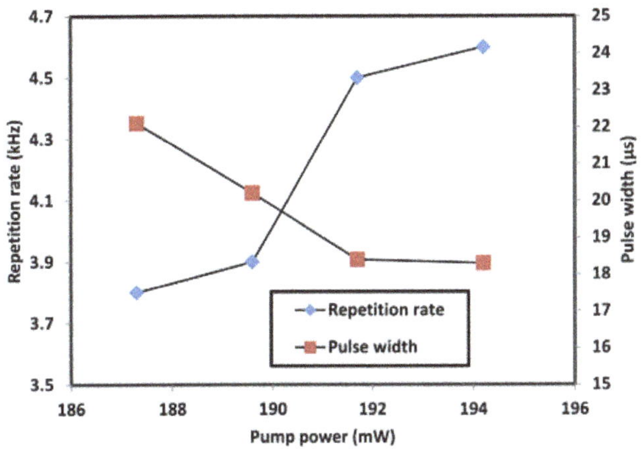

Fig. 9 *Repetition rate and pulse width as a function of pump power*

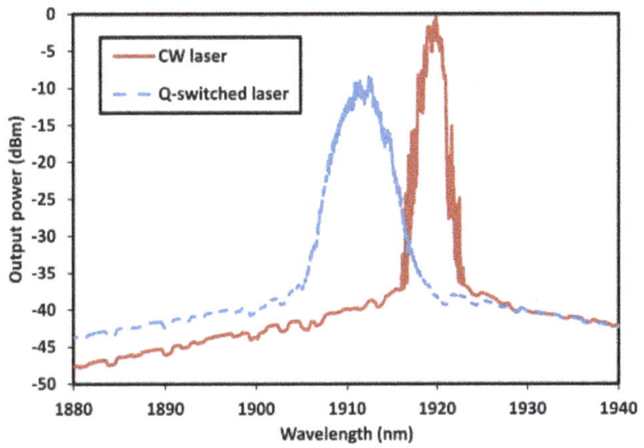

Fig. 11 *Optical spectra of the TDFL with CW and Q-switching modes of operation*

could be improved by reducing the insertion loss of the SA and optimising the laser cavity. Fig. 8 shows the typical oscilloscope trace of the pulse envelop at the pump power of 191.7 mW. As seen in the figure, the full-width at half-maximum (FWHM) or pulse width was obtained at 18.4 μs.

Fig. 9 shows the repetition rate and the pulse width of the proposed Q-switched TDFL against the pump power. The repetition rate has a monotonically increasing, near-linear relationship with the pump power level, which is consistent with other reported results of the SWCNT-based fibre lasers [8]. When the pump power is tuned from 187.3 to 194.2 mW, the pulse train repetition rate varies from 3.8 to 4.6 kHz. On the other hand, the pulse width is inversely proportional to the pump power, where the pulse duration becomes shorter as the pump power increases. The shortest pulse width of 18.3 μs is achieved at the maximum pump power of 194.2 mW. The pulse width is expected to decrease further if the pump power can be augmented beyond 194.2 mW as long as it is still kept below the damage threshold of the MWCNT-PVA-based SA. Shortening the total cavity length of the fibre laser is another alternative to obtain a shorter pulse. Fig. 10 shows the output power and pulse energy as functions of pump power. It is found that both output pump power and pulse energy increase with the pump power. At the maximum pump power of 194.2 mW, the average pump power and pulse energy of the Q-switched laser are obtained at 0.58 mW and 126.1 nJ.

5 Q-switching performance with 1552 nm pumping

Aside from 802 nm pumping, the TDFL can also be pumped by 1552 nm light to create a population inversion between $^3F_4$ and $^3H_6$ energy levels and generates laser at 1900 nm region. Here, the performance of the Q-switched TDFL is investigated using a 1552 nm pump as the pump source based on the similar setup of Fig. 3. The total cavity length is measured to be around 11.6 m because of the increment of both TDF and WDM fibre lengths. Without the SA in the cavity, the TDFL starts to operate in CW mode at threshold pump power of 256 mW. As the SA is incorporated into the cavity, a stable self-started Q-switching pulse train is obtained at the slightly higher pump power of 302.2 nm than the CW operation. Fig. 11 compares the optical spectrum of the Q-switched TDFL with the CW one, which was obtained by removing the SA. Both lasers operate at a longer wavelength compared with the previous TDFL configured with 802 nm pumping because of the longer TDF length used. The longer TDF provides a higher population inversion and thus shifts the operating wavelength to a longer wavelength region. The Q-switched TDFL operates at wavelength of 1910.8 nm, which is slightly shorter than the CW laser owing to the insertion loss in the SA. It has a broad FWHM of 3.0 nm because of the self-phase modulation (SPM) effect in the ring cavity and SNR of ~30 dB.

Fig. 12 shows the oscilloscope trace of the typical Q-switched pulse train at 1552 nm pump power of 382.1 mW. It is observed that the Q-switching operation is stable for the TDFL where neither amplitude variation nor timing jitter was notable in the pulse train. The spacing between two pulses in Fig. 12 is measured to be around 46.0 μs, which can be translated to repetition rate of 21.7 kHz. At this pump power, the pulse width and average pump power were measured to be 7.9 μs and 2.2 mW, respectively, and thus the pulse energy is calculated to be 103.4 nJ. Fig. 13 shows how repetition rate and pulse width are related to the pump power

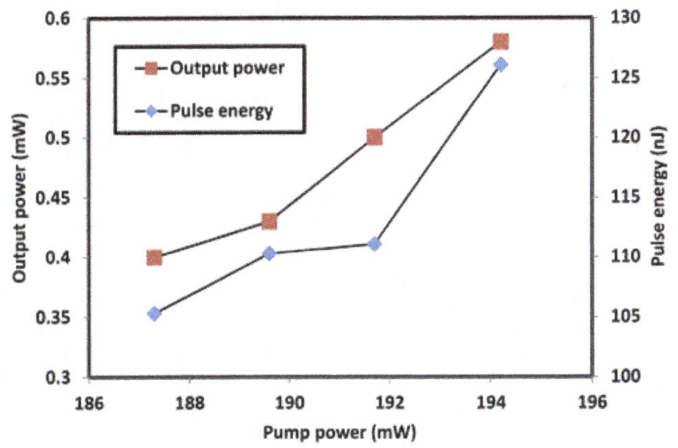

Fig. 10 *Average output power and pulse energy as a function of pump power*

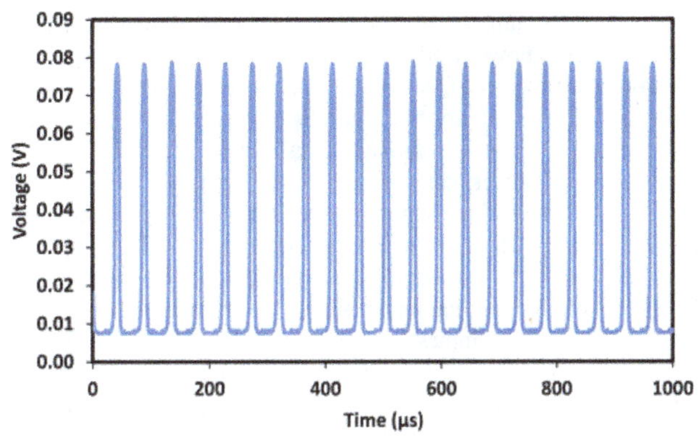

Fig. 12 *Typical pulse train for the proposed TDFL at 1552 nm pump power of 382.1 mW*

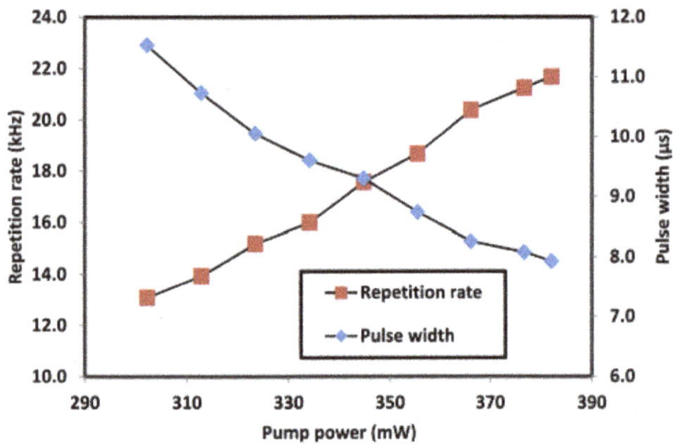

Fig. 13 *Repetition rate and pulse width as a function of 1552 nm pump power*

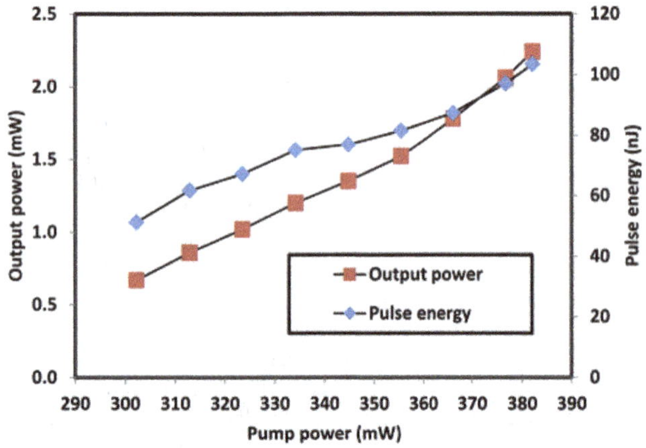

Fig. 14 *Average output power and pulse energy as a function of 1552 nm pump power*

for the TDFL. In the laser, the dependence of the pulse repetition rate can be seen to increase almost linearly with the pump power, whereas the pulse width decreases also almost linearly with the pump power. This agrees well with the previous result on 802 nm pumping. By varying the pump power from 302.2 to 382.1 mW, the pulse repetition rate of the Q-switched TDFL configured with 1552 nm pumping can be widely tuned from 13.1 to 21.7 kHz, whereas the corresponding pulse width reduces from 11.5 to 7.9 µs. The Q-switching pulse becomes unstable and disappeared as the pump power increases above 382.1 mW.

Fig. 14 shows the average output power and pulse energy of the proposed Q-switched TDFL as a function of 1552 nm pump power. It is observed that both output power and pulse energy increase with the increment of the pump power from 302.2 to 382.1 mW. Further increasing the pump power leads to randomising the pulses and the pulse train of the passively Q-switched laser became unstable and strong amplitude variation appeared. It is predicted that the fluctuation is caused by the lower damage threshold of the MWCNT under higher pump power, considering the fact that the laser will completely stop lasing if the pump power is further increased. It is also observed that the Q-switching operation of the laser can be resumed back as the pump power is reduced within 302.2–382.1 mW. At the maximum pump power of 382.1 mW, the TDFL produces the highest average output power and pulse energy of 2.2 mW and 103.4 nJ, respectively. These results indicate that both 802 and 1552 nm pumping schemes can be used for generating Q-switching pulse train in the TDFL. It is also shown that the MWCNTs-PVA SA has a big potential for superior Q-switching and saturable absorption compared with conventional light absorbing components when carefully employed in an appropriate laser system. The fabrication of the SA is also simple and thus the cost of the laser should be low. The simple and low cost laser is suitable for applications in metrology, environmental sensing and biomedical diagnostics.

6 Conclusion

A stable passive Q-switched TDFL operating at 1900 nm region is successfully demonstrated using 802 and 1552 nm pumping schemes, in conjunction with MWCNTs-SA. The SA employs MWCNT film, which is fabricated by mixing the MWCNTs homogeneous solution into a PVA polymer solution. With 1552 nm pumping, the average output power, repetition rate and pulse width are obtained at 2.2 mW, 21.7 kHz and 7.9 µs, respectively, when the pump power is fixed at 382.1 mW. As compared with the 802 nm pumping, these performances are better, which can be accounted from the higher pump power applied to the TDF. The Q-switched TDFL configured by 802 nm pump gives the pulse width, repetition rate and output power of 18.3 µs, 4.6 kHz and 0.58 mW, respectively, when the pump power is 194.2 mW. The highest pulse energy of 126.1 nJ is also obtained at this pump power.

7 Acknowledgment

This work is financially supported by Ministry of Education and University of Malaya under various grant schemes (Grant Numbers: PG139-2012B, ER012-2012A, RP008D-13AET and RP008C-13AET).

8 References

[1] El-Sherif A.F., King T.A.: 'High-energy, high-brightness Q-switched Tm^{3+}-doped fiber laser using an electro-optic modulator', *Opt. Commun.*, 2003, **218**, (4–6), pp. 337–344

[2] Jiang M., Ma H.F., Ren Z.Y., ET AL.: 'A graphene Q-switched nanosecond Tm-doped fiber laser at 2 µm', *Laser Phys. Lett.*, 2013, **10**, (5), p. 055103

[3] Wang Q., Geng J., Jiang Z., Luo T., Jiang S.: 'Mode-locked Tm–Ho-codoped fiber laser at 2.06 µm', *IEEE Photonics Technol. Lett.*, 2011, **23**, pp. 682–684

[4] Kivistö S., Koskinen R., Paajaste J., Jackson S.D., Guina M., Okhotnikov O.G.: 'Passively Q-switched Tm3+, Ho3+-doped silica fiber laser using a highly nonlinear saturable absorber and dynamic gain pulse compression', *Opt. Express*, 2008, **16**, (26), p. 22058

[5] Solodyankin M.A., Obraztsova E.D., Lobach A.S., ET AL.: 'Mode-locked 1.93 µm thulium fiber laser with a carbon nanotube absorber', *Opt. Lett.*, 2008, **33**, (12), pp. 1336–1338

[6] Kieu K., Wise F.: 'Soliton thulium-doped fiber laser with carbon nanotube saturable absorber', *IEEE Photonics Technol. Lett.*, 2009, **21**, (3), pp. 128–130

[7] Hasan T., Sun Z., Wang F., ET AL.: 'Nanotube–polymer composites for ultrafast photonics', *Adv. Mater.*, 2009, **21**, (38-39), pp. 3874–3899

[8] Ahmad F., Harun S., Nor R., Zulkepely N., Ahmad H., Shum P.: 'A passively mode-locked erbium-doped fiber laser based on a single-wall carbon nanotube polymer', *Chin. Phys. Lett.*, 2013, **30**, (5), p. 054210

[9] Harun S.W., Saidin N., Zen D.I.M., ET AL.: 'Self-starting harmonic mode-locked thulium-doped fiber laser with carbon nanotubes saturable absorber', *Chin. Phys. Lett.*, 2013, **30**, (9), p. 094204, doi: i10.1088/0256–307x/30/9/094204

[10] Costa S., Borowiak-Palen E., Kruszynska M., Bachmatiuk A., Kalenczuk R.: 'Characterization of carbon nanotubes by Raman spectroscopy', *Mater. Sci. Poland*, 2008, **26**, (2), pp. 433–441

[11] Dresselhaus M.S., Dresselhaus G., Saito R., Jorio A.: 'Raman spectroscopy of carbon nanotubes', *Phys. Rep.*, 2005, **409**, (2), pp. 47–99

[12] Zhang L., Wang Y.G., Yu H.J., ET AL.: 'Passive mode-locked Nd: YVO4 laser using a multi-walled carbon nanotube saturable absorber', *Laser Phys.*, 2011, **21**, pp. 1382–1386

[13] Ramadurai K., Cromer C.L., Lewis L.A., ET AL.: 'High-performance carbon nanotube coatings for high-power laser radiometry', *J. Appl. Phys.*, 2008, **103**, p. 013103

[14] Banhart F.: 'Irradiation effects in carbon nanostructures', *Rep. Prog. Phys.*, 1999, **62**, p. 1181

21

Energy detection UWB system based on pulse width modulation

Song Cui, Fuqin Xiong

Department of Electrical and Computer Engineering, Cleveland State University, 2121 Euclid Avenue, Cleveland, OH 44115, USA
E-mail: s.cui99@csuohio.edu

Abstract: A new energy detection ultra-wideband system based on pulse width modulation is proposed. The bit error rate (BER) performance of this new system is slightly worst than that of a pulse position modulation (PPM) system in additive white Gaussian noise channels. In multipath channels, this system does not suffer from cross-modulation interference as PPM, so it can achieve better BER performance than PPM when cross-modulation interference occurs. In addition, when synchronisation errors occur, this system is more robust than PPM.

1 Introduction

Ultra-wideband (UWB) impulse radio has received more and more attention as a promising technology for many applications, such as short-range and high-speed wireless Internet, covert communications, ground penetrating radar, through-wall image and localisation [1]. In UWB systems, the data are carried by sub-nanosecond pulses and each of these short pulses will generate a large number of multipath components in multipath channels. These multipath components have fine time resolution, so they can be resolved and combined in a Rake receiver [2]. However, the Rake receiver needs a large number of fingers to capture enough signal energy to demodulate the received signals, so this leads to a very complex structure of the receiver and greatly increase the computational burden of channel estimation [2, 3]. In a Rake receiver, each finger includes a correlator and these correlators need extremely accurate synchronisation to align the received signals with the template signals to perform correlation. The acceptable synchronisation error is much smaller than one pulse duration, and a small synchronisation error can severely degrade the system performance [3].

Non-coherent technologies have been developed to avoid the challenges in Rake receivers. Energy detection (ED) has been a conventional non-coherent technology in the communication field for many years. In recent years, ED is applied to UWB systems for on-off keying (OOK) [4–7] and pulse position modulation (PPM) systems [5, 6, 8]. Although ED is a sub-optimal technology, it has many advantages. Its receiver structure is very simple and channel estimation is not required. In addition, ED does not need as accurate synchronisation as a Rake receiver because ED does not use correlators. Now, ED is attracting more and more researchers in the field of UWB.

In this paper, a new ED receiver based on pulse width modulation (PWM) is proposed. Although the popular modulation methods in UWB are PPM and pulse amplitude modulation, PWM has been proved to be a suitable scheme for UWB systems [9]. The receiver in [9] uses a Rake receiver, so we develop the ED receiver in this paper. The bit error rate (BER) performance of this new ED PWM system is slightly worst than that of an ED PPM system in additive white Gaussian noise (AWGN) channels. However, the BER performance of PWM can surpass that of PPM in multipath channels since PWM does not suffer cross-modulation interference (CMI) as PPM. In addition, PWM is more robust to synchronisation errors than PPM, and the BER performance of PWM can be better than PPM when synchronisation errors occur.

The left of this paper is organised as follows. Section 2 introduces the system model. Section 3 analyses the system performance in AWGN channel. Section 4 analyses the system performance in multipath channel. Section 5 analyses the system performance in the presence of synchronisation errors. In Section 6, the numerical results are analysed. Section 7 is the conclusion of this paper.

2 System model

2.1 *System model of PWM*

In a PWM system, the modulation is achieved by transmitting pulses with different widths to denote bit 0 or 1. The model of the transmitter in [9] is used here. However, we only research the case of single-user communication, and a bit is assumed to transmit only once. Therefore the equation of the transmitter in [9] is simplified to

$$s_{\mathrm{PWM}}(t) = \sum_j \sqrt{E_{\mathrm{p}}} p_{\mathrm{b}_j}(t - \mathrm{j}T_{\mathrm{f}}) \tag{1}$$

where T_{f} denotes the frame period and $p_{\mathrm{b}_j}(t)$ denotes the pulse waveform for *j*th transmitted data bit b_j. The data bit has a binary value of either 0 or 1. When bit 0 or 1 is transmitted, the transmitted pulse waveform is $p_0(t)$ or $p_1(t)$, respectively, where $p_0(t)$ and $p_1(t)$ are amplitude-normalised pulse waveforms with different widths. The pulse energy is adjusted by E_{p} and the energies of $p_0(t)$ and $p_1(t)$ are different. Therefore we define the energies of the pulses for bits 0 and 1, E_0 and E_1, by $E_i = E_{\mathrm{p}} \int_{-T_{\mathrm{p}_i}/2}^{T_{\mathrm{p}_i}/2} [p_i(t)]^2 \, \mathrm{d}t$, $(i = 0,\ 1)$, where T_{p_i} is the pulse width. The ratio E_0/E_1 depends on which pulse waveforms are chosen [9]. In this paper, we use the second-order derivative of the Gaussian pulse [10]

$$p(t) = (1 - 4\pi t^2/\alpha_i^2)\exp(-2\pi t^2/\alpha_i^2) \tag{2}$$

where α_i is the shape factor. The pulse width T_{p_i} is set to $2.4\alpha_i$ and the detailed method to choose the pulse width for a specific α_i value can be found in [10]. After the second-order derivative of the Gaussian pulse is chosen, the modulation is achieved by using different α_i values for bits 0 and 1. Increasing α_i will increase T_{p_i} and thus decrease the bandwidth. We choose $\alpha_1 = 2\alpha_0$ to achieve modulation, where α_1 and α_0 are the shape factors for $p_1(t)$ and $p_0(t)$, respectively. Therefore we have $T_{\mathrm{p}_1} = 2T_{\mathrm{p}_0}$, where T_{p_1} and T_{p_0} denote the width of $p_1(t)$ and $p_0(t)$, respectively. Based on these assignments, the relationship between E_0 and E_1 is

$$E_0 = 0.5E_1 \tag{3}$$

A simple proof of (3) is given as follows: the values of E_0 and E_1 are $E_0 = E_p \int_{-1.2\alpha_0}^{1.2\alpha_0} (1 - 4\pi t^2/\alpha_0^2)^2 e^{-4\pi t^2/\alpha_0^2} dt$ and $E_1 = E_p \int_{-1.2\alpha_1}^{1.2\alpha_1} (1 - 4\pi t^2/\alpha_1^2)^2 e^{-4\pi t^2/\alpha_1^2} dt$, respectively. Since $\alpha_1 = 2\alpha_0$, we let $\tau = t/2$. And then it is straightforward to use the new variable τ to convert E_1 to $E_1 = 2E_p \int_{-1.2\alpha_0}^{1.2\alpha_0} (1 - 4\pi\tau^2/\alpha_0^2)^2 e^{-4\pi\tau^2/\alpha_0^2} d\tau = 2E_0$.

The design idea of the receiver originates from the spectral characteristics of the Gaussian pulse. The Fourier transform X_f and centre frequency f_c of the kth-order derivative of the Gaussian pulse are [10]

$$X_f \propto f^k \exp(-\pi f^2 \alpha_i^2/2) \tag{4}$$

$$f_c = \sqrt{k}/(\alpha_i\sqrt{\pi}) \tag{5}$$

where f is the frequency. Using (4), the spectra of $p_0(t)$ and $p_1(t)$ are plotted in Fig. 1. The bandwidth of $p_0(t)$ is almost twice that of $p_1(t)$. The centre frequency of $p_0(t)$ is f_{c0}. Two ideal filters, Filters 1 and 2, are also shown in Fig. 1. The passband of Filter 1 is $[0, f_{c0}]$ and that of Filter 2 is $[f_{c0}, 2f_{c0}]$. When $p_0(t)$ is transmitted, both Filters 1 and 2 pass about half of its energy. If we subtract the signal energies captured by Filters 1 and 2, the result is ~ 0. When $p_1(t)$ is transmitted, Filter 1 passes almost all of the energy of $p_1(t)$, but Filter 2 rejects the energy of $p_1(t)$. Therefore, if we subtract the energies, the result is $\sim E_1$. We can determine the transmitted bit is 0 or 1 by measuring the difference of signal energies captured by the two filters. This inspires us to design the receiver as in Fig. 2. To make sure that our design idea of the receiver is based on strict theoretical support rather than direct observable results, we use MAPLE software to perform numerical calculation to obtain the following equations

$$E_{01} - E_{02} \simeq 0 \tag{6}$$

$$E_{01} + E_{02} \simeq E_0 \tag{7}$$

$$E_{11} - E_{12} \simeq E_1 \tag{8}$$

$$E_{11} + E_{12} \simeq E_1 \tag{9}$$

where E_{01}, E_{02}, E_{11} and E_{12} denote the signal energies captured by the two filters. The first subscript denotes the transmitted bit is 0 or 1 and the second subscript means Filter 1 or 2. The detailed procedure to obtain these equations is given in Appendix. Equations (6) and (8) show the adequate theory support for our design idea.

The receiver in Fig. 2 has two branches, and each branch is a conventional ED receiver. The only difference between the two branches is the passbands of the filters. The passbands of Filters 1 and 2 are like that in Fig. 1. The signal arriving at the receiver is denoted by $s(t)$, the AWGN is denoted by $n(t)$, and the sum of $s(t)$ and $n(t)$ is denoted by $r(t)$. The integration interval $T \le T_f$. The decision statistic is given by $Z = Z_1 - Z_2$, where Z_1 and Z_2 are the outputs of branches 1 and 2, respectively. Finally, Z is compared with threshold γ to determine the transmitted bit. If $Z \ge \gamma$, the transmitted bit is 1, otherwise it is 0.

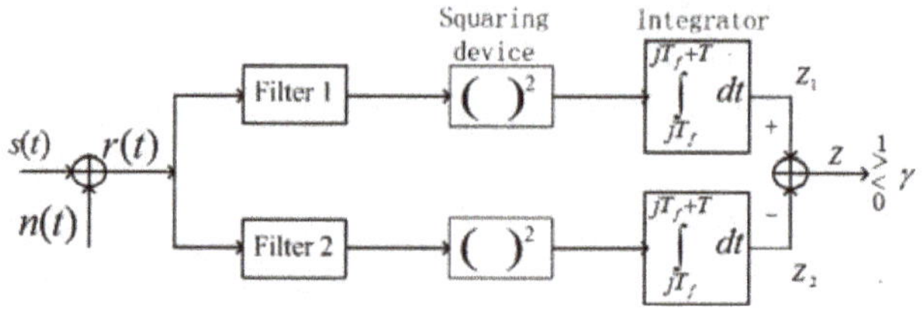

Fig. 2 *Receiver structure of an ED PWM system*

We will compare the performance of this new system with the existing ones in the following, and the models of these systems are simply stated as follows: the OOK system does not transmit a signal when data are 0, so it has difficulty to achieve synchronisation, especially when a stream of zeros is transmitted [11]. Therefore we only compare PPM with our new system in this paper. In Section 2.2, the system model of PPM is depicted.

2.2 System model of PPM

The transmitted signal of a PPM system is [12]

$$s(t)_{\text{PPM}} = \sum_j \sqrt{E_p} p(t - jT_f - \delta b_j) \tag{10}$$

where δ is called the modulation index and the pulse shift amount is determined by δb_j. The frame period is denoted by T_f, $p(t)$ is the pulse waveform with normalised-energy and E_p denotes signal energy. The receiver of an ED PPM system includes a conventional ED receiver, and the decision statistic Z is obtained as [13]

$$Z = Z_1 - Z_2 = \int_{jT_f}^{jT_f+T} r^2(t)\,dt - \int_{jT_f+\delta}^{jT_f+\delta+T} r^2(t)\,dt \tag{11}$$

where $T \le \delta$ denotes the length of integration interval. The decision threshold of PPM is $\gamma = 0$. If $Z \ge \gamma = 0$, the transmitted bit is 0, otherwise it is 1.

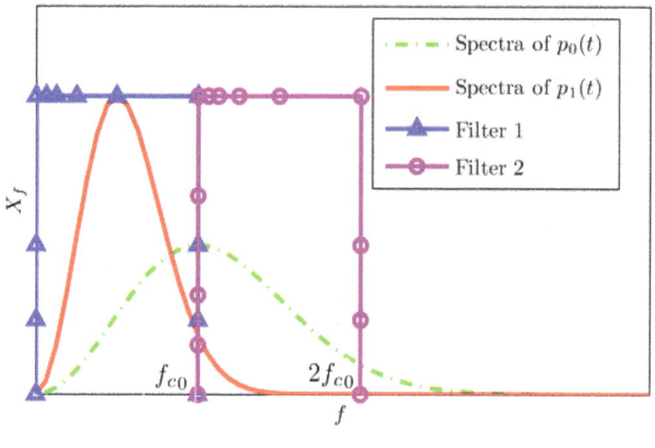

Fig. 1 *Spectral curves of $p_0(t)$ and $p_1(t)$ and the passbands of Filters 1 and 2*

3 BER performance in AWGN channels

3.1 PWM performance in AWGN channels

In Fig. 2, the outputs of conventional energy detectors, Z_1 and Z_2, are chi-square variables with approximately a degree of $2TW$ [14], where W denotes the bandwidth of the filters and T denotes the integration time. When the value of $2TW$ is large, the chi-square variable is approximated as a Gaussian variable. This popular method is called Gaussian approximation in ED communication systems [5, 13, 15–17]. The mean value and variance of this Gaussian variable are [18]

$$\mu = N_0TW + E \tag{12}$$

$$\sigma^2 = N_0^2TW + 2N_0E \tag{13}$$

where μ and σ^2 are the mean value and variance, and $N_0/2$ denotes the double-sided power spectral density of AWGN. The signal energy captured by the filter is denoted by E. If the spectrum of the signal does not fall in the passband of filter, then $E=0$. In Fig. 2, when bit 0 is transmitted, the signal energy almost distributes equally in the passbands of Filters 1 and 2. The probability density functions (pdfs) of Z_1 and Z_2 are $Z_1 \sim N(N_0TW + E_{01},\ N_0^2TW + 2N_0E_{01})$ and $Z_2 \sim N(N_0TW + E_{02},\ N_0^2TW + 2N_0E_{02})$. Since $Z = Z_1 - Z_2$, we have $Z \sim N(E_{01} - E_{02},\ 2N_0^2TW + 2N_0(E_{01} + E_{02}))$. Using (6) and (7), the pdf of Z is

$$H_0{:}Z \sim N(0,\ 2N_0^2TW + 2N_0E_0) \tag{14}$$

When bit 1 is transmitted, the signal energy almost passes through Filter 1 entirely. The pdf of Z_1 and Z_2 can be expressed as $Z_1 \sim N(N_0TW + E_{11},\ N_0^2TW + 2N_0E_{11})$ and $Z_2 \sim N(N_0TW + E_{12},\ N_0^2TW + 2N_0E_{12})$. The pdf of Z is $Z \sim N(E_{11} - E_{12},\ 2N_0^2TW + 2N_0(E_{11} + E_{12}))$. Using (8) and (9), the pdf of Z becomes

$$H_1{:}Z \sim N(E_1,\ 2N_0^2TW + 2N_0E_1) \tag{15}$$

Since E_0 and E_1 have different values, we will denote them using average bit energy E_b. Assuming bits 0 and 1 are randomly transmitted at the same probability, we obtain $E_b = (1/2)(E_0 + E_1)$ [5, 6]. From (3), we have $E_0 = 0.5E_1$, so E_0 and E_1 can be expressed as

$$E_0 = \frac{2}{3}E_b \tag{16}$$

$$E_1 = \frac{4}{3}E_b \tag{17}$$

Substituting (16) and (17) into (14) and (15), respectively, we obtain

$$H_0{:}Z \sim N\left(0,\ 2N_0^2TW + \frac{4}{3}N_0E_b\right) \tag{18}$$

$$H_1{:}Z \sim N\left(\frac{4}{3}E_b,\ 2N_0^2TW + \frac{8}{3}N_0E_b\right) \tag{19}$$

We follow the method in [15] to derive BER using (18) and (19). Firstly, we calculate the BER when bits 0 and 1 are transmitted as follows

$$P_0 = \int_\gamma^\infty f_0(x)\,dx = \int_\gamma^\infty \frac{1}{\sqrt{2\pi}\sigma_0} e^{-\left((x-\mu_0)^2/2\sigma_0^2\right)}\,dx \tag{20}$$

$$P_1 = \int_{-\infty}^\gamma f_1(x)\,dx = \int_{-\infty}^\gamma \frac{1}{\sqrt{2\pi}\sigma_1} e^{-\left((x-\mu_1)^2/2\sigma_1^2\right)}\,dx \tag{21}$$

where $f_0(x)$ and $f_1(x)$ denote the pdfs of Z when bits 0 and 1 are transmitted, respectively, and γ denotes the decision threshold. From (18) and (19), it is straightforward to obtain $\mu_0 = 0$, $\sigma_0^2 = 2N_0^2TW + \frac{4}{3}N_0E_b$, $\mu_1 = (4/3)E_b$, $\sigma_1^2 = 2N_0^2TW + \frac{8}{3}N_0E_b$. Substituting these parameter values into (20) and (21), and then expressing P_0 and P_1 in terms of the complementary error function $Q(\cdot)$, we obtain

$$P_0 = Q\left(\gamma/\sqrt{2N_0^2TW + \frac{4}{3}N_0E_b}\right) \tag{22}$$

$$P_1 = Q\left(\left(\frac{4}{3}E_b - \gamma\right)/\sqrt{2N_0^2TW + \frac{8}{3}N_0E_b}\right) \tag{23}$$

The optimal threshold is obtained by setting $P_0 = P_1$ [5, 15, 17], and then we have

$$\frac{\gamma}{\sqrt{2N_0^2TW + \frac{4}{3}N_0E_b}} = \frac{((4/3)E_b - \gamma)}{\sqrt{2N_0^2TW + (8/3)N_0E_b}} \tag{24}$$

Solving (24), the optimal threshold is

$$\gamma = \frac{((4/3)E_b)\sqrt{2N_0^2TW + (4/3)N_0E_b}}{\sqrt{2N_0^2TW + (8/3)N_0E_b} + \sqrt{2N_0^2TW + (4/3)N_0E_b}} \tag{25}$$

The total BER is $P_e = 0.5(P_0 + P_1)$. Since $P_0 = P_1$, we have $P_e = P_0$. Substituting (25) into (22), the total BER of PWM in AWGN channels is

$$P_e = Q\left(\frac{(4/3)(E_b/N_0)}{\sqrt{2TW + (8/3)(E_b/N_0)} + \sqrt{2TW + (4/3)(E_b/N_0)}}\right) \tag{26}$$

3.2 PPM performance in AWGN channels

The BER equation of ED PPM has been derived in [5], and its expression is as follows

$$P_e = Q\left(\frac{E_b/N_0}{\sqrt{2TW + 2E_b/N_0}}\right) \tag{27}$$

4 BER performance in multipath channels

In this section, the BER performances of PPM and PWM in multipath channels are researched. The IEEE 802.15.4a channel model [19] is used in this paper. The signal convolves with the channel impulse response in multipath channels and becomes

$$r(t) = s(t) \otimes h(t) + n(t) \tag{28}$$

where $h(t)$ denotes the channel impulse response, $n(t)$ is AWGN and the convolution operation is denoted by $\otimes$.

4.1 PPM performance in multipath channels

The BER performance of PPM in multipath channels was analysed in our previous publication [17], so we briefly summarise it as follows. Fig. 3 shows the frame structures of PPM in multipath channels. In Fig. 3, δ denotes modulation index, T_0 and T_1 are the time intervals reserved for multipath components of bits 0 and 1, respectively. The relationship of the above three parameters is $\delta = T_0 = T_1$. The value of δ must be designed appropriately. A too large value will waste the transmission time and reduce the data rate. However, if it is less than the maximum channel spread D,

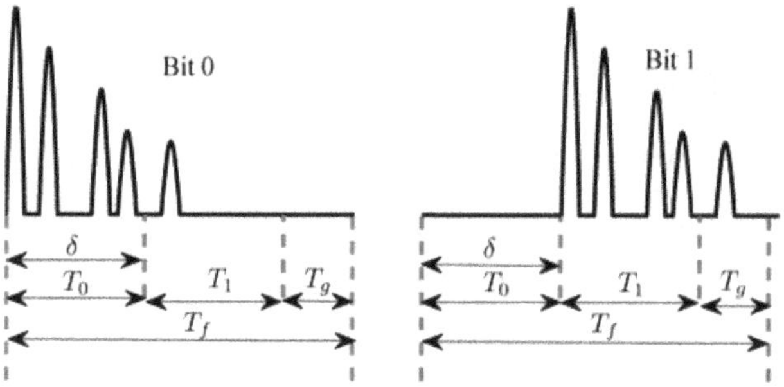

Fig. 3 *PPM frame structures in multipath channels*

the CMI will occur [13, 16, 17, 20]. In fact, it is difficult to guarantee that the value of δ is always appropriate, so the effect of CMI on system performance cannot be neglected. When CMI occurs, the system performance will degrade considerably. Even increasing the transmitting power will not improve the performance because of the proportional increase of interference [17, 20]. Synchronisation error is assumed to be 0 in analysis. When δ is less than the maximum channel spread D, some multipath components of bit 0 extend into the interval T_1 and cause CMI. However, the multipath components of bit 1 do not cause CMI and some of them just extend into the guard interval T_g, which is used to prevent inter-frame interference (IFI). The frame period is $T_f = T_0 + T_1 + T_g$. To achieve as high a data rate as possible and prevent IFI simultaneously, the frame period is set to $T_f = \delta + D$ [13, 17].

The BER of PPM in multipath channels is derived in [17]

$$P_e = \frac{1}{2}Q\left(\frac{(\beta_a - \beta_b)(E_b/N_0)}{\sqrt{2TW + 2(\beta_a + \beta_b)(E_b/N_0)}}\right) + \frac{1}{2}Q\left(\frac{\beta_a(E_b/N_0)}{\sqrt{2TW + 2\beta_a(E_b/N_0)}}\right) \tag{29}$$

where β_a and β_b are defined as follows. When bit 0 is transmitted, $\beta_a = E_{T_0}/E_b$ and $\beta_b = E_{T_1}/E_b$. The two variables, E_{T_0} and E_{T_1}, denote the captured signal energies in the integration intervals T_0 and T_1, respectively. Under this conversion, E_{T_0} and E_{T_1} are expressed as $\beta_a E_b$ and $\beta_b E_b$, respectively. The values of β_a and β_b are in the range [0, 1]. When bit 1 is transmitted, $E_{T_0} = 0$ and $E_{T_1} = \beta_a E_b$. When there is no CMI, $\beta_a = 1$ and $\beta_b = 0$, (29) reduces to (27).

4.2 *PWM performance in multipath channels*

Fig. 4 is the frame structure of PWM in multipath channels. CMI does not occur as it does in PPM systems. To compare PWM with PPM under the same energy capture condition, the integration interval T_0 of PWM has the same length as the T_0 of PPM. In addition, synchronisation is assumed to be perfect as in PPM. The guard interval is T_g, and the frame period is set to $T_f = T_0 + T_g = D$. This will achieve the maximum data rate and prevent IFI simultaneously. This frame structure is applied to both bits 0 and 1. We assume λ denotes the ratio of the captured signal energy in interval T_0 to the total signal energy at each branch. When bit 0 is transmitted, the signal energy distributes almost equally to two branches, that is $E_{01} = E_{02} = E_b/3$. Therefore the captured signal energy E_{T_0} in two branches are all $\lambda E_b/3$, and the resultant pdf of Z is

$$H_0{:}Z \sim N\left(0,\ 2N_0^2TW + N_0\lambda\frac{4}{3}E_b\right) \tag{30}$$

When bit 1 is transmitted, the signal energy almost entirely distributes to Branch 1, and the signal energy of Branch 2 is ~0. Therefore we have $E_{11} = 4\lambda E_b/3$ and $E_{12} = 0$. The pdf of Z is

$$H_1{:}Z \sim N\left(\lambda\frac{4}{3}E_b,\ 2N_0^2TW + N_0\lambda\frac{8}{3}E_b\right) \tag{31}$$

Using (20) and (21), and following the method in Section 3.1, we obtain the decision threshold γ and BER as follows

$$\gamma = \frac{((4/3)\lambda E_b)\sqrt{2N_0^2TW + (4/3)N_0\lambda E_b}}{\sqrt{2N_0^2TW + (8/3)N_0\lambda E_b} + \sqrt{2N_0^2TW + (4/3)N_0\lambda E_b}} \tag{32}$$

$$P_e = Q\left(\frac{(4/3)\lambda(E_b/N_0)}{\sqrt{2TW + (8/3)\lambda(E_b/N_0)} + \sqrt{2TW + (4/3)\lambda(E_b/N_0)}}\right) \tag{33}$$

where γ is not a constant and it is changed by the captured energy in interval T_0. When $T_0 = D$, the integrators capture all signal energy and then $\lambda = 1$. Equations (32) and (33) reduce to (25) and (26), respectively.

5 BER performance in the presence of synchronisation errors

5.1 *PPM performance in the presence of synchronisation errors*

The BER performance of PPM in the presence of synchronisation errors is analysed in our publication [17]. A brief summary is as follows. In Fig. 5, the PPM frame structures in the presence of synchronisation errors ε are shown. Since only the effect of synchronisation errors is analysed, so the modulation index is set to $\delta = D = T_0 = T_1$ to avoid CMI. The BER performances of PPM is analysed in the range $\varepsilon \in [0, D/2]$. The frame length is set to $T_f = 2D + T_g$, where the value of the guard interval T_g is set to $D/2$, the

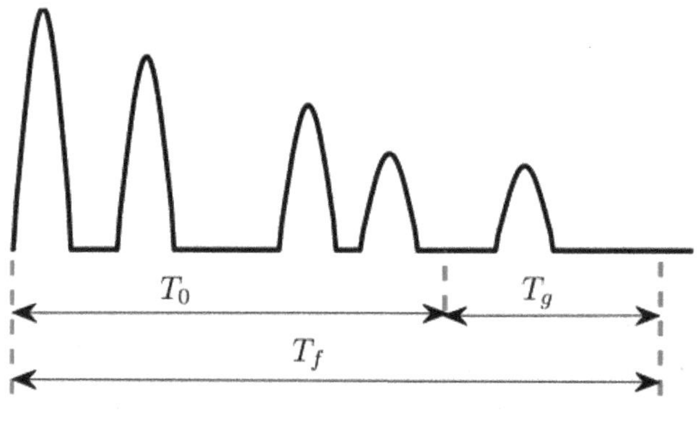

Fig. 4 *PWM frame structures in multipath channels*

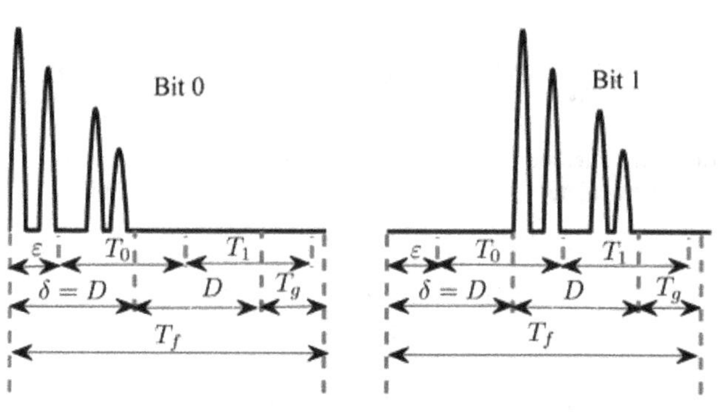

Fig. 5 *PPM frame structures in the presence of synchronisation errors*

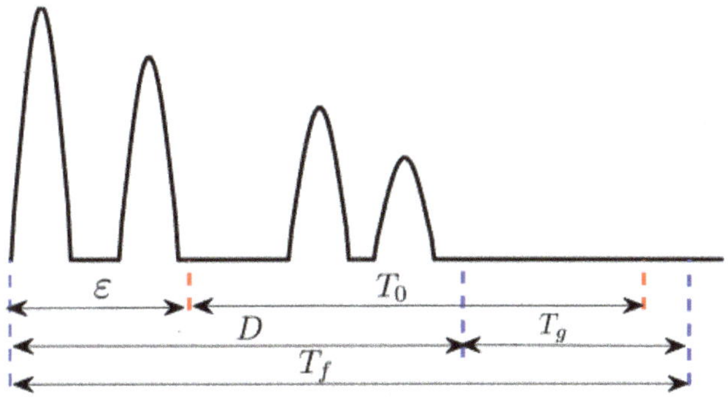

Fig. 6 *PWM frame structure in the presence of synchronisation errors*

maximum synchronisation error used in analysis. Under this frame length design, IFI is effectively avoided.

The BER of PPM in the presence of synchronisation errors is [17]

$$P_e = \frac{1}{2}Q\left(\frac{\eta E_b/N_0}{\sqrt{2TW + 2\eta E_b/N_0}}\right) + \frac{1}{2}Q\left(\frac{(2\eta - 1)E_b/N_0}{\sqrt{2TW + 2E_b/N_0}}\right) \quad (34)$$

where η is defined as follows. When bit 0 is transmitted, $\eta = E_{T_0}/E_b$ and $E_{T_1} = 0$. When bit 1 is transmitted, $\eta = E_{T_1}/E_b$ and $E_{T_0} = (1-\eta)E_b$.

5.2 *PWM performance in the presence of synchronisation errors*

Fig. 6 depicts the PWM frame structure in the presence of synchronisation errors ε. The integration interval $T_0 = D$ is the same as that of PPM. The frame length is $T_f = T_g + D$, where $T_g = D/2$ as in Section 5.1. From Fig. 6, the pdfs of Z are

$$H_0{:}Z \sim N\left(0,\ 2N_0^2TW + \frac{4}{3}N_0\rho E_b\right) \quad (35)$$

$$H_1{:}Z \sim N\left(\frac{4}{3}\rho E_b,\ 2N_0^2TW + \frac{8}{3}N_0\rho E_b\right) \quad (36)$$

where ρ denotes the ratio of the captured signal energy in T_0 to the total signal energy at each branch. Using (20) and (21), and following the method in Section 3.1, the decision threshold γ and total BER are

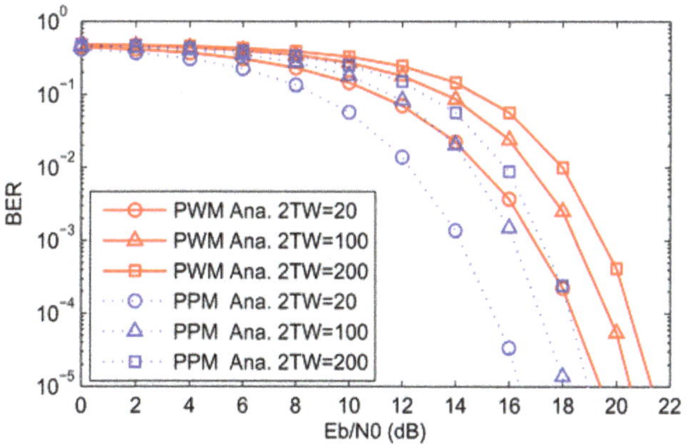

Fig. 8 *Comparison of BER performance of PWM and PPM for different 2TW values*

$$\gamma = \frac{((4/3)\rho E_b)\sqrt{2N_0^2TW + (4/3)N_0\rho E_b}}{\sqrt{2N_0^2TW + (8/3)N_0\rho E_b} + \sqrt{2N_0^2TW + (4/3)N_0\rho E_b}} \quad (37)$$

$$P_e = Q\left(\frac{(4/3)\rho(E_b/N_0)}{\sqrt{2TW + (8/3)\rho(E_b/N_0)} + \sqrt{2TW + (4/3)\rho(E_b/N_0)}}\right) \quad (38)$$

where γ is also an adaptive threshold. When $\varepsilon = 0$, no synchronisation errors occur and the integrators capture all signal energy. Therefore we have $\rho = 1$ and then (37) and (38) reduce to (25) and (26), respectively.

6 Numerical results and analysis

Fig. 7 shows the BER curves of PWM for different $2TW$ values in AWGN channels. In simulation, the bandwidth of the filters is 3 GHz, the shape factors for bits 0 and 1 are 0.25×10^{-9} and 0.5×10^{-9}, respectively, and the corresponding pulse durations are 0.6 and 1.2 ns. The analytical BER curves are obtained directly from (26). From Fig. 7, it is observed that the simulated and analytical curves match better when $2TW$ is increased. The reason is that the Gaussian approximation is more accurate under large $2TW$ values [15, 17]. After the bandwidth W is chosen, we only can change $2TW$ by changing the integration time T. Therefore the Gaussian approximation is more accurate when T is larger. However, the large T degrades BER performance since the integrator captures more noise energy [17]. After the UWB signal travels through the multipath channel, the large number of multipath

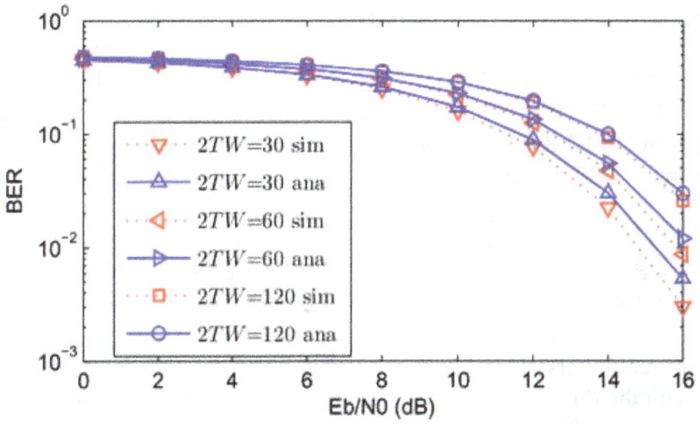

Fig. 7 *BER performance of PWM for different 2TW values in AWGN channels*

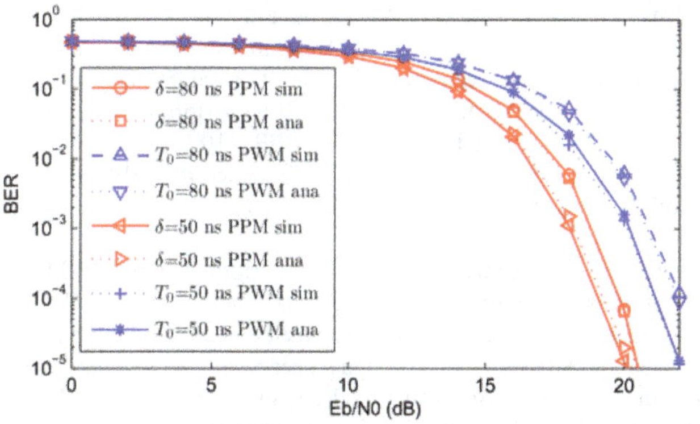

Fig. 9 *Comparison of BER performance of PWM and PPM in multipath channels ($\delta = 80$ and 50 ns)*

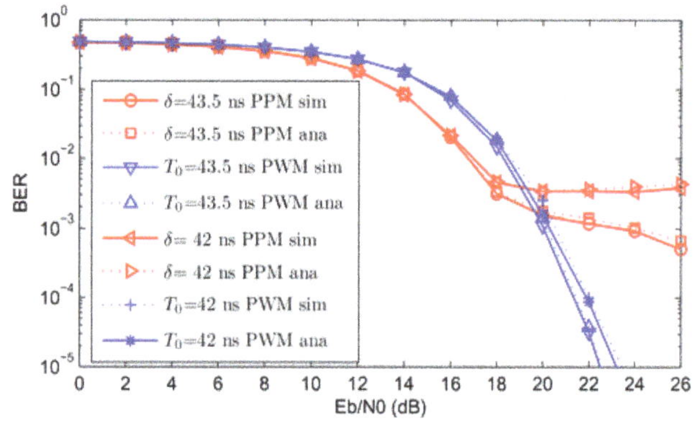

Fig. 10 *Comparison of BER performance of PWM and PPM in multipath channels (δ = 43.5 and 42 ns)*

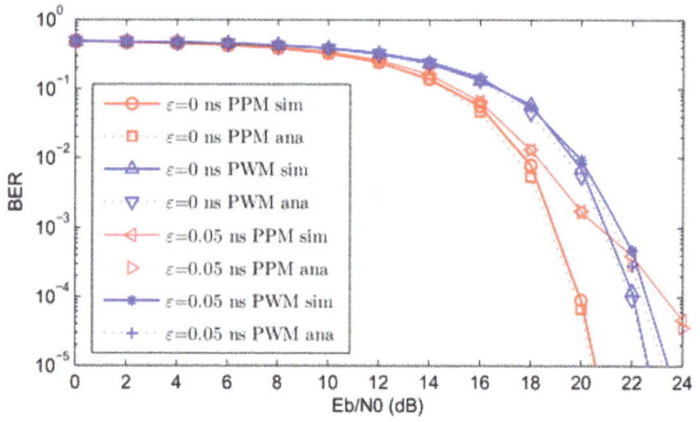

Fig. 11 *Comparison of BER performance of PWM and PPM in the presence of synchronisation errors (ε = 0 and 0.05 ns)*

components cause a very long channel delay. To capture the enough signal energy, the integration interval must be very long and this leads to a very large $2TW$ value.

Fig. 8 shows the analytical BER curves of PWM and PPM for different $2TW$ values. Since (26) has been proved an accurate BER equation for PWM above and (27) was also proved to be accurate for PPM [5], we use analytical BER curves to compare the BER performances of PWM and PPM in AWGN channels. In Fig. 8, When $2TW = 20$, PPM achieves 2.7 dB improvement over PWM at BER = 10^{-3}. When $2TW = 100$ and 200, the improvements are 2.3 and 2.2 dB, respectively. An ED PPM system exhibits better BER performance than an ED PWM system in AWGN channels.

Figs. 9 and 10 show the BER performance comparisons of PWM and PPM in multipath channels. In simulation, the shape factor of the pulses for PPM is 0.5×10^{-9} and the bandwidth of the filter is 3 GHz, the same as PWM. The CM1 model [19] is used in simulation. Synchronisation is perfect, and the maximum channel spread D is truncated to 80 ns. The frame length is designed using the method mentioned in Section 4, so IFI is avoided in simulation. In this paper, $\delta = T_0 = T_1$ for PPM, and the T_0 of PWM equals the T_0 of PPM. In the following, when a value of δ is given, it implies that T_0 and T_1 also have the same value. Therefore we only mention δ in the following. The analytical BER curves of PPM and PWM are obtained directly from (29) and (33), respectively. In these two equations, we need to know the values of parameters β_a, β_b and λ. There is no mathematical formula to calculate the captured energy as a function of the length of the integration interval for IEEE 802.15.4a channel. Therefore we use the statistic method in [17] to obtain values for the above parameters. Firstly, the MATLAB code in [19] is used to generate realisations of the channel impulse response $h(t)$. Then, we calculate the ratio of energy in a specific time interval to the total energy of a channel realisation to obtain the values for these parameters. These values are substituted into (29) and (33) to achieve the analytical BER. Both the simulated and the analytical BER are obtained by averaging over 100 channel realisations. In Fig. 9, when $\delta = 80$ ns, no CMI occurs and PPM achieves better BER performance than PWM. The improvement is ~2.1 dB at BER = 10^{-3}. When δ = 50 ns, PPM still achieves better BER performance than PWM in spite of the slight CMI and the improvement is ~2 dB at BER = 10^{-3}.

However, we can see from Fig. 9 that the performances of PWM and PPM are both improved compared with when $\delta = 80$ ns. The reason is that the multipath components existing in the time interval between 50 and 80 ns include low signal energy and the integrators capture more noise energy than signal energy in this interval. In Fig. 10, when $\delta = 43.5$ ns, PWM achieves better BER performance than PPM and the improvement is ~4 dB at BER = 10^{-3}. When $\delta = 42$ ns, PWM requires an increase of E_b/N_0 ~0.2 dB to maintain

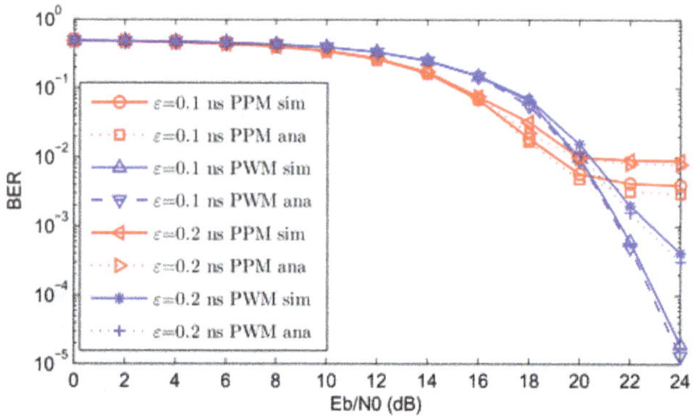

Fig. 12 *Comparison of BER performance of PWM and PPM in the presence of synchronisation errors (ε = 0.1 and 0.2 ns)*

BER = 10^{-3}. However, PPM cannot achieve this BER level and exhibits a BER floor. The BER performance of PPM cannot be improved by increasing the transmitted power because of the proportional increase of CMI [17, 20]. Unlike PPM, however, PWM still achieves a good BER performance when the signal transmitting power is increased.

Figs. 11 and 12 are comparisons of BER performance when synchronisation errors occur. In simulation, the modulation index δ is set to the maximum channel spread $D = 80$ ns, so no CMI occurs in simulation. The frame structure is designed by following the method mentioned in Section 5, so IFI is avoided in simulation. The analytical BER curves are obtained directly from (34) and (38), and the values for parameters η and ρ in (34) and (38) are obtained by using the statistic method similar to the one described above. Both the simulated and analytical BERs are obtained by averaging over 100 channel realisations. In Fig. 11, when $\varepsilon = 0$ ns, no synchronisation error occurs, and PPM achieves better BER performance than PWM. The improvement is ~2.1 dB at BER = 10^{-3}. When $\varepsilon = 0.05$ ns, PPM still achieves ~0.5 dB improvement at BER = 10^{-3}. However, there is a crossing point between the BER curves of PWM and PPM. The BER performance of PWM actually has surpassed PPM if we compare them using better BER values. In Fig. 12, when $\varepsilon = 0.1$ and 0.2 ns, the BER curves of PPM exhibit BER floors because of a severe synchronsation error, but PWM still achieves a good BER. When we compare these synchronisation error values to the maximum channel delay, they are really very small. Even it is still small when we compare them with the duration of a single pulse. Under so small synchronisation errors, PPM has been degraded severely. However, PWM still can achieve good BER performance. Therefore the robustness of PWM is very significant.

The reason the BER performance of PWM is better than PPM in the presence of CMI or synchronisation errors can be explained as follows: in a PPM system, modulation is achieved by shifting the pulse position, and the orthogonality of the signals is achieved in time domain. When CMI or synchronisation errors occur, this orthogonality is easily destroyed [17]. And this results in the energy cancellation between T_0 and T_1, thereby the euclidean distance is reduced dramatically and the BER performance is severely degraded. In a PWM system, when bit 0 is transmitted, the mean values in (30) and (35) are always 0. The euclidean distance is not affected by small T_0 values or synchronisation errors when bit 0 is transmitted. When bit 1 is transmitted, although the mean values in (31) and (36) are reduced, but the phenomena of energy cancellation does not occur in a PWM system.

7 Conclusion

A new ED UWB system based on pulse width modulation is proposed in this paper. The BER performance of this new system is compared with PPM in AWGN channels, multipath channels and in the presence of synchronisation errors. In AWGN channels, the BER performance of PPM is slightly better than PWM. However, in multipath channels, PPM suffers from CMI if the integration interval is shorter than maximum channel spread. This causes the degradation of BER performance of PPM, so the BER performance of PWM can surpass that of PPM in multipath channels. In addition, when synchronisation errors occur, PWM is more robust and achieves better BER performance than PPM. If we choose PWM other than PPM, it will lower the requirement of synchronisation accuracy and we can choose cheap synchroniser to reduce the cost.

8 References

[1] Yang L., Giannakis G.B.: 'Ultra-wideband communications: an idea whose time has come', *IEEE Signal Process. Mag.*, 2004, **21**, (6), pp. 26–54

[2] Carbonelli C., Mengali U.: 'M-PPM noncoherent receivers for UWB applications', *IEEE Trans. Wirel. Commun.*, 2006, **5**, (8), pp. 2285–2294

[3] He N., Tepedelenlioglu C.: 'Performance analysis of non-coherent UWB receivers at different synchronization levels', *IEEE Trans. Wirel. Commun.*, 2006, **5**, (6), pp. 1266–1273

[4] Paquelet S., Aubert L.M.: 'An energy adaptive demodulation for high data rates with impulse radio'. IEEE Radio and Wireless Conf., Atlanta, USA, September 2004, pp. 323–326

[5] Dubouloz S., Denis B., de Rivaz S., Ouvry L.: 'Performance analysis of LDR UWB non-coherent receivers in multipath environments'. IEEE Int. Conf. Ultra-Wideband, Zurich, Switzerland, September 2005, pp. 491–496

[6] Witrisal K., Leus G., Janssen G., ET AL.: 'Noncoherent ultra-wideband systems', *IEEE Signal Process. Mag.*, 2009, **26**, (4), pp. 48–66

[7] Mu D., Qiu Z.: 'Weighted non-coherent energy detection receiver for UWB OOK system'. IEEE the Ninth Int. Conf. Signal Processing, Beijing, China, October 2008, pp. 1846–1849

[8] Amico A., Mengali U., Arias-de-reyna E.: 'Energy-detection UWB receivers with multiple energy measurements', *IEEE Trans. Wirel. Commun.*, 2007, **6**, (7), pp. 2652–2659

[9] Wang F., Xu C., Ji X., Zhang Y.: 'Performance analysis of time-hopping pulse width modulation impulse radio'. The Fourth Int. Conf. Wireless Communication, Networking and Mobile Computing, Dalian, China, October 2008, pp. 1–5

[10] Benedetto M.-G.D., Giancola G.: 'Understanding ultra wide band radio fundamentals' (Prentice-Hall, Upper Saddle River, NJ, USA, 2004)

[11] Nekoogar F.: 'Ultra-wideband communications: fundamentals and applications' (Prentice-Hall, Upper Saddle River, NJ, USA, 2005)

[12] Win M.Z., Scholtz R.A.: 'Impulse radio: how it works', *IEEE Commun. Lett.*, 1998, **2**, (2), pp. 36–38

[13] Cheng X., Guan Y.: 'Mitigation of cross-modulation interference in UWB energy detector receiver', *IEEE Commun. Lett.*, 2009, **13**, (6), pp. 375–377

[14] Urkowitz H.: 'Energy detection of unknown deterministic signals', *Proc. IEEE*, 1967, **55**, (4), pp. 523–531

[15] Humblet P., Azizoglu M.: 'On the bit error rate of lightwave systems with optical amplifiers', *J. Lightwave Technol.*, 1991, **9**, (11), pp. 1576–1582

[16] Celebi H., Arslan H.: 'Cross-modulation interference and mitigation technique for ultrawideband PPM signaling', *IEEE Trans. Veh. Technol.*, 2008, **57**, (2), pp. 847–858

[17] Cui S., Xiong F.: 'UWB system based on energy detection of derivatives of the Gaussian pulse', *Eurasip J. Wirel. Commun. Netw.*, 2011, 206, DOI: 10.1186/1687-1499-2011-206

[18] Mills R.F., Prescotte G.E.: 'A comparison of various radiometer detection models', *IEEE Trans. Aerosp. Electron. Syst.*, 1996, **32**, (1), pp. 467–474

[19] Molisch A.F., Balakrishnan K., Cassioli D., ET AL.: IEEE 802.15.4a channel model-final report. pp. 1–40. Available at http://www.ieee802.org/15/pub/04/15-04-0662-02-004a-channel-model-final-report-r1.pdf, accessed November 2013

[20] Arslan H.: 'Cross-modulation interference reduction for pulse-position modulation UWB signals'. IEEE 64th Vehicular Technology Conf., Montreal, Canada, September 2006, pp. 1–5

9 Appendix

We use mathematical tool MAPLE to calculate the energy distribution relationships of E_{01}, E_{02}, E_{11} and $E_{1,2}$ as follows

$$\begin{aligned} E_{01} &= \int_0^{\sqrt{2}/(\alpha_0\sqrt{\pi})} [X_f]^2\,\mathrm{d}f \\ &= \int_0^{\sqrt{2}/(\alpha_0\sqrt{\pi})} \left(f^2 \mathrm{e}^{-\pi f^2\alpha_0^2/2}\right)^2 \mathrm{d}f \end{aligned} \tag{39}$$

$$\begin{aligned} E_{02} &= \int_{\sqrt{2}/(\alpha_0\sqrt{\pi})}^{2\sqrt{2}/(\alpha_0\sqrt{\pi})} [X_f]^2\,\mathrm{d}f \\ &= \int_{\sqrt{2}/(\alpha_0\sqrt{\pi})}^{2\sqrt{2}/(\alpha_0\sqrt{\pi})} \left(f^2 \mathrm{e}^{-\pi f^2\alpha_0^2/2}\right)^2 \mathrm{d}f \end{aligned} \tag{40}$$

$$E_0 = \int_0^{\infty} [X_f]^2\,\mathrm{d}f = \int_0^{\infty} \left(f^2 \mathrm{e}^{-\pi f^2\alpha_0^2/2}\right)^2 \mathrm{d}f \tag{41}$$

$$\begin{aligned} E_{11} &= \int_0^{\sqrt{2}/(\alpha_0\sqrt{2})} [X_f]^2\,\mathrm{d}f \\ &= \int_0^{\sqrt{2}/(\alpha_0\sqrt{\pi})} \left(f^2 \mathrm{e}^{-\pi f^2\alpha_1^2/2}\right)^2 \mathrm{d}f \end{aligned} \tag{42}$$

$$\begin{aligned} E_{12} &= \int_{\sqrt{2}/(\alpha_0\sqrt{\pi})}^{2\sqrt{2}/(\alpha_0\sqrt{\pi})} [X_f]^2\,\mathrm{d}f \\ &= \int_{\sqrt{2}/(\alpha_0\sqrt{\pi})}^{2\sqrt{2}/(\alpha_0\sqrt{\pi})} \left(f^2 \mathrm{e}^{-\pi f^2\alpha_1^2/2}\right)^2 \mathrm{d}f \end{aligned} \tag{43}$$

$$E_1 = \int_0^{\infty} [X_f]^2\,\mathrm{d}f = \int_0^{\infty} \left(f^2 \mathrm{e}^{-\pi f^2\alpha_1^2/2}\right)^2 \mathrm{d}f \tag{44}$$

where $\sqrt{2}/\left(\alpha_0\sqrt{2}\right)$ is the value of the centre frequency f_{c0} of $p_0(t)$. After substituting the values of α_0 and $\alpha_1 = 2\alpha_0$ into these equations, we obtain

$$\begin{aligned} E_{01} - E_{02} &\simeq 0.451E_0 - 0.543E_0 \\ &= -0.09E_0 = -0.06E_b \simeq 0 \end{aligned} \tag{45}$$

$$\begin{aligned} E_{01} + E_{02} &\simeq 0.451E_0 + 0.543E_0 \\ &= 0.994E_0 \simeq E_0 \end{aligned} \tag{46}$$

where $E_{01} = 0.451E_0$ and $E_{02} = 0.543E_0$ can be obtained by calculating E_{01}/E_0 and E_{02}/E_0, respectively. And E_b is the average bit energy. From (16), we know $E_0 = (2/3)E_b$ and then we use (2/3) E_b to replace E_0 in (45). The reason we round off $0.06E_b$ to 0 in (45) can be explained as follows: UWB signals are transmitted in a very low power, so $0.06E_b$ is a very small value, and when we evaluate system BER performance in terms of E_b/N_0, this $0.06E_b/N_0$ is very small compared with E_b/N_0. Therefore it is reasonable to round off $0.06E_b$ to 0. Similarly, we can obtain the relationship of E_{11} and E_{12} as

$$\begin{aligned} E_{11} - E_{12} &\simeq 0.993E_1 - 0.0068E_1 \\ &= 0.986E_1 \simeq E_1 \end{aligned} \tag{47}$$

$$\begin{aligned} E_{11} + E_{12} &\simeq 0.993E_1 + 0.0068E_1 \\ &= 0.9998E_1 \simeq E_1 \end{aligned} \tag{48}$$

These results are all verified by different values of α_0 and $\alpha_1 = 2\alpha_0$.

22

Stand-alone excitation synchronous wind power generators with power flow management strategy

Tzuen-Lih Chern[1], Ping-Lung Pan[1], Yu-Hsiang Chern[2], Ji-Xian Huang[1], Jyh-Horng Chou[2], Whei-Min Lin[1], Chih-Chiang Cheng[1]

[1]Department of Electrical Engineering, National Sun Yat-sen University, Kaohsiung, Taiwan
[2]Graduate Institute of Electrical Engineering, National Kaohsiung First University of Science and Technology, Kaohsiung, Taiwan
E-mail: tlchen@ee.nsysu.edu.tw

Abstract: This study presents a stand-alone excitation synchronous wind power generator (SESWPG) with power flow management strategy (PFMS). The rotor speed of the excitation synchronous generator tracks the utility grid frequency by using servo motor tracking technologies. The automatic voltage regulator governs the exciting current of generator to achieve the control goals of stable voltage. When wind power is less than the needs of the consumptive loading, the proposed PFMS increases motor torque to provide a positive power output for the loads, while keeping the generator speed constant. Conversely, during the periods of wind power greater than output loads, the redundant power of generator production is charged to the battery pack and the motor speed remains constant with very low power consumption. The advantage of the proposed SESWPG is that the generator can directly output stable alternating current (AC) electricity without using additional DC–AC converters. The operation principles with software simulation for the system are described in detail. Experimental results of a laboratory prototype are shown to verify the feasibility of the system.

1 Introduction

Wind power is environmentally friendly and inexhaustible. As such, it has been considered one of most clean substitute energy sources during past decades. In global markets, two main categories of wind power generator (WPG), doubly fed induction generator and permanent magnet (PM) synchronous generator, are widely used for grid connection and stand-alone applications. Compared with common renewable energy systems like photovoltaic electricity generator as an isolated electricity source, the stand-alone WPG system has easy installation, continuous operation ability and smaller occupation futures that are suitable for the areas where it is difficult to access public power grids or the roofs on urban building for serving major or auxiliary power supplier [1, 2]. The turbine speed of stand-alone wind generator is susceptible to the wind random dynamic, which results unpredictable voltage and frequency drift [3]. Moreover, stand-alone WPGs are often used at a non-ideal wind site. When wind speed is insufficient for driving the wind generator, the wind generator must be backed up by other form of electricity supplier in order to keep the electricity uninterrupted. An unavoidable voltage fluctuation and current impact occurs during power bus switching periods; this phenomenon greatly affects the power quality of the generator. For the above issues, conventional generators use variable speed variable frequency technologies. Rectifier and alternating current (AC) converter technologies for AC/DC transformation are widely used to obtain constant frequency and voltage for a commercial loading. The voltage flicker because of unstable wind speed can be avoided by using voltage control systems [4–15]. However, the use of multi-stage power transfer technologies will cause unavoidable power losses from the rectifier's conducting resistance and high-frequency power switches, which not only increases power consumption but decreases the system reliability.

In this paper, a novel converter-less stand-alone excitation synchronous wind power generator (SESWPG) with its control framework is proposed. The presented system has constant frequency and voltage output. The main structure consists of an excitation synchronous generator, a PM synchronous servo motor, a battery pack, a servo driver and signal processor. The wind and servo motor powers are combined and transmitted to the excitation synchronous generator via coaxial configuration. The automatic voltage regulator (AVR) controls the excitation field current to achieve the stable output voltage. According to the wind speed variation, the servo motor provides a compensatory power to force the generator speed to be constant. The additional power of servo motor finally becomes electricity, and outputs to the loads. The proposed robust integral servo motor control strategy employs phase tracking function, which can reduce the generator output frequency and phase shifting from the wind disturbances. The proposed power flow management strategy (PFMS) first aims to govern the system input and output power balance and then achieve maximum power tracking. By judging the measured servo motor power consumption, the system will obtain input wind information. If wind power is less than the generator consumer loading, servo motor provides a positive power to the generator. When wind power is larger than the consumer loading, a near negative servo motor power is detected and the battery charger is activated immediately. The result is that battery pack absorbs the excess wind power, therefore no power is wasted. Based on physical theorems, a mathematical model of the proposed system is established to examine the performance of the control function for the designed framework. The detailed structure and experimental results will be discussed in the later sections.

2 Power flow and speed

Keeping power flow balance between the inputs and outputs is the basic design concept of the system. The main designed principles related to whole system's power flow are described as follows. To simplify the analysis procedures, one assumes that the power losses of all power transmission devices are considered in the ideal state, thus the mechanical power losses of the wind turbine, the servo motor and the excitation synchronous generator are ignored. The power flow of the major dynamic components in the system is illustrated as Fig. 1*a*, where P_W, P_M, P_G, P_L and P_B are the powers of wind, servo motor, excitation synchronous generator, output load and battery pack. In the proposed system, the power flow equations can be defined as

$$T_G\omega_G = T_W\omega_W + T_M\omega_M \tag{1}$$

The wind power and motor power are integrated and transferred to the generator. Generator converts kinetic energy to electricity, and outputs to the load P_{Load} and battery P_B can be described as

$$T_G\omega_G = P_G = P_{Load} + P_B \tag{2}$$

Fig. 1*b* shows the corresponding coaxial configuration of the proposed system. The input-end of wind generator rotor shaft receives a rotating torque from the speed increasing gear box, and the tail-end of the generator rotor shaft is coupled with servo motor. It is obvious that the inputted power of the excitation synchronous generator is the sum of the wind power and servo motor powers. Owing to the compact structure, power losses in energy deliver procedure can be minimised.

3 Structure of the proposed wind power generator

The framework of the proposed system is shown as Fig. 1*c*. Based on coaxial configuration, the wind turbine provides kinetic energy to the excitation synchronous generator. The excitation synchronous generator converts mechanical power into electricity and outputs to the isolated load and battery. In the natural environment, the wind power dynamics are varying and random, and the shaft speeds and the frequencies of the excitation synchronous generator are affected by wind dynamics. To protect against the disturbances of wind power, the servo motor is added to the stabilised rotary speeds and frequencies of the generator. When the system enters operation status, the signals relating to the output frequencies, output voltages and the encoder of rotary speeds are sent to the controllers which are built in a microprocessor control unit (MCU). The power control strategies are written in an MCU, and the design of control loops is expected to adjust the currents of the excitement field and control the servo motor precisely. Constant frequencies and voltages of the synchronous generator can thus be achieved. Although inputted wind power is sufficient, through rectifiers and buck converters in the unit 1 of power flow management, the excitation synchronous generator can provide battery pack with surplus electricity aside from providing the load requirements. As such, the entire electrical power outputted by the excitation synchronous generator can be fully utilised. The battery voltage is increased from lower to higher level through boost type converter to provide the driver power of the servo motor. If necessary, battery pack releases the stored electrical power, the excitation synchronous generator obtains kinetics from the servo motor for stabilising the rotary speed and the frequency of output voltage, which is able to assist the electricity production rate of generator as well. Therefore the power losses of the system are very low. When the battery electricity is deficient, one solution for driving the servo motor is providing electricity from the utility grid.

The block diagram of the proposed stand-alone WPG system is shown as Fig. 1*d*. In this system, the generator output frequency is set at 60 Hz. To obtain higher quality of AC power, the generator output frequency must be kept constant. The mechanical angle θ_M of the servo motor shaft is detected by using rotary encoder. By software processing, the rotary speed information ω_M of the servo motor is obtained. The phase/frequency synchronisation strategy compares the reference phase angle and frequency with the generator's feedback signals, and produces the position command θ_{cmd} with pulse-type signals to the servo motor driver. Using the coaxial configuration, the speed ω_W and torque T_W of the wind turbine, and the speed ω_M and torque T_M of the servo motor are integrated to drive the generator. The AVR is the second control factor of the generator, except servo motor. It is available to adjust the control signal I_f of the magnetic field current to stabilise the output voltages of the excitation synchronous generator and produce the electrical power to the loads.

Although wind power is sufficient for load demand, the power flow management system forces the partial power into the battery pack. As shown in Fig. 1*d*, partial power of generator AC output voltages is converted to DC voltages by rectifiers. A buck-type power converter is designed to determine the charging current I_B of the batteries which are conducted by pulse width modulation (PWM). The power flow management unit compiles the information of the servo motor including three-phase currents I_a, I_b, I_c, rotary speeds ω_M and the rotary angle θ_M to calculate the power of the servo motor. Thereafter, it is available to determine the charging current for battery pack. When inputted wind power is insufficient, the stored power can be released from the battery pack. Through boost type converters, a high-level DC voltage flow into DC bus drives the servo motor.

4 Power flow management unit

Figs. 1*d* and 2 are the block diagrams of power flow management unit and detailed control loops, respectively. The design of PFMS operating principles are mainly according to (1) and (2). In practical dynamics, the generator absorbs wind power and transfers the aerodynamic into electricity power for the loads and battery. Assume that the generator has constant loading power within rated power of the generator. The controllers estimate the magnitude of wind power by measuring motor power and wind speed gauge. In Fig. 2, the MCU senses motor power as the feedback signal for power flow control. The three-phase input currents I_a, I_b and I_c and voltages V_a, V_b and V_c of the servo motor are detected and fed back to the MCU. By some calculations, the quantity of the servo motor power consumed P_M is obtained. When the measured motor power is negative, it means that the output power of the generator is larger than the load demand. At this moment, the control system charges the surplus power into batteries. On the contrary, the battery charging is stopped. During charged operation, the charge current is determined by the power deviation of power command and measured motor power. In normal operation, the value of the power command P_{M-cmd} of the proportional–integral (PI) controller is set to be zero. For fast convergence, a forward loop of PI controller is designed to determine the quantity of charger power. The parameters K_{P1} and K_{P2} are proportional gains, and K_{I1} and K_{I2} are integral control gains of two PI controllers, respectively. Multiplying the battery voltage V_B and the charging current I_B yields a feedback charging power. PI control loops for buck-type converter are built after the forward loop controller. It compares the signal ΔP and the feedback battery power to generate PWM signal. The additional K_B is a constant gain for the buck converter. With PWM technology, the charging currents of batteries can be regulated by buck converter.

This paper presents a power flow management unit to deal with power distribution for two sub-functions when the system suffers wind power variation. Assuming that the system output load is constant, there exist two operating possibility conditions. The relevant control strategies for two conditions are detailed as below.

4.1 Wind power is sufficient

By referring to Fig. 2, when input wind power is sufficient, (1) is satisfied. Aside from providing electrical power to the load end, the excitation synchronous generator can also charge surplus power into the battery pack through the rectifiers and buck converters in the power flow management unit. Therefore the power outputted by the excitation synchronous generator can be entirely utilised

$$\begin{aligned} &P_W > P_{Load} \\ &P_W + P_M = P_{Load} + P_B,\ P_B > 0,\ P_M \sim 0 \end{aligned} \tag{3}$$

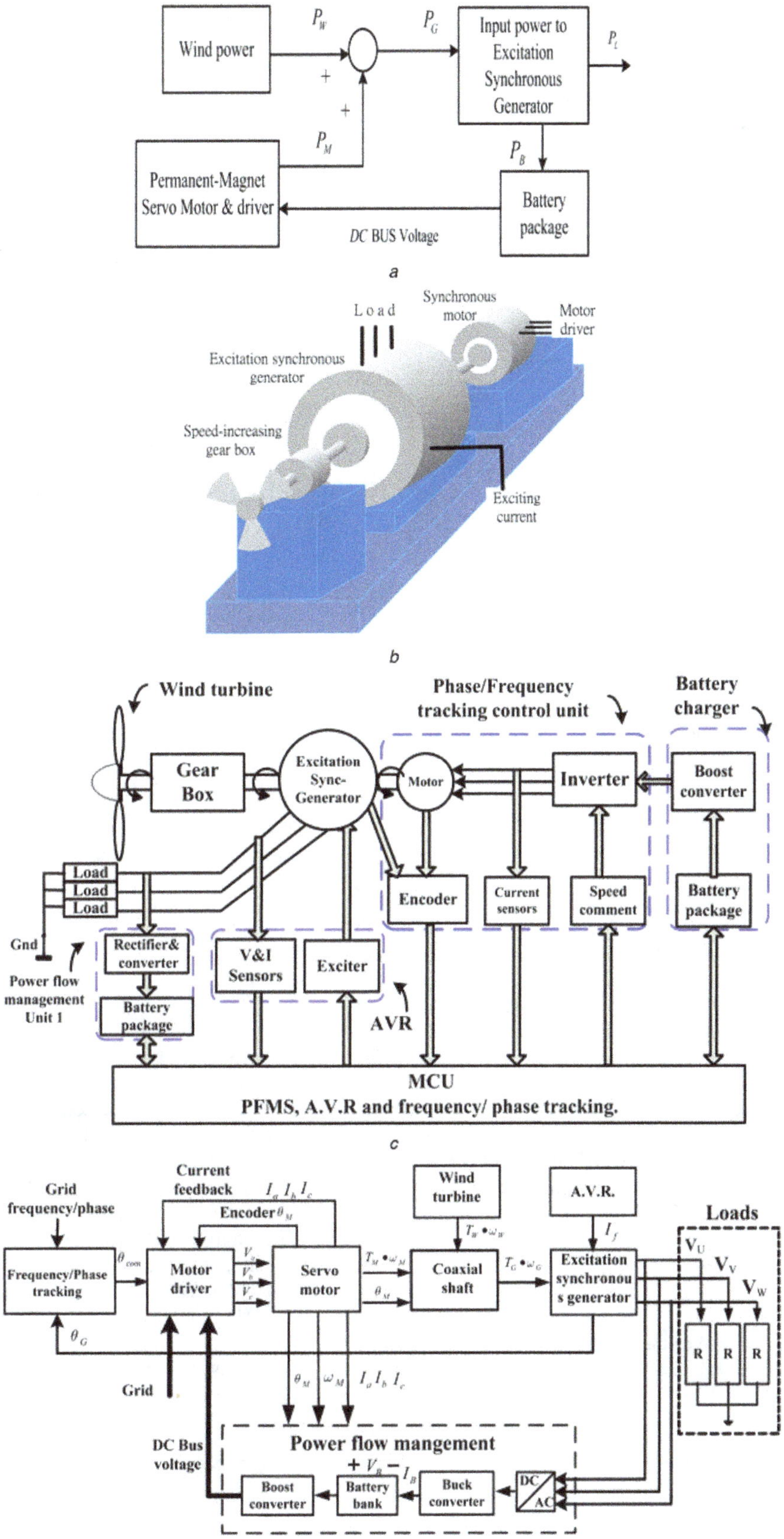

Fig. 1 *Proposed WPG system*
a Power flow
b Coaxial configuration
c Framework
d Block diagram

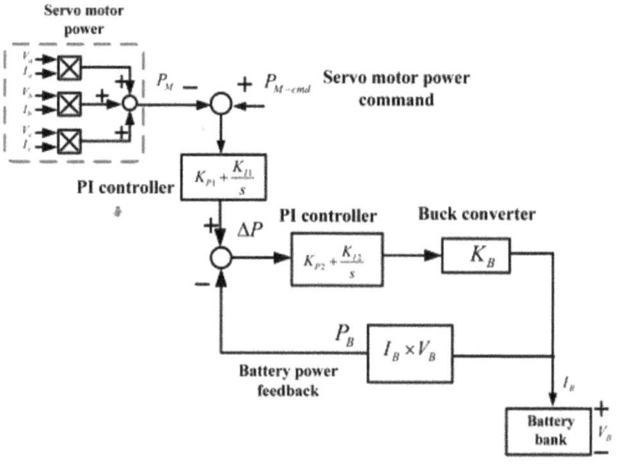

Fig. 2 *Control loops of power flow management unit with battery charger*

4.2 *Wind power is deficient*

By referring to Fig. 2, deficient wind power input affects the rotary speed of the excitation synchronous generator, with the frequencies of output electricity supply changeable. Therefore this control method adopts the servo control function of the servo motor to stabilise the rotary speed of the excitation synchronous generator and the frequencies of output electricity supply. At this moment

$$\begin{aligned} &P_{\mathrm{W}} < P_{\mathrm{Load}} \\ &P_{\mathrm{W}} + P_{\mathrm{M}} = P_{\mathrm{Load}} + P_{\mathrm{B}},\ P_{\mathrm{B}} = 0,\ P_{\mathrm{M}} > 0 \end{aligned} \tag{4}$$

5 AVR and excitation synchronous generator

Excitation synchronous generator has been widely applied in various electricity machines to provide electricity for living requirements [16–18]. In this paper, the excitation synchronous generator behaviours in a wind generator under different dynamic conditions will be clearly defined by using computer simulation. Fig. 3*a* depicts the control block diagram of the proposed system. The three-phase winding of the generator is connected to a consumer load. Through the reaction to the electric field and magnetic force effects, it is possible for a stator to generate the power corresponding to the back-electromotive force (EMF) of a generator. Based on the nature of the conductor, the correlation between the voltage and the rotary speed of back-EMF can be designated as

$$E = \ell \cdot \omega_{\mathrm{G}} \times B \tag{5}$$

E is the back-EMF voltage of the excitation synchronous generator stator, ℓ is the effective length of the conductance magnet, ω_g is the rotor speed and B is the strength of the magnetic field

$$B = \frac{\mu N}{\ell} \cdot I_{\mathrm{f}} \tag{6}$$

Here μ is the permeability coefficient of the conductance magnet, N is the turn number of winding and I_{f} is the current of the rotor. Combining (5) and (6) can give

$$E = \mu N \cdot \omega_{\mathrm{G}} \cdot I_{\mathrm{f}} \tag{7}$$

where

$$I_{\mathrm{f}}(s) = \frac{E_{\mathrm{f}}(s)}{L_{\mathrm{f}} s + R_{\mathrm{f}}} \tag{8}$$

$E_{\mathrm{f}}(s)$ denotes input voltage across the exciter fields, L_{f} is the equivalent inductance value of the exciter windings and R_{f} is the equivalent resistance of the exciter windings. In this paper, a three-phase excitation synchronous generator is proposed. In view of the distribution of magnetic fields in each phase of a generator, the back-EMF of each phase in the excitation generator must further multiply the sine wave signal sin θ_{G} of a corresponding angle. Therefore the back-EMF of a three-phase generator is

$$\begin{cases} E_U = \mu N \cdot \omega_{\mathrm{G}} \cdot I_{\mathrm{f}} \sin(\theta_{\mathrm{G}}) \\ E_V = \mu N \cdot \omega_{\mathrm{G}} \cdot I_{\mathrm{f}} \sin\left(\theta_{\mathrm{G}} + \frac{2}{3}\pi\right) \\ E_W = \mu N \cdot \omega_{\mathrm{G}} \cdot I_{\mathrm{f}} \sin\left(\theta_{\mathrm{G}} + \frac{4}{3}\pi\right) \end{cases} \tag{9}$$

The generator back torque T_{G} can be determined as follows

$$T_{\mathrm{G}} = \mu N I_{\mathrm{f}} \left[I_U \sin(\theta_{\mathrm{G}}) + I_V \sin\left(\theta_{\mathrm{G}} + \frac{2}{3}\pi\right) + I_{\mathrm{W}} \sin\left(\theta_{\mathrm{G}} + \frac{4}{3}\pi\right) \right] \tag{10}$$

The block diagram of the three-phase excitation synchronous generator, including the excitation input current and the generator output according to (5)–(10), is shown in Fig. 3*a*.

Fig. 3*b* shows the control block of AVR for the generator. When the field winding of synchronous generator is excited, through feeding back the output voltages with root mean square V_{rms} measured from the excitation synchronous generator and comparing the feedback signal with the voltage command V_{cmd}, we have a voltage error for regulating the output voltage. This voltage error is further processed through a PI controller, and an excitement gain K_{f}, to generate the control signals for the excitement field in the excitation synchronous generator. Therefore it is possible to regulate the output voltage of the excitation synchronous generator by aiming to stabilise output voltages.

6 Servo motor controller design

To reduce the influence of wind fluctuations on the generator, the transient and dynamic responses of the servo motor controller must meet robustness requirements. Thus, the robust integral structure control (RISC) method was chosen to ensure the voltage phase and the frequency in phase with grid. Among general electrical motors, the three-phase PM synchronous motor has the highest efficiency and lowest maintenance needs. It is therefore chosen to provide the controllable power for the servo control structure. This study constructed the analysis model based on the electrical circuit, motor torque and mechanical theorems. Fig. 4*a* shows the block diagram of the three-phase PM synchronous motor [19–21]. The nominal parameters of the model are listed in Table 1. According to (1), the wind power, generator power and servo motor powers can be transferred to three torque functions. The electromagnetic torque of servo motor can be expressed as

$$T_{\mathrm{M}} = K_{\mathrm{t}} \cdot \left[I_a \sin\theta_r + I_b \sin\left(\theta_r + \frac{2}{3}\pi\right) + I_c \sin\left(\theta_r + \frac{4}{3}\pi\right) \right] \tag{11}$$

where P is the number of motor poles. The mechanical torque, shaft velocity and rotor position are related by

$$T_{\mathrm{M}} + (T_{\mathrm{W}} - T_{\mathrm{G}}) = J\left(\frac{2}{P}\right)\frac{\mathrm{d}\omega_r}{\mathrm{d}t} + B_m\left(\frac{2}{P}\right)\omega_r \tag{12}$$

$$\theta_r = \int \omega_r \mathrm{d}t \tag{13}$$

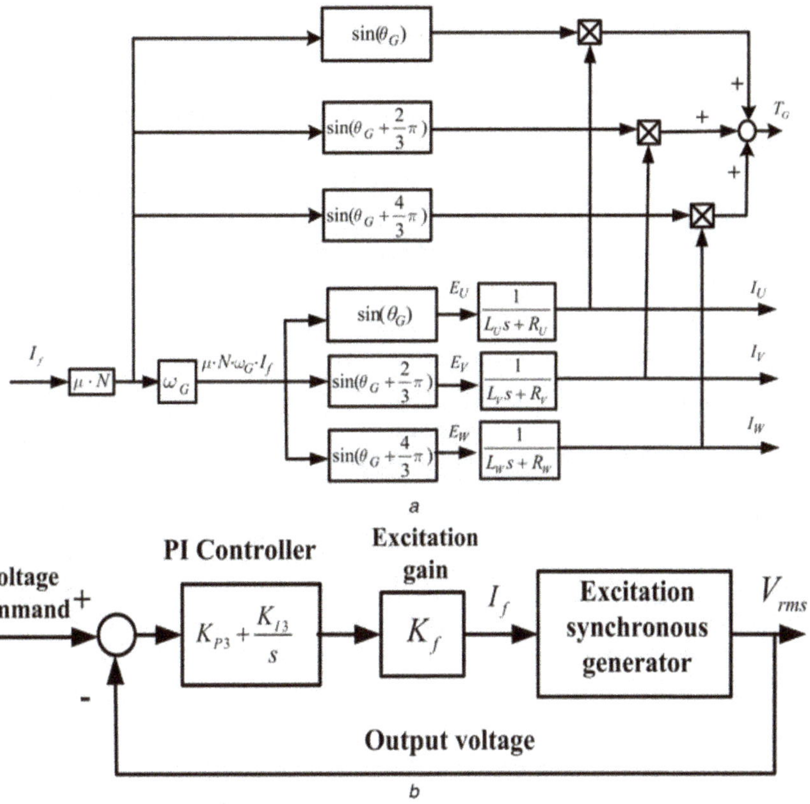

Fig. 3 *Block diagram of*
a Excitation synchronous generator
b Automatic voltage regulation

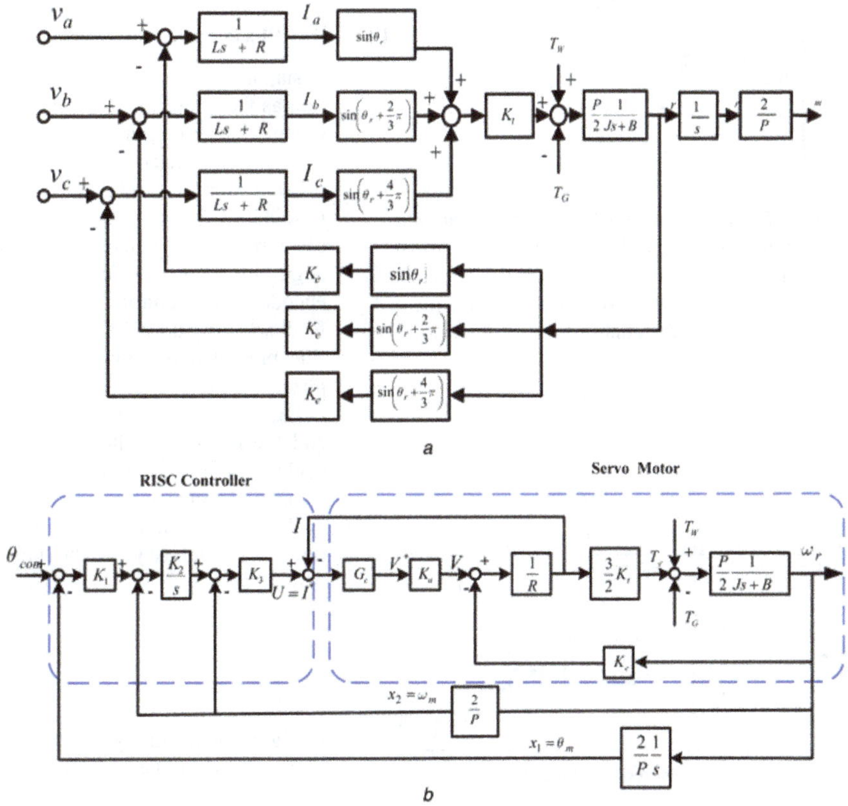

Fig. 4 *Block diagram of*
a PM synchronous motor
b Servo motor position control loop

Table 1 Specifications of excitation synchronous generator

Items	Values
rated power output	1 kW
rated voltage output	AC 220 V
phase	three-phase
pole	4-pole
stator phase resistance (R_U, R_V, R_W)	0.17 Ω
stator phase inductance (L_U, L_V, L_W)	4.3 mH
product of coefficient of conductance magnet and winding turn ($\mu \bullet N$)	0.04

Fig. 4*b* shows the position control structure; it includes the RISC controller and servo motor transfer function. The three-phase PM synchronous motor can be simplified as a first-order mathematical model. The symbols of the figure are listed in Table 1. The well-known motor current feedback controller can avoid instantaneous current stress to the servo driver. This technology is applied to the servo motor control to increase the control performance. The outer-loop of the RISC controller is designed to achieve a fast and accurate servo tracking response under load disturbances and plant parameter variations [22]. The parameters K_1 and K_3 are proportional gains and K_2 is integral gain.

The dynamics of PM synchronous motor can be described as below

$$\begin{cases} \dot{x}(t) = x_2(t) \\ \dot{x}_2(t) = -a_1x_1(t) - a_2x_2(t) + bU(t) - T_L(t) \end{cases} \tag{14}$$

where

$$\begin{cases} a_1 = 0 \\ a_2 = \dfrac{B}{J} + \dfrac{(3/2)K_tK_e}{JR + JG_cK_a} \\ b = \dfrac{(3/2)GcK_aK_t}{JR + JG_cK_a} \\ T_L(t) = \dfrac{1}{J}\left(T_W - T_G\right) \end{cases} \tag{15}$$

The RISC is one of state feedback control which combines an integral controller and the series state feedback information of the plant. The state equations of the servo motor control system are expressed as (14). Symbols a_i, $i = 1, 2$ and b are the state variables of the plant, T_L is the total disturbance and U is the control input of the system.

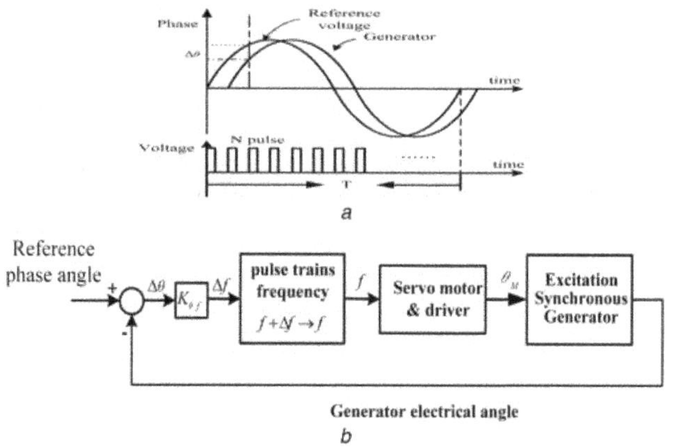

Fig. 5 *Phase tracking control strategy*
a Phase difference calculation
b Phase tracking controller

For the servo motor control system, the control function U can be expressed as below

$$U(s) = K_1K_2K_3\frac{\theta_{cmd}(s) - x_1(s)}{s} - K_2K_3x_1 - K_3x_2 \tag{16}$$

The transfer function of the system is

$$\frac{x_1(s)}{\theta_{cmd}(s)} = \frac{K_1K_2K_3b}{s^3 + (a_2 + K_3b)s^2 + (a_1 + K_2K_3b)s + K_1K_2K_3b} \tag{17}$$

Designing the system characteristic function to lay on the stable plane, one can obtain (18)

$$(s + \lambda_1)(s + \lambda_2)(s + \lambda_3) = 0 \tag{18}$$

where λ_1, λ_2 and λ_3 are the selected close loop poles of the system. Then the characteristic function of (17) can be rewritten as

$$s^3 + (\lambda_1 + \lambda_2 + \lambda_3)s^2 + (\lambda_1\lambda_2 + \lambda_1\lambda_3 + \lambda_2\lambda_3)s + \lambda_1\lambda_2\lambda_3 = 0 \tag{19}$$

The system control gains K_1, K_2 and K_3 can be determined by the pole-zero placement technique

$$\begin{cases} K_3 = \dfrac{(\lambda_1 + \lambda_2 + \lambda_3) - a_2}{b} \\ K_2 = \dfrac{(\lambda_1\lambda_2 + \lambda_2\lambda_3 + \lambda_2\lambda_3) - a_1}{K_3b} \\ K_1 = \dfrac{\lambda_1\lambda_2\lambda_3}{K_2K_3b} \end{cases} \tag{20}$$

7 Phase tracking control strategy

A standard grid sinusoidal waveform with voltage phase information as the reference command is required to ensure the generator output voltage is similar to the grid before the excitation synchronous generator system connects to the consumer loads. To do this, a phase tracking control strategy is introduced. Firstly, additional sinusoidal waveform signals are sent to the MCU controller. As the servo motor and generator are coupled with same shaft, electrical angle for both devices can be obtained by detecting motor encoder. The controller compares the reference command with feedback rotor position signals, and a phase deviation is obtained. The presented position control loop will rapidly track the tiny phase error to obtain high-quality electricity.

Fig. 5*a* describes the phase deviation $\Delta\theta$ between the generator and the reference voltage. The time period over one sinusoidal cycle is defined as T. In MCU, a constant frequency pulse-type signal is created to count the pulse numbers in one sinusoidal cycle time. Assuming there are N pulse signals over one cycle time, the frequency of the pulse numbers in every period is defined as

$$f = \frac{N}{T} \tag{21}$$

In Fig. 5*b*, while the system has a phase deviation $\Delta\theta$, the deviation frequency Δf can be expressed as below

$$\Delta f = K_{\theta f} \times \Delta\theta \tag{22}$$

where $K_{\theta f}$ is a constant gain. The new pulse frequency f can be obtained as below

$$f + \Delta f \rightarrow f \tag{23}$$

The MCU will generate pulse trains of frequency command f to drive the servo motor, so that the generator can lock the generator frequency and the phase command.

8 Simulation and experimental verifications

To confirm the functionality of the generator design, a simulation model of the WPG framework with an excitation synchronous generator and its corresponding sub-systems are built by using MATLAB/SIMULINK® and MATLAB/SIMPOWER® software. The sub-systems include the wind power input, servo motor phase tracking control, maximum power tracking control, AVR and excitation synchronous generator. The detailed generator parameters of the excitation synchronous generator proposed in this paper are listed as Tables 1 to 3. It is clear from the parameter tables that if the excitation synchronous generator is planned to output signals of electricity supply with the frequency identical to that of the urban electricity supply then, in a 4-pole generator, the rotary speed must be rated at 1800 rpm. As such, it guarantees the output electricity of generator is constantly rated at 60 Hz. According to simulation results, the optimal control parameters of the simulation were applied to a practical system.

In Fig. 6*a*, a 660 W constant loading is set for simulations. The wind power is set to be 660 W during the beginning 5 s in the simulation. At the fifth second, the wind power variations begin. A sinusoidal wind disturbance with $660\ \text{W} \pm 100\sin((2\pi/5)t)$ W amplitude is suddenly applied for inspecting the power tracking ability. During 0–5 s, the wind power for driving the generator is sufficient and stable for the 660 W load. Thus, the battery charger

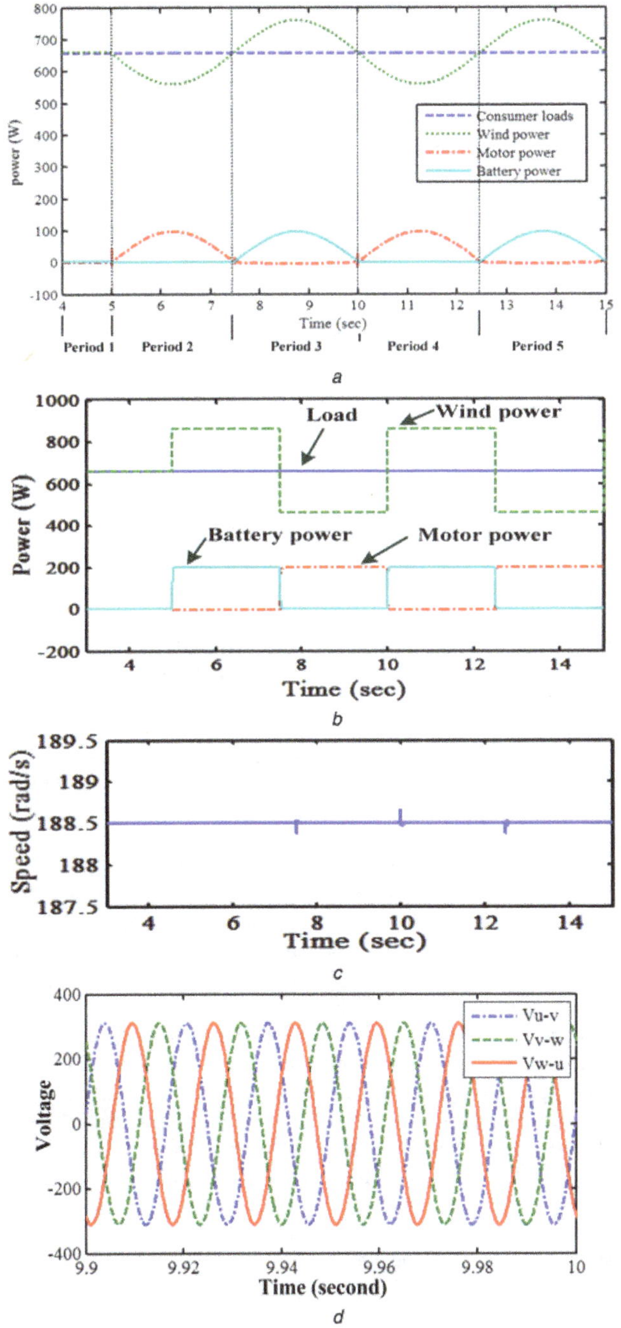

Fig. 6 *Simulation results*
a Power curves at a sinusoidal wind disturbance
b Power curves at a step wind disturbance
c Speed response at a step wind disturbance
d Three line voltage of the synchronous generator

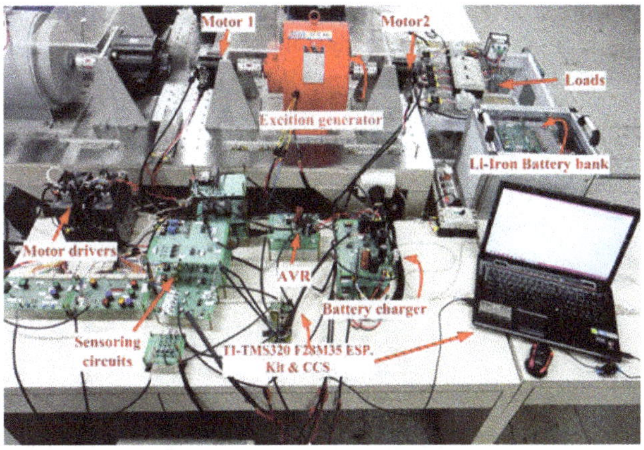

Fig. 7 *Photograph of experimental setup*

Table 2 Specifications of servo motor

Items	Values
torque constant (K_t)	1.09 Nt − m/A
back EMF constant (K_e)	60 V/krpm
rotor inertia (J)	3.3×10^{-4} kg m^2
stator wining resistance (R)	1.5 Ω
stator wining inductance (L)	8 mH

Table 3 PFMS, AVR and servo driver controller parameters

Items	Values
proportional gain (K_1)	5
integral gain (K_2)	20
proportional gain (K_3)	5
current control gain (G_C)	10
PWM amplifier gain (K_a)	10
proportional gain (K_{p1})	20
integral gain (K_{I1})	1
proportional gain (K_{P2})	2
integral gain (K_{I2})	25
excitation gain (K_f)	0.4
proportional gain (K_{P3})	0.8
integral gain (K_{I3})	1

is not triggered. During this interval, the servo motor unnecessarily provides torque to the system. During the time period 5–7.5 s in Fig. 6*a*, the inputted wind power is changed, and the wind power is not large enough to drive the generator for maintaining constant load and speed. Therefore batteries release the stored electrical energy to provide enough power for the servo motor driver to keep the loading power and shaft speed stable. Conversely, within the period 7.5–10 s, the inputted wind power is larger than the AC loading power. The electrical power of the generator not only supplies the AC load but also starts to charge the batteries. To evaluate the system transient response, input wind with step changes is applied in Fig. 6*b*. Fig. 6*c* shows that a slight speed change occurs at the time period of the sudden wind change. Fig. 6*d* shows the voltage waveform transient response of three-phases excitation synchronous generator. It indicates that the voltage and frequency are constant under wind disturbances.

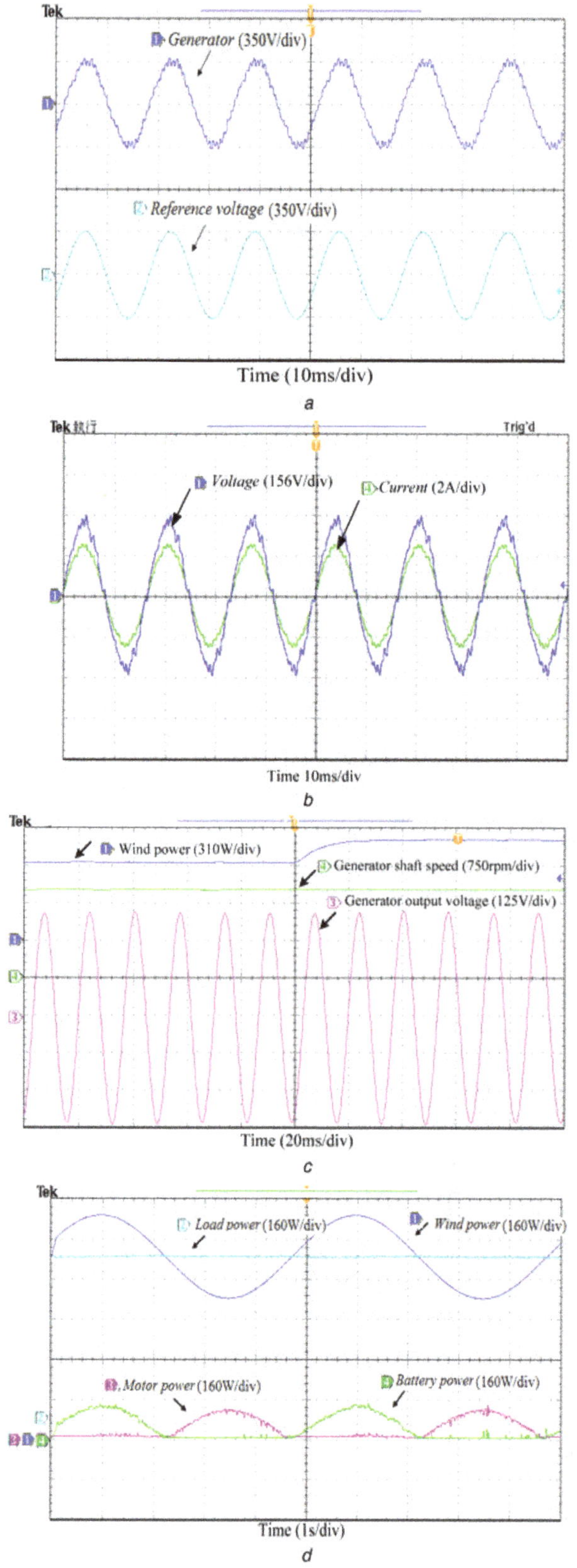

Fig. 8 *Experimental results*
a Line voltage waveforms of grid and generator
b Line voltage and current waveforms of generator
c Transient responses for wind power, line voltage and generator shaft speed
d Power curves at a sinusoidal wind disturbance

To demonstrate the validity of the proposed control system, the experiment has been carried out for a 1 kW SESWPG as shown in Fig. 7. The control system is implemented on a platform consisting of an MCU Texas Instruments F28M35®. There are two motors in the experiment system. Motor 1 employs torque control to simulate natural wind power, and motor 2 is the servo motor to complete phase-locked function. The system parameters including the SESWPG, servo motor and the control parameters are listed in Tables 1–3. The output voltage is a three phase of 220 V_{rms}/60 Hz phase to phase.

According to generator output status, the testing procedure for the experimental system is divided into two steps. In the first step, the system performs phase tracking control. While generator is entering synchronal status, a constant loading is connected to ensure a stable output. In the second step, a sinusoidal wind disturbance with $720\ \text{W} \pm 150\sin((2\pi/5)t)\ \text{W}$ amplitude is suddenly applied for observing the power tracking condition. The experimental results measured from the practical SESWPG are shown in Fig. 8. Ensuring the phase tracking is an essential work. Firstly, a preset wind torque accelerates the generator shaft speed up to ~1800 rpm, and then the phase tracking control is started to regulate the generator output waveform in phase with grid voltage. Fig. 8*a* shows the generator and reference voltages which indicate that the generator voltage is in phase with the reference voltage. The generator voltage and current waveforms are shown in Fig. 8*b*. In this case, three resistances with wye connection as the independent load were connected to the generator.

The expected functions of the proposed PFMS are the produced power of generator following wind power fluctuation; meanwhile output voltage waveform remains constant and stable. Experimental process assumes that the input of the system has a 660 W stable wind power at the time beginning in Fig. 8*c*. At the fourth second, the wind power is increased from 660 W up to 860 W. It is clear that the motor speed and line voltage remained constant as shown in Fig. 8*c*.

For observing the transient responses of wind power, generator power, motor power and battery power, experiment waveforms are measured in Fig. 8*d*. A sinusoidal wind power with $740\ \text{W} \pm 150\sin((2\pi/5)t)\ \text{W}$ amplitude is added for observing the power tracking condition. Although wind power is less than the demand of the AC loading, the proposed PFMS increases motor torque to provide a positive power output for the loads while keeping the generator speed constant. Conversely, when the wind power is greater than output loads, the redundant power of generator is charged to the battery pack and the motor speed remains constant with very low power consumption. In Fig. 8*d*, there was 80 W difference between wind power and generator power. This phenomenon illustrates the existing energy consumptions from mechanical friction and inertia, because the system is accelerated to 1800 rpm and some are the result of generator conversion efficiency.

9 Conclusion

In this paper, an SESWPG with power flow management strategy is presented. The excitation synchronous generator and servo motor are coupled on a coaxial structure in which the generator frequency and voltage are kept constant by the proposed control schemes. Although wind speed is less than generator demand, the proposed PFMS increases motor torque such that the shortfall of power is replenished. Conversely, during the periods where wind speed is greater than the generator demand, the redundant power of the generator is charged to the battery pack and the motor speed remains constant with very low power consumption. The system operation principles and design considerations are analysed and described in detail. With software simulation and laboratory test, the steady state and transient responses under serious conditions are shown for confirming the feasibility. The experimental results show that the proposed system can provide stable AC power for an off-grid loading without using power inverters. Owing to the advantages of converter-less and coaxial configuration, the power converter devices are greatly reduced, moreover the system reliability is increased as well.

10 Acknowledgment

This study was supported by the National Science Council of Taiwan, through grant number NSC 102-3113-P-110-005.

11 References

[1] Barote L., Marinescu C., Cirstea M.N.: 'Control structure for single-phase stand-alone wind-based energy sources', *IEEE Trans. Ind. Electron.*, 2013, **60**, (2), pp. 764–772

[2] Sharma S., Singh B.: 'Control of permanent magnet synchronous generator-based stand-alone wind energy conversion system', *IEEE Trans. Power Electron.*, 2012, **5**, (8), pp. 1519–1526

[3] Munoz-Aguilar R.S., Dòria-Cerezo A., Fossas E., Cardoner R.: 'Sliding mode control of a stand-alone wound rotor synchronous generator', *IEEE Trans. Ind. Electron.*, 2011, **58**, (10), pp. 4888–4897

[4] Iwanski G., Koczara W.: 'DFIG-based power generation system with UPS function for variable-speed applications', *IEEE Trans. Ind. Electron.*, 2008, **55**, (8), pp. 3047–3054

[5] Singh B., Sharma S.: 'Design and implementation of four-leg voltage-source-converter-based VFC for autonomous wind energy conversion system', *IEEE Trans. Ind. Electron.*, 2012, **59**, (12), pp. 4694–4703

[6] Di Gerlando A., Foglia G., Iacchetti M.F., Perini R.: 'Axial flux PM machines with concentrated armature windings: design analysis and test validation of wind energy generators', *IEEE Trans. Ind. Electron.*, 2011, **58**, (9), pp. 3795–3805

[7] Zhang S., Tseng K.-J., Vilathgamuwa D.M., Nguyen T.D., Wang X.-Y.: 'Design of a robust grid interface system for PMSG-based wind turbine generators', *IEEE Trans. Ind. Electron.*, 2011, **58**, (1), pp. 316–328

[8] Bhende C.N., Mishra S., Malla S.G.: 'Permanent magnet synchronous generator-based standalone wind energy supply system', *IEEE Trans. Sust. Energy*, 2011, **2**, (4), pp. 361–373

[9] Geng H., Xu D., Wu B., Yang G.: 'Active damping for PMSG-based WECS with dc-link current estimation', *IEEE Trans. Ind. Electron.*, 2011, **58**, (4), pp. 1110–1119

[10] Bu F., Huang W., Hu Y., Shi K.: 'An excitation-capacitor-optimized dual stator-winding induction generator with the static excitation controller for wind power application', *IEEE Trans. Energy Convers.*, 2011, **26**, (1), pp. 122–131

[11] Rajaei A., Mohamadian M., Varjani A.Y.: 'Vienna-rectifier-based direct torque control of PMSG for wind energy application', *IEEE Trans. Ind. Electron.*, 2013, **60**, (7), pp. 2919–2929

[12] Alepuz S., Calle A., Busquets-Monge S., Kouro S., Wu B.: 'Use of stored energy in PMSG rotor inertia for low-voltage ride-through in back-to-back NPC converter-based wind power systems', *IEEE Trans. Ind. Electron.*, 2013, **60**, (5), pp. 1787–1796

[13] Razzaghi R., Davarpanah M., Sanaye-Pasand M.: 'A novel protective scheme to protect small-scale synchronous generators against transient instability', *IEEE Trans. Ind. Electron.*, 2013, **60**, (4), pp. 1659–1667

[14] Geng H., Xu D., Wu B., Yang G.: 'Active damping for PMSG-based WECS with DC-link current estimation', *IEEE Trans. Ind. Electron.*, 2011, **58**, (4), pp. 1110–1119

[15] Wang J., Qu R., Liu Y.: 'Comparison study of superconducting generators with multiphase armature windings for large-scale direct-drive wind turbines', *IEEE Trans. Appl. Supercond.*, 2013, **23**, (3), p. 5201005

[16] Colli V.D., Marignetti F., Attaianese C.: 'Analytical and multiphysics approach to the optimal design of a 10 MW DFIG for direct-drive wind turbines', *IEEE Trans. Ind. Electron.*, 2012, **59**, (7), pp. 2791–2799

[17] Cardenas R., Pena R., Alepuz S., Asher G.: 'Overview of control systems for the operation of DFIGs in wind energy applications', *IEEE Trans. Ind. Electron.*, 2013, **60**, (7), pp. 2776–2798

[18] Xie K., Jiang Z., Li W.: 'Effect of wind speed on wind turbine power converter reliability', *IEEE Trans. Energy Convers.*, 2012, **27**, (1), pp. 96–104

[19] Stipetic S., Kovacic M., Hanic Z., Vrazic M.: 'Measurement of excitation winding temperature on synchronous generator in rotation using infrared thermography', *IEEE Trans. Ind. Electron.*, 2012, **59**, (5), pp. 2288–2298

[20] Chen Y.-T., Chiu C.-L., Jhang Y.-R., Tang Z.-H., Liang R.-H.: 'A driver for the single-phase brushless DC fan motor with hybrid winding structure', *IEEE Trans. Ind. Electron.*, 2013, **60**, (10), pp. 4369–4375

[21] Chern T.L., Chang J., Chang G.K.: 'DSP-based integral variable structure model following control for brushless DC motor drives', *IEEE Trans. Trans. Power Electron.*, 1997, **12**, (1), pp. 53–63

[22] Chern T.L., Wu Y.C.: 'Design of brushless DC position servo systems using integral variable structure approach', *IEE Proc., B.*, 1993, **140–1**, pp. 27–34

23

Performance analysis of automatic generation control of interconnected power systems with delayed mode operation of area control error

Janardan Nanda, Dushyant Sharma, Sukumar Mishra

Department of Electrical Engineering, Indian Institute of Technology Delhi, Hauz Khas, New Delhi, India
E-mail: dushyantnitrkl@gmail.com

Abstract: This study presents automatic generation control (AGC) of interconnected power systems comprising of two thermal and one hydro area having integral controllers. Emphasis is given to a delay in the area control error for the actuation of the supplementary controller and to examine its impact on the dynamic response against no delay which is usually the practice. Analysis is based on 50% loading condition in all the areas. The system performance is examined considering 1% step load perturbation. Results reveal that delayed mode operation provides a better system dynamic performance compared with that obtained without delay and has several distinct merits for the governor. The delay is linked with reduction in wear and tear of the secondary controller and hence increases the life of the governor. The controller gains are optimised by particle swarm optimisation. The performance of delayed mode operation of AGC at other loading conditions is also analysed. An attempt has also been made to find the impact of weights for different components in a cost function used to optimise the controller gains. A modified cost function having different weights for different components when used for controller gain optimisation improves the system performance.

1 Introduction

In interconnected power systems power is exchanged between utilities over tie lines by which they are connected. The desired operating conditions include nominal frequency, voltage profile and load flow configuration. These quantities are controlled by controlling the active and reactive powers.

Automatic generation control (AGC) plays an important role in maintaining desired system frequency and tie line flow during normal operating conditions and also under small perturbations. Many investigations in the area of AGC of isolated and interconnected power systems have been reported in the past, mostly pertaining to suitable design of secondary controller. Various controller optimisation techniques for design of secondary controller such as classical, optimal, bacterial foraging, genetic algorithm, artificial neural networks, fuzzy logic, particle swarm optimisation (PSO) and so on have been reported in the literature [1–20].

However, so far in the studies that have been reported, there is no mention in the literature regarding the effect of any delay in the area control error (ACE) to actuate the secondary controller of a governor and its impact on dynamic responses of the system. Generally, the ACE command that goes to the speeder gear to regulate the real power generation is present all the time. What happens to the function of the governor and the corresponding dynamic responses if the ACE is just delayed before it actuates the supplementary controller? The motivation of delaying the ACE is that once there is a delay in ACE, the total working of the secondary controller from the instant of initiation of the perturbation (i.e., from the instant when frequency deviates from the desired value) to the steady state will be reduced and consequently the life expectancy of the controller will be increased. The governor secondary control (regulator) works only when it is actuated by ACE. If ACE is absent for some interval of time, then the governor secondary control does not work for that interval of time and thus the additional mechanical movements of governor settings and associated linkage mechanisms and so on of the governor, because of secondary control, remain absent, thus reducing wearing and tearing of the equipment for the interval. These aspects are examined by the authors for a power system consisting of two thermal and one hydro system having generation rate constraints (GRCs) and working under different loading conditions.

The frequency control device is required to be fast acting and this aspect is very well taken care of by the fast acting inherent or primary control of the governor. It may be mentioned here that a governor has two controls, one the inherent or the primary control which is pretty fast, and another the slow secondary or supplementary control and this control is actuated by the ACE. Furthermore, any undesired triggering of under frequency relays is taken care of by fast acting primary control, thus giving us scope to appropriately delay the ACE to actuate the secondary control.

The main concept is to explore the allowable delay to actuate the supplementary controller in order to take advantage of reduced wear and tear and more life for the governor while maintaining good dynamic response.

Sensitivity analysis is also performed to see if the selected delay is acceptable for different loadings by observing the response at different loading conditions. Moreover, the robustness of the controller gains has been examined for changes in the delay in the ACE.

It has been shown in the literature that optimum proportional integral (PI) controller gives same behaviour as an optimum integral (I) controller in operation of AGC of interconnected power systems.

In view of the above discussion, the authors have investigated the AGC problem of a three area system provided with I controllers and with a time delay circuit for the ACE.

Delay in the secondary control loop provides reduced wear and tear in the governor valve assembly linkages and also mitigates the possibility of the controller entering into saturation which leads to undesirable responses.

2 System investigated

Investigations have been carried out on a three area system consisting of two thermal and one hydro system considering GRC. A step load perturbation (SLP) of 1% of rated capacity of each area has been considered in all the areas simultaneously which is justified as all the areas in general undergo load perturbations at all points of time. For the load frequency control, I controller is considered. All details for the system are given in Table A1 in Appendix.

A typical value of GRC for thermal units as 3% per minute for both ramp-up and ramp-down is considered which is the case in most power plants in India. Modern turbines with GRC of 10%

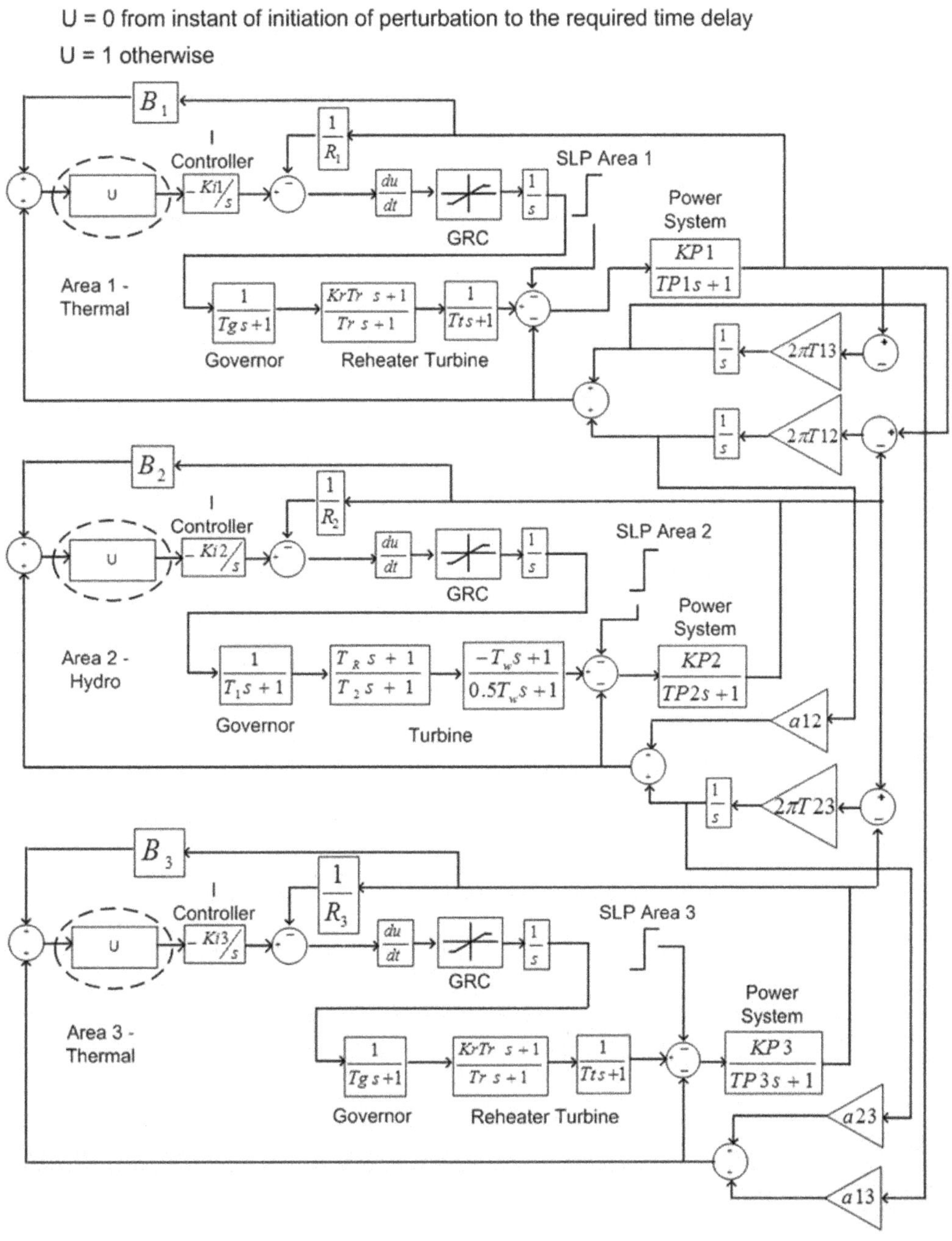

Fig. 1 *Transfer function block diagram of three area power system with delayed ACE*

per minute are now operational with super critical boiler parameters. System response with GRC of 10% per minute is also observed.

In normal mode of operation, ACE goes to supplementary controller all the time following a load change in the electrical power system. There is continuous change in load in the electrical network and hence ACE is generated all the time. In delayed mode of operation, the ACE is allowed to go to supplementary controller only after some time delay from the occurrence of the load change in power system as shown in Fig. 1 (encircled in dotted black).

The power system parameters are taken from [4, 10, 19, 20]. Results are obtained by giving an SLP of 1% in all the areas. The analysis carried out is based on 50% nominal loading condition in all the areas. GRC has been taken as 3% per minute and 10% per minute for the system investigated. Speed regulation parameter is taken as 4%. Detailed model of operation of AGC of interconnected power system is depicted in Fig. 1.

The system is analysed for various delays in the ACE and the changes in frequency (ΔF) and tie line power (ΔP_{tie}) of the three areas are observed corresponding to each delay.

Controller gains are optimised using integral square error (ISE) criterion, where the objective is to minimise the cost function given by

$$J = \int_{0}^{T_{\text{simulation}}} (\Delta F_1^2 + \Delta F_2^2 + \Delta F_3^2 + \Delta P_{\text{tie12}}^2 + \Delta P_{\text{tie23}}^2 + \Delta P_{\text{tie31}}^2)\,dt \quad (1)$$

Fig. 2 shows the modulation of ACE of the three areas for the system under study for a time delay of 0, 5, 10, 15 and 20 s with GRC of 3% per minute in the thermal areas and 1% SLP in all the three areas. 0 s delay is the normal without delay operation while for other time delays the ACE is absent to the controller for a time periods of 0–5, 0–10, 0–15 and 0–20 s.

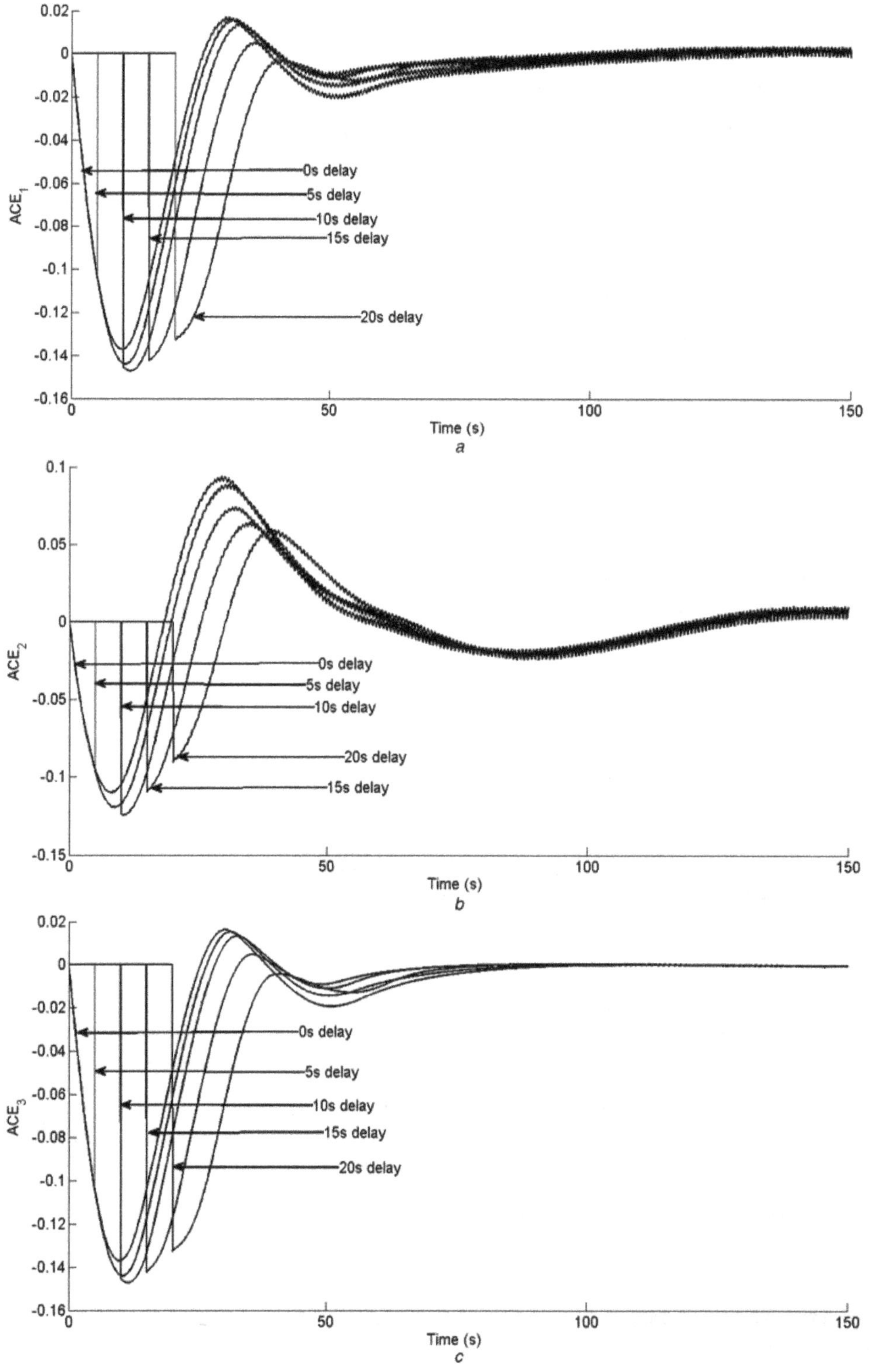

Fig. 2 *Modulated ACE of the three areas for different time delays for 1% SLP in all the areas considering GRC of 3% per minute in the thermal areas*
a Area 1
b Area 2
c Area 3

3 Results and analysis

3.1 Responses for equal load perturbations in the three areas

Figs. 3–5 show dynamic responses for frequency and tie line power change in the three areas for 1% SLP considering GRC of 3 or 10% per minute in the thermal systems and a speed regulation parameter (R) of 4%. Optimum I controller has been used considering different delays of ACE in steps of 5 s. The controller gains are optimised by using PSO. The optimised controller gains for the three areas for 3% GRC are given in Table A2 in Appendix.

It is observed in Figs. 3 and 4 (for 3% per minute GRC in the thermal area) the frequency deviation responses show slight deterioration in terms of peak undershoot but appreciable improvement is observed in peak overshoot and settling time when a delay is applied in the controller. The peak overshoot for a normal (without delay) operation is about 0.06 Hz in area 1 (Fig. 3*a*)

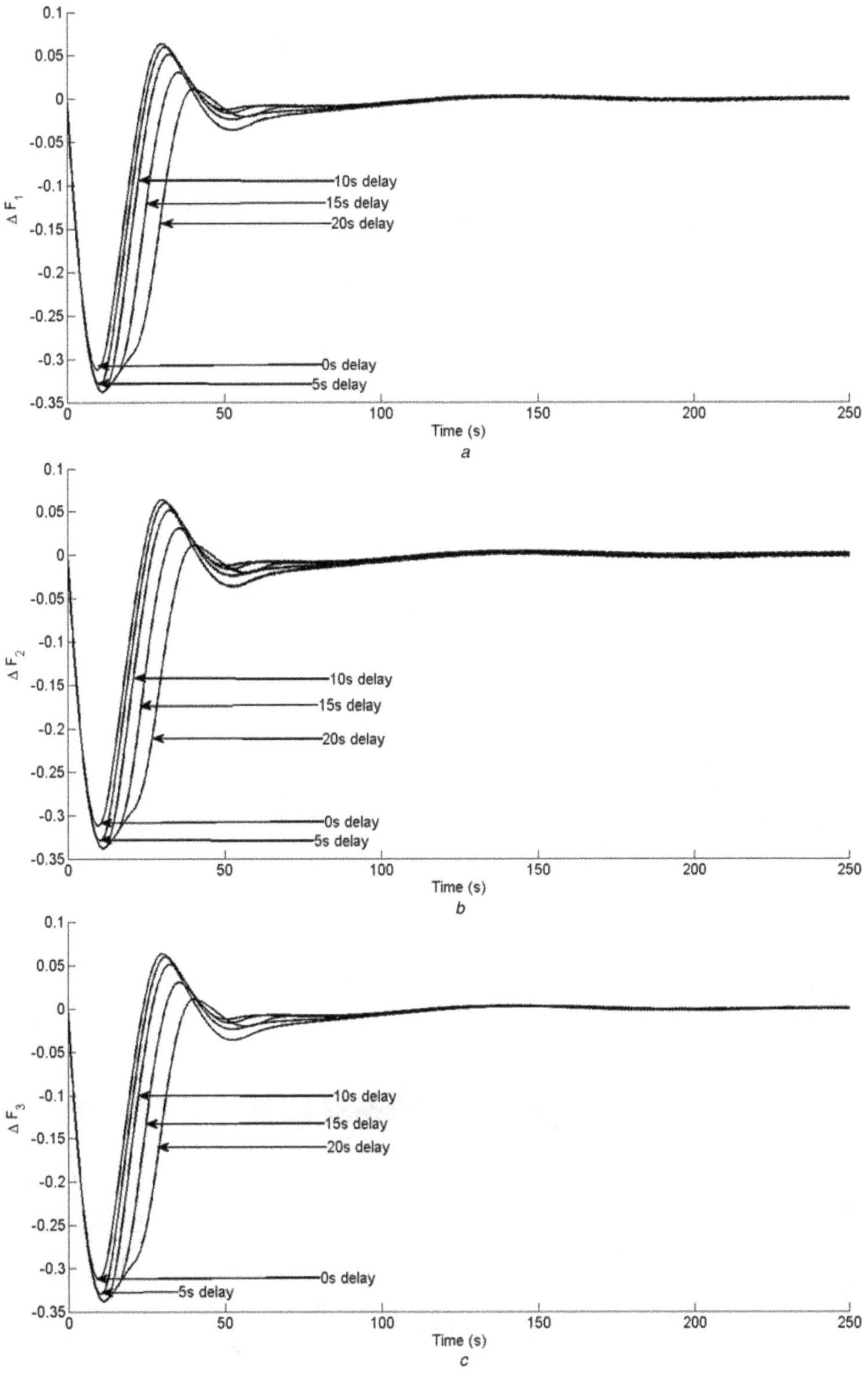

Fig. 3 *Dynamic responses for frequency deviations in the three areas at various time delays for 1% SLP in all areas considering GRC of 3% per minute in the thermal areas and a speed regulation parameter (R) of 4%*
a Area 1
b Area 2
c Area 3

which reduces with increase in delay and is found to be even <0.01 Hz when delay is 20 s. The frequency oscillations are also reduced when a delay is introduced. Undershoot of 0.04 Hz in the second oscillation is also maximum under normal (without delay operation). Similar results are seen in the other areas as well (Figs. 3*b* and *c*). Looking at the responses for change in tie line power flow (Fig. 4), we can see that there is considerable improvement in peak deviation for an operation of the system with delay introduced in the AGC operation while the settling time is almost same. The same is true for all the three tie lines. As it can be seen in Fig. 4, the peak deviation in tie line power flow is least for 10 and 15 s delay, and we can clearly say that a delay of

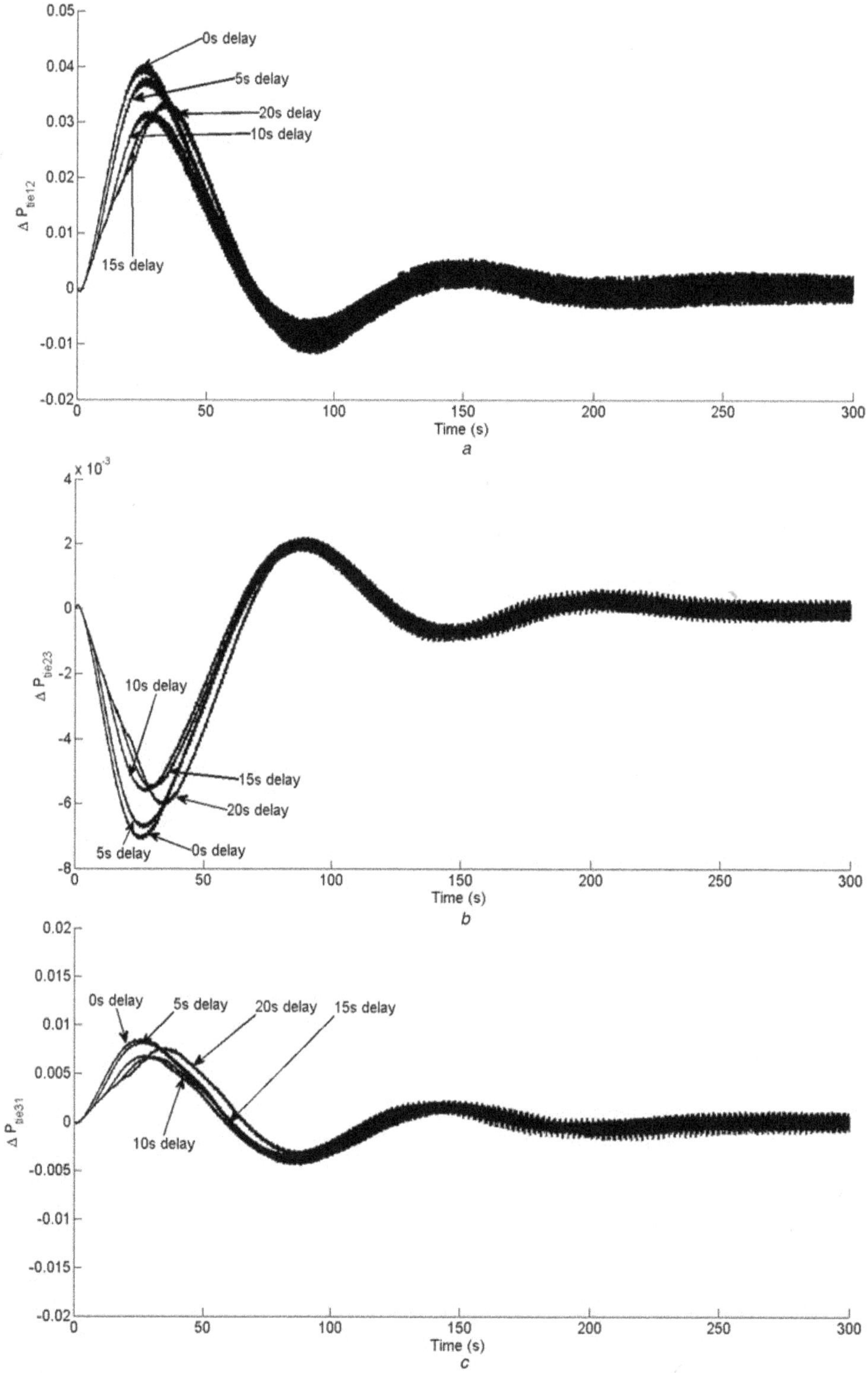

Fig. 4 *Dynamic responses for tie power deviations at various time delays for 1% SLP in all areas considering GRC of 3% per minute in the thermal areas and a speed regulation parameter (R) of 4%*
a ΔP_{tie12}
b ΔP_{tie23}
c ΔP_{tie31}

about 10–15 s is welcome for ACE actuation owing to improvement in settling time and peak deviations in the responses of frequency and tie line power changes.

The deviations in peak undershoot in frequency response are small as a result of the governor primary or inherent control. Moreover, as the secondary controller is slow acting because of I action, its contribution during the small time duration when ACE is absent is small and hence slight deviations are seen in peak undershoot.

Similar to the results observed in Fig. 3, the results with GRC of 10% per minute (Fig. 5) show improved responses up to an ACE delay of 10 s and then the deviations in the tie line power increase with further delay in the ACE actuation. Fig. 5*a* shows the frequency deviation in area 1 and Fig. 5*b* shows the tie power deviation in line connecting area 1 and area 2. Similar results are obtained for other areas and lines as well. Although the frequency and the tie power response for 10 and 15 s delay are quite similar

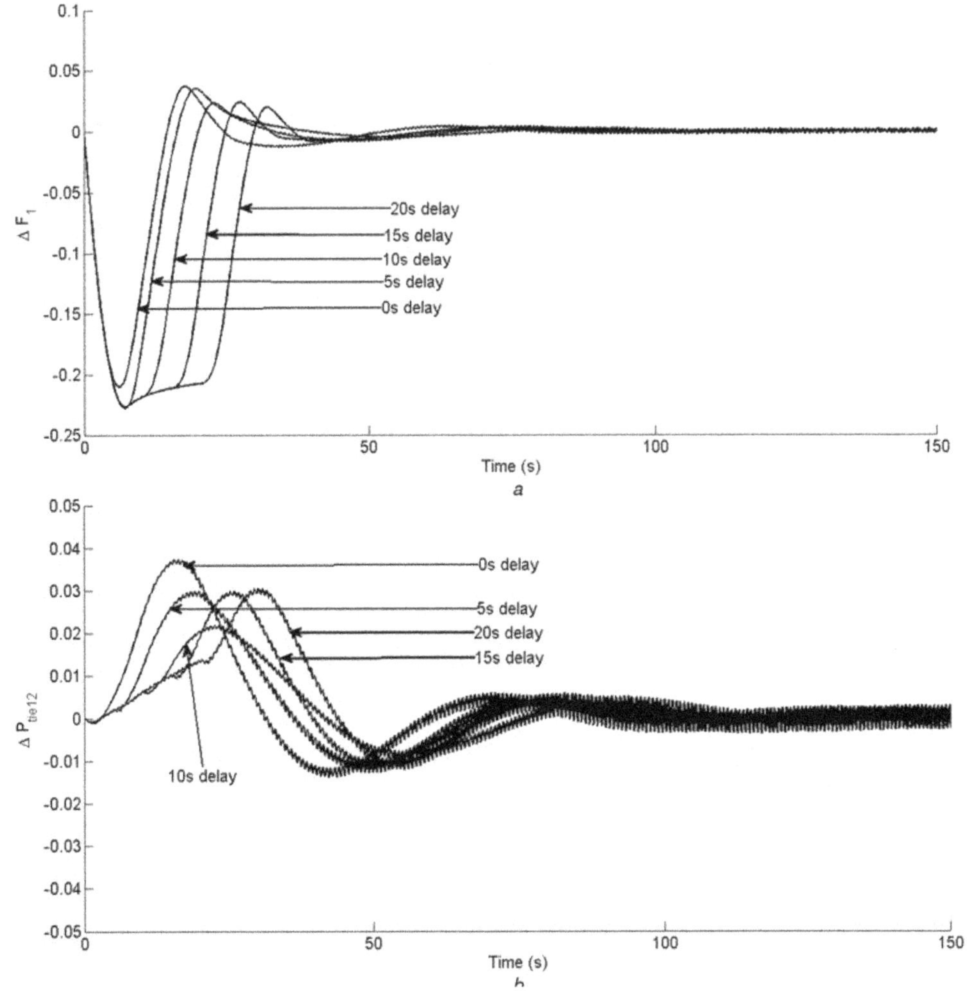

Fig. 5 *Dynamic responses for frequency and tie line power change at various time delays for 1% SLP in all areas considering GRC of 10% per minute in the thermal areas and a speed regulation parameter (R) of 4%*
a ΔF_1
b ΔP_{tie12}

as in Figs. 3–4; there is considerable deterioration in tie power response when delay is increased from 10 to 15 s (Fig. 5*b*). Thus a higher value of GRC reduces the permissible ACE delay for retaining the quality of dynamic responses. It can thus be recommended that irrespective of 3 or 10% per minute GRC, an ACE delay of about 10 s can be comfortably accepted in practice without hampering dynamic responses. The choice of appropriate time delay stays with the utility. A larger delay may also be accepted depending on the acceptance of the deviations and the settling times.

3.2 *Responses with un-optimised gains*

We have seen that a delay in ACE can be acceptable in a power system. The studies performed so far have been made by optimising gains for each delay. For existing systems the gains may have been tuned for in-practice operation (i.e. without delay operation). Even with these gains a delay in the controller actuation shows better response as shown in Fig. 6. Taking $R1 = R2 = R3 = 4\%$ and taking a time delay of 10 s, the frequency and tie power deviations response obtained are compared with normal without delay operation at gains optimised for normal operation (i.e. 0 s delay).

Looking at the responses it can be observed that with 10 s delay the responses are improved (similar to Figs. 3 and 4). Fig. 6a shows that other than a slight deterioration in peak undershoot the response for delayed mode operation is better as there are less oscillations and peak overshoot is reduced. The settling time remains the same for both the cases. In Fig. 6b, it is observed that the peak deviation is reduced to 0.032 pu in delayed mode operation from 0.04 pu in normal mode of operation.

Therefore from the above responses it is eminent that a 10 s delay (even without optimising the gains) can be accepted for the system investigated without hampering the system performance in terms of peak deviation or settling time.

3.3 *Responses for multiple load changes*

The delayed mode AGC operation is also analysed when there are multiple load changes within the period of ACE delay. A 1% load perturbation is followed by 2% load change at 4 s. Delayed mode operation at 10 s is compared with normal without delay operation and the result is shown in Fig. 7. The results obtained earlier hold good in this case too. There is a slight deterioration in frequency undershoot in delayed mode operation. The peak undershoot for 0 s delay is −1.16 Hz which is slightly increased to −1.19 Hz. The settling time remains the same for the two operating scenarios. Thus even for multiple load changes, the delayed mode operation of ACE can be accepted. Tie power deviations and frequency deviations in other areas also match the previous findings.

4 Sensitivity analysis

The system investigated has been analysed for 50% nominal loading condition in all the areas. To check the acceptance of the

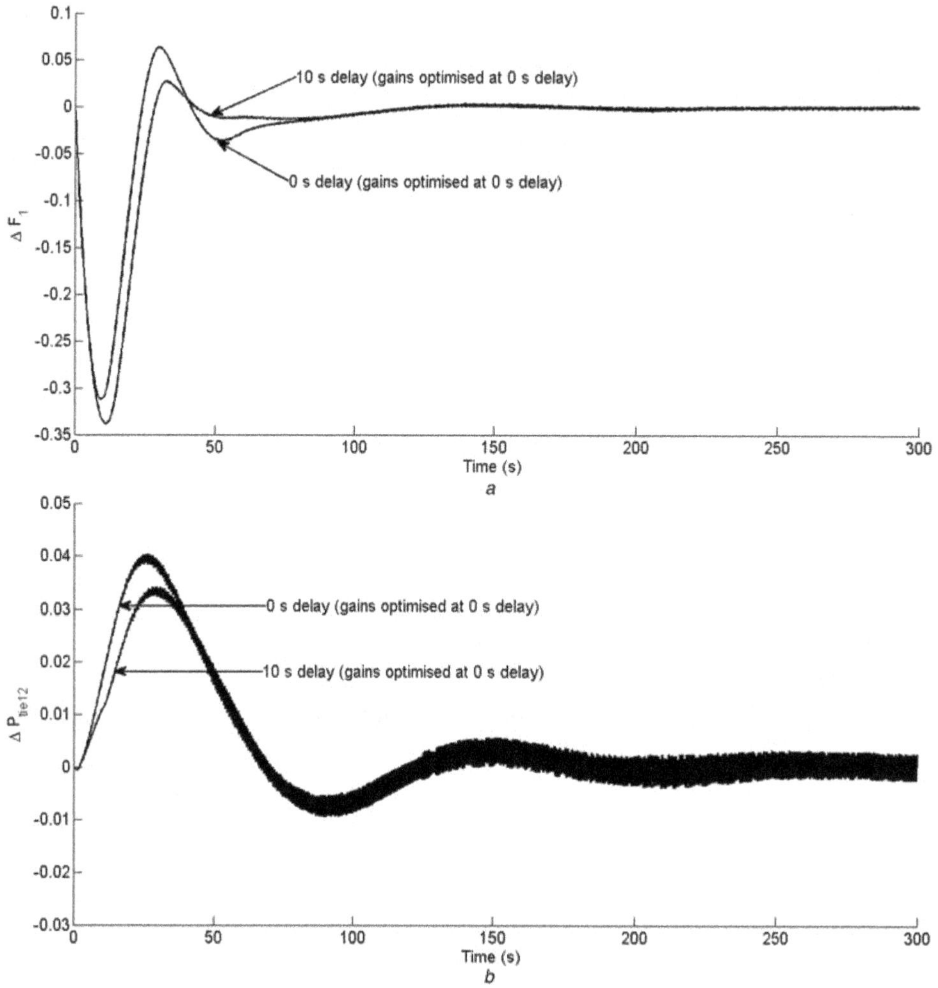

Fig. 6 *Comparison of dynamic responses for frequency and tie line power change in the three areas for 1% SLP in all areas considering GRC of 3% per minute in the thermal areas at 0 s delay (normal operation) and 10 s delay at gains optimised at 0 s delay*
a ΔF_1
b ΔP_{tie12}

delay at different loading conditions, the system is analysed at 80% and 30% loadings for $R = 4\%$, GRC = 3% per minute in the thermal areas and 1% SLP in all the areas. Similar to Section 3.2 the gains are optimised only at one condition which is the nominal 50% loading and 0 s delay. The performance with 80% loading for 10 s delay is compared with performance without delay in Fig. 8, and the performance with 30% loading for 10 s delay is compared with performance without delay in Fig. 9. The frequency deviation in area 1 and tie power deviation in one line is shown in these figures; the others show similar results.

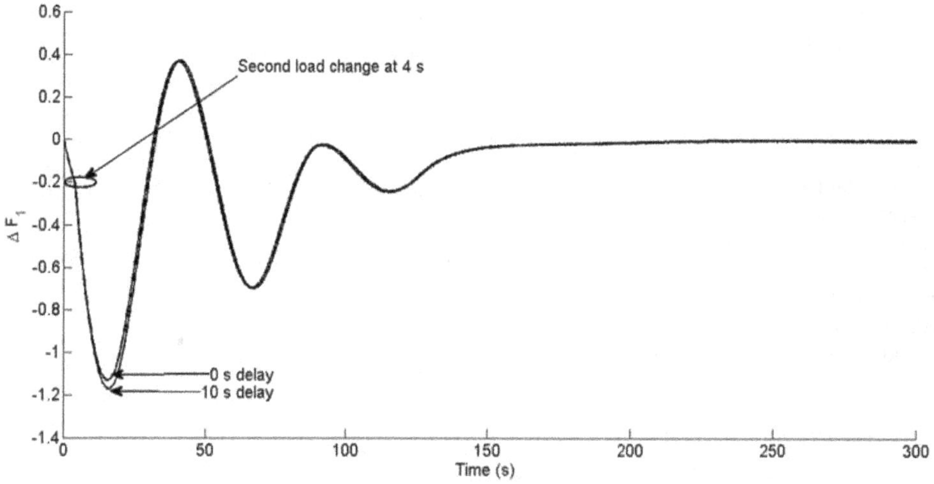

Fig. 7 *ΔF_1 when a delay of 10 s is used in AGC operation and there are multiple load changes within 10 s*

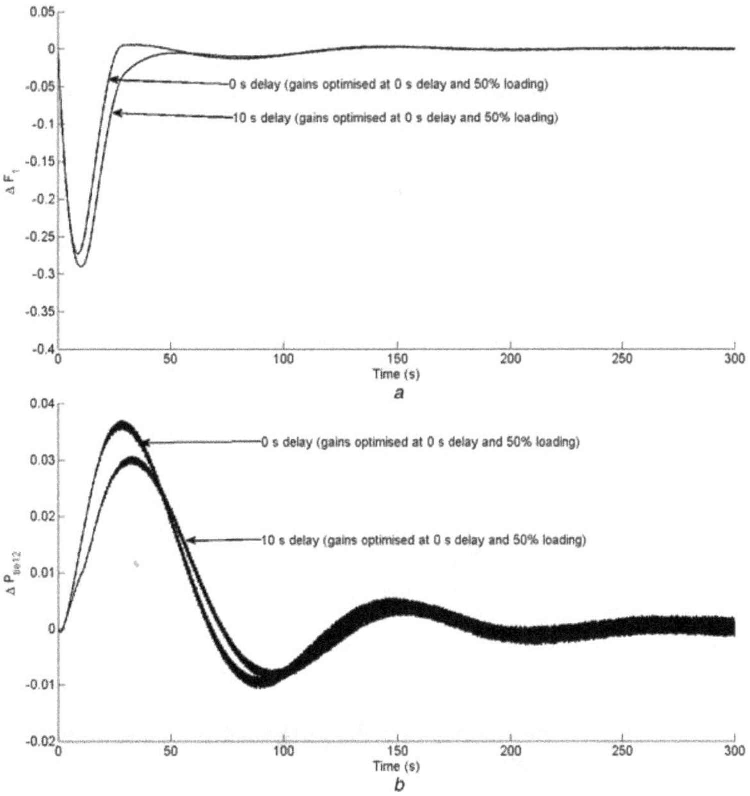

Fig. 8 *Comparison of dynamic responses for frequency and tie line power change in the three areas for 1% SLP in all areas considering GRC of 3% per minute in the thermal areas at 0 s delay and 10 s delay at 80% loading with gains optimised at nominal 50% loading and 0 s delay*
a ΔF_1
b ΔP_{tie12}

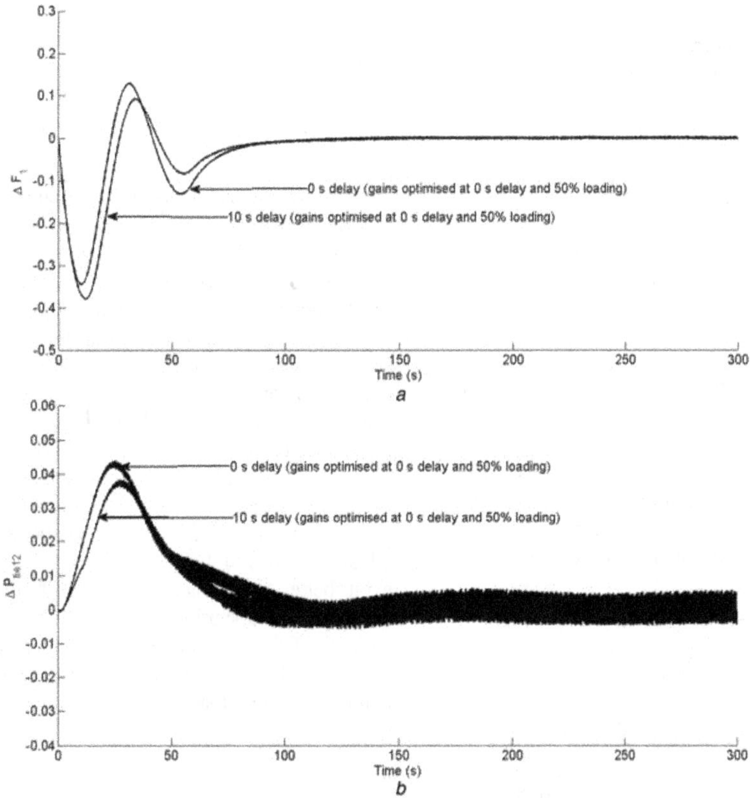

Fig. 9 *Comparison of dynamic responses for frequency and tie line power change in the three areas for 1% SLP in all areas considering GRC of 3% per minute in the thermal areas at 0 s delay and 10 s delay at 30% loading with gains optimised at nominal 50% loading and 0 s delay*
a ΔF_1
b ΔP_{tie12}

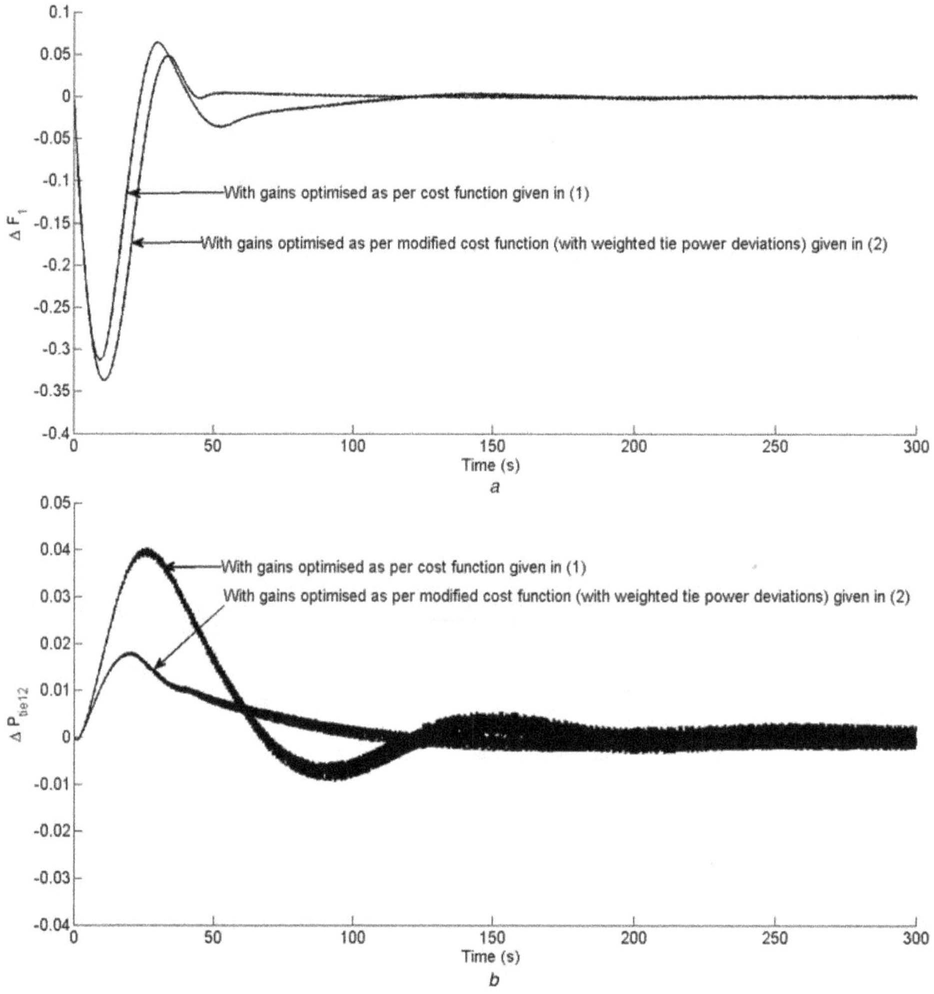

Fig. 10 *Comparison of dynamic responses for frequency and tie line power change in the three areas for 1% SLP in all areas considering GRC of 3% per minute in the thermal areas at 0 s delay (normal operation) at gains optimised by different cost functions*
a ΔF_1
b ΔP_{tie12}

The results in both the cases are similar as obtained in the analysis so far. The delayed mode AGC operation enhances the frequency as well as tie power deviations, except for slight deterioration in peak undershoot in the case of frequency response.

5 Cost function optimisation

An attempt has been made to change the cost function used for optimisation in order to achieve better system performance. As per the normal cost function given in (1), the contributions of tie line power change in the total cost are much less as compared with the contributions of the frequency deviations. Therefore it is justified to put more weights to the tie line deviations in the cost function so that their contributions are also reflected. Such a cost function is given in (2) below

$$J = \int_0^{T_{\text{simulation}}} (\Delta F_1^2 + \Delta F_2^2 + \Delta F_3^2 + w_1 \Delta P_{\text{tie12}}^2 + w_2 \Delta P_{\text{tie23}}^2 + w_3 \Delta P_{\text{tie31}}^2)\mathrm{d}t \quad (2)$$

Controller gain optimised using such a cost function will enhance the tie power responses of the system. The weights $w1$, $w2$ and $w3$ are to be judiciously decided and optimised depending on the contributions in the original case. In this paper, we have made this analysis by setting all the three weights $w1$, $w2$ and $w3$ to 50.

The system response is compared with original response (obtained by choosing cost function given in (1)) in Fig. 10. Responses are shown for one area and one tie. Similar results are observed with other areas and lines as well. From Fig. 10 the results obtained at two different cost functions show that the frequency deviation response shows slight deterioration in peak undershoot but considerable improvement is observed in peak overshoot, oscillations and settling time. The settling time with original cost function is 130 s which reduces to 80 s with modified cost function. Peak overshoot in tie power with original cost function is 0.04 pu which is reduced to just 0.018 pu in the latter case. This shows that more participation of the tie power deviations in the cost will lead to improved responses. In the above analysis, the weights are all set to 50 and further optimisation of the weights may give even better performance and can further enhance the delayed mode AGC operation performance. With properly optimised weights of the cost function and then properly selecting gains accordingly, the performance of delayed mode operation of secondary controller can be made more acceptable.

6 Conclusions

A properly chosen delay in the ACE to actuate the secondary controller of the governor with optimum I controller provides dynamic responses better than without delay operation of ACE prevalent in practice. A delay in the ACE should be recommended, as wear and tear of the moving parts and linkages of governing system is reduced significantly and thus increasing the life of the governor.

Sudden loading and unloading which may lead to the mal-operation of the safety valves can be avoided to a great extent in the presence of delayed ACE.

The optimum I gains with nominal loading condition and operating conditions are robust and can be used under wide changes in the loading conditions and time delays.

Selection of optimum weights in the cost function is also an important aspect in the AGC performance. Controller gains optimised using properly weighted cost function gives better tie line and frequency responses.

7 References

[1] Concordia C., Kirchmayer L.K.: 'Tie-line power & frequency control of electric power systems', *AIEE Trans., III-A*, 1953, **72**, pp. 562–572

[2] Concordia C., Kirchmayer L.K.: 'Tie-line power & frequency control of electric power systems: part II', *AIEE Trans., III-A*, 1954, **73**, pp. 133–146

[3] Kirchmayer L.K.: 'Economic control of interconnected systems' (Wiley, New York, 1959)

[4] Elgerd O.I.: 'Electric energy systems theory an introduction' (Tata McGraw-Hill, 1983)

[5] Nanda J., Kaul B.L.: 'Automatic generation control of an interconnected power system', *Proc. Inst. Electr. Eng.*, 1978, **125**, (5), pp. 385–390

[6] Elgerd O.I., Fosha C.E.: 'Optimum megawatt-frequency control of multiarea electric energy systems', *IEEE Trans. Power Appl. Syst.*, 1970, **PAS-89**, (4), pp. 556–563

[7] Fosha C.E., Elgerd O.I.: 'The megawatt-frequency control problem – a new approach via optimal control theory', *IEEE Trans. Power Appl. Syst.*, 1970, **PAS-89**, (4), pp. 563–577

[8] Leum M.: 'The development and field experience of a transistor electric governor for hydro turbines', *IEEE Trans. Power Appar. Syst.*, 1966, **PAS-85**, pp. 393–402

[9] Kothari M.L., Satsangi P.S., Nanda J.: 'Sampled data automatic generation control of interconnected reheat thermal system considering generation rate constraint', *IEEE Trans. Power Appar. Syst.*, 1981, **PAS-100**, (5), pp. 2334–2342

[10] Nanda J., Kothari M.L., Satsangi P.S.: 'Automatic generation control of an interconnected hydrothermal system in continuous and discrete modes considering generation rate constraints', *Control Theory and Applications, Proc. Inst. Electr. Eng.*, 1983, **130**, (1), pp. 17–27, pt. D

[11] Hari L., Kothari M.L., Nanda J.: 'Optimum selection of speed regulation parameter for automatic generation control in discrete mode considering generation rates constraint', *Proc. Inst. Electr. Eng.*, 1991, **138**, (5), pp. 401–406

[12] Beaufays F., Abdel-Magid Y., Widrow B.: 'Application of neural network to load frequency control in power systems', *Neural Netw.*, 1994, **7**, (1), pp. 183–194

[13] Abdel-Magid Y.L., Dawoud M.M.: 'Tuning of AGC of interconnected reheat thermal systems with genetic algorithms'. IEEE Int. Conf. on Systems, Man and Cybernetics, Vancouver, BC, 1995, vol. **3**, pp. 2622–2627

[14] Djukanovic M., Novicevic M., Sobajic D.J., Pao Y.P.: 'Conceptual development of optimal load frequency control using artificial neural networks and fuzzy set theory', *Int. J. Eng. Intell. Syst. Electr. Eng. Commun.*, 1995, **3**, (2), pp. 95–108

[15] Chown G.A., Hartman R.C.: 'Design & experience of fuzzy logic controller for automatic generation control (AGC)', *IEEE Trans. Power Syst.*, 1998, **13**, (3), pp. 965–970

[16] Abdel-Magid Y.L., Abido M.A.: 'AGC tuning of interconnected reheat thermal systems with particle swarm optimization'. Proc. Tenth IEEE Int. Conf. on Electronics, Circuits and Systems, ICECS, 2003, vol. **1**, pp. 376–379

[17] Ghoshal S.P., Goswami S.K.: 'Application of GA based optimal integral gains in fuzzy based active power-frequency control of non-reheat and reheat thermal generating systems', *Electr. Power Syst. Res.*, 2003, **67**, (2), pp. 79–88

[18] Ghoshal S.P.: 'Application of GA/GA-SA based fuzzy automatic generation control of a multi-area thermal generating system', *Electr. Power Syst. Res.*, 2004, **70**, (2), pp. 115–127

[19] Nanda J., Mangla A., Suri S.: 'Some new findings on automatic generation control of an interconnected hydrothermal system with conventional controllers', *IEEE Trans. Energy Convers.*, 2006, **21**, (1), pp. 187–194

[20] Nanda J., Mishra S., Saikia L.C.: 'Maiden application of bacterial foraging-based optimization technique in multiarea automatic generation control', *IEEE Trans. Power Syst.*, 2009, **24**, (2), pp. 602–609

8 Appendix

See Tables A1 and A2.

Table A1 Nominal system parameters

Description	Symbol	Value
type of units in each area	area 1	thermal
	area 2	hydro
	area 3	thermal
rated capacity of each area	P_{r1}	2000 MW
	P_{r2}	1000 MW
	P_{r3}	4000 MW
nominal frequency	f	60 Hz
inertia constant of the three areas	H_1, H_2, H_3	5 s
thermal unit hydraulic governor time constant	T_g	0.08 s
thermal unit reheater steam	T_r	10 s
turbine parameters	T_t	0.3 s
	K_r	0.5
hydro unit mechanical governor time constant	T_1	48.75 s
hydro unit turbine parameters	T_R	5 s
	T_2	0.513 s
	T_w	1 s
system loading		50% of rated capacity of each area
nominal speed regulation parameters	$R1$, $R2$, $R3$	2.4 Hz/pu MW
load frequency parameter	$D1$, $D2$, $D3$	0.00833 pu MW/Hz
power system gains	$KP1$, $KP2$, $KP3$	120
power system time constants	$TP1$, $TP2$, $TP3$	20
synchronising coefficient	$T12$, $T23$, $T13$	0.544
GRC thermal	ramp-up	3% per minute and 10% per minute
	ramp-down	3% per minute and 10% per minute
GRC hydro	ramp-up	270% per minute
	ramp-down	360% per minute

Table A2 Controller gains at different time delays for GRC in thermal areas = 3% per minute, R = 4% and loading = 50%

Time delay (s)	Controller gains		
	$Ki1$	$Ki2$	$Ki3$
0	0.0635	0.1288	0.0657
5	0.0761	0.1233	0.0666
10	0.0882	0.1253	0.0865
15	0.1153	0.1403	0.1115
20	0.1413	0.1545	0.1193

Permissions

All chapters in this book were first published in TJE, by The Institution of Engineering and Technology (TIET); hereby published with permission under the Creative Commons Attribution License or equivalent. Every chapter published in this book has been scrutinized by our experts. Their significance has been extensively debated. The topics covered herein carry significant findings which will fuel the growth of the discipline. They may even be implemented as practical applications or may be referred to as a beginning point for another development.

The contributors of this book come from diverse backgrounds, making this book a truly international effort. This book will bring forth new frontiers with its revolutionizing research information and detailed analysis of the nascent developments around the world.

We would like to thank all the contributing authors for lending their expertise to make the book truly unique. They have played a crucial role in the development of this book. Without their invaluable contributions this book wouldn't have been possible. They have made vital efforts to compile up to date information on the varied aspects of this subject to make this book a valuable addition to the collection of many professionals and students.

This book was conceptualized with the vision of imparting up-to-date information and advanced data in this field. To ensure the same, a matchless editorial board was set up. Every individual on the board went through rigorous rounds of assessment to prove their worth. After which they invested a large part of their time researching and compiling the most relevant data for our readers.

The editorial board has been involved in producing this book since its inception. They have spent rigorous hours researching and exploring the diverse topics which have resulted in the successful publishing of this book. They have passed on their knowledge of decades through this book. To expedite this challenging task, the publisher supported the team at every step. A small team of assistant editors was also appointed to further simplify the editing procedure and attain best results for the readers.

Apart from the editorial board, the designing team has also invested a significant amount of their time in understanding the subject and creating the most relevant covers. They scrutinized every image to scout for the most suitable representation of the subject and create an appropriate cover for the book.

The publishing team has been an ardent support to the editorial, designing and production team. Their endless efforts to recruit the best for this project, has resulted in the accomplishment of this book. They are a veteran in the field of academics and their pool of knowledge is as vast as their experience in printing. Their expertise and guidance has proved useful at every step. Their uncompromising quality standards have made this book an exceptional effort. Their encouragement from time to time has been an inspiration for everyone.

The publisher and the editorial board hope that this book will prove to be a valuable piece of knowledge for researchers, students, practitioners and scholars across the globe.

List of Contributors

Sara Hashemi and Keivan Navi
Department of Electrical and Computer Engineering, Shahid Beheshti University, G.C., Tehran, Iran

Sinisa Djurović, Damian S. Vilchis-Rodriguez and Alexander Charles Smith
School of Electrical and Electronic Engineering, The University of Manchester, Power Conversion Group, Sackville Street Building, M13 MPL Manchester, UK

Ahmed S. Morsy and Ayman S. Abdelkhalik
Electrical Power and Machines Department, Faculty of Engineering, Alexandria University, Alexandria, Egypt

Shehab Ahmed
Electrical & Computer Engineering, Texas A&M University at Qatar, Doha 23874, Qatar

Ahmed Mohamed Massoud
Electrical Engineering Department, Qatar University, Doha, Qatar

Kyle Holzer and Jeffrey S. Walling
University of Utah PERFIC Laboratory, Salt Lake City, UT 84112, USA

Daniel Castanheira, Adão Silva and Atílio Gameiro
DETI, Instituto de Telecomunicações, University of Aveiro, Aveiro, Portugal

Lei Sun, Chi Tung Ko, Marco Ho, Ka Nang Leung, Chiu Sing Choy and Kong Pang Pun
Department of Electronic Engineering, Chinese University of Hong Kong, Shatin 852, Hong Kong, People's Republic of China

Wai Tung Ng
Department of Electrical & Computer Engineering, University of Toronto, 10 King's College Road, Toronto, Ontario, Canada M5S 3G4

Li Peng, Hao Li and Xin Li
Department of Electronics and Information Engineering, Huazhong University of Science and Technology, Wuhan 430074, People's Republic of China

Pingliang Zeng
Power System Department, China Electric Power Research Institute, Beijing 100192, People's Republic of China

Emmanuel Chifuel Manasseh
Department of Electronics and Telecommunications Engineering, School of Computational and Communication Science and Engineering, Nelson Mandela African Institution of Science and Technology, P.O. Box 447, Arusha, Tanzania

Shuichi Ohno and Toru Yamamoto
Department of System Cybernetics, Graduate school of Engineering, Hiroshima University, 1-4-1 Kagamiyama, Higashi-Hiroshima, 739-8527, Japan

Aloys Mvuma
Department of Telecommunications and Communications Networks, School of Informatics, College of Informatics and Virtual Education, University of Dodoma, P.O. Box 490, Dodoma, Tanzania

Pankaj Kumar Das
Department of Electronics and Communication Engineering, Sant Longowal Institute of Engineering and Technology, Longowal-148106, India

Manoj Kumar Majumder and Brajesh Kumar Kaushik
Department of Electronics and Communication Engineering, Indian Institute of Technology Roorkee, Roorkee - 247667, India

Chitrakant Sahu and Jawar Singh
Department of Electronics and Communication Engineering, PDPM Indian Institute of Information Technology, Design and Manufacturing Jabalpur, Madhya Pradesh, India

Fei Yuan
Department of Electrical and Computer Engineering, Ryerson University, Toronto, ON, Canada

Konstantinos O. Oureilidis and Charis S. Demoulias
Department of Electrical and Computer Engineering, Aristotle University, Thessaloniki, 54124, Greece

Xu Xuefei and Liao Guisheng
National Laboratory of Radar Signal Processing, Xidian University, Xi'an 710071, People's Republic of China

Ahmed A. Hakeem, Ahmed Abbas Elserougi and Ayman Samy Abdel-Khalik
Department of Electrical Engineering, Alexandria University, Alexandria, Egypt

Shehab Ahmed
ECEN Department, Texas A&M University at Qatar, Doha, Qatar

Ahmed Mohamed Massoud
Department of Electrical Engineering, Alexandria University, Alexandria, Egypt
Electrical Department, Qatar University, Doha, Qatar

Sin Jin Tan, Zian Cheak Tiu and Sulaiman Wadi Harun
Department of Electrical Engineering, Faculty of Engineering, University of Malaya, 50603 Kuala Lumpur, Malaysia
Photonics Research Center, University of Malaya, 50603 Kuala Lumpur, Malaysia

Harith Ahmad
Photonics Research Center, University of Malaya, 50603 Kuala Lumpur, Malaysia

Liechen Li
Science and Technology on Microwave Imaging Laboratory, Institute of Electronics, Chinese Academy of Sciences, Beijing 100190, People's Republic of China
University of Chinese Academy of Sciences, Beijing 100049, People's Republic of China

Daojing Li
Science and Technology on Microwave Imaging Laboratory, Institute of Electronics, Chinese Academy of Sciences, Beijing 100190, People's Republic of China

Bo Liu
China Academy of Space Technology, Beijing 100094, People's Republic of China

Qingjuan Zhang
Patent Examination Cooperation Center Jiangsu Center of the Patent Office, SIPO, Suzhou 215011, People's Republic of China

William J. B. Heffernan
Electric Power Engineering Centre, University of Canterbury, Christchurch 8140, New Zealand

Neville R. Watson and Jeremy D. Watson
Electrical & Computer Engineering Department, University of Canterbury, Private Bag 4800, Christchurch 8140, New Zealand

Shyam Babu Mahato, Ben Allen and Enjie Liu
Centre for Wireless Research (CWR), University of Bedfordshire, Luton LU1 3JU, UK

Tien Van Do
Department of Telecommunication, Budapest University of Technology and Economics, Budapest, Hungary

Jie Zhang
Centre for Wireless Network Design (CWiND), Department of Electronic and Electrical Engineering, University of Sheffield, Sheffield, UK

Dr. Hans Gustat
Dept. Circuit Design, IHP - Institute for High-Performance Microelectronics, Im Technologiepark 25, D-15236 Frankfurt (Oder)

Norazlina Saidin
Photonics Research Centre, University of Malaya, 50603 Kuala Lumpur, Malaysia
Department of Electrical Engineering, Faculty of Engineering, University of Malaya, 50603 Kuala Lumpur, Malaysia
Department of Electrical and Computer Engineering, International Islamic University Malaysia, Jalan Gombak, 53100 Kuala Lumpur, Malaysia

Dess I. M. Zen
Photonics Research Centre, University of Malaya, 50603 Kuala Lumpur, Malaysia
Department of Electrical and Electronic Engineering, National Defense University of Malaysia, Kem Sungai Besi, 57000 Kuala Lumpur, Malaysia

Fauzan Ahmad, Hazlihan Haris and Harith Ahmad
Photonics Research Centre, University of Malaya, 50603 Kuala Lumpur, Malaysia

Muhammad T. Ahmad
Photonics Research Centre, University of Malaya, 50603 Kuala Lumpur, Malaysia
Faculty of Electronic and Computer Engineering, Universiti Teknikal Malaysia Melaka, 76100 Durian Tunggal, Melaka, Malaysia

Anas A. Latiff
Faculty of Electronic and Computer Engineering, Universiti Teknikal Malaysia Melaka, 76100 Durian Tunggal, Melaka, Malaysia

Kaharudin Dimyati
Department of Electrical and Electronic Engineering, National Defense University of Malaysia, Kem Sungai Besi, 57000 Kuala Lumpur, Malaysia

Sulaiman W. Harun
Photonics Research Centre, University of Malaya, 50603 Kuala Lumpur, Malaysia
Department of Electrical Engineering, Faculty of Engineering, University of Malaya, 50603 Kuala Lumpur, Malaysia

Song Cui and Fuqin Xiong
Department of Electrical and Computer Engineering, Cleveland State University, 2121 Euclid Avenue, Cleveland, OH 44115, USA

Tzuen-Lih Chern, Ping-Lung Pan, Ji-Xian Huang, Whei-Min Lin and Chih-Chiang Cheng
Department of Electrical Engineering, National Sun Yat-sen University, Kaohsiung, Taiwan

Yu-Hsiang Chern and Jyh-Horng Chou
Graduate Institute of Electrical Engineering, National Kaohsiung First University of Science and Technology, Kaohsiung, Taiwan

Janardan Nanda, Dushyant Sharma and Sukumar Mishra
Department of Electrical Engineering, Indian Institute of Technology Delhi, Hauz Khas, New Delhi, India

www.ingramcontent.com/pod-product-compliance
Lightning Source LLC
Chambersburg PA
CBHW080300230326
41458CB00097B/5239
* 9 7 8 1 6 8 2 8 5 0 1 1 4 *